Characters of Solvable Groups

GRADUATE STUDIES IN MATHEMATICS 189

Characters of Solvable Groups

I. Martin Isaacs

AMERICAN MATHEMATICAL SOCIETY
Providence, Rhode Island

2010 *Mathematics Subject Classification.* Primary 20C15;
Secondary 20C20, 20D10, 20D35, 20F16, 20F14.

For additional information and updates on this book, visit
www.ams.org/bookpages/gsm-189

Library of Congress Cataloging-in-Publication Data

Names: Isaacs, I. Martin, 1940– author.
Title: Characters of solvable groups / I. Martin Isaacs.
Other titles: Solvable groups
Description: Providence, Rhode Island : American Mathematical Society, [2018] | Series: Graduate studies in mathematics ; volume 189 | Includes bibliographical references and index.
Identifiers: LCCN 2017047819 | ISBN 9781470434854 (alk. paper)
Subjects: LCSH: Solvable groups. | Group theory. | AMS: Group theory and generalizations – Representation theory of groups – Ordinary representations and characters. msc | Group theory and generalizations – Representation theory of groups – Modular representations and characters. msc | Group theory and generalizations – Abstract finite groups – Solvable groups, theory of formations, Schunck classes, Fitting classes, π-length, ranks. msc | Group theory and generalizations – Abstract finite groups – Subnormal subgroups. msc | Group theory and generalizations – Special aspects of infinite or finite groups – Solvable groups, supersolvable groups. msc | Group theory and generalizations – Special aspects of infinite or finite groups – Derived series, central series, and generalizations. msc
Classification: LCC QA177 .I8325 2018 | DDC 512/.2—dc23
LC record available at `https://lccn.loc.gov/2017047819`

∞ The paper used in this book is acid-free and falls within the guidelines
established to ensure permanence and durability.
Visit the AMS home page at `http://www.ams.org/`

10 9 8 7 6 5 4 3 2 1   23 22 21 20 19 18

To Deborah

Contents

Preface

When character theory was developed at the end of the 19th century and in the early years of the 20th, it was viewed primarily as a powerful technique for proving theorems about finite groups. For example, Burnside's $p^a q^b$-theorem and Frobenius's theorem on the existence of what we now call Frobenius kernels were early triumphs of the use of character theory as a tool for proving group-theory theorems. Indeed, characters continued to play an essential role in proving theorems about groups for another 50 years or so, culminating in the Feit–Thompson odd-order theorem.

Another thread in the development of character theory is the establishment of connections between the characters and the subgroup structure of a group. Examples of this are Brauer's theorem on induced characters, which asserts that every character of G is a $\mathbb{Z}$-linear combination of characters induced from certain nilpotent subgroups, and also the Itô–Michler theorem, which gives a necessary and sufficient condition in terms of irreducible character degrees for G to have a normal abelian Sylow p-subgroup.

Research on the connections between characters and subgroups continues unabated, and there are still a number of intriguing open questions in this area. For example, consider the still-unresolved McKay conjecture, which posits that for each finite group G and each prime number p, the numbers of irreducible characters with degrees not divisible by p of G and of the normalizer of a Sylow p-subgroup of G are equal.

Within the last half century or so, there has been considerable attention paid to the character theory of groups such as solvable and p-solvable groups that have an abundance of normal subgroups. One reason for this is that groups of this type often serve to motivate and to "test" more general questions. For example, the Itô–Michler theorem was initially proved (by

Itô) for solvable groups; it was easily generalized to p-solvable groups, and then it was proved by Michler for arbitrary finite groups. Another example is provided by the McKay conjecture. I gave the first proof of a case of this conjecture by establishing it for solvable groups of odd order, and more generally, the conjecture was proved by Okuyama and Wajimais for all p-solvable groups. Other outstanding general problems, such as the Alperin weight conjecture and the Feit question on fields of character values, have also been settled for solvable groups.

In addition to its role in testing conjectures, there is another aspect of the many-normal-subgroups branch of character theory that I find fascinating. There are a number of surprising theorems that hold for such groups that are simply false in the general case. For example, if H is a subgroup of G and θ is an irreducible character of H, an obvious necessary condition for θ to be the restriction of a character of G is that $\theta(x)$ and $\theta(y)$ must be equal whenever x and y are elements of H that are conjugate in G. Remarkably, if H is a Hall subgroup and G is solvable, this condition is also sufficient, but if G is not solvable, the sufficiency can fail even if H is a Sylow subgroup of G. Another example is this: if H is an arbitrary subgroup of a group G and χ is an irreducible character of G, there is little that can be said about the degrees of the irreducible constituents of the restriction of χ to H. If G is solvable of odd order, however, there is always some constituent of the restriction with degree dividing the degree of χ.

Despite its title, this book is not exclusively about solvable groups. When discussing this material informally, I often write "solvable groups" as a convenient (but not necessarily accurate) shorthand for the more awkward and less precise phrase "many-normal-subgroups groups". But of course in formal settings such as the statements of theorems, the assumptions are specific and precisely stated.

For many years, most of my research has been in solvable group character theory. My work, together with that of many others, has resulted in a substantial accumulation of theory and theorems, very little of which has previously appeared in book form. My goal in this text is to provide the student and researcher with enough of the theory to demonstrate its power and beauty and to provide a foundation for further study. In order to keep the exposition as accessible as possible, however, I was forced to omit some important and interesting results.

I have included a number of applications, including proofs of solvable-group cases of various outstanding general conjectures. Perhaps this is a good place to explain my belief that one test of a theory is that it should have the power to answer questions that it did not ask. By an "application", therefore, I mean a result that is external to the theory. For example, in

Chapter 2, factorization theory and π-special characters are used to prove the solvable case of the Feit conjecture, although on its face the Feit conjecture has no connection with factorization or with π-special characters.

I mentioned that I have tried to keep the exposition in this book accessible, but some preparation is certainly necessary. I expect that readers will be comfortable with basic group theory and character theory: the group theory that might be acquired in a strong first-year graduate algebra course and the character theory in the first half of my character theory book should be sufficient. I prefer to explain as much as possible in the text without invoking external references, but occasionally I do refer to one of my earlier books: [**CTFG**], *Character theory of finite groups*, AMS Chelsea Publishing, Providence, RI, 2006; [**FGT**], *Finite group theory*, Graduate Studies in Mathematics, 92, American Mathematical Society, 2008; and [**AGC**], *Algebra: A graduate course*, Graduate Studies in Mathematics, 100, American Mathematical Society, 2009. There are also numerous external references at the end of each chapter in the Notes and References sections. In those sections, I have attempted to document the history of the material in the chapter, but let me take this opportunity to apologize to anyone who feels that I have not sufficiently acknowledged his or her contribution; any such omissions were inadvertent.

Several mathematicians have contributed to this effort. First, I want to thank my friend and frequent collaborator Gabriel Navarro. Not only did he point out numerous typographical errors and unclear presentations as he read through each chapter, but also, and much more importantly, Gabriel is responsible for a number of the ideas in this book, ideas that came from his work or that were inspired by our collaboration and by his frequent email communications with me. I also want to thank my former student, Mark Lewis. Mark independently started working on a book on solvable group character theory, and he was kind enough to share a preliminary manuscript with me. A number of the ideas in that document influenced my thinking, especially on the character-correspondence material. Finally, I offer thanks to Ping Jin, who found an embarrassingly large number of typos in the chapters he read. Of course, none of these people is responsible for any of the typographical or mathematical errors that may remain.

I. Martin Isaacs
Berkeley, California
August 2017

Part 1

π-Theory

Chapter 1

π-Separable Groups and Character Theory Background

1A

We include under the rubric "π-theory" the study of theorems that relate the set $\mathrm{Irr}(G)$ of irreducible characters of a finite solvable group G to some set π of prime numbers. But despite the word "solvable" appearing in the previous sentence as well as in the title of this book, many of the results of π-theory do not actually require that our groups should be solvable; what we generally need is that from the "perspective" of the prime set π, our groups should resemble solvable groups. To be more precise, it will usually suffice that our groups are "π-separable", and in this introductory section, we review the definition and basic properties of this class of groups.

In fact, solvable groups are guaranteed to be π-separable for all prime sets π. We could, therefore, have limited our study to solvable groups, and had we done so, the essence of the theory would have survived. (Readers who are so inclined can thus ignore the extra generality, and they can simply pretend that our π-separable groups are actually solvable.) Alternatively, by appealing to the Feit–Thompson odd-order theorem and expending only a little extra effort, we can obtain the more general π-separable results, and that is the course we follow.

Given a prime set π, we say that a finite group G is a π-group if every prime divisor of the order $|G|$ of G lies in π. We generally write π' to denote the complement of the set π in the set of all prime numbers, and thus a

π'-group is one for which no prime in π divides the order. We say that G is π-**separable** if there exist normal subgroups $N_i \lhd G$ with

$$1 = N_0 \subseteq N_1 \subseteq \cdots \subseteq N_r = G\,,$$

and such that for $0 < i \leq r$, each of the factor groups N_i/N_{i-1} is either a π-group or a π'-group. It should be clear that G is π-separable if and only if it is π'-separable, and it is almost as easy to see that every subgroup and every homomorphic image of a π-separable group is π-separable.

Recall now that $\mathbf{O}_\pi(G)$ is the unique largest normal π-subgroup of G and $\mathbf{O}_{\pi'}(G)$ is the unique largest normal π'-subgroup of G. If G is π-separable and nontrivial, therefore, then either $\mathbf{O}_\pi(G) > 1$ or $\mathbf{O}_{\pi'}(G) > 1$ (or both), and thus G has a nontrivial characteristic subgroup K that is either a π-group or a π'-group. Also, if $K < G$, then G/K is π-separable, and so it has a nontrivial characteristic π or π'-subgroup. Continuing like this, we see that if G is π-separable, then it must have a chain of *characteristic* subgroups with factors that are π-groups or π'-groups. In other words, in the previous paragraph, where we defined π-separability, we could have required the subgroups N_i to be characteristic and not merely normal.

What if we weaken the condition on the N_i in our definition, and instead of assuming that each of these subgroups is normal in G, suppose we assume only that each is normal in the next? We suppose, therefore, that we have the subnormal series

$$1 = N_0 \lhd N_1 \lhd \cdots \lhd N_r = G\,,$$

where each factor is either a π-group or a π'-group. In this situation, and assuming that G is nontrivial, we see that either $\mathbf{O}^\pi(G) < G$ or $\mathbf{O}^{\pi'}(G) < G$. (Recall that $\mathbf{O}^\pi(G)$ is the unique smallest normal subgroup of G whose quotient is a π-group, and similarly for $\mathbf{O}^{\pi'}(G)$.) In either case, therefore, G has a proper characteristic subgroup L such that G/L is a π-group or a π'-group. By intersecting L with the given subnormal series in G, we see that L has a corresponding subnormal series, and thus if L is nontrivial, it has a proper characteristic subgroup whose factor group is either a π-group or a π'-group. Also, since this subgroup is characteristic in a characteristic subgroup of G, it is itself characteristic in G. Continuing like this, we see that if G has a subnormal series with π- and π'-factors, then in fact it has a characteristic series with π- and π'-factors, and thus by definition G is π-separable. In other words, in our original definition of π-separability, it would have been sufficient to start with a subnormal series in place of the given normal series.

We offer one additional perspective on the definition of π-separability. We have seen that G is π-separable if and only if it has a subnormal series in which each factor is a π-group or a π'-group. Recall that a **composition**

series of G is a subnormal series in which each factor is a simple group. Observe that a subnormal series in a finite group can always be refined (by deleting repeats and then inserting additional subgroups if necessary) to a composition series. It follows that G is π-separable if and only if each factor in some composition series (or equivalently, by the Jordan–Hölder theorem, in *every* composition series) is either a π-group or a π'-group.

We recall now that a **chief series** of G is a normal series in which each factor N_i/N_{i-1} is a minimal normal subgroup of G/N_{i-1}. If $N \triangleleft G$ and M/N is a minimal normal subgroup of G/N, then, assuming that G is finite, it should be clear that M/N is one of the factors of some chief series for G, and for that reason, we refer to M/N as a **chief factor** of G.

The hypothesis that a group G is π-separable is often exploited by using the fact that each chief factor M/N of G is either a π-group or a π'-group. To see why this is so, observe first that M/N is π-separable, and thus M/N has a nontrivial characteristic subgroup K/N, where K/N is either a π-group or a π'-group. Also, since K/N is characteristic in the normal subgroup M/N, it follows that $K/N \triangleleft G/N$. By assumption, however, M/N is minimal normal in G/N, so $K/N = M/N$, and thus the chief factor M/N is either a π-group or a π'-group, as claimed. In particular, if M is minimal normal in a π-separable group G, then M is a chief factor of G, and hence M is either a π-group or a π'-group. Also, if N is maximal normal in G, then G/N is a chief factor, and thus G/N is either a π-group or a π'-group.

It is also true that if G is π-separable and H is a maximal (but not necessarily normal) subgroup of G, then the index $|G : H|$ must be either a π-number or a π'-number. To see this, let $L = \mathrm{core}_G(H)$, the largest normal subgroup of G contained in H. Then $L < G$, so we can choose a chief factor K/L of G. Since $K > L$, we see that K is not contained in H, so $H < KH \subseteq G$. By assumption, H is maximal in G, and thus $KH = G$,and we have $|G : H| = |KH : H| = |K : K \cap H|$, and this divides $|K : L|$ because $L \subseteq K \cap H$. It follows that $|G : H|$ is a π-number or a π'-number, as claimed.

To relate π-separable groups to solvable groups, recall that by definition a group G is **solvable** if it has a normal series with abelian factors, or equivalently, if its derived series

$$G \supseteq G' \supseteq G'' \supseteq \cdots$$

eventually reaches the identity. If G is solvable, then every composition factor of G is solvable and simple, and thus has prime order. (And it is easy to see that the converse holds too; if every composition factor of G has prime order, then G is solvable.) If π is an arbitrary set of primes, then obviously, for each prime p, either $p \in \pi$ or $p \in \pi'$, and thus each composition factor of a solvable group is either a π-group or a π'-group, and it follows (as we have already mentioned) that solvable groups are π-separable for all prime sets π.

Conversely, suppose that G is π-separable for all π. (In fact, it suffices to assume that G is π-separable for all singleton prime sets π.) If S is a composition factor of G and p is a prime divisor of $|S|$, then, taking $\pi = \{p\}$, we see that S is not a π'-group, and so S must be a π-group; in other words, S is a p-group. Now S is a simple p-group, so it has prime order, and since all composition factors of G have prime order, it follows that G is solvable.

One sometimes sees in the literature the phrase "π-solvable". Although we shall seldom use this concept, we mention for completeness that to say that G is **π-solvable** means that it is π-separable, and that in addition every composition factor of G that is a π-group has prime order. Note that if $\pi = \{p\}$ is a singleton set, then every simple π-group has prime order, and in this case a π-separable group is automatically π-solvable. For historical reasons, such groups are referred to as being **p-solvable**, and we will follow that convention. To repeat, therefore, "p-solvable" and "$\{p\}$-separable" are the same thing.

Recall that a **Hall π-subgroup** H of a group G (if it exists) is a π-subgroup such that the index $|G : H|$ is a π'-number or, equivalently, $|H|$ is the full "π-part" of $|G|$. A celebrated result of P. Hall asserts that if G is solvable, then G actually has Hall π-subgroups for all prime sets π, and in fact, G is solvable if and only if such subgroups exist for all π. Hall also showed that in a solvable group G and for every prime set π, all Hall π-subgroups of G are conjugate and that every π-subgroup is contained in some Hall π-subgroup. (Thus in solvable groups, analogues of the Sylow theorems hold with *sets* of primes replacing individual primes.)

As we suggested earlier, π-separable groups can be thought of as groups that look solvable from the perspective of the prime set π. In particular, we prove the following.

Theorem 1.1. *Suppose that G is π-separable for some set π of primes. Then*

(a) *G has a Hall π-subgroup.*

(b) *Every two Hall π-subgroups of G are conjugate in G.*

(c) *If $U \subseteq G$ is a π-subgroup, then there exists a Hall π-subgroup H of G with $U \subseteq H$.*

The key to the proof of Theorem 1.1 is the Schur–Zassenhaus theorem. (This result appears as Theorems 3.8 and 3.12 in [**FGT**].)

Theorem 1.2 (Schur–Zassenhaus). *Let $N \triangleleft G$, and assume that $|N|$ and $|G:N|$ are relatively prime. The following then hold.*

(a) *There exists a subgroup $H \subseteq G$ such that $NH = G$ and $N \cap H = 1$.*

(b) *If either N or G/N is solvable, then all subgroups H as in* (a) *are conjugate in G.*

A subgroup H such that $NH = G$ and $N \cap H = 1$ is said to be a **complement** for N in G. Under the assumption that $|N|$ and $|G/N|$ are relatively prime, the Schur–Zassenhaus theorem asserts that such complements exist, and it is easy to see (using the coprimeness) that the complements for N in G are exactly the subgroups of G having order equal to $|G:N|$.

If we assume the additional solvability hypothesis in Theorem 1.2(b), the Schur–Zassenhaus theorem also tells us that all complements for N in G are conjugate. Note, however, that the coprimeness hypothesis guarantees that at least one of $|N|$ and $|G:N|$ must be odd, so by the Feit–Thompson odd-order theorem, at least one of N or G/N is solvable, and thus the complements for N in G are conjugate unconditionally. Unfortunately, there is no known direct proof of this fact, independent of the odd-order theorem.

To prove Theorem 1.1, we first establish the following, which is a special case.

Corollary 1.3. *Let $N \triangleleft G$, where N is a π'-group and G/N is a π-group. Then G has Hall π-subgroups, and all such subgroups are conjugate in G. Also, if $U \subseteq G$ is a π-subgroup, then U is contained in some Hall π-subgroup of G.*

Proof. By the Schur–Zassenhaus theorem (and the odd-order theorem) there exist complements for N in G, and all such complements are conjugate in G. The complements for N in G are those subgroups of G having order $|G:N|$ and index $|N|$, and so they are exactly the Hall π-subgroups of G. All of these Hall π-subgroups are thus conjugate in G.

Observe now that U is a complement for N in NU. Also, if H is a complement for N in G, then since $N \subseteq NU \subseteq G = NH$, Dedekind's lemma yields $NU = N(NU \cap H)$. Since $N \cap (NU \cap H) \subseteq N \cap H = 1$, we see that $NU \cap H$ is a complement for N in NU, and thus by the conjugacy part of the Schur–Zassenhaus theorem applied in the group NU, the complements U and $NU \cap H$ are conjugate. We can thus write $U = (NU \cap H)^g \subseteq H^g$ for some element $g \in G$, and it follows that U is contained in the Hall π-subgroup H^g. ∎

Proof of Theorem 1.1. Observe that (a) follows from (c) by taking U to be the trivial subgroup, and thus we need to prove only (b) and (c). We proceed by induction on $|G|$.

Let N be minimal normal in G so that, as we have seen, N is either a π-group or a π'-group. To prove (c), let $U \subseteq G$ be a π-subgroup, and note that UN/N is a π-subgroup of G/N. By the inductive hypothesis, (c) holds in the group G/N, so G/N has a Hall π-subgroup V/N containing UN/N. We thus have $U \subseteq UN \subseteq V$.

Now $|V| = |N||V/N|$, so if N is a π-group, then $|V|$ is a product of two π-numbers. In this case, V is a π-group, and since $|G : V| = |(G/N) : (V/N)|$ is a π'-number, it follows that V is a Hall π-subgroup of G and (c) holds. Continuing to assume that N is a π-group, let H be an arbitrary Hall π-subgroup of G. Then HN is a π-group, and so $|HN|$ is at most the order of the Hall π-subgroup H. We deduce that $H = HN \supseteq N$, and thus H/N is a Hall π-subgroup of G/N. By the inductive hypothesis, (b) holds in G/N, and thus H/N and V/N are conjugate, and it follows that H and V are conjugate in G.

We can now assume that N is a π'-group, so we can apply Corollary 1.3 in the group V. We deduce that V has a Hall π-subgroup K containing U and that all Hall π-subgroups of V are conjugate. Now K is a π-group and $|G : K| = |G : V||V : K|$ is a product of two π'-numbers, and it follows that K is a Hall π-subgroup of G.

Finally, suppose that H is an arbitrary Hall π-subgroup of G. Then HN/N is easily seen to be a Hall π-subgroup of G/N, so by the inductive hypothesis, HN/N and V/N are conjugate in G/N. We can thus write $H^g \subseteq (HN)^g = V$ for some element $g \in G$. Since H^g is a Hall π-subgroup of G contained in V, it is actually a Hall π-subgroup of V. We have established that all Hall π-subgroups of V are conjugate, however, and it follows that H^g is conjugate to K, and thus H and K are conjugate in G, as required.

We close this section with Glauberman's lemma, which we will need in later chapters. Although this result appears as Corollary 13.10 of [**CTFG**] and also as Theorem 3.26 of [**FGT**], we present the proof here because it provides an excellent demonstration of the use of both parts of the Schur–Zassenhaus theorem.

Lemma 1.4 (Glauberman). *Suppose that a group A acts via automorphisms on a group G, where $|A|$ and $|G|$ are relatively prime. Suppose also that both A and G act on some set Ω, where G acts transitively, and assume that the following compatibility condition holds for all $\alpha \in \Omega$, $g \in G$, and $a \in A$:*

$$(\alpha \cdot g) \cdot a = (\alpha \cdot a) \cdot g^a \,.$$

Then

(a) *A fixes some point in Ω.*

(b) *The set of A-fixed points in Ω is an orbit under $\mathbf{C}_G(A)$.*

Proof. Let Γ be the semidirect product of G by A, constructed using the given action of A on G, and view A and G as subgroups of Γ. Thus $G \lhd \Gamma$ and the conjugation action of A on G in Γ is the original action, and every element of Γ is uniquely of the form ag, with $a \in A$ and $g \in G$.

We define an action of Γ on Ω by setting $\alpha{\cdot}(ag) = (\alpha{\cdot}a){\cdot}g$. To see that this is truly an action, let $a, b \in A$ and $g, h \in G$, and compute

$$(\alpha{\cdot}(ag)){\cdot}(bh) = (((\alpha{\cdot}a){\cdot}g){\cdot}b){\cdot}h = (((\alpha{\cdot}a){\cdot}b){\cdot}g^b){\cdot}h = \alpha{\cdot}(abg^bh) = \alpha{\cdot}(agbh)\,,$$

where the second equality follows from the compatibility condition.

Fix a point $\alpha \in \Omega$, and note that Γ acts transitively on Ω because G acts transitively. Writing G_α and Γ_α to denote the stabilizers of α in G and in Γ, we have

$$|\Gamma : \Gamma_\alpha| = |\Omega| = |G : G_\alpha|\,,$$

and this is a π-number, where π is the set of prime divisors of $|G|$. Now G is a normal Hall π-subgroup of Γ, and it follows that Γ_α also has a normal Hall π-subgroup. By the Schur–Zassenhaus theorem, therefore, Γ_α has a Hall π'-subgroup H, and we see that H is a full Hall π'-subgroup of Γ.

Since A is also a Hall π'-subgroup of Γ, it follows from the conjugacy part of the Schur–Zassenhaus theorem that $A = H^x$ for some element $x \in \Gamma$. Since H fixes α, we see that A fixes $\alpha{\cdot}x$, and this proves (a).

The compatibility condition shows that if A fixes $\beta \in \Omega$, then A also fixes $\beta{\cdot}t$, for elements $t \in \mathbf{C}_G(A)$. Then $\mathbf{C}_G(A)$ permutes the fixed points of A in Ω, and to prove (b), it remains to show that this action is transitive.

Let β and γ be A-fixed points, and choose $g \in G$ such that $\beta{\cdot}g = \gamma$. Then the set Λ of all elements of G that carry β to γ is the coset gC, where $C = G_\gamma$.

Now C acts transitively by right multiplication on Λ, and since β and γ are fixed by A, it is easy to see that the set Λ and the subgroup C are permuted by the given action of A on G. Also, $|A|$ and $|C|$ are relatively prime, and it is easy to see that the compatibility condition is satisfied for the actions of A on C and on Λ and the action of C on Λ. By part (a) of the lemma, there is an A-fixed point $x \in \Lambda$, and thus $x \in \mathbf{C}_G(A)$. Also, by the definition of Λ, we have $\beta{\cdot}x = \gamma$, and this shows that $\mathbf{C}_G(A)$ acts transitively on the set of A-fixed points in Ω. ∎

1B

There are two positive integers that are naturally associated with each irreducible character χ of an arbitrary finite group G. One of these, of course, is the degree $\chi(1)$, and we recall that $\chi(1)$ divides $|G|$. The other integer to which we refer also divides $|G|$: it is the "determinantal order" $o(\chi)$, which we are about to define.

Let $\mathcal{X}$ be a matrix representation of G that corresponds to the character χ. Thus $\mathcal{X}$ is a homomorphism from G into the matrix group $GL(n, \mathbb{C})$, where $n = \chi(1)$, and for each element $g \in G$, the trace of the matrix $\mathcal{X}(g)$ is $\chi(g)$. The function $\lambda : G \to \mathbb{C}$, defined by $\lambda(g) = \det(\mathcal{X}(g))$, is clearly a homomorphism, or in other words, it is a linear character of G. We would like to write $\lambda = \det(\chi)$, but to justify this, we must observe that λ does not depend on the particular matrix representation $\mathcal{X}$ that we chose. If $\mathcal{Y}$ is another matrix representation of G yielding χ, we know that $\mathcal{Y}$ is similar to $\mathcal{X}$, and this means that there is some invertible $n \times n$ matrix M such that $\mathcal{Y}(g) = M^{-1}\mathcal{X}(g)M$ for all $g \in G$. Since $\mathcal{Y}(g)$ and $\mathcal{X}(g)$ are similar matrices, they have equal determinants, and thus the values of $\lambda(g)$ computed using $\mathcal{X}$ and $\mathcal{Y}$ are equal. This justifies the notation $\lambda = \det(\chi)$.

Caution is appropriate here because the linear character $\det(\chi)$ cannot be determined by examining the character table of G. For example, the dihedral group D_8 and the quaternion group Q_8 have identical character tables, and each of these groups has a unique irreducible character χ of degree 2, but $\lambda = \det(\chi)$ is the principal character for Q_8, and it is nonprincipal for D_8. (In fact, it is easy to see that $\lambda(x) = -1$ if x is a noncentral involution in D_8.)

Finally, we define the **determinantal order** $o(\chi)$ to be the order of $\det(\chi)$ as an element of the multiplicative group of linear characters of G. Since this group has order $|G : G'|$, we see that $o(\chi)$ divides $|G : G'|$, and thus $o(\chi)$ divides $|G|$, as we promised.

An important connection between character degrees, determinantal orders, and the normal structure of a group is the following, to which most of the remainder of this section is devoted. ∎

Theorem 1.5 (Gallagher). *Let $N \lhd G$, where G/N is a π'-group, and suppose that $\theta \in \mathrm{Irr}(N)$ is invariant in G. If both $\theta(1)$ and $o(\theta)$ are π-numbers, then θ extends to G, and there is a unique extension χ such that $o(\chi)$ is a π-number. In fact, $o(\chi) = o(\theta)$.*

This result is proved more than once in [**CTFG**], where the proofs are scattered over several chapters. The proof we present here ultimately relies on the theory of projective representations and not, as in Gallagher's original argument, on Brauer's theorem on induced characters. (As we shall see, if G/N is solvable, the proof is much more elementary, so this is one of the places where less work would be required if we were willing to limit ourselves to solvable groups.)

First, we *assume* that an extension exists in Theorem 1.5, and we prove that in that situation, there is a unique extension with the appropriate determinantal order. We mention that this unique extension of θ is called the **canonical extension**.

Lemma 1.6. *Assume the hypotheses of Theorem* 1.5, *and suppose that* θ *extends to* G. *Then* θ *has a unique extension* $\chi \in \mathrm{Irr}(G)$ *such that* $o(\chi)$ *is a* π*-number. In fact,* $o(\chi) = o(\theta)$.

To prove this, we need another important and well-known result of Gallagher: his correspondence theorem. This appears as Corollary 6.17 of [**CTFG**], and we state it here without proof. In Chapter 2, however, we will need a generalization (Theorem 2.5) of the Gallagher correspondence. Although that result also appears in [**CTFG**], it is probably less well known, and we will provide a proof.

In the following, we write $\mathrm{Irr}(G|\theta)$ to denote the set of irreducible characters of G that "lie over" a given irreducible character θ of some subgroup N of G. In other words, the members of $\mathrm{Irr}(G|\theta)$ are the irreducible constituents of the induced character θ^G, or equivalently, they are the characters $\chi \in \mathrm{Irr}(G)$ such that θ is a constituent of the restriction χ_N. (In this situation, we sometimes say instead that θ "lies under" χ.) If $N \triangleleft G$ and χ "lies over" θ, where $\theta \in \mathrm{Irr}(N)$ is invariant in the conjugation action of G, then in fact χ_N is actually a multiple of θ. This is a consequence of Clifford's theorem, which we discuss in more detail in Theorem 1.17, below.

Theorem 1.7 (Gallagher correspondence)**.** *Let* $N \triangleleft G$, *and suppose* $\psi \in \mathrm{Irr}(G)$ *and that* $\psi_N = \theta$ *is irreducible. Then the map* $\beta \mapsto \beta\psi$ *is a bijection from* $\mathrm{Irr}(G/N)$ *onto* $\mathrm{Irr}(G|\theta)$.

Proof of Lemma 1.6. Write $m = o(\theta)$, so that by hypothesis, m is a π-number. If χ is any extension of θ, then $\det(\chi)$ extends $\det(\theta)$, and so $\det(\chi)$ has order at least m. Our goal is to show that for some choice of χ, this order is exactly m, but that for no extension of θ other than χ is the order a π-number.

Let ψ be an arbitrary extension of θ to G. By the Gallagher correspondence, every irreducible character of G that lies over θ has the form $\beta\psi$ for some uniquely determined character β of G/N. Also, since $(\beta\psi)_N = \beta(1)\theta$, we see that the extensions of θ to G are exactly the characters of the form $\beta\psi$ for *linear* characters β of G/N, and so we must compute the determinantal orders of these characters. Writing $n = \psi(1)$ and $\mu = \det(\psi)$, we observe that

$$\det(\beta\psi) = \beta^{\psi(1)} \det(\psi) = \beta^n \mu\,.$$

Since $\psi_N = \theta$, we see that $\mu_N = \det(\theta)$, and thus

$$(\mu^m)_N = (\mu_N)^m = (\det(\theta))^m = 1_N ,$$

and μ^m can be viewed as a linear character of G/N. By hypothesis, $n = \psi(1) = \theta(1)$ is a π-number, so mn is a π-number. Since the group of linear characters of G/N is a π'-group, it follows that the map $\beta \mapsto \beta^{mn}$ maps this group onto itself, so there exists a linear character γ of G/N such that $\gamma^{mn} = \mu^{-m}$.

We have

$$\det(\gamma\psi)^m = (\gamma^n\mu)^m = \gamma^{nm}\mu^m = 1_G ,$$

and thus $o(\gamma\psi)$ divides m. Since this order cannot be less than m, we have equality, as wanted.

Finally, to prove the uniqueness assertion, we can change notation and assume that ψ is an extension of θ with determinantal order m. We assume also that $o(\beta\psi)$ is a π-number, where β is a linear character of G/N, and we work to show that $\beta = 1_G$. As before, we have $\det(\beta\psi) = \beta^n\mu$, but now we have $o(\mu) = m$, so $\beta^n = \mu^{-1}\det(\beta\psi)$ is a product of two linear characters whose orders are π-numbers. Also, n is a π-number, and thus $o(\beta)$ must be a π-number. Now β is a character of G/N, so its order divides the π'-number $|G/N|$, and hence $o(\beta) = 1$. Since β is linear, it is principal, as required. ∎

If G/N is solvable, it is comparatively easy to prove the existence of an extension in Theorem 1.5, and we do this first. The general result is considerably deeper, and our proof will rely on the solvable case and on the theory of character-triple isomorphisms, which we will review. We need a preliminary lemma, which appears as Corollary 6.20 in [**CTFG**].

Lemma 1.8. *Let $N \triangleleft G$, where $|G:N|$ is prime, and let $\theta \in \mathrm{Irr}(N)$. Then either θ has exactly $|G:N|$ distinct conjugates in G or else every member of $\mathrm{Irr}(G|\theta)$ is an extension of θ.*

In particular, Lemma 1.8 guarantees that if θ is invariant in G, then θ extends to G. In fact, this conclusion holds even if we weaken the assumption that G/N has prime order, and we assume only that G/N is cyclic. This more general result is an easy corollary of the theory of character-triple isomorphisms, but for the lemma as stated, there is an entirely elementary proof.

Proof of Lemma 1.8. The stabilizer of θ contains N, so it is either N or G, and in the first case θ has exactly $|G:N|$ distinct conjugates. We can assume, therefore, that θ is invariant in G, and we work to show that if $\chi \in \mathrm{Irr}(G|\theta)$, then $\chi_N = \theta$.

Since θ is invariant in G, we have seen that Clifford's theorem guarantees that $\chi_N = e\theta$ for some integer e, and our goal is to show that $e = 1$. If χ vanishes on $G - N$, it follows from the definition of the character inner product that

$$|N|e^2 = |N|[\chi_N, \chi_N] = |G|[\chi, \chi] = |G|\,,$$

and so $e^2 = |G : N|$. Since $|G : N|$ is prime, this is impossible, and hence there exists an element $g \in G - N$ such that $\chi(g) \neq 0$. It follows that the characters $\beta\chi$ are distinct as β runs over the linear characters of G/N, and they are irreducible since χ is irreducible and the β are linear. Each of the characters $\beta\chi$ lies over θ, and thus θ^G has at least $|G : N|$ distinct irreducible constituents, each of which has degree $\chi(1) = e\theta(1)$. Then

$$|G : N|\theta(1) = \theta^G(1) \geq |G : N|\chi(1) = |G : N|e\theta(1)\,,$$

and so $1 \geq e$. We conclude that $e = 1$, as wanted. ■

Corollary 1.9. *Theorem* 1.5 *holds in the case where G/N is solvable.*

Proof. We proceed by induction on $|G : N|$. There is nothing to prove if $N = G$, so we can assume that the solvable group G/N is nontrivial, and hence there exists a prime-index subgroup $M/N \lhd G/N$. By the inductive hypothesis, θ has a unique canonical extension $\xi \in \mathrm{Irr}(M)$, and the uniqueness guarantees that ξ is invariant in G. Since $|G : M|$ is prime, we conclude by Lemma 1.8 that ξ extends to G, and thus θ extends to G. The result then follows by Lemma 1.6. ■

To handle the case where G/N is not necessarily solvable, we will use the following general extendibility criterion.

Theorem 1.10. *Let $\theta \in \mathrm{Irr}(N)$, where $N \lhd G$, and assume that for each prime divisor p of $|G/N|$, there exists a Sylow p-subgroup P/N of G/N such that θ extends to P. Then θ extends to G.*

Assuming Theorem 1.10 for the moment, the proof of Theorem 1.5 is easy.

Proof of Theorem 1.5. Given a prime divisor p of G/N, let P/N be a Sylow p-subgroup of G/N. Then P/N is a solvable π'-group, so θ extends to P by Corollary 1.9. It follows by Theorem 1.10 that θ extends to G, and the result follows by Lemma 1.6. ■

Our proof of Theorem 1.10 proceeds in two steps: we prove the result when N is central in G, and then we use character-triple isomorphisms to reduce the general problem to that case.

Lemma 1.11. *Assume the hypotheses of Theorem* 1.10, *and assume in addition that $N \subseteq \mathbf{Z}(G)$. Then θ extends to G.*

Proof. Since N is abelian, θ is linear, so we can write $\theta = \prod \mu_i$, where each factor is a linear character having prime-power order. This can be done, moreover, so that the characters μ_i all lie in the subgroup generated by θ in the group of linear characters of N. By hypothesis, θ extends to the various groups P as in the statement of Theorem 1.10, so the same holds for all powers of θ, and thus the characters μ_i satisfy the hypotheses of the lemma. If we can show each μ_i has an extension to G, then the product of these extensions will be the desired extension of θ, and thus it is no loss to assume that $o(\theta)$ is some prime power q.

Let p be the prime divisor of q. Let P/N be a Sylow p-subgroup of G/N such that θ has an extension $\varphi \in \mathrm{Irr}(P)$, and note that φ is linear. Each irreducible constituent of φ^G lies over φ, and hence it lies over θ. Since N is central, the restrictions of the constituents of φ^G to N are multiples of θ, and it follows that

$$(\varphi^G)_N = \varphi^G(1)\theta = |G : P|\varphi(1)\theta = |G : P|\theta \,.$$

Now let $\lambda = \det(\varphi^G)$, and observe that

$$\lambda_N = \det(\varphi^G)_N = \det((\varphi^G)_N) = \det(|G : P|\theta) = \theta^{|G:P|} \,.$$

Since $|G : P|$ and q are relatively prime, there exists an integer m such that $m|G : P| \equiv 1 \bmod q$, and we have

$$(\lambda^m)_N = \theta^{m|G:P|} = \theta \,,$$

where the last equality holds since $o(\theta) = q$. Thus λ^m is the desired extension of θ. ∎

To complete the proof of Theorem 1.5, we need some theory from Chapter 11 of [**CTFG**], which we review here. If $N \triangleleft G$ and $\theta \in \mathrm{Irr}(N)$ is invariant in G, we refer to the object (G, N, θ) as a **character triple**, and our first goal is to define what it means for two character triples to be "isomorphic".

We begin by introducing some nomenclature that we hope will make this somewhat complicated definition a little more transparent. Given a character triple (G, N, θ), we refer to a subgroup X such that $N \subseteq X \subseteq G$ as an **intermediate** subgroup, and we say that a (possibly reducible) character ξ of an intermediate subgroup is **admissible** if its restriction to N is a multiple of θ. (Equivalently, ξ is admissible precisely when every irreducible constituent of ξ lies in $\mathrm{Irr}(X|\theta)$.)

Suppose now that (G, N, θ) and (G^*, N^*, θ^*) are character triples. To say that they are isomorphic, we require first that $G/N \cong G^*/N^*$, and we fix a particular isomorphism $\tau : G/N \to G^*/N^*$. Given an intermediate subgroup X of (G, N, θ), we write X^* to denote the intermediate subgroup of (G^*, N^*, θ^*) defined by the equation $X^*/N^* = \tau(X/N)$. The map $X \mapsto X^*$

is thus a bijection from the set of intermediate subgroups of (G, N, θ) onto the set of intermediate subgroups of (G^*, N^*, θ^*).

Next, we assume that for each intermediate subgroup X of (G, N, θ), there exists a bijection $\mathrm{Irr}(X|\theta) \to \mathrm{Irr}(X^*|\theta^*)$. These maps on irreducible characters are then extended linearly so that for every admissible character ξ of an intermediate subgroup X of (G, N, θ), there is a uniquely defined admissible character ξ^* of X^*, and we observe that this defines a bijection from the set of all admissible characters of intermediate subgroups of (G, N, θ) onto the set of all admissible characters of intermediate subgroups of (G^*, N^*, θ^*). Since the maps $\xi \mapsto \xi^*$ are defined as linear extensions of bijections on irreducible characters, it is clear that if ξ and η are admissible characters of an intermediate subgroup X, then $(\xi + \eta)^* = \xi^* + \eta^*$, and we refer to this as the "additivity property". It should also be clear that $[\xi^*, \eta^*] = [\xi, \eta]$, and this is called the "isometry property".

The isomorphism $\tau : G/N \to G^*/N^*$ together with the $*$-maps on admissible characters are said to form a **character-triple isomorphism** from (G, N, θ) to (G^*, N^*, θ^*) if the various maps respect character restriction. To be more precise, we require that whenever $N \subseteq X \subseteq Y \subseteq G$ and ξ is an admissible character of Y, we have

$$(\xi_X)^* = (\xi^*)_{X^*} .$$

We refer to this as the "restriction property", and we mention that a consequence that can be proved using Frobenius reciprocity is the corresponding "induction property", which asserts that

$$(\eta^Y)^* = (\eta^*)^{Y^*},$$

when η is an admissible character of X.

A key observation is that character-triple isomorphisms respect character extendibility.

Lemma 1.12. *Let (G, N, θ) and (G^*, N^*, θ^*) be isomorphic character triples, and suppose that $N \subseteq U \subseteq G$ and $\psi \in \mathrm{Irr}(U|\theta)$. Then $\psi^* \in \mathrm{Irr}(U^*|\theta^*)$ and*

$$\frac{\psi(1)}{\theta(1)} = \frac{\psi^*(1)}{\theta^*(1)} .$$

In particular, θ is extendible to U if and only if θ^* is extendible to U^*.

Proof. By definition, the $*$-map carries irreducible characters to irreducible characters, so we certainly have $\psi^* \in \mathrm{Irr}(U^*|\theta^*)$. We can write $\psi_N = e\theta$ for some integer e, and thus

$$e\theta^* = (e\theta)^* = (\psi_N)^* = (\psi^*)_{N^*} ,$$

where the third equality follows from the restriction property. Thus

$$\frac{\psi(1)}{\theta(1)} = e = \frac{\psi^*(1)}{\theta^*(1)},$$

as required. Now ψ is an extension of θ if and only if $e = 1$, and this happens precisely when ψ^* is an extension of θ^*. ■

If $N \triangleleft G$ and we wish to show that some G-invariant character $\theta \in \mathrm{Irr}(N)$ extends to G, it suffices by Lemma 1.12 to find a character triple (G^*, N^*, θ^*) isomorphic to (G, N, θ) and to prove that θ^* extends to G^*. At first, this may not seem to help much, but in fact, it is always possible to choose an isomorphic triple (G^*, N^*, θ^*) such that $N^* \subseteq \mathbf{Z}(G^*)$, and consequently this technique can be very powerful. The key result here is the following, which appears in [**CTFG**] as Theorem 11.28. The proof depends on the theory of projective representations, and we will not reproduce it here.

Theorem 1.13. *Given an arbitrary character triple (G, N, θ), there exists an isomorphic character triple (G^*, N^*, θ^*) such that θ^* is faithful and $N^* \subseteq \mathbf{Z}(G^*)$.*

It should be noted that we have stated our definition of an isomorphism of character triples somewhat differently from the way it appears in Definition 11.23 of [**CTFG**]. In that definition, there are four conditions on the various maps. Conditions (a) and (b) are the additive and isometry properties, and, as we have observed, these are automatically satisfied in our formulation. Condition (c) of Definition 11.23 is our restriction property. The proof of Theorem 11.28 of [**CTFG**] actually establishes that two further properties can be assumed to hold. One of these is (d) in Definition 11.23 and the other appears as the definition of a "strong" character-triple isomorphism in Problem 11.13 of [**CTFG**]. However, we need neither of these now, so we will defer further discussion of them.

We now prove Theorem 1.10, and this will complete the proof of Gallagher's theorem.

Proof of Theorem 1.10. By Theorem 1.13, there exists a character triple (G^*, N^*, θ^*) isomorphic to (G, N, θ) and such that $N^* \subseteq \mathbf{Z}(G^*)$. By hypothesis, we know that for each prime divisor p of $|G/N| = |G^*/N^*|$, there exists a Sylow p-subgroup P/N of G/N such that θ extends to P. Then P^*/N^* is a Sylow p-subgroup of G^*/N^*, and by Lemma 1.12 we see that θ^* extends to P^*. By Lemma 1.11, therefore, θ^* extends to G^*, and thus by Lemma 1.12 again, θ extends to G. ■

Another application of character-triple isomorphisms, and specifically Theorem 1.13, is the following strong form of Lemma 1.8.

Corollary 1.14. *Let $N \triangleleft G$, where G/N is cyclic, and let $\theta \in \mathrm{Irr}(N)$ be invariant in G. Then θ extends to G.*

Proof. Let (G^*, N^*, θ^*) be a character triple isomorphic to (G, N, θ), where $N^* \subseteq \mathbf{Z}(G^*)$. Now $G^*/N^* \cong G/N$ and hence G^*/N^* is cyclic. It follows that G^* is abelian, and thus the members of $\mathrm{Irr}(G^*|\theta^*)$ are linear, and we conclude that θ^* extends to G^*. By Lemma 1.12, therefore, θ extends to G. ∎

We need one more theorem that connects character degrees and determinantal orders with normal subgroups.

Theorem 1.15. *Let $N \triangleleft G$ and $\chi \in \mathrm{Irr}(G)$, and let $\theta \in \mathrm{Irr}(N)$ be a constituent of χ_N. The following then hold.*

(a) *$\chi(1)$ divides $|G : N|\theta(1)$.*

(b) *$o(\chi)$ divides $|G : N|o(\theta)$.*

An immediate consequence of Theorem 1.15(a) is the following.

Corollary 1.16. *Let $N \triangleleft G$, and suppose that $\chi \in \mathrm{Irr}(G)$ has degree relatively prime to $|G : N|$. Then χ_N is irreducible.*

Proof. Let θ be an irreducible constituent of χ_N, so $\theta(1) \leq \chi(1)$, and we want to prove equality. But $\chi(1)$ divides $|G : N|\theta(1)$ by Theorem 1.15, and since $\chi(1)$ and $|G : N|$ are relatively prime, we conclude that $\chi(1)$ divides $\theta(1)$. This yields $\chi(1) \leq \theta(1)$, and the result follows. ∎

To prove Theorem 1.15, we recall Clifford's theorem and the Clifford correspondence from Chapter 6 of [**CTFG**].

Theorem 1.17 (Clifford). *Let $N \triangleleft G$ and suppose $\chi \in \mathrm{Irr}(G)$. Then there exist an integer e and a G-orbit $\{\theta_i \mid 1 \leq i \leq t\}$ of distinct irreducible characters of N such that $\chi_N = e\sum \theta_i$.*

Theorem 1.18 (Clifford correspondence). *Let $N \triangleleft G$ and $\theta \in \mathrm{Irr}(N)$, and let $T = G_\theta$ be the stabilizer of θ in G. Then $\psi \mapsto \psi^G$ is a bijection from $\mathrm{Irr}(T|\theta)$ onto $\mathrm{Irr}(G|\theta)$. Also, if $\psi \in \mathrm{Irr}(T|\theta)$ and $\psi^G = \chi \in \mathrm{Irr}(G|\theta)$ are corresponding characters, then $[\psi_N, \theta] = [\chi_N, \theta]$.*

Observe that if θ is one of the t characters θ_i in Theorem 1.17, then $\chi(1) = et\theta(1)$, and if $T = G_\theta$ as in Theorem 1.18, then $t = |G : T|$. Also, the integer e in Theorem 1.17 is equal to

$$[\chi_N, \theta] = [\psi_N, \theta] = \frac{\psi(1)}{\theta(1)},$$

where ψ is the Clifford correspondent of χ with respect to θ. (The second equality holds because θ is invariant in T, and thus ψ_N is a multiple of θ.)

Proof of Theorem 1.15. In the notation of Clifford's theorem, θ is one of the characters θ_i, and we have $\chi(1) = et\theta(1)$. To prove (a), therefore, we must show that et divides $|G : N|$, or equivalently, that e divides $|T : N|$, where $T = G_\theta$. If ψ is the Clifford correspondent of χ with respect to θ, we have seen that $e = \psi(1)/\theta(1)$, so our task is to compute this ratio of degrees.

Now (T, N, θ) is a character triple, and we have seen that if (T^*, N^*, θ^*) is any isomorphic triple, then $\psi(1)/\theta(1) = \psi^*(1)/\theta^*(1)$. We can choose an isomorphic triple such that $N^* \subseteq \mathbf{Z}(T^*)$, and thus $\theta^*(1) = 1$, and we have $e = \psi^*(1)$. Since the degree of an irreducible character of a finite group always divides the index of the center, we deduce that e divides $|T^* : N^*| = |T : N|$, as desired. (This fact about character degrees is a consequence of Theorem 3.12 of [**CTFG**].)

For (b), write $\lambda = \det(\chi)$ and $\mu_i = \det(\theta_i)$, where the characters θ_i are as in Clifford's theorem. Then

$$\lambda_N = (\prod_{i=1}^{t} \mu_i)^e ,$$

and since all θ_i are conjugate to θ in G, each of these characters has determinantal order equal to $o(\theta)$. Then $(\mu_i)^{o(\theta)} = 1_N$, and we have

$$(\lambda^{o(\theta)})_N = (\lambda_N)^{o(\theta)} = 1_N .$$

We can thus view $\lambda^{o(\theta)}$ as a character of G/N, and hence $(\lambda^{o(\theta)})^{|G:N|} = 1_G$. It follows that $o(\chi) = o(\lambda)$ divides $o(\theta)|G : N|$, proving (b). ∎

1C

Suppose that $N \triangleleft G$, where G is an arbitrary finite group, and let $\theta \in \mathrm{Irr}(N)$ be invariant in G. If θ is extendible to G, it follows by the Gallagher correspondence (Theorem 1.7) that $|\mathrm{Irr}(G|\theta)| = |\mathrm{Irr}(G/N)|$, so in this situation, the number of irreducible characters of G that lie over θ is equal to the number of conjugacy classes of G/N. More generally, if θ is G-invariant but is not necessarily extendible, a theorem of Gallagher asserts that the number of irreducible characters of G that lie over θ is equal to the number of certain "θ-good" conjugacy classes of G/N. We begin our discussion of Gallagher's theorem with the relevant definition.

Write $\overline{G} = G/N$, and fix an element $\overline{g} \in \overline{G}$. (As usual, the overbar denotes the natural homomorphism $G \to \overline{G}$, and thus $\overline{g}$ is another name for the coset Ng.) We say that $\overline{g}$ is a **θ-good** element of $\overline{G}$ if for every choice of $\overline{x} \in \overline{G}$ centralizing $\overline{g}$, the character θ is extendible to $\langle N, g, x\rangle$. Note that the subgroup $\langle N, g, x\rangle$ is uniquely determined by the cosets $\overline{g}$ and $\overline{x}$, and observe that $\overline{x}$ centralizes $\overline{g}$ if and only if $\langle N, g, x\rangle/N$ is abelian or, equivalently, $[g, x] \in N$. If $\overline{g}$ is θ-good, then clearly, the same is true about every conjugate of $\overline{g}$ in $\overline{G}$, and we say that a conjugacy class of $\overline{G}$ is θ-good if its elements are θ-good.

Theorem 1.19 (Gallagher's character count). *Let $N \lhd G$, and let $\theta \in \mathrm{Irr}(N)$ be G-invariant. Then $|\mathrm{Irr}(G|\theta)|$ is equal to the number of θ-good classes of G/N.*

To prove Theorem 1.19, we will use a character-triple isomorphism to reduce the problem to the case where N is central and θ is faithful. In that situation, the following characterization of the θ-good elements of $\overline{G} = G/N$ will be useful.

Lemma 1.20. *Let $N \subseteq \mathbf{Z}(G)$, and suppose that $\theta \in \mathrm{Irr}(N)$ is faithful. Then an element $\overline{g} \in \overline{G} = G/N$ is θ-good if and only if $[g,x] = 1$ whenever $x \in G$ and $[g,x] \in N$.*

Proof. Suppose first that $\overline{g}$ is θ-good. If $[g,x] \in N$, then $\overline{x}$ centralizes $\overline{g}$, and thus by assumption, θ extends to some character λ of $\langle N, g, x\rangle$. In particular, $\lambda(1) = \theta(1) = 1$, so λ is linear, and thus $[g,x] \in N \cap \ker(\lambda) = \ker(\theta) = 1$.

Conversely, suppose that $[g,x] = 1$ for all elements $x \in G$ such that $[g,x] \in N$. If $\overline{x}$ centralizes $\overline{g}$, then $[g,x] \in N$, and thus by hypothesis, $[g,x] = 1$. Then g and x commute, and since N is central, $\langle N, g, x\rangle$ is an abelian group. It follows that θ extends to $\langle N, g, x\rangle$, and thus $\overline{g}$ is θ-good, as required. ∎

Our next result is quite general.

Lemma 1.21. *Let $\chi \in \mathrm{Irr}(G)$. Then*

$$\sum_{x,y\in G} \chi([x,y]) = \frac{|G|^2}{\chi(1)}\,.$$

Proof. Let $\mathcal{X}$ be a representation that affords the character χ. Given an element $x \in G$, let K be the conjugacy class of x, and write $\hat{K}$ to denote the sum of the elements of K in the group algebra $\mathbb{C}[G]$. Then $\hat{K}$ is central in $\mathbb{C}[G]$, so by Schur's lemma, $\mathcal{X}(\hat{K})$ is a scalar matrix, and we can write $\mathcal{X}(\hat{K}) = \omega I$, where $\omega \in \mathbb{C}$ and I is an identity matrix. Taking traces, we have $|K|\chi(x) = \chi(1)\omega$, and we obtain the familiar formula $\omega = \chi(x)|K|/\chi(1)$.

As y runs over G, each element of K occurs $|\mathbf{C}_G(x)|$ times in the form x^y, and thus working in $\mathbb{C}[G]$ we have

$$\sum_{y\in G}[x,y] = \sum_{y\in G} x^{-1}x^y = x^{-1}\sum_{y\in G} x^y = |\mathbf{C}_G(x)|x^{-1}\hat{K}\,.$$

Then

$$\sum_{y\in G} \mathcal{X}([x,y]) = |\mathbf{C}_G(x)|\mathcal{X}(x^{-1})\mathcal{X}(\hat{K}) = |\mathbf{C}_G(x)|\omega\mathcal{X}(x^{-1}),$$

and taking traces yields

$$\sum_{y\in G} \chi([x,y]) = |\mathbf{C}_G(x)|\omega\chi(x^{-1}) = \frac{|\mathbf{C}_G(x)|\chi(x)|K|\chi(x^{-1})}{\chi(1)} = \frac{|G||\chi(x)|^2}{\chi(1)},$$

where the last equality holds because $|\mathbf{C}_G(x)||K| = |G|$. Summing over $x \in G$ and using the fact that $\sum |\chi(x)|^2 = |G|$, we obtain

$$\sum_{x,y\in G} \chi([x,y]) = \frac{|G|^2}{\chi(1)},$$

as required. ■

We need one more preliminary result.

Lemma 1.22. *Let (G,N,θ) and (G^*,N^*,θ^*) be isomorphic character triples, where $\tau : G/N \to G^*/N^*$ is the associated group isomorphism. Then τ defines a bijection from the set of θ-good elements of G/N onto the set of θ^*-good elements of G^*/N^*.*

Proof. Fix an element Ng of G/N. Then $\tau(Ng)$ is a coset of N^* in G^*, so we can write $\tau(Ng) = N^*g^*$, where g^* is some (not uniquely determined) element of G^*. Now if $H = \langle N, g, x\rangle$ for some element $x \in G$, then the group H/N is generated by Ng and one additional element; thus, the isomorphic group $H^*/N^* = \tau(H/N)$ is generated by N^*g^* and one other element. We can thus write $H^* = \langle N^*, g^*, y\rangle$ for some element $y \in G^*$, and hence the correspondence $H \mapsto H^*$ carries the set of subgroups of the form $H = \langle N, g, x\rangle$ for $x \in G$ onto the set of subgroups $\langle N^*, g^*, y\rangle$ for $y \in G^*$. Also H/N and H^*/N^* are isomorphic, so H/N is abelian if and only if H^*/N^* is abelian. Lemma 1.12 tells us that θ extends to H if and only if θ^* extends to H^*, and it follows that Ng is θ-good if and only if N^*g^* is θ^*-good. ■

Proof of Theorem 1.19. If we replace (G,N,θ) by an isomorphic character triple, then by the definition of a character-triple isomorphism, there is no change in the number of irreducible characters of G that lie over θ, and by Lemma 1.22 there is also no change in the number of θ-good classes of G/N. By Theorem 1.13, there exists a character triple (G^*,N^*,θ^*) isomorphic to (G,N,θ) and such that N^* is central in G^* and θ^* is faithful; hence, replacing (G,N,θ) by (G^*,N^*,θ^*) allows us to assume that N is central and θ is faithful.

Now write $n = |\operatorname{Irr}(G|\theta)|$. If $\chi \in \operatorname{Irr}(G|\theta)$, then since $N \subseteq \mathbf{Z}(G)$, we have $\chi_N = \chi(1)\theta$, and thus $[\theta^G, \chi] = [\theta, \chi_N] = \chi(1)$, and we have $\theta^G = \sum \chi(1)\chi$, where the sum runs over the n characters $\chi \in \mathrm{Irr}(G|\theta)$. Using Lemma 1.21 now, we compute that

$$\sum_{x,y \in G} \theta^G([x,y]) = \sum_{\chi \in \mathrm{Irr}(G|\theta)} \chi(1) \sum_{x,y \in G} \chi([x,y]) = n|G|^2 .$$

For elements $x \in G$, write $C_x = \{y \in G \mid [x,y] \in N\}$, so C_x is the full preimage of $\mathbf{C}_{\overline{G}}(\overline{x})$ in G, and in particular C_x is a subgroup.

Now θ^G vanishes on $G - N$ and $\theta^G(z) = |G : N|\theta(z)$ for elements $z \in N$. It follows that in the sum above we need consider only pairs of elements $x, y \in G$ such that $[x,y] \in N$, and thus we have

$$\begin{aligned} n|G|^2 = \sum_{\substack{x,y \in G \\ [x,y] \in N}} \theta^G([x,y]) &= |G : N| \sum_{\substack{x,y \in G \\ [x,y] \in N}} \theta([x,y]) \\ &= |G : N| \sum_{x \in G} \sum_{y \in C_x} \theta([x,y]) . \end{aligned}$$

We argue next that if $\overline{x}$ is θ-good, then the inner sum on the right above is equal to $|\mathbf{C}_G(x)|$, and that it is zero, otherwise. First, if $\overline{x}$ is θ-good, then by Lemma 1.20, we have $[x,y] = 1$ for all $y \in C_x$, and in this case, the inner sum is equal to $|C_x| = |\mathbf{C}_G(x)|$, as wanted. If $\overline{x}$ is not θ-good, then the map $y \mapsto [x,y]$ is a homomorphism from C_x onto a nontrivial subgroup $M \subseteq N$, and thus each element of M occurs equally often as y runs over C_x. Since θ is faithful, we see that $\sum_{z \in M} \theta(z) = 0$, and it follows that the inner sum is zero in this case, as claimed.

Now consider the elements $x \in G$ such that $\overline{x}$ lies in some fixed θ-good class of $\overline{G}$. There are $|N|$ elements x that map to $\overline{x}$, and each such element x has $|G : \mathbf{C}_G(x)|$ conjugates in G. For each of these elements, we have seen that the inner sum above is equal to $|\mathbf{C}_G(x)|$, and thus the total contribution to the double sum from elements corresponding to a given θ-good class is $|N||G|$. Writing m to denote the total number of θ-good classes in $\overline{G}$, we conclude that $n|G|^2 = m|G : N||N||G|$, and thus $n = m$, as required. ∎

An immediate consequence of Theorem 1.19, which does not seem to be at all obvious from first principles, is the following.

Corollary 1.23. *Let $N \lhd G$ and $\theta \in \mathrm{Irr}(N)$, and suppose that θ is G-invariant. Then $|\mathrm{Irr}(G|\theta)| \leq |\mathrm{Irr}(G/N)|$.*

Proof. We know by Theorem 1.19 that $|\mathrm{Irr}(G|\theta)|$ is the number of θ-good classes of G/N. Since $|\mathrm{Irr}(G/N)|$ is the total number of classes of G/N, the inequality is obvious. ∎

By the Gallagher correspondence, equality occurs in Corollary 1.23 if θ extends to G, but equality can hold even if θ is not extendible. To see this, suppose that $N \subseteq \mathbf{Z}(G) \cap G'$, where N has prime order and the identity is the only element of N that is a commutator in G. If $\theta \in \operatorname{Irr}(N)$ is faithful, Lemma 1.20 guarantees that every element of G/N is θ-good. Then $|\operatorname{Irr}(G|\theta)| = |\operatorname{Irr}(G/N)|$, but θ cannot extend to G because $N \subseteq G'$ so N is contained in the kernel of every linear character of G. In fact, it is not very hard to find specific examples of this phenomenon.

We close with an application that is occasionally useful. If $N \triangleleft G$ and $\theta \in \operatorname{Irr}(N)$ is invariant in G, then θ is said to be **fully ramified** with respect to G/N if $|\operatorname{Irr}(G|\theta)| = 1$.

Theorem 1.24. *Let $N \triangleleft G$, and suppose that $\theta \in \operatorname{Irr}(N)$ is fully ramified with respect to G/N. Then every prime divisor of $|G : N|$ divides either $\theta(1)$ or $o(\theta)$ or both.*

We need the following.

Lemma 1.25. *Let $N \triangleleft G$ and $\theta \in \operatorname{Irr}(N)$, and suppose that θ is G-invariant. If $\overline{g} \in G/N$ has order relatively prime to $\theta(1)o(\theta)$, then $\overline{g}$ is θ-good.*

Proof. Let $K = \langle N, g \rangle$, so $|K : N| = o(\overline{g})$, and hence $|K : N|$ is relatively prime to $\theta(1)o(\theta)$. By Theorem 1.5, therefore, θ has a unique canonical extension $\hat{\theta} \in \operatorname{Irr}(K)$.

Now let $H = \langle N, g, x \rangle$, where $x \in G$, and assume that H/N is abelian. Then $K \triangleleft H$ and H/K is cyclic. Also, the uniqueness of $\hat{\theta}$ guarantees that it is invariant in H, and it follows by Corollary 1.14 that $\hat{\theta}$ extends to H. Then θ extends to H, so by definition $\overline{g}$ is θ-good. ∎

Proof of Theorem 1.24. Suppose that p is a prime divisor of $|G : N|$ that does not divide $\theta(1)o(\theta)$, and let $\overline{g}$ be an element of order p in G/N. By Lemma 1.25, both $\overline{g}$ and the identity $\overline{1}$ are θ-good, and thus G/N has at least two θ-good classes. By Theorem 1.19, therefore, $|\operatorname{Irr}(G|\theta)| \geq 2$, and this is a contradiction since, by hypothesis, $|\operatorname{Irr}(G|\theta)| = 1$. ∎

Problems

1.1. Suppose that G is π-separable, and let p and q be primes, where $p \in \pi$ and $q \notin \pi$. Show that G has a Hall $\{p, q\}$-subgroup.

1.2. Let A act on G via automorphisms, where $|A|$ and $|G|$ are relatively prime.

(a) Show that for each prime number p, there exists an A-invariant Sylow p-subgroup of G. Also, show that $C = \mathbf{C}_G(A)$ acts transitively by conjugation on the set of A-invariant Sylow p-subgroups of G.

(b) If P is maximal (with respect to containment) among A-invariant p-subgroups of G, show that P is a Sylow p-subgroup of G.

(c) Let P be an A-invariant Sylow p-subgroup of G. Show that $P \cap C$ is a Sylow p-subgroup of C, where $C = \mathbf{C}_G(A)$.

1.3. Let A act on G via automorphisms, where $|A|$ and $|G|$ are relatively prime, and suppose that $G = XY$, where X and Y are A-invariant subgroups. Show that $C = (C \cap X)(C \cap Y)$, where $C = \mathbf{C}_G(A)$.

1.4. Let A act on G via automorphisms, where $|A|$ and $|G|$ are relatively prime, and suppose that $H \subseteq G$ is an A-invariant subgroup. Show that $|H : H \cap C|$ divides $|G : C|$, where $C = \mathbf{C}_G(A)$.

Hint: Show for each prime p that the p-part of $|H : H \cap C|$ is at most the p-part of $|G : C|$.

1.5. Let A act on G via automorphisms, and let $\chi \in \mathrm{Irr}(G)$ be A-invariant. Suppose that N is an A-invariant normal subgroup of G such that $|A|$ and $|G : N|$ are relatively prime. Show that χ lies over some A-invariant character θ of N. Also, show that θ is unique if $\mathbf{C}_{G/N}(A) = 1$. Finally, if the action of A on G/N is trivial, show that every irreducible character of N lying under χ is A-invariant.

1.6. Let $\theta \in \mathrm{Irr}(N)$, where $N \triangleleft G$, and write $\theta^G = \sum e_\chi \chi$, where the sum runs over the irreducible characters χ of G and the coefficients e_χ are nonnegative integers. Show that $\sum (e_\chi)^2 = |T : N|$, where T is the stabilizer of θ in G.

1.7. Let $N \triangleleft G$, and suppose that $\theta \in \mathrm{Irr}(N)$ is invariant in G, and let $\chi \in \mathrm{Irr}(G|\theta)$.

(a) Show that θ is fully ramified with respect to G/N if and only if

$$\left(\frac{\chi(1)}{\theta(1)}\right)^2 = |G : N|.$$

(b) Suppose that θ is fully ramified with respect to G/N. If P/N is a Sylow subgroup of G/N. Show that θ is fully ramified with respect to P/N.

1.8. Let $\chi \in \mathrm{Irr}(G)$, and suppose that $|G|/\chi(1)$ is a π-number, where π is a set of primes.

(a) Let ψ be an irreducible constituent of χ_N, where $N \triangleleft G$. Show that $|N|/\psi(1)$ is a π-number.

(b) Assume that G is π-separable, and let $N = \mathbf{O}_\pi(G)$. Assume that χ_N is a multiple of some character $\theta \in \mathrm{Irr}(N)$. Show that G is a π-group.

Notes and References

The first clear statement of the Schur–Zassenhaus theorem seems to be in Zassenhaus's group theory text [**77**], where it is attributed to Schur. It appears that the existence of a complement (at least in the case where the normal subgroup is abelian) may be implicit in Schur's work, but the rest of the theorem is probably due to Zassenhaus himself.

Glauberman's lemma, which is Lemma 1.4(a), appears in [**26**]. Although Lemma 1.4(b) is an easy consequence of Lemma 1.4(a), it does not seem to be stated explicitly in Glauberman's paper.

Gallagher's theorem, Theorem 1.5, is the essential content of [**24**]. In that paper, the result is proved using Brauer's theorem on induced characters. (See Chapter 8 of [**CTFG**].) The Gallagher correspondence (Theorem 1.7) appears as Theorem 2 of the same paper.

Character-triple isomorphisms were introduced by Isaacs in [**34**].

Clifford's theorem, Theorem 1.17, appears in [**10**], although the author says that much of the content of his paper was already known to Frobenius. It is not clear where Theorem 1.18, the Clifford correspondence, was first stated explicitly.

Gallagher's character count, Theorem 1.19, appears in [**25**].

Problem 1.7(b) can be found in [**15**].

Chapter 2

π-Special Characters

2A

If G is a π-group for some set π of primes, then for all irreducible characters χ of G, it is clear that both $\chi(1)$ and $o(\chi)$ are π-numbers. It would be pleasant if there were some sort of converse for this statement. In other words, if $\chi \in \mathrm{Irr}(G)$ and both $\chi(1)$ and $o(\chi)$ are π-numbers, we might hope that χ would behave (in some respects, at least) as if its domain G were a π-group. In fact, if G is π-separable, then under some additional hypotheses, this is exactly what does happen.

For every π-separable group G, we will define a subset $\mathcal{X}_\pi(G)$ of $\mathrm{Irr}(G)$ such that the degrees and determinantal orders of all members of $\mathcal{X}_\pi(G)$ are π-numbers. In addition, we want our set $\mathcal{X}_\pi(G)$ to "respect" the normal subgroup structure of G. More precisely, we require that for every character $\chi \in \mathcal{X}_\pi(G)$ and every normal subgroup N of G, the irreducible constituents of the restriction χ_N should lie in $\mathcal{X}_\pi(N)$.

By Clifford's theorem, we know that if $N \lhd G$ and $\chi \in \mathrm{Irr}(G)$, then the degrees of the irreducible constituents of χ_N divide $\chi(1)$. If $\chi(1)$ is a π-number, therefore, the degrees of the irreducible constituents of χ_N are also π-numbers; so from the point of view of character degrees, the requirement that the set $\mathcal{X}_\pi(G)$ should respect the normal structure of G imposes no additional burden.

In general, the determinantal order of an irreducible constituent of χ_N need not divide $o(\chi)$, and for that reason it is necessary to impose additional determinantal-order requirements on a character χ before we can admit it to the set $\mathcal{X}_\pi(G)$. In fact, it should be clear that not only must we consider restrictions to normal subgroups of G, but also to their normal subgroups, and so forth. The requirement should thus be stated in terms of subnormal subgroups.

If G is π-separable, we say that a character $\chi \in \mathrm{Irr}(G)$ is **π-special** if $\chi(1)$ is a π-number and $o(\theta)$ is a π-number for every irreducible constituent θ of the restriction χ_S for every subnormal subgroup $S \lhd\lhd G$. (We stress that we are defining π-special characters *only* for π-separable groups.) The set of π-special characters of G is denoted $\mathcal{X}_\pi(G)$, and it should be clear from the foregoing discussion that if $\chi \in \mathcal{X}_\pi(G)$ and θ is an irreducible constituent of χ_S, where $S \lhd\lhd G$, then $\theta \in \mathcal{X}_\pi(S)$.

We shall see that (as promised) the π-special characters of a π-separable group G behave in some ways as though G were a π-group. (In fact, it is tempting to say that a π-special character of G "thinks" that G is a π-group.) We will mention just one example of this phenomenon now: the values of a π-special character always lie in a field generated over the rationals by π-roots of unity. (This is proved as Corollary 2.13.)

We begin our discussion of π-special characters with a few much more elementary observations. First, it is clear that the principal character 1_G of a π-separable group G is π-special, and of course it is also π'-special. In fact, 1_G is the only character that is both π-special and π'-special. To see this, consider a character $\chi \in \mathcal{X}_\pi(G) \cap \mathcal{X}_{\pi'}(G)$. Then $\chi(1)$ is simultaneously a π-number and a π'-number, and thus $\chi(1) = 1$, and χ is linear. Also, $o(\chi)$ is both a π-number and a π'-number, so $o(\chi) = 1$, and since χ is linear, it is principal.

Now let $N \lhd G$ be a π'-subgroup, so every irreducible character of N is π'-special. If $\chi \in \mathrm{Irr}(G)$ is π-special, then the irreducible constituents of χ_N are π-special, and it follows that all of these irreducible constituents are principal. Then $N \subseteq \ker(\chi)$, and in particular $\mathbf{O}_{\pi'}(G)$ is contained in the kernel of every π-special character of G. Similarly, of course, $\mathbf{O}_\pi(G)$ is contained in the kernel of every π'-special character of G.

Next, recall that if $N \lhd G$, it is customary to identify the members of $\mathrm{Irr}(G/N)$ with the set of characters $\chi \in \mathrm{Irr}(G)$ such that $N \subseteq \ker(\chi)$. It is easy to see that if G is π-separable, then under this identification, the π-special characters of G/N correspond to those π-special characters of G with kernels containing N.

Using this observation, we can characterize the π-separable groups for which all irreducible characters are π-special.

Corollary 2.1. *Let G be π-separable. Then all members of* $\mathrm{Irr}(G)$ *are π-special if and only if G is a π-group.*

Proof. If G is a π-group, then clearly, every irreducible character is π-special. Conversely, we suppose that $\mathcal{X}_\pi(G) = \mathrm{Irr}(G)$, and we proceed by induction on $|G|$ to show that G must be a π-group. We can assume that $|G| > 1$, and we let N be minimal normal in G. Since all members of $\mathrm{Irr}(G/N)$ are π-special, the inductive hypothesis implies that G/N is a π-group, and thus it suffices to show that N is a π-group.

Recall that a minimal normal subgroup of G must be either a π'-group or a π-group. If N is a π'-group, then $N \subseteq \ker(\chi)$ for all $\chi \in \mathcal{X}_\pi(G) = \mathrm{Irr}(G)$, and since $N > 1$, this is a contradiction. It follows that N is a π-group, as wanted. ∎

We turn now to some of the deeper properties of π-special characters. Perhaps the most surprising fact about these characters is the following product theorem of D. Gajendragadkar.

Theorem 2.2 (Gajendragadkar's product theorem). *Let G be π-separable, and suppose that $\alpha, \beta \in \mathrm{Irr}(G)$ are π-special and π'-special, respectively. Then $\alpha\beta$ is irreducible. Also, if $\alpha\beta = \mu\nu$, where μ is π-special and ν is π'-special, then $\mu = \alpha$ and $\nu = \beta$.*

We mention that it is an open question whether or not a product of two *faithful* irreducible characters of a noncyclic solvable group can be irreducible. In the situation of Gajendragadkar's theorem, however, if G is nontrivial, it is impossible for both α and β to be faithful because at least one of $\mathbf{O}_{\pi'}(G)$ or $\mathbf{O}_\pi(G)$ must be nontrivial, and these subgroups are contained in $\ker(\alpha)$ and $\ker(\beta)$, respectively.

We begin work now toward a proof of Gajendragadkar's product theorem.

Lemma 2.3. *Let $\chi \in \mathrm{Irr}(G)$, where G is π-separable, and suppose that $\chi(1)$ and $o(\chi)$ are π-numbers. Then χ is π-special if and only if the irreducible constituents of χ_M are π-special for every maximal normal subgroup M of G.*

Proof. Suppose first that for every maximal normal subgroup M of G, the irreducible constituents of χ_M are π-special. Since $\chi(1)$ is a π-number, we see that to show that χ is π-special, it suffices to show that the determinantal orders of the irreducible constituents of χ_S are π-numbers for all $S \lhd\lhd G$. By hypothesis, $o(\chi)$ is a π-number, so there is nothing to prove if $S = G$, and we need consider only proper subnormal subgroups. Each such subgroup S, however, is contained in some maximal normal subgroup M of G, so if θ is an irreducible constituent of χ_S, then θ lies under some irreducible constituent ψ of χ_M. By assumption, ψ is π-special, and since $S \lhd\lhd M$, it follows that $o(\theta)$ is a π-number, as required.

Conversely, if χ is π-special, then the irreducible constituents of χ_M are π-special for all subnormal subgroups M of G, and in particular, this holds for maximal normal subgroups. ■

We digress briefly to recall three basic facts about character induction that will be needed in the proofs of the next few results. (They will also be used frequently thereafter, often without explicit reference.) First, suppose that $H \subseteq G$ and that α and β are characters (or, more generally, arbitrary class functions) of H and G, respectively. Then Frobenius reciprocity asserts that $[\alpha^G, \beta] = [\alpha, \beta_H]$. In this situation, we also have $(\alpha\beta_H)^G = \alpha^G\beta$. Finally, suppose that $G = HK$, where $K \subseteq G$, and let $D = H \cap K$. If α is a class function of H, then $(\alpha^G)_K = (\alpha_D)^K$. These three equalities follow easily from the definition of an induced class function as given in Chapter 5 of [**CTFG**]. (We should mention also that the third of these formulas is perhaps the most useful special case of a more general result due to G. Mackey.)

The following is our principal tool for proving that a character is π-special.

Theorem 2.4. *Let $N \lhd G$, where G is π-separable, and let $\theta \in \mathrm{Irr}(N)$ be π-special.*

(a) *If G/N is a π-group, then every member of $\mathrm{Irr}(G|\theta)$ is π-special.*

(b) *If G/N is a π'-group and θ is invariant in G, then θ has a canonical extension $\chi \in \mathrm{Irr}(G)$ and χ is the unique π-special character in $\mathrm{Irr}(G|\theta)$. If θ is not invariant in G, then no member of $\mathrm{Irr}(G|\theta)$ is π-special.*

Proof. We proceed by induction on $|G|$. For (a), we assume that G/N is a π-group, and we let $\chi \in \mathrm{Irr}(G|\theta)$. By hypothesis, θ is π-special, so $\theta(1)$ and $o(\theta)$ are π-numbers, and since Theorem 1.15 guarantees that $\chi(1)$ divides $|G : N|\theta(1)$ and $o(\chi)$ divides $|G : N|o(\theta)$, we see that $\chi(1)$ and $o(\chi)$ are π-numbers. By Lemma 2.3, therefore, it suffices to show that the irreducible constituents of χ_M are π-special for all proper normal subgroups M of G. (Of course, we need only consider *maximal* normal subgroups M, but maximality contributes nothing to this part of the argument.)

Let $D = M \cap N$, and let φ be an irreducible constituent of θ_D. Since $D \lhd N$ and θ is π-special, we see that φ is π-special. Also, φ lies under χ, so it must lie under some irreducible constituent ψ of χ_M. Now $D \lhd M$ and $M/D = M/(N \cap M) \cong NM/N$, which is a subgroup of G/N, and thus M/D is a π-group. By the inductive hypothesis applied in the group M with respect to the normal subgroup D and the character $\varphi \in \mathrm{Irr}(D)$, we deduce that ψ is π-special. By Clifford's theorem, all irreducible constituents of χ_M

are conjugate to ψ, so they are all π-special, and this completes the proof of (a).

For (b), we assume that G/N is a π'-group. By Clifford's theorem, the degree of every member in $\mathrm{Irr}(G|\theta)$ is divisible by the π'-number $|G:T|$, where $T = G_\theta$ is the stabilizer of θ. If one of these characters is π-special, therefore, then $|G:T| = 1$, and θ must be invariant.

Now assume that θ is G-invariant. Since $\theta(1)$ and $o(\theta)$ are π-numbers, Gallagher's theorem applies, and we deduce that θ has a unique canonical extension $\chi \in \mathrm{Irr}(G|\theta)$. We must show that χ is π-special and that no member of $\mathrm{Irr}(G|\theta)$ other than χ is π-special.

We handle the second of these tasks first, so suppose that $\psi \in \mathrm{Irr}(G|\theta)$ is π-special. Then $\psi(1)$ is a π-number, and since $|G:N|$ is a π'-number, Corollary 1.16 tells us that ψ_N is irreducible, and thus $\psi_N = \theta$. Then ψ is an extension of θ, and since we are assuming that ψ is π-special, $o(\psi)$ is a π-number. By definition, therefore, ψ is the canonical extension of θ, and hence $\psi = \chi$, as required.

To prove that χ is π-special, suppose first that $\mathbf{O}^\pi(G) < G$, and write $M = \mathbf{O}^\pi(G)$. Then $|G:M|$ is a π-number, and since $|G:N|$ is a π'-number, it follows that $MN = G$. Let $D = M \cap N$, so $D \triangleleft M$ and $M/D \cong G/N$ is a π'-group.

Choose an irreducible constituent φ of θ_D, and note that φ is π-special since θ is π-special and $D \triangleleft N$. Now φ lies under χ, so it lies under some irreducible constituent ψ of χ_M. Also, $\psi(1)$ divides $\chi(1)$ by Clifford's theorem, so $\psi(1)$ is a π-number, and thus ψ_D is irreducible by Corollary 1.16. We conclude that $\psi_D = \varphi$, and thus φ is invariant in M. Since $\varphi(1)$ and $o(\varphi)$ are π-numbers, Gallagher's theorem guarantees the existence of a canonical extension ξ of φ to M. (But note that we do not know that $\xi = \psi$.) By the inductive hypothesis applied in the group M, we conclude that ξ is π-special, and thus by part (a) of the theorem, all irreducible constituents of ξ^G are π-special. We have

$$(\xi^G)_N = (\xi^{MN})_N = (\xi_D)^N = \varphi^N ,$$

and since θ is an irreducible constituent of φ^N, it follows that θ lies under at least one of the irreducible constituents of ξ^G. All of these constituents are π-special, and yet we know that χ is the only member of $\mathrm{Irr}(G|\theta)$ that can possibly be π-special. It follows that χ is π-special, as desired.

It remains to complete the proof of (b) in the case where $\mathbf{O}^\pi(G) = G$. To do this, we will appeal to Lemma 2.3, so we choose a maximal normal subgroup M of G, and we let ψ be an irreducible constituent of χ_M. It suffices to show that ψ is π-special.

Since M is maximal normal in G, we know that G/M must be either a π-group or a π'-group, but since $\mathbf{O}^{\pi}(G) = G$, it follows that G/M is a π'-group. As before, let $D = N \cap M$, so that $D \lhd G$. Now $|G : D|$ is a π'-number since it divides $|G : N||G : M|$, and thus χ_D is irreducible by Corollary 1.16, and in particular χ_M is irreducible. Then $\chi_M = \psi$, and we write $\psi_D = \varphi$, so φ is irreducible. Also, φ lies under θ, and thus φ is π-special. Since $\psi = \chi_M$, we see that $o(\psi)$ divides $o(\chi)$, which is a π-number, and thus ψ is the canonical extension of φ to M. By the inductive hypothesis, ψ is π-special, and the result now follows by Lemma 2.3. ■

In order to prove Gajendragadkar's theorem, we need a tool that can show that a product of two characters is irreducible. The result we use is the following generalization of the Gallagher correspondence (Theorem 1.7). It appears as Theorem 6.16 of [**CTFG**].

Theorem 2.5. *Let $N \lhd G$, suppose $\psi \in \mathrm{Irr}(G)$, and suppose that $\psi_N = \theta$ is irreducible. Let $\varphi \in \mathrm{Irr}(N)$, where φ is invariant in G, and assume that $\varphi\theta$ is irreducible. Then the map $\beta \mapsto \beta\psi$ is a bijection from $\mathrm{Irr}(G|\varphi)$ onto $\mathrm{Irr}(G|\varphi\theta)$.*

Observe that if $\varphi = 1_N$, then the assumptions in Theorem 2.5 that φ is invariant and that $\varphi\theta$ is irreducible are automatically satisfied. Also $\mathrm{Irr}(G|\varphi) = \mathrm{Irr}(G/N)$, so in this case the assertion of Theorem 2.5 is exactly Theorem 1.7.

Proof of Theorem 2.5. Since φ is G-invariant, we have $(\varphi^G)_N = |G : N|\varphi$, and thus the irreducibility of φ and Frobenius reciprocity yield

$$[\varphi^G, \varphi^G] = [\varphi, (\varphi^G)_N] = [\varphi, |G : N|\varphi] = |G : N| .$$

Also $\varphi\theta$ is invariant in G and irreducible, so exactly the same reasoning yields

$$[(\varphi\theta)^G, (\varphi\theta)^G] = |G : N| .$$

Now write

$$\varphi^G = \sum_{\beta} b_\beta \beta ,$$

where β runs over $\mathrm{Irr}(G|\varphi)$ and the coefficients b_β are positive integers. Observe that this yields

$$\sum_{\beta} (b_\beta)^2 = [\varphi^G, \varphi^G] = |G : N| .$$

Next, we compute

$$(\varphi\theta)^G = (\varphi\psi_N)^G = \varphi^G\psi = \sum_{\beta} b_\beta \beta\psi .$$

Clearly, for $\beta, \gamma \in \mathrm{Irr}(G|\varphi)$, we have $[\beta\psi, \gamma\psi] \geq \delta_{\beta,\gamma}$, where the "Kronecker delta" $\delta_{\beta,\gamma}$ is 1 if $\beta = \gamma$ and is 0 otherwise. Since the coefficients b_β and b_γ are positive, this yields

$$\begin{aligned}\sum_\beta (b_\beta)^2 = |G : N| = [(\varphi\theta)^G, (\varphi\theta)^G] &= \sum_{\beta,\gamma} b_\beta b_\gamma [\beta\psi, \gamma\psi] \\ &\geq \sum_{\beta,\gamma} b_\beta b_\gamma \delta_{\beta,\gamma} \\ &= \sum_\beta (b_\beta)^2 ,\end{aligned}$$

and thus equality holds everywhere. In particular, $[\beta\psi, \gamma\psi] = \delta_{\beta,\gamma}$, and thus as β runs over $\mathrm{Irr}(G|\varphi)$, the characters $\beta\psi$ are distinct and irreducible, as required. Finally, the equation

$$(\varphi\theta)^G = \sum_\beta b_\beta \beta\psi$$

shows that every member of $\mathrm{Irr}(G|\varphi\theta)$ is one of the characters $\beta\psi$, and so our map is surjective. ∎

We can now prove Gajendragadkar's product theorem.

Proof of Theorem 2.2. We can assume that $G > 1$, and we proceed by induction on $|G|$. We will prove the two parts of the theorem simultaneously, so we assume that both the irreducibility assertion and the uniqueness of factorization hold for all smaller groups.

Let $\alpha, \beta \in \mathrm{Irr}(G)$ be π-special and π'-special, respectively, and let N be maximal normal in G, so that G/N is either a π-group or a π'-group. Because of the symmetry between π and π', we can assume that in fact G/N is a π-group, and thus, since $\beta(1)$ is a π'-number, Theorem 1.15 guarantees that β_N is irreducible. Write $\theta = \beta_N$, and let φ be an arbitrary irreducible constituent of α_N. Then φ is π-special and θ is π'-special, and so by the inductive hypothesis on irreducibility applied in N, we see that $\varphi\theta \in \mathrm{Irr}(N)$.

Let $T = G_\varphi$ be the stabilizer of φ in G, and observe that T also stabilizes $\varphi\theta$. In fact, T must be the full stabilizer of $\varphi\theta$ in G because if $g \in G$ and $\varphi\theta = (\varphi\theta)^g = \varphi^g\theta^g$, then since φ^g is π-special and θ^g is π'-special, the inductive hypothesis on uniqueness yields $\varphi^g = \varphi$, and thus $g \in T$, as required.

Let $\eta \in \mathrm{Irr}(T|\varphi)$ be the Clifford correspondent of α with respect to φ, so $\eta^G = \alpha$. (See Theorem 1.18.) Also, let $\psi = \beta_T$ so $\psi_N = \theta$, and in particular, ψ is irreducible. We have

$$\alpha\beta = \eta^G\beta = (\eta\beta_T)^G = (\eta\psi)^G ,$$

and we observe that $\eta\psi \in \mathrm{Irr}(T|\varphi\theta)$ by Theorem 2.5. (We definitely are not appealing to the inductive hypothesis here to establish the irreducibility of $\eta\psi$ because we do not know that $T < G$.)

Since T is the stabilizer of $\varphi\theta$ and $\eta\psi$ lies over $\varphi\theta$, the Clifford correspondence tells us that $\alpha\beta = (\eta\psi)^G$ is irreducible, as required.

Now suppose that $\alpha\beta = \mu\nu$, where μ is π-special and ν is π'-special. Then the irreducible constituents of μ_N and ν_N are π-special and π'-special, respectively, and by the inductive hypothesis on irreducibility, the product of each irreducible constituent of μ_N with each irreducible constituent of ν_N is irreducible. It follows that every irreducible constituent of $(\mu\nu)_N$ is one of these products. Since $\mu\nu = \alpha\beta$ lies over the irreducible character $\varphi\theta$, we deduce that $\varphi\theta$ must be a product of an irreducible constituent of μ_N with an irreducible constituent of ν_N. By the inductive hypothesis on uniqueness, it follows that φ lies under μ and θ lies under ν.

Since G/N is a π-group and θ is π'-special, it follows by Theorem 2.4(b) that at most one π'-special character of G can lie over θ, and thus $\nu = \beta$. Since φ lies under μ, we can consider the Clifford correspondent ξ of μ with respect to φ, so $\xi \in \mathrm{Irr}(T|\varphi)$ and $\xi^G = \mu$. Reasoning as before, we see that $\xi\psi$ is the Clifford correspondent of $\mu\nu = \alpha\beta$ with respect to $\varphi\theta$, so by the uniqueness of the Clifford correspondence, we must have $\xi\psi = \eta\psi$. Finally, by the injectivity of the correspondence in Theorem 2.5, we deduce that $\xi = \eta$, and thus $\alpha = \eta^G = \xi^G = \mu$. ∎

If G is π-separable, we say that a character $\chi \in \mathrm{Irr}(G)$ is **π-factored** if $\chi = \alpha\beta$ for some π-special character α and π'-special character β of G. Note that linear characters are automatically π-factored since every element of an arbitrary finite abelian group can be decomposed as the product of a π-element and a π'-element. Also, it is trivial that π-special and π'-special characters are π-factored. (If χ is π-special, then $\chi = \alpha\beta$, where $\alpha = \chi$ and $\beta = 1_G$, and similarly if χ is π'-special.)

If $\chi \in \mathrm{Irr}(G)$ is π-factored and $S \lhd\lhd G$, then every irreducible constituent of χ_S is π-factored, and in fact if $\chi = \alpha\beta$, where α is π-special and β is π'-special, then each irreducible constituent of χ_S has the π-factored form $\mu\nu$, where μ is an irreducible constituent of α_S and ν is an irreducible constituent of β_S. This is clear since $\chi_S = \alpha_S\beta_S$, and by Gajendragadkar's theorem, the product of each irreducible constituent of α_S with each irreducible constituent of β_S is irreducible.

Lemma 2.6. *Suppose that $\chi = \alpha\beta$, where α and β are π-special and π'-special characters, respectively, of a π-separable group G. Then* $\ker(\chi) = \ker(\alpha) \cap \ker(\beta)$.

Proof. If $x \in \ker(\alpha) \cap \ker(\beta)$, then

$$\chi(x) = \alpha(x)\beta(x) = \alpha(1)\beta(1) = \chi(1),$$

and thus $x \in \ker(\chi)$, and we have $\ker(\alpha) \cap \ker(\beta) \subseteq \ker(\chi)$. To prove the reverse containment, let $N = \ker(\chi)$. Then the principal character 1_N is an irreducible constituent of χ_N, and thus by the remarks preceding the statement of the lemma, we can write $1_N = \mu\nu$, where μ is a constituent of α_N and ν is a constituent of β_N. It follows that μ and ν are linear, and since they are inverses of each other in the group of linear characters of N, they have equal orders. But $o(\mu)$ is a π-number and $o(\nu)$ is a π'-number, so $o(\mu) = 1 = o(\nu)$, and thus μ and ν are principal. It follows that $N \subseteq \ker(\alpha) \cap \ker(\beta)$, as needed. ∎

It is easy to see from Lemma 2.6 that if $N \triangleleft G$ and $\chi \in \operatorname{Irr}(G/N)$, then χ is π-factored when viewed as a character of G/N if and only if it is π-factored when viewed as a character of G. This enables us to characterize those π-separable groups G for which *every* irreducible character is π-factored.

Theorem 2.7. *Let G be π-separable. Then every irreducible character of G is π-factored if and only if G has a normal Hall π-subgroup and a normal Hall π'-subgroup.*

We need the following lemma, whose proof uses the elementary fact that if C is a proper subgroup of a finite group H, then H cannot be the union of the conjugates of C. (In fact, by Problem 1A.7 of [**FGT**], the number of elements of H that are not in any conjugate of C is at least $|C|$, but that is more information than we need.)

Lemma 2.8. *Suppose that a π-group K acts via automorphisms on a π'-group H, and assume either that every conjugacy class of H is K-invariant or that every irreducible character of H is K-invariant. Then the action of K on H is trivial.*

Proof. Brauer's permutation lemma (Theorem 6.32 of [**CTFG**]) guarantees that each automorphism of H fixes equal numbers of irreducible characters and conjugacy classes. If all irreducible characters of H are K-invariant, therefore, then all classes of H are also K-invariant. We can thus assume that in fact, all of the classes of H are K-invariant.

To show that K acts trivially on H, it suffices to prove that each Sylow subgroup of K acts trivially, and thus it is no loss to assume that K is a p-group for some prime p, where, of course, p does not divide $|H|$. Since each class X of H is K-invariant, it is a union of K-orbits, and since K is a p-group, each of these orbits has p-power size. Now $|X|$ divides $|H|$, which is not divisible by p, so some K-orbit in X has size not divisible by p. This orbit must have size 1, and hence the class X contains an element of the subgroup $C = \mathbf{C}_H(K)$.

Since every conjugacy class of H contains an element of C, it follows that H is the union of the conjugates of C in H. This can only happen, however, if $C = H$, and this means that K acts trivially on H. ∎

Proof of Theorem 2.7. First, suppose that H and K are a normal Hall π-subgroup and a normal Hall π'-subgroup, respectively, of G. We can then write $G = H \times K$, and by general character-theoretic properties of direct products, every irreducible character χ of G has the form $\chi = \alpha\beta$, where $\alpha \in \operatorname{Irr}(G/K)$ and $\beta \in \operatorname{Irr}(G/H)$. Since G/K is a π-group, α is a π-special character of G/K, and thus α is π-special as a character of G. Similarly, β is a π'-special character of G, and thus χ is π-factored.

Conversely now, we assume that every member of $\operatorname{Irr}(G)$ is π-factored, and we proceed by induction to show that G has a normal Hall π-subgroup and a normal Hall π'-subgroup. We can assume $G > 1$, so we can choose a minimal normal subgroup N of G, and since G is π-separable, N must be either a π or a π'-group. By symmetry, it is no loss to assume that N is a π-group.

The hypothesis that every irreducible character of G is π-factored is inherited by G/N, so by the inductive hypothesis, G/N has a normal Hall π-subgroup H/N, and we see that H is a normal Hall π-subgroup of G. Let K be a Hall π'-subgroup of G, so K acts by conjugation on H. To prove that $K \triangleleft G$, it suffices to prove that this action is trivial, so by Lemma 2.8 it is enough to show that every irreducible character of H is K-invariant.

Let $\theta \in \operatorname{Irr}(H)$, and let $\chi \in \operatorname{Irr}(G|\theta)$. By hypothesis, χ is π-factored, so we can write $\chi = \alpha\beta$, where α is π-special and β is π'-special. Since H is a normal π-subgroup, $H \subseteq \ker(\beta)$, and thus $\chi_H = (\alpha\beta)_H = \beta(1)\alpha_H$. As θ is an irreducible constituent of χ_H, we see that θ must be a constituent of α_H, and thus α is a π-special character of G lying over θ. Since G/H is a π'-group, Theorem 2.4 guarantees that $\theta = \alpha_H$, so θ is G-invariant, and thus θ is K-invariant, as required. ∎

Next, we address the question of how rare π-factored characters are among the irreducible characters of π-separable groups. We shall see that they are common enough so that if $\chi \in \operatorname{Irr}(G)$ where G is π-separable, then although χ itself may not be π-factored, there necessarily exist a subgroup $H \subseteq G$ and a π-factored character $\psi \in \operatorname{Irr}(H)$ such that $\chi = \psi^G$.

Recall that an irreducible character χ of a finite group G is said to be **primitive** if there is no character η of a proper subgroup of G such that $\chi = \eta^G$. If $\chi \in \operatorname{Irr}(G)$ is not primitive, therefore, it is induced from some character η of a proper subgroup H. If η is not primitive, it is induced from some character ξ of a still smaller subgroup, and in that case $\chi = \eta^G = (\xi^H)^G = \xi^G$. Continuing like this, we see that every irreducible character of a finite group is induced from a primitive character of some subgroup, and we will justify the assertion of the previous paragraph by showing that all primitive characters of π-separable groups are π-factored.

Actually, more is true. An irreducible character χ of a finite group G is **quasi-primitive** if for every normal subgroup $N \lhd G$, the restriction χ_N is homogeneous. (This means that χ_N is a multiple of an irreducible character.) If χ is primitive, it is automatically quasi-primitive because if θ is an irreducible constituent of χ_N, then by the Clifford correspondence, χ is induced from some character of the stabilizer G_θ of θ, and by the assumed primitivity of χ, we must have $G_\theta = G$. Then θ is G-invariant, and we see by Clifford's theorem that $\chi_N = e\theta$ for some integer e. Of course, the assumption that χ is quasi-primitive says little if G has few normal subgroups, but a theorem of T. Berger asserts that for solvable groups, quasi-primitivity is actually equivalent to primitivity.

Theorem 2.9. *Quasi-primitive characters of π-separable groups are π-factored. More generally, if $\chi \in \mathrm{Irr}(G)$ and*

$$1 = N_0 \subseteq N_1 \subseteq \cdots \subseteq N_r = G$$

is any normal series such that each factor N_i/N_{i-1} is either a π-group or a π'-group and each restriction χ_{N_i} is homogeneous, then χ is π-factored.

Proof. Since G is π-separable, G has a normal series $\{N_i\}$ in which each factor is either a π-group or a π'-group. Let $\chi \in \mathrm{Irr}(G)$, and assume that each restriction χ_{N_i} is homogeneous. (In particular, this will be the case if χ is quasi-primitive.)

Let θ_i be the unique irreducible constituent of χ_{N_i}, so θ_i is invariant in G. We show by induction on i that θ_i is π-factored, and it will follow that, in particular, $\theta_r = \chi$ is π-factored, as required. To start the induction, observe that θ_0 is the principal character of the trivial group, and so it is obviously π-factored.

For the inductive step, it suffices to consider two consecutive subgroups in our normal series, and for notational simplicity, we call these M and N, where $M \lhd N$ and N/M is either a π-group or a π'-group. We have an N-invariant character $\theta \in \mathrm{Irr}(M)$, where θ is π-factored, and we complete the proof by showing that every member of $\mathrm{Irr}(N|\theta)$ is π-factored.

Write $\theta = \alpha\beta$, where α is π-special and β is π'-special, and observe that since θ is invariant in N, the uniqueness of factorization in Gajendragadkar's theorem guarantees that both α and β are N-invariant. Because of the symmetry between π and π', it is no loss to assume that N/M is a π'-group, and thus α has a π-special canonical extension $\psi \in \mathrm{Irr}(N)$ by Theorem 2.4(b).

Now

$$\theta^N = (\alpha\beta)^N = \psi\beta^N ,$$

and by Theorem 2.4(a), all of the irreducible constituents of β^N are π'-special. Then θ^N is a sum of products of ψ with certain π'-special characters of N, and since ψ is π-special, it follows by Gajendragadkar's theorem that these products are irreducible. Each irreducible constituent of θ^N is thus one of these products and, hence, is π-factored, as required. ∎

We mentioned that one can almost say that a π-special character χ of a π-separable group G "thinks" that G is a π-group. If H is a Hall π-subgroup of G, therefore, χ thinks that H is the whole group, and thus χ_H should be irreducible. This is not much of a proof, of course, but nevertheless, the assertion is true. In fact, Gajendragadkar's restriction theorem below includes this and says even more.

Theorem 2.10 (Gajendragadkar's restriction theorem)**.** *Suppose that G is π-separable, and let $H \subseteq G$, where $|G : H|$ is a π'-number. Then restriction defines an injective map from $\mathcal{X}_\pi(G)$ into $\mathcal{X}_\pi(H)$. In particular, this holds if H is a Hall π-subgroup of G.*

The first part of the following general lemma is needed for our proof of Theorem 2.10. Although the second part will not be used now, we present both parts because the two assertions are in some sense "dual" to each other, and the two proofs are quite parallel.

Lemma 2.11. *Let $N \lhd G$ and $H \subseteq G$, and suppose that $G = NH$. Write $D = N \cap H$, and let $\varphi \in \mathrm{Irr}(D)$ and $\theta \in \mathrm{Irr}(N)$.*

(a) *Suppose that $\theta_D = \varphi$, and assume that the stabilizer of φ in H stabilizes θ. Then restriction defines a bijection $\mathrm{Irr}(G|\theta) \to \mathrm{Irr}(H|\varphi)$.*

(b) *Suppose that $\varphi^N = \theta$, and assume that the stabilizer of θ in H stabilizes φ. Then induction defines a bijection $\mathrm{Irr}(H|\varphi) \to \mathrm{Irr}(G|\theta)$.*

Proof. Let $\mathcal{O}$ be the G-orbit of θ in $\mathrm{Irr}(N)$, and let T be the stabilizer of θ in G, so $|\mathcal{O}| = |G : T|$. Similarly, if $\mathcal{U}$ is the H-orbit of φ in $\mathrm{Irr}(D)$ and S is the stabilizer of φ in H, we have $|\mathcal{U}| = |H : S|$. Also, we observe that H acts transitively on $\mathcal{O}$ because $NH = G$, and thus $|\mathcal{O}| = |H : T \cap H|$.

For (a), we are assuming that $\theta_D = \varphi$, and thus restriction to D maps the H-orbit $\mathcal{O}$ of θ onto the H-orbit $\mathcal{U}$ of φ, and thus $|\mathcal{O}| \geq |\mathcal{U}|$. We are also assuming that $S \subseteq T \cap H$, so $|\mathcal{U}| = |H : S| \geq |H : T \cap H| = |\mathcal{O}|$, and it follows that $|\mathcal{O}| = |\mathcal{U}|$, and thus restriction to D defines a bijection from $\mathcal{O}$ onto $\mathcal{U}$.

Let $\chi \in \mathrm{Irr}(G|\theta)$, and let ψ be an irreducible constituent of χ_H. By Clifford's theorem, $\chi_N = e\sum\mathcal{O}$ for some integer e, and thus $\chi_D = e\sum\mathcal{U}$. Then ψ lies over at least one of the characters in $\mathcal{U}$, and we deduce by Clifford's theorem applied in the group H that $\psi_D = f\sum\mathcal{U}$ for some integer f. It follows that $f \leq e$ and that if equality holds, then $\chi_D = \psi_D$, and thus $\chi_H = \psi$.

Now χ is a constituent of ψ^G, so

$$e = [\chi_N, \theta] \leq [(\psi^G)_N, \theta] = [(\psi_D)^N, \theta] = [\psi_D, \theta_D] = [\psi_D, \varphi] = f\,,$$

and it follows that $e = f$. Then $\chi_H = \psi$, and so restriction defines a map from $\mathrm{Irr}(G|\theta)$ to $\mathrm{Irr}(H|\varphi)$, as wanted. Also, we conclude that

$$[\chi_N, \theta] = [(\psi^G)_N, \theta]\,,$$

and thus χ is the unique irreducible constituent of ψ^G that lies over θ, and it follows that our restriction map is injective.

To see that this restriction map is surjective, let $\psi \in \mathrm{Irr}(H|\varphi)$ be arbitrary. Then

$$0 < [\psi_D, \varphi] = [(\psi_D)^N, \theta] = [(\psi^G)_N, \theta]\,,$$

and thus some irreducible constituent χ of ψ^G lies over θ. We know that χ_H is irreducible, however, and thus $\chi_H = \psi$, as required.

For (b) we are assuming that $\varphi^N = \theta$, and thus induction to N maps the H-orbit $\mathcal{U}$ onto the H-orbit $\mathcal{O}$, and thus $|\mathcal{U}| \geq |\mathcal{O}|$. We are also assuming that $T \cap H \subseteq S$, and thus $|\mathcal{O}| = |H : T \cap H| \geq |H : S| = |\mathcal{U}|$, and it follows that $|\mathcal{O}| = |\mathcal{U}|$, and thus induction to N defines a bijection from $\mathcal{U}$ onto $\mathcal{O}$.

Let $\psi \in \mathrm{Irr}(H|\varphi)$, and let χ be an irreducible constituent of ψ^G. By Clifford's theorem, $\psi_D = e\sum\mathcal{U}$ for some integer e, and thus

$$(\psi^G)_N = (\psi_D)^N = e\sum\mathcal{O}\,.$$

Then χ lies over at least one of the characters in $\mathcal{O}$, and so by Clifford's theorem, $\chi_N = f\sum\mathcal{O}$ for some integer f. It follows that $f \leq e$ and that if equality holds, then $\psi^G = \chi$.

Now ψ is a constituent of χ_H, and so

$$e = [\psi_D, \varphi] \leq [\chi_D, \varphi] = [\chi_N, \varphi^N] = [\chi_N, \theta] = f\,,$$

and it follows that $f = e$. Then $\psi^G = \chi$, and thus induction defines a map from $\mathrm{Irr}(H|\varphi)$ to $\mathrm{Irr}(G|\theta)$, as wanted. Also, $[\psi_D, \varphi] = [\chi_D, \varphi]$, so ψ is the unique irreducible constituent of χ_H that lies over φ, and it follows that the map is injective.

To see that this induction map is surjective, let $\chi \in \mathrm{Irr}(G|\theta)$ be arbitrary. Then χ lies over φ, so there exists $\psi \in \mathrm{Irr}(H)$ lying over φ and under χ. We know that ψ^G is irreducible, however, and thus $\psi^G = \chi$, as required. ■

Proof of Theorem 2.10. First assume that $H \lhd G$. If $\chi \in \mathcal{X}_\pi(G)$, then $\chi(1)$ is a π-number, so χ_H is irreducible by Corollary 1.16. Also, since $H \lhd G$ and χ_H lies under the π-special character χ, it follows that χ_H is π-special. Restriction thus defines a map $\mathcal{X}_\pi(G) \to \mathcal{X}_\pi(H)$, and this map is injective by Theorem 2.4(b).

For the general case, we can assume that $H < G$, and we proceed by induction on $|G|$. Let N be a maximal normal subgroup of G, and write $D = N \cap H$, so $|N : D| = |NH : H|$, and thus $|N : D|$ divides the π'-number $|G : H|$. Then $|N : D|$ is a π'-number, and since $N < G$, it follows by the inductive hypothesis that restriction defines an injection $\mathcal{X}_\pi(N) \to \mathcal{X}_\pi(D)$.

Suppose $|G : N|$ is a π'-number. By the first paragraph of the proof, restriction defines an injection $\mathcal{X}_\pi(G) \to \mathcal{X}_\pi(N)$, so by composing with the restriction map $\mathcal{X}_\pi(N) \to \mathcal{X}_\pi(D)$, we see that restriction defines an injection $\mathcal{X}_\pi(G) \to \mathcal{X}_\pi(D)$.

Now let $\chi \in \mathcal{X}_\pi(G)$, and write $\varphi = \chi_D$, so $\varphi \in \mathcal{X}_\pi(D)$. Let $\psi = \chi_H$ and observe that $\psi_D = \varphi$, so ψ is an extension of φ to H, and since $D \lhd H$ and $|H : D| = |NH : N|$ is a π'-number, we see by Theorem 2.4(b) that to show that ψ is π-special, it suffices to check that ψ is the canonical extension of φ to H. This is the case, however, because $o(\psi)$ divides the π-number $o(\chi)$ since $\chi_H = \psi$. It follows that restriction defines a map $\mathcal{X}_\pi(G) \to \mathcal{X}_\pi(H)$, and this map is injective since following it by the restriction map to D yields an injective map.

We can now suppose that $|G : N|$ is a π-number, and thus $NH = G$ and $|N : D| = |G : H|$, and this is a π'-number. Given $\chi \in \mathcal{X}_\pi(G)$, let $\theta \in \mathrm{Irr}(N)$, where θ lies under χ. Then $\theta \in \mathcal{X}_\pi(N)$, and so $\theta_D \in \mathcal{X}_\pi(D)$ by the inductive hypothesis, and we write $\varphi = \theta_D$.

Since the restriction map $\mathcal{X}_\pi(N) \to \mathcal{X}_\pi(D)$ is injective, it follows that the stabilizer of φ in H stabilizes θ, and hence we can apply Lemma 2.11(a) to deduce that restriction defines a bijection $\mathrm{Irr}(G|\theta) \to \mathrm{Irr}(H|\varphi)$. In particular, χ_H is irreducible, and we write $\psi = \chi_H$. Then ψ lies over the π-special character φ, and since $D \lhd H$ and $|H : D| = |G : N|$ is a π-number, it follows by Theorem 2.4(a) that ψ is π-special.

We know now that restriction defines a map $\mathcal{X}_\pi(G) \to \mathcal{X}_\pi(H)$, and it remains to show that this map is injective. We suppose, therefore, that $\xi \in \mathcal{X}_\pi(G)$ and $\xi_H = \psi$, and we work to show that $\xi = \chi$. Now ξ lies over φ, so some irreducible constituent μ of ξ_N lies over φ. Also μ is π-special since it lies under ξ, and since μ lies over φ, it follows by the injectivity of the restriction map $\mathcal{X}_\pi(N) \to \mathcal{X}_\pi(D)$ that $\mu = \theta$, and thus $\xi \in \mathrm{Irr}(G|\theta)$. We conclude by the injectivity of the restriction map $\mathrm{Irr}(G|\theta) \to \mathrm{Irr}(H|\varphi)$ that $\xi = \chi$, as required. ■

It is natural to ask if there is any reasonable way to describe the image of Gajendragadkar's restriction map $\mathcal{X}_\pi(G) \to \mathrm{Irr}(H)$, where H is a Hall π-subgroup of a π-separable group G. An obvious necessary condition for a (not necessarily irreducible) character ψ of a subgroup H to be the restriction of a character of G is that $\psi(x) = \psi(y)$ whenever $x, y \in H$ are conjugate in G. We prove in Theorem 3.6 that if G is π-separable and H is a Hall π-subgroup, then this condition is also sufficient. In fact, we shall see that the image of the restriction map $\mathcal{X}_\pi(G) \to \mathrm{Irr}(H)$ is exactly the set of irreducible characters of the Hall π-subgroup H that satisfy this condition (see the discussion following Corollary 3.15).

Next, we present a little application of the part of the theory of π-special characters that we have developed thus far. First, however, we should explain that in this book, what we mean by an "application" is any result that can be proved using the theory, but which does not explicitly refer to the theory in its statement.

Recall that a character $\chi \in \mathrm{Irr}(G)$ is said to be **monomial** if $\chi = \lambda^G$, where λ is a linear character of some (not necessarily proper) subgroup of G. Monomial characters arise often in the character theory of solvable groups, and, indeed, a group for which every irreducible character is monomial is necessarily solvable. (Recall that a group with this property is said to be an **M-group**, and the fact that M-groups are solvable is a theorem of K. Taketa, which appears as Corollary 5.13 of [**CTFG**], and which is also proved here as Theorem 9.1.) There are still a number of open questions concerning monomial characters and M-groups. For example, although it is known that a normal subgroup of an M-group need not be an M-group, it is unknown if this can happen if the normal subgroup has odd order or odd index.

One of the things that make monomial characters difficult to work with is the fact that if $\chi \in \mathrm{Irr}(G)$ is monomial and $\chi = \psi^G$, where ψ is a character of some subgroup of G, then ψ may not be monomial. In order to study this phenomenon, we say that a character $\chi \in \mathrm{Irr}(G)$ is **supermonomial** if whenever $\chi = \psi^G$, then ψ is monomial. Equivalently, χ is supermonomial if and only if every primitive character that induces χ from a subgroup of G is linear.

As an application of the theory of π-special characters, we can give a condition sufficient to guarantee that a character is supermonomial. (Other sufficient conditions are presented in Chapter 9.)

Theorem 2.12. *Let $\chi \in \mathrm{Irr}(G)$, where G is π-separable and $\chi(1)$ is a π-number, and let $H \subseteq G$ be a Hall π-subgroup.*

(a) *If χ_H has a linear constituent, then χ is supermonomial.*

(b) *If χ_H has a principal constituent, then the multiplicative order of every linear character inducing χ is a π'-number.*

Proof. For (a), we assume that $\chi = \gamma^G$, where γ is a primitive character of some subgroup $U \subseteq G$, and our goal is to show that γ is linear. Since γ is primitive, Theorem 2.9 guarantees that we can write $\gamma = \alpha\beta$, where α is π-special and β is π'-special, and thus

$$\chi(1) = \gamma^G(1) = |G : U|\gamma(1) = |G : U|\alpha(1)\beta(1)\,.$$

Since $\chi(1)$ is a π-number and $\beta(1)$ is a π'-number, it follows that β is linear. It also follows that $|G : U|$ is a π-number, and since $|G : H|$ is a π'-number, we have $UH = G$.

Let $D = U \cap H$, so $|U : D| = |G : H|$, which is a π'-number. Now α is π-special, so it follows by Gajendragadkar's restriction theorem (Theorem 2.10) that α_D is irreducible, and thus $\gamma_D = \alpha_D\beta_D$ is irreducible because β is linear.

By assumption, χ_H has a linear constituent λ, and thus

$$0 < [\chi_H, \lambda] = [(\gamma^G)_H, \lambda] = [(\gamma_D)^H, \lambda] = [\gamma_D, \lambda_D]\,,$$

so λ_D is a linear constituent of the irreducible character γ_D. Then $\gamma_D = \lambda_D$, and hence γ is linear, proving (a).

For (b), we assume that λ is the principal character of H. Then $\gamma_D = \lambda_D = 1_D$, and hence $D \subseteq \ker(\gamma)$. It follows that the multiplicative order of γ divides the π'-number $|U : D|$, and this establishes (b). ∎

Recall now that by definition, the cyclotomic field $\mathbb{Q}_n$ is the subfield of the complex numbers generated over the rationals by a primitive nth root of unity. Recall also that all character values of a finite group G lie in the cyclotomic field $\mathbb{Q}_{|G|}$. The following consequence of Gajendragadkar's restriction theorem gives additional information about the values of π-special characters.

Corollary 2.13. *Let χ be a π-special character of a π-separable group G. Then all values of χ lie in the cyclotomic field $\mathbb{Q}_{|H|}$, where H is a Hall π-subgroup of G.*

Proof. All values of χ lie in $\mathbb{Q}_{|G|}$, so it suffices to show that χ is invariant under the Galois group $\mathcal{G} = \mathrm{Gal}(\mathbb{Q}_{|G|}/\mathbb{Q}_{|H|})$. For elements $h \in H$, it is certainly true that $\chi(h) \in \mathbb{Q}_{|H|}$, and thus for $\sigma \in \mathcal{G}$, we have $(\chi^\sigma)_H = (\chi_H)^\sigma = \chi_H$.

Since χ is π-special and the application of the field automorphism σ cannot affect either the degrees or the determinantal orders of characters of subgroups of G, it follows that χ^σ is also π-special. The injectivity of the restriction map $\mathcal{X}_\pi(G) \to \mathrm{Irr}(H)$ thus guarantees that $\chi = \chi^\sigma$, as required. ■

For the final result in this section, we need the basic (but not quite trivial) fact that if a and b are arbitrary positive integers, then $\mathbb{Q}_a \cap \mathbb{Q}_b = \mathbb{Q}_d$, where d is the greatest common divisor of a and b (see Theorem 20.12 of [**AGC**]). In particular, if a and b are relatively prime, then $\mathbb{Q}_a \cap \mathbb{Q}_b = \mathbb{Q}_1 = \mathbb{Q}$, the rational number field.

Corollary 2.14. *Let χ be a π-special character of a π-separable group G, and let $K \subseteq G$ be a π'-subgroup. Then χ_K is rational valued.*

Proof. All values of χ_K lie in $\mathbb{Q}_{|K|}$, and by Corollary 2.13 these values also lie in $\mathbb{Q}_{|H|}$, where H is a Hall π-subgroup of G. But $\mathbb{Q}_{|K|} \cap \mathbb{Q}_{|H|} = \mathbb{Q}$ since $|K|$ and $|H|$ are relatively prime, and the result follows.

2B

In this section, we obtain some applications of the theory of π-special characters to the character theory of solvable groups, and in particular we prove the solvable case of a conjecture of W. Feit. (The general conjecture remains unresolved.)

Since a solvable group G is π-separable for every set π of primes, a given character $\chi \in \mathrm{Irr}(G)$ can be π-special for several different sets π. It should be clear, however, that there is a unique *smallest* set $\pi = \pi(\chi)$ such that χ is π-special. Observe that χ is π-special for a set π of primes if and only if $\pi \supseteq \pi(\chi)$.

For our next result, we need to be able to compute the determinant of the product of two (not necessarily irreducible) characters. The relevant formula is

$$\det(\xi\eta) = \det(\xi)^{\eta(1)} \det(\eta)^{\xi(1)} .$$

To see why this is so, we can work one element at a time, so we can assume that the group is cyclic. We can thus write

$$\xi = \sum_{i=1}^{\xi(1)} \lambda_i \qquad \text{and} \qquad \eta = \sum_{j=1}^{\eta(1)} \mu_j ,$$

where the summands λ_i and μ_j are linear characters. Then

$$\xi\eta = \sum_{i,j} \lambda_i\mu_j ,$$

and thus

$$\det(\xi\eta) = \prod_{i,j} \lambda_i \mu_j \,.$$

In this product, λ_i occurs $\eta(1)$ times and μ_j occurs $\xi(1)$ times, and this yields

$$\det(\xi\eta) = \Big(\prod_i \lambda_i\Big)^{\eta(1)} \Big(\prod_j \mu_j\Big)^{\xi(1)} = \det(\xi)^{\eta(1)} \det(\eta)^{\xi(1)} \,,$$

as wanted. ∎

Theorem 2.15. *Let G be solvable, and suppose $\alpha, \beta \in \operatorname{Irr}(G)$, where $\pi(\alpha)$ and $\pi(\beta)$ are disjoint. Then $\alpha\beta$ is irreducible and $\pi(\alpha\beta) = \pi(\alpha) \cup \pi(\beta)$.*

Proof. Write $\sigma = \pi(\alpha)$ and $\tau = \pi(\beta)$. Then $\tau \subseteq \sigma'$, and thus α is σ-special and β is σ'-special, and it follows by Gajendragadkar's product theorem that $\alpha\beta$ is irreducible.

Let σ_0 be the set of prime divisors of $\alpha(1)$, and let τ_0 be the set of prime divisors of $\beta(1)$. Then σ_0 and τ_0 are subsets of σ and τ, respectively, and clearly, $\sigma_0 \cup \tau_0$ is exactly the set of primes that divide the degree of $\alpha\beta$.

Similarly, let σ_1 and τ_1 be the sets of all primes dividing determinantal orders of irreducible constituents of α_S and β_S, respectively, as S runs over all subnormal subgroups of G. Since $\sigma = \sigma_0 \cup \sigma_1$ and $\tau = \tau_0 \cup \tau_1$, it suffices to show that $\sigma_1 \cup \tau_1$ is exactly the set of primes dividing the determinantal orders of irreducible constituents of $(\alpha\beta)_S$ for subnormal subgroups $S \lhd\lhd G$.

We have seen that a consequence of Gajendragadkar's product theorem is the fact that the irreducible constituents of $(\alpha\beta)_S$ are exactly the characters of the form $\mu\nu$, where μ is an irreducible constituent of α_S and ν is an irreducible constituent of β_S. Now $\det(\mu\nu) = \det(\mu)^{\nu(1)} \det(\nu)^{\mu(1)}$, so $o(\mu\nu)$ divides $o(\mu)o(\nu)$, and thus every prime dividing $o(\mu\nu)$ lies in $\sigma_1 \cup \tau_1$.

Conversely, we must show that every prime p in $\sigma_1 \cup \tau_1$ divides the order of some irreducible constituent of the restriction of $\alpha\beta$ to a subnormal subgroup. We can assume that $p \in \sigma_1$, and we choose $S \lhd\lhd G$ and an irreducible constituent μ of α_S so that p divides $o(\mu)$. Let ν be any irreducible constituent of β_S. Now $\det(\mu\nu)$ is a product of two linear characters: $\det(\mu)^{\nu(1)}$ and $\det(\nu)^{\mu(1)}$. The prime p does not divide the exponent $\nu(1)$ in the first factor because $\nu(1)$ is a τ-number and $p \in \sigma$, and thus the order of the first factor is divisible by p. The other factor of $\det(\mu\nu)$ is a power of $\det(\nu)$, and so its order is a τ-number, and hence is not divisible by p. Since $\det(\mu\nu)$ is a product of a linear character with order divisible by p with a linear character whose order is not divisible by p, it follows that $o(\mu\nu)$ is divisible by p, as required. ∎

For notational simplicity when we deal with singleton sets, we write "p-special" rather than "$\{p\}$-special", and we write p'-special instead of $\{p\}'$-special.

Corollary 2.16. *Let G be solvable, and let π be a finite set of primes. For each member $p \in \pi$, let α_p be a p-special character of G. Then*

$$\chi = \prod_{p\in\pi} \alpha_p$$

is irreducible and $\pi(\chi) \subseteq \pi$. Furthermore, if χ is also a product of p-special characters β_p for $p \in \pi$, then $\alpha_p = \beta_p$ for all p.

Proof. We can assume that $|\pi| > 1$, and we proceed by induction on $|\pi|$. Choose $q \in \pi$ and let $\tau = \pi - \{q\}$. Write $\psi = \prod \alpha_p$, where p runs over τ, so by the inductive hypothesis, ψ is irreducible and $\pi(\psi) \subseteq \tau$. Since $q \notin \tau$ and $\pi(\psi) \subseteq \tau$, it follows by Theorem 2.15 that $\chi = \alpha_q\psi$ is irreducible and $\pi(\chi) = \pi(\alpha_q) \cup \pi(\psi) \subseteq \{q\} \cup \tau = \pi$, as wanted.

Now suppose that $\chi = \prod \beta_p$, where p runs over π and each character β_p is p-special. Then $\chi = \beta_q\xi$, where $\xi = \prod \beta_p$ for $p \in \tau$. By the inductive hypothesis, ξ is irreducible and $\pi(\xi) \subseteq \tau$. We have

$$\alpha_q\psi = \chi = \beta_q\xi\,,$$

and both α_q and β_q are q-special. Also, both ψ and ξ are q'-special since they are τ-special and $\tau \subseteq \{q\}'$. It follows from the uniqueness in Gajendragadkar's factorization theorem that $\alpha_q = \beta_q$ and $\psi = \xi$. Finally, since $\psi = \xi$, the inductive hypothesis yields $\alpha_p = \beta_p$ for all $p \in \tau$. ■

If G is solvable, we say that $\chi \in \mathrm{Irr}(G)$ is **fully factored** if we can write $\chi = \prod \alpha_p$, where p runs over some set π of primes and each factor α_p is p-special. (We allow π to be empty, and in that case our interpretation is that $\chi = 1_G$, the principal character.) We stress that in a full factorization there is at most one factor for each prime p.

Theorem 2.17. *Let $\chi \in \mathrm{Irr}(G)$ be primitive, where G is solvable. Then $\chi = \prod \alpha_p$, where p runs over $\pi(\chi)$ and α_p is p-special. In particular, χ is fully factored.*

Proof. We proceed by induction on $|\pi(\chi)|$. We can certainly assume that $\pi(\chi)$ is nonempty, and we choose $q \in \pi(\chi)$. Since χ is primitive, Theorem 2.9 allows us to write $\chi = \mu\nu$, where μ is q-special and ν is q'-special. Also, $\pi(\chi) = \pi(\mu) \cup \pi(\nu)$ by Theorem 2.15. Now $\pi(\nu) \subseteq q'$, so $q \notin \pi(\nu)$, and we conclude that $\pi(\nu) < \pi(\chi)$. Also, since $\pi(\mu) \subseteq \{q\}$, we see that $\pi(\chi) = \{q\} \cup \pi(\nu)$.

We argue next that ν is primitive. Otherwise, we can write $\nu = \gamma^G$, where $\gamma \in \mathrm{Irr}(J)$ and $J < G$. Then

$$\chi = \mu\nu = \mu\gamma^G = (\mu_J\gamma)^G ,$$

and this contradicts the assumed primitivity of χ.

Since $\pi(\nu) < \pi(\chi)$, we can apply the inductive hypothesis to write $\nu = \prod \alpha_p$, where p runs over $\pi(\nu)$ and α_p is p-special. Since $q \notin \pi(\nu)$ and $\pi(\chi) = \{q\} \cup \pi(\nu)$, the result follows by taking $\alpha_q = \mu$. ■

We mention that the characters α_p in Theorem 2.17 must all be nonprincipal. Otherwise, we could delete one of the factors, and then Corollary 2.16 would imply that $\pi(\chi)$ is contained in a proper subset of $\pi(\chi)$.

Next, as an easy application, we prove a result of B. Huppert. We need a bit of basic character theory: If $\chi \in \mathrm{Irr}(G)$, then $\chi(1)^2 \le |G : \mathbf{Z}(G)|$. To see this, observe that $\chi_{\mathbf{Z}(G)} = \chi(1)\lambda$ for some linear character λ of $\mathbf{Z}(G)$, and thus χ occurs with multiplicity $\chi(1)$ as a constituent of λ^G. It follows that $\chi(1)^2 \le \lambda^G(1) = |G : \mathbf{Z}(G)|$, as wanted.

Theorem 2.18. *Let $\chi \in \mathrm{Irr}(G)$, where G is solvable, and assume that χ is primitive. Then $\chi(1)^2$ divides $|G : \mathbf{Z}(G)|$.*

Proof. Given a prime p, let p^a be the p-part of $\chi(1)$. (In other words, p^a is the full power of p dividing $\chi(1)$.) It suffices to show for every choice of p that p^{2a} is at most the p-part of $|G : \mathbf{Z}(G)|$ or, equivalently, $p^{2a} \le |P : P \cap \mathbf{Z}(G)|$, where P is a Sylow p-subgroup of G.

Since χ is primitive, it is fully factored by Theorem 2.17, and thus we can write $\chi = \prod \alpha_q$, where q runs over some set of distinct primes and α_q is q-special. Then $\alpha_q(1)$ is a power of q, and since $\chi(1) = \prod \alpha_q(1)$, we see that for each prime divisor p of $\chi(1)$, the p-part of $\chi(1)$ is exactly $\alpha_p(1)$.

By Gajendragadkar's restriction theorem, $(\alpha_p)_P$ is irreducible, and thus

$$p^{2a} = \alpha_p(1)^2 \le |P : \mathbf{Z}(P)| \le |P : P \cap \mathbf{Z}(G)| ,$$

as required. ■

We need a fact about fields of values of fully factored characters. If χ is a character of a group G, we write $\mathbb{Q}(\chi)$ to denote the field generated over the rational numbers $\mathbb{Q}$ by all the values of χ, and similarly, if $\mathcal{X}$ is a set of characters of G, we write $\mathbb{Q}(\mathcal{X})$ to denote the field generated over $\mathbb{Q}$ by all values of all the members of $\mathcal{X}$. Note that fields generated by values of characters of G are subfields of the cyclotomic field $\mathbb{Q}_{|G|}$, and since the Galois group $\mathrm{Gal}(\mathbb{Q}_{|G|}/\mathbb{Q})$ is abelian, all of these character-value fields are Galois extensions of $\mathbb{Q}$.

Lemma 2.19. *Let G be solvable, and suppose that $\chi = \prod \alpha_p$, where p runs over some set π of distinct primes and $\alpha_p \in \mathrm{Irr}(G)$ is p-special for each member $p \in \pi$. Then*

$$\mathbb{Q}(\chi) = \mathbb{Q}(\{\alpha_p \mid p \in \pi\}),$$

the field generated over $\mathbb{Q}$ by the values of all of the characters α_p.

Proof. Write $E = \mathbb{Q}(\{\alpha_p \mid p \in \pi\})$. Since χ is the product of the α_p, it is clear that $\mathbb{Q}(\chi) \subseteq E$, and we consider an arbitrary element σ of the Galois group $\mathrm{Gal}(E/\mathbb{Q}(\chi))$. Then

$$\chi = \chi^\sigma = \prod_{p \in \pi} (\alpha_p)^\sigma,$$

and since $(\alpha_p)^\sigma$ is p-special, it follows from the uniqueness in Corollary 2.16 that $(\alpha_p)^\sigma = \alpha_p$ for all p. Thus σ fixes all values of the characters α_p, and hence σ is the identity automorphism of E. We conclude that $\mathrm{Gal}(E/\mathbb{Q}(\chi))$ is trivial, and so $E = \mathbb{Q}(\chi)$. ∎

We are now ready to discuss the Feit conjecture. If $\chi \in \mathrm{Irr}(G)$, then in addition to the two integers $\chi(1)$ and $o(\chi)$ that we have been considering, there are several other numbers that are naturally associated with χ. One of these, which we call $f(\chi)$, is defined to be the smallest positive integer f such that all of the values of χ lie in the cyclotomic field $\mathbb{Q}_f$. Since we know that the values of χ lie in $\mathbb{Q}_{|G|}$, it is clear that $f(\chi)$ exists and that $f(\chi) \le |G|$. In fact, it is a consequence of the following lemma that $f(\chi)$ divides $|G|$.

Lemma 2.20. *Let $\chi \in \mathrm{Irr}(G)$, and suppose that all values of χ lie in $\mathbb{Q}_m$ for some integer m. Then $f(\chi)$ divides m.*

Proof. Write $f = f(\chi)$. Then all values of χ lie in $\mathbb{Q}_f \cap \mathbb{Q}_m = \mathbb{Q}_d$, where d is the greatest common divisor of f and m. It follows from the definition of $f(\chi)$ that $f \le d$, and since d divides f, we have $f = d$. Then f divides m, as claimed. ∎

The Feit conjecture asserts that if $\chi \in \mathrm{Irr}(G)$, where G is an arbitrary finite group, then G contains an element of order $f(\chi)$. The next result shows that to prove this, we need consider only primitive characters.

Corollary 2.21. *Suppose $\chi \in \mathrm{Irr}(G)$ and $\chi = \psi^G$ for some character ψ of a subgroup $J \subseteq G$. If J contains an element of order $f(\psi)$, then G contains an element of order $f(\chi)$. In particular, if the Feit conjecture holds for all proper subgroups of G and $\chi \in \mathrm{Irr}(G)$ is imprimitive, then G contains an element of order $f(\chi)$.*

Proof. Since $\chi = \psi^G$, it follows from the definition of an induced character that each value of χ lies in every subfield of $\mathbb{C}$ that contains all values of ψ. In particular, if $e = f(\psi)$, then $\chi(g) \in \mathbb{Q}_e$ for all $g \in G$, and it follows by Lemma 2.20 that $f(\chi)$ divides e. By assumption, there exists an element $x \in J$ with order e, so some power of x has order $f(\chi)$, and this completes the proof of the first assertion.

Now suppose that the Feit conjecture holds for all proper subgroups of G, and let $\chi \in \mathrm{Irr}(G)$ be imprimitive. Then $\chi = \psi^G$ for some (necessarily irreducible) character ψ of a subgroup $J < G$. By assumption, the Feit conjecture holds for J, so there exists $x \in J$ with order $f(\psi)$. By the first part of the proof, therefore, G contains an element of order $f(\chi)$. ■

Theorem 2.22. *The Feit conjecture holds for every solvable group.*

Proof. Let G be a solvable group having the smallest possible order such that the Feit conjecture fails for G. Then there exists a character $\chi \in \mathrm{Irr}(G)$ such that no element of G has order $f = f(\chi)$. Also, since the conjecture holds for all proper subgroups of G, it follows by Corollary 2.21 that χ must be primitive, and thus Theorem 2.17 guarantees that χ is fully factored. We can thus write $\chi = \prod \alpha_p$, where p runs over some set π of distinct primes, and α_p is p-special.

By Lemma 2.19, the values of each the characters α_p lie in $\mathbb{Q}(\chi) \subseteq \mathbb{Q}_f$, and thus $f(\alpha_p)$ divides f by Lemma 2.20. Also, since α_p is p-special, it follows by Corollary 2.13 that the values of α_p lie in $\mathbb{Q}_{|P|}$, where P is a Sylow p-subgroup of G, and thus $f(\alpha_p)$ is a power of p. We conclude that $f(\alpha_p)$ divides f/q for every prime divisor q of f different from p. In particular, if q divides f and $q \neq p$, then the values of α_p lie in $\mathbb{Q}_{f/q}$.

If q is a prime divisor of f, then $f/q < f$, so some value of χ does not lie in $\mathbb{Q}_{f/q}$. Let E_q be the field generated over $\mathbb{Q}_{f/q}$ by the values of χ, and observe that $\mathbb{Q}_{f/q} < E_q \subseteq \mathbb{Q}_f$. By elementary Galois theory, we see that there exists an automorphism τ_q of $\mathbb{Q}_f$ such that τ_q acts nontrivially on E_q and trivially on $\mathbb{Q}_{f/q}$. In particular, for each prime divisor q of f, we have $\chi^{\tau_q} \neq \chi$. Since τ_q acts trivially on $\mathbb{Q}_{f/q}$ and this field contains the values of α_p for $p \neq q$, we have

$$(\alpha_p)^{\tau_q} = \alpha_p \qquad \text{if} \quad p \neq q\,.$$

Now τ_q does not fix χ, so it fails to fix at least one of the factors α_p, and hence $p = q$. It follows that $q \in \pi$ for every prime divisor q of f, and for every such prime,

$$\alpha_q^{\tau_q} \neq \alpha_q\,.$$

By assumption, G has no element of order f, and so if $x \in G$, the greatest common divisor d of f and $o(x)$ is a proper divisor of f, and thus d divides f/q for some prime divisor q of f.

Since the values of α_p lie in $\mathbb{Q}_f$, we have

$$\alpha_q(x) \in \mathbb{Q}_f \cap \mathbb{Q}_{o(x)} = \mathbb{Q}_d \subseteq \mathbb{Q}_{f/q},$$

and thus $\alpha_q(x)$ is fixed by τ_q. For each element $x \in G$, therefore, there exists a prime divisor q of f such that the function $\alpha_q - \alpha_q^{\tau_q}$ vanishes at x. The product of these functions is thus identically zero, and so we can write

$$0 = \prod_q \left(\alpha_q - \alpha_q^{\tau_q}\right),$$

where q runs over the prime divisors of f. Expanding this product, we obtain

$$0 = \sum_\Delta (-1)^{|\Delta|} \prod_q \alpha_q^{\sigma_q},$$

where Δ runs over all subsets of the set of prime divisors of f and

$$\sigma_q = \begin{cases} \tau_q & \text{if } q \in \Delta, \\ 1 & \text{if } q \notin \Delta. \end{cases}$$

Since each character $\alpha_q^{\sigma_q}$ is q-special, it follows by Theorem 2.32 that except for the signs, each of the products in the above sum is an irreducible character of G. Also, since $\alpha_q^{\tau_q}$ and α_q are different for all q, we see by Corollary 2.16 that these irreducible characters are distinct. Distinct irreducible characters are linearly independent, however, and this is the desired contradiction. ∎

For our final application of the theory of π-special characters, we prove the following.

Theorem 2.23. *Let G be solvable, and let $\chi \in \mathrm{Irr}(G)$ be primitive. Suppose that ψ is a character of G that agrees with χ on all elements that have prime-power order in G. Then $\psi = \chi$.*

We need the following standard fact.

Lemma 2.24. *Let χ be a character of some group G, and let p be a prime. If $\chi(x) = 0$, where $x \in G$ has p-power order, then p divides $\chi(1)$.*

Proof. Let P be a prime ideal containing p in the ring of algebraic integers, and note that $P \cap \mathbb{Q}$ is an ideal of $\mathbb{Z}$ containing p, and thus $P \cap \mathbb{Q} = p\mathbb{Z}$. If ε is an mth root of unity, where m is a power of p, then

$$(\varepsilon - 1)^m \equiv \varepsilon^m - 1 = 0 \mod P,$$

and since P is a prime ideal, it follows that $\varepsilon - 1 \in P$. Then $\varepsilon \equiv 1 \bmod P$, and since $\chi(x)$ is a sum of $\chi(1)$ roots of unity, each of p-power order, it follows that $0 = \chi(x) \equiv \chi(1) \bmod P$. Then $\chi(1) \in P \cap \mathbb{Q} = p\mathbb{Z}$, as required. ■

Proof of Theorem 2.23. Since $\psi(1) = \chi(1)$, we see that ψ and χ agree on the trivial subgroup of G, and thus we can choose a normal subgroup N of G maximal with the property that $\psi_N = \chi_N$. Our goal is to show that $N = G$, so we assume that $N < G$, and we work to obtain a contradiction. Let M/N be a chief factor of G, and observe that since G is solvable, M/N is a p-group for some prime p.

By assumption, χ is primitive, and thus χ_N is a multiple of some character $\varphi \in \mathrm{Irr}(N)$. Also, χ is $\{p\}$-factored by Theorem 2.9, and thus φ is $\{p\}$-factored, and we can write $\varphi = \alpha\beta$, where α is p-special and β is p'-special. Now φ is invariant in G, so β is invariant in G by the uniqueness of the factorization, and in particular β is invariant in M. Since β is p'-special and M/N is a p-group, it follows by Theorem 2.4(b) that β has a p'-special extension $\gamma \in \mathrm{Irr}(M)$.

Now $\varphi^M = (\alpha\beta)^M = \alpha^M\gamma$, and the irreducible constituents of α^M are p-special by Theorem 2.4(a). It follows that every member of $\mathrm{Irr}(M|\varphi)$ is a product of some p-special character of M with γ. In particular, since $\chi_N = \psi_N$ is a multiple of φ, the irreducible constituents of χ_M and ψ_M have factorizations of this form, and so we can write $\chi_M = \xi\gamma$ and $\psi_M = \eta\gamma$, where all irreducible constituents of ξ and η are p-special.

Let P be a Sylow p-subgroup of M, and observe that by hypothesis,

$$\xi_P\gamma_P = \chi_P = \psi_P = \eta_P\gamma_P\,.$$

Since p does not divide $\gamma(1)$, Lemma 2.24 guarantees that all values of γ on P are nonzero, and thus we can cancel γ_P to deduce that $\xi_P = \eta_P$. By Gajendragadkar's restriction theorem, the restrictions to P of distinct p-special characters of M are distinct and irreducible, and since all irreducible constituents of ξ and η are p-special, it follows that $\xi = \eta$. Then $\chi_M = \xi\gamma = \eta\gamma = \psi_M$, and this contradicts the maximality of N. ■

Corollary 2.25. *Let $\chi \in \mathrm{Irr}(G)$ be primitive, where G is solvable. Then $\mathbb{Q}(\chi)$ is generated by the values of χ at elements of prime power order in G.*

Proof. Let $\mathbb{Q} \subseteq F \subseteq \mathbb{Q}(\chi)$, where F is the subfield generated by the values of χ at the elements of prime power order in G. If $\sigma \in \mathrm{Gal}(\mathbb{Q}(\chi)/F)$. Then χ^σ and χ agree on all elements of prime-power order, and since χ is primitive, Theorem 2.23 implies that $\chi^\sigma = \chi$. Then σ is the identity field automorphism, and it follows that $F = \mathbb{Q}(\chi)$, as wanted.

2C

In this section, we discuss some of the ways in which the prime 2 plays an exceptional role in the theory of π-special characters. For example, suppose that $H \subseteq G$, where G is π-separable, and let $\chi \in \mathrm{Irr}(G)$ be π-special. We have seen that if $|G : H|$ is a π'-number, then χ_H is irreducible, and in fact, χ_H is π-special. One might ask if any part of this theorem survives if we drop the assumption that $|G : H|$ is a π'-number. In this generality, of course, χ_H is not necessarily irreducible, but we might guess that if χ_H is *assumed* to be irreducible, then in fact, it must be π-special. (This seems plausible because the degree and determinantal order of χ_H are certainly π-numbers.) Unfortunately, even with the assumption that χ_H is irreducible, it is not necessarily true that it is π-special. A counterexample occurs in the solvable group $G = GL(2,3)$ of order 48, which has a faithful $\{2\}$-special character χ with degree 2. If H is the normalizer in G of a Sylow 3-subgroup T, it is easy to see that χ_H is irreducible, but this character cannot be $\{2\}$-special since T is not contained in its kernel.

If $|G|$ is odd or, more generally, if $2 \notin \pi$, then it is true that an irreducible restriction of a π-special character must be π-special. In this section, we prove the odd-order case of this result, and we develop some of the theory needed in the more general situation, but we defer the proof of the full theorem until later, after we have developed some additional theory in Chapter 7.

For odd-order groups, the proof of this restriction theorem is comparatively easy because in that case there is a useful necessary and sufficient condition for a character $\chi \in \mathrm{Irr}(G)$ to be π-special. (By the Feit–Thompson theorem, odd-order groups are solvable, so there is no need to make an explicit assumption that G is π-separable in the following.) ∎

Theorem 2.26. *Assume that $|G|$ is odd, and let π be an arbitrary set of primes. Then a character $\chi \in \mathrm{Irr}(G)$ is π-special if and only if $\chi(1)$ is a π-number and the values of χ lie in some cyclotomic field $\mathbb{Q}_m$, where m is a π-number.*

Observe that in the language of the previous section, the condition on the values of χ in this theorem is equivalent to saying that $f(\chi)$ is a π-number.

Proof of Theorem 2.26. Since the necessity of the condition is immediate from Corollary 2.13, it is enough to show for an odd-order group G that if the degree of an irreducible character χ of G is a π-number and the values of χ are contained in $\mathbb{Q}_m$, where m is a π-number, then χ is π-special. In fact, we prove something a bit stronger.

Factor $|G| = ab$, where a is a π-number and b is a π'-number, and recall from Galois theory that since a and b are relatively prime, the Galois group $\mathrm{Gal}(\mathbb{Q}_{ab}/\mathbb{Q}_a)$ is isomorphic (via restriction) to $\mathrm{Gal}(\mathbb{Q}_b/\mathbb{Q})$. It follows that $\mathrm{Gal}(\mathbb{Q}_{ab}/\mathbb{Q}_a)$ contains an element σ that acts on $\mathbb{Q}_b$ like complex conjugation, and of course σ acts like the identity map on $\mathbb{Q}_a$. Also, since σ^2 fixes all elements of $\mathbb{Q}_a$ and $\mathbb{Q}_b$, we have $\sigma^2 = 1$. (We refer to σ as the **magic field automorphism** for π and G, and we observe that if $H \subseteq G$, then the restriction of σ to $\mathbb{Q}_{|H|}$ is exactly the magic field automorphism for π and H.)

If m is a π-number, then $\mathbb{Q}_m \cap \mathbb{Q}_{|G|} \subseteq \mathbb{Q}_a$, so a character of G with values in $\mathbb{Q}_m$ is fixed by the magic field automorphism σ. To complete the proof, we show by induction on the (odd) group order that if $\chi \in \mathrm{Irr}(G)$, where $\chi(1)$ is a π-number and $\chi^\sigma = \chi$, then χ is π-special. We can certainly assume that G is nontrivial, so we can choose a maximal normal subgroup M of G. Then G/M is either a π-group or a π'-group, and we consider these cases separately.

Suppose first that G/M is a π-group. By assumption, $\chi^\sigma = \chi$, and thus σ permutes the set of irreducible constituents of χ_M. The cardinality of this set divides $|G/M|$, and hence it is odd, and since $\sigma^2 = 1$, it follows that σ must fix some irreducible constituent θ of χ_M, and of course $\theta(1)$ divides $\chi(1)$, so $\theta(1)$ is a π-number. Clearly, θ is fixed by the restriction of σ to $\mathbb{Q}_{|M|}$, and since this restriction is the magic field automorphism for M, the inductive hypothesis guarantees that θ is π-special. We conclude by Theorem 2.4(a), that χ is π-special.

Now assume that G/M is a π'-group, and write $\theta = \chi_M$, so that θ is irreducible and G-invariant, and $\theta(1) = \chi(1)$ is a π-number. Also, θ is fixed by σ, and hence it is fixed by the magic automorphism for M. By the inductive hypothesis, θ is π-special, and so by Theorem 2.4(b), it has a π-special canonical extension $\xi \in \mathrm{Irr}(G)$.

By the Gallagher correspondence, we can write $\chi = \beta\xi$, where β is some uniquely determined linear character of G/M. Now

$$(\xi^\sigma)_M = (\xi_M)^\sigma = \theta^\sigma = \theta\,,$$

so ξ^σ is an extension of θ. Also $o(\xi^\sigma) = o(\xi)$, so in fact $\xi^\sigma = \xi$, the canonical extension. Then

$$\beta\xi = \chi = \chi^\sigma = (\beta\xi)^\sigma = \beta^\sigma\xi\,,$$

and we conclude by the uniqueness of β that $\beta^\sigma = \beta$. Since G/M is a π'-group and β is linear, the values of β are π'-roots of unity, and thus they are mapped to their complex conjugates by σ. Then $\beta = \beta^\sigma = \overline{\beta}$, and hence β is real valued. As G/M has odd order, we conclude that β is the principal character, and we have $\chi = \beta\xi = \xi$, so χ is π-special. ∎

Corollary 2.27. *Let $H \subseteq G$, where $|G|$ is odd, and suppose that $\chi \in \mathrm{Irr}(G)$ is π-special and that χ_H is irreducible. Then χ_H is π-special.*

Proof. Clearly, $\mathbb{Q}(\chi_H) \subseteq \mathbb{Q}(\chi)$, and since χ is π-special, we know that $\mathbb{Q}(\chi) \subseteq \mathbb{Q}_m$ for some π-number m. Since χ_H has π-degree, the result is an immediate consequence of Theorem 2.26. ∎

There is an equally easy result about induced characters for odd-order groups.

Corollary 2.28. *Let $\psi \in \mathrm{Irr}(H)$, where $H \subseteq G$ and $|G|$ is odd, and suppose that ψ^G is irreducible. If ψ is π-special and $|G : H|$ is a π-number, then ψ^G is π-special.*

Proof. We see from the definition of an induced character that $\mathbb{Q}(\psi^G) \subseteq \mathbb{Q}(\psi)$, and since ψ is π-special, we have $\mathbb{Q}(\psi) \subseteq \mathbb{Q}_m$ for some π-number m. Since $\psi^G(1) = |G : H|\psi(1)$ is a π-number, the result follows by Theorem 2.26. ∎

It may not be obvious, but the converse of Corollary 2.28 is also true, although an entirely different technique of proof seems to be needed to establish it. As in Corollary 2.28, assume that $|G|$ is odd, and let $\psi \in \mathrm{Irr}(H)$, where $H \subseteq G$ and ψ^G is irreducible. If ψ^G is π-special, we will show that ψ is π-special and $|G : H|$ is a π-number. (Of course, the second assertion is obvious because $|G : H|$ divides the π-number $\psi^G(1)$.) In fact, something much stronger is true: It is not actually necessary to assume that ψ^G is irreducible. If we assume only that each irreducible constituent of ψ^G is π-special, we can still conclude that ψ is π-special and that $|G : H|$ is a π-number. (And note that in this generality the second assertion is no longer obvious.)

A version of the converse of Corollary 2.28 is valid even without assuming that $|G|$ is odd. If $2 \in \pi$, the result holds as stated: ψ is π-special. If $2 \notin \pi$, a suitably modified form of the conclusion is true. In that case, ψ may not be π-special, but there exists a certain linear character $\delta_{(G,H)}$ of H such that $\delta_{(G,H)}\psi$ is π-special.

As we shall see, $\delta_{(G,H)}$ depends on the embedding of H in G, but not on the character ψ. Also, the linear character $\delta_{(G,H)}$ is a **sign character**, which means that its values are ± 1 or, equivalently, that its square is principal. (Note that a sign character of an odd-order group is necessarily principal because the order of a linear character divides the order of the group.)

We refer to $\delta_{(G,H)}$ as the π-**standard** sign character of H with respect to G. Unfortunately, it will take some effort to define it and to establish its essential properties. Before we proceed with that, however, we offer a formal statement of the theorem that we have been discussing.

Theorem 2.29. *Let $\psi \in \mathrm{Irr}(H)$, where $H \subseteq G$ and G is π-separable, and assume that every irreducible constituent of ψ^G is π-special. Then $|G:H|$ is a π-number and $\psi(1)$ is a π-number. Also, $\lambda\psi$ is π-special, where $\lambda \in \mathrm{Irr}(H)$ is principal if $2 \in \pi$, and $\lambda = \delta_{(G,H)}$ if $2 \notin \pi$.*

Note that if $|G|$ is odd, it is irrelevant whether or not the prime 2 is included in the set π. This is so because in that case, H can have no nontrivial sign character, and thus $\delta_{(G,H)} = 1_H$, so λ is principal unconditionally.

Sign characters are necessary in Theorem 2.29 and related results because the determinantal order does not behave well for induced characters. More specifically, if $H \subseteq G$ and ψ is a character of H, then $o(\psi^G)$ may not divide $o(\psi)$, although, as we shall see, $o(\psi^G)$ always divides $2o(\psi)$. (The group G of order 2 provides an easy example showing that the extra factor of 2 may be necessary: if ψ is the principal character of the trivial subgroup, then $o(\psi) = 1$ but $o(\psi^G) = 2$.)

This difficulty can sometimes be avoided by finding an appropriate sign character δ of H such that $o(\psi^G)$ divides $o(\delta\psi)$ for all characters ψ of H, and in this situation we say that δ has the **induction determinant** property for H in G. Equivalently, a sign character δ of H has the induction determinant property for H in G if and only if $o((\delta\psi)^G)$ divides $o(\psi)$ for all characters ψ of H. (These two formulations are equivalent because each can be obtained from the other by substituting $\delta\psi$ for ψ and observing that δ^2 is principal.)

To study determinants of induced characters, we begin with some observations about permutation characters and representations. Let G act on a set Ω, and recall that if χ is the corresponding permutation character of G, then by definition, $\chi(g)$ is the number of points in Ω that are fixed by g. The associated **permutation sign character** of G is the function σ_Ω whose value at an element $g \in G$ is 1 if g induces an even permutation on Ω and -1 if g induces an odd permutation. We will show (somewhat informally) that $\sigma_\Omega = \det(\chi)$.

A permutation representation $\mathcal{P}$ of G corresponding to the action of G on Ω can by constructed by defining the matrix $\mathcal{P}(g)$ as follows. The rows and columns of $\mathcal{P}(g)$ are indexed by Ω, and, for $s, t \in \Omega$, the entry of $\mathcal{P}(g)$ at position (s,t) is 1 if $s{\cdot}g = t$, and it is 0, otherwise. The trace of $\mathcal{P}(g)$ is thus the number of ones on the diagonal, and this is the number of points $s \in \Omega$ fixed by g. Then $\mathcal{P}$ is a representation of G yielding the permutation character χ, and thus $(\det(\chi))(g) = \det(\mathcal{P}(g))$.

Since each row and column of $\mathcal{P}(g)$ contains exactly one nonzero entry (equal to 1), we can rearrange the rows of $\mathcal{P}(g)$ to form the $m \times m$ identity matrix I, where $m = |\Omega|$. It follows that $\det(\mathcal{P}(g)) = \pm\det(I) = \pm 1$. The permutation of the rows needed to convert $\mathcal{P}(g)$ to the identity matrix is exactly the permutation of Ω induced by g, and this determines the sign. We thus have $\det(\mathcal{P}(g)) = \sigma_\Omega(g)$, and hence $\det(\chi) = \sigma_\Omega$, as claimed. An important special case is where Ω is the set of right cosets of a subgroup $H \subseteq G$, where G acts by right multiplication. It is easy to see in this situation that $\chi = (1_H)^G$, and thus $\det((1_H)^G)$ is a sign character, and in particular, $o((1_H)^G)$ divides 2.

We have just seen that if ψ is the principal character of a subgroup $H \subseteq G$, then $o(\psi^G)$ divides $2o(\psi)$. The following generalizes this divisibility assertion for an arbitrary character ψ of H, and it produces a sign character with the induction determinant property whenever $|G : H|$ is odd.

Lemma 2.30. *Let ψ be an arbitrary character of a subgroup $H \subseteq G$. Then*

(a) *$o(\psi^G)$ divides $2o(\psi)$.*

(b) *Assume that $|G : H|$ is odd, and let $\delta = \det(((1_H)^G)_H)$. Then δ is a sign character of H and $o(\psi^G)$ divides $o(\delta\psi)$, and thus δ has the induction determinant property in G.*

Proof. Let $m = |G : H|$, and let T be a set of representatives for the right cosets of H in G. Then $|T| = m$, and we can view T as a set of labels for the right cosets of H. The action of G (by right multiplication) on these cosets can thus be interpreted as an action of G on T. Let $\mathcal{P}$ be the corresponding permutation representation, with rows and columns indexed by the elements of T. Then the entry of $\mathcal{P}(g)$ at position (s, t) is 1 if $Hsg = Ht$ and, otherwise, the entry at position (s, t) is 0.

Next, we choose a representation $\mathcal{Y}$ of H yielding the character ψ, and then we use $\mathcal{Y}$ and the permutation representation $\mathcal{P}$ to construct a representation $\mathcal{X}$ of G that yields the character ψ^G. Given an element $g \in G$, we form the matrix $\mathcal{X}(g)$ by replacing each entry in the $m \times m$ matrix $\mathcal{P}(g)$ with a $d \times d$ matrix, where $d = \psi(1)$. Each zero entry in $\mathcal{P}(g)$ is replaced with a zero matrix, and an entry equal to 1 in position (s, t) of $\mathcal{P}(g)$ is replaced by the matrix $\mathcal{Y}(sgt^{-1})$. (This makes sense because $Hsg = Ht$, and thus $sgt^{-1} \in H$.) Then $\mathcal{X}(g)$ is a $dm \times dm$ matrix, and it is not hard to check that $\mathcal{X}$ is indeed a representation of G that yields the character ψ^G.

To compute the determinant of $\mathcal{X}(g)$, we permute its rows while keeping track of the relevant signs. We know that some permutation of the rows of $\mathcal{P}(g)$ will convert $\mathcal{P}(g)$ to the identity matrix, and we use essentially the same row permutation on $\mathcal{X}(g)$, but instead of moving individual rows, we move d consecutive rows at a time. What results from this manipulation is a matrix with $d \times d$ blocks along the diagonal and zeros elsewhere, where each of the diagonal blocks is a matrix of the form $\mathcal{Y}(h)$, where $h = sgt^{-1} \in H$

for appropriate $s, t \in T$. The determinant $\varepsilon(g)$ of this block-diagonal matrix is the product of the determinants of the diagonal blocks, and each of those has the form $\det(\mathcal{Y}(h)) = (\det(\psi))(h)$. It follows that $\varepsilon(g)$ is a root of unity with order dividing $o(\psi)$. Then

$$(\det(\psi^G))(g) = \det(\mathcal{X}(g)) = \pm\varepsilon(g),$$

and it follows that $o(\psi^G)$ divides $2o(\psi)$, establishing (a).

Now assume that $m = |G : H|$ is odd. Let σ be the permutation sign character of the action of G on the right cosets of H in G, so that as we have seen, $\sigma = \det((1_H)^G)$, and thus $\delta = \sigma_H$ is a sign character. Also, $\sigma(g) = \det(\mathcal{P}(g))$ for all elements $g \in G$.

Next, we determine the ambiguous sign in our formula for $\det(\mathcal{X}(g))$. Since the row permutations that we used to put $\mathcal{X}(g)$ into block diagonal form move d consecutive rows at a time, the sign has the form τ^d, where τ is the sign of of the underlying permutation of the m rows of $\mathcal{P}(g)$ that were used to convert $\mathcal{P}(g)$ to the identity matrix. Then $\tau = \det(\mathcal{P}(g)) = \sigma(g)$, and thus we have $\det(\psi^G) = \sigma^d\varepsilon$, or equivalently, $\sigma^d \det(\psi^G) = \varepsilon$. (Note that this shows that ε is a linear character of G.)

Now $\psi^G(1) = md$, and since m is odd and σ is a sign character, we have $\sigma^{md} = \sigma^d$. Recall that $\delta = \sigma_H$, so

$$\det((\delta\psi)^G) = \det(\sigma\psi^G) = \sigma^{md}\det(\psi^G) = \sigma^d \det(\psi^G) = \varepsilon,$$

and we have $o((\delta\psi)^G) = o(\varepsilon)$, and thus $o((\delta\psi)^G)$ divides $o(\psi)$ for all characters ψ of H. Assertion (b) now follows by substituting $\delta\psi$ for ψ. ■

Suppose now that G is π-separable, where $2 \notin \pi$, and let $H \subseteq G$, where $|G : H|$ is a π-number. Then $|G : H|$ is odd, and we know by Lemma 2.30(b) that the sign character $\delta = \det(((1_H)^G)_H)$ has the induction determinant property. We shall see that in the case where H is a maximal subgroup of G, this character δ is equal to the π-standard sign character $\delta_{(G,H)}$. In general, however, $\delta_{(G,H)}$ is not equal to $\det(((1_H)^G)_H)$, but nevertheless, $\delta_{(G,H)}$ has the induction determinant property, even when H is not maximal.

To define the π-standard sign character $\delta_{(G,H)}$, we begin with some notation. If G acts via automorphisms on a group X, we write σ_X to denote the permutation sign character corresponding to the action of G on the set X. Continuing to assume that G is π-separable and $2 \notin \pi$, we define a character $\nu_{(G)}$ as the product of certain sign characters of G of the form σ_X. To be specific, we consider a normal series for G with factors that are π-groups and π'-groups, and we define $\nu_{(G)}$ to be the product of the permutation sign characters σ_F, as F runs over those factors of the given series that are π-groups. (Thus, although it is not reflected in the notation, the definition of $\nu_{(G)}$ depends on the set π as well as on the group G.)

Our first task is to justify the notation by showing that if $2 \notin \pi$, then $\nu_{(G)}$ is independent of the particular normal series that was used in its construction. To do this, we need the following.

Lemma 2.31. *Let G act on an odd-order group X, and suppose that Y is a G-invariant normal subgroup of X. Then the permutation sign characters σ_X, σ_Y, and $\sigma_{X/Y}$ of G are defined, and we have $\sigma_X = \sigma_Y \sigma_{X/Y}$.*

Proof. Since the kernel of a sign character is a normal subgroup with 2-power index, we see that in order to prove that two sign characters of G are equal, it suffices to show that they agree on a Sylow 2-subgroup of G. It is thus no loss to assume that G is a 2-group.

Writing α and β to denote the permutation characters of the actions of G on Y and on X/Y, respectively, we argue that $\alpha\beta$ is the permutation character associated with the action of G on X. For elements $g \in G$, we have $\alpha(g) = |\mathbf{C}_Y(g)|$ and $\beta(g) = |\mathbf{C}_{X/Y}(g)|$, so we must show that $|\mathbf{C}_X(g)| = |\mathbf{C}_Y(g)||\mathbf{C}_{X/Y}(g)|$.

Let $C/Y = \mathbf{C}_{X/Y}(g)$, and consider a coset S of Y in C. Since g acts trivially on C/Y, the coset S is fixed by g, and thus S is a union of $\langle g \rangle$-orbits, each of which has 2-power size. As $|S| = |Y|$ is odd, at least one of these orbits must have size 1, so S contains an element of $\mathbf{C}_X(g)$. Now $\mathbf{C}_X(g) \subseteq C$, and it follows that $C = Y\mathbf{C}_X(g)$, and we have

$$|Y||\mathbf{C}_{X/Y}(g)| = |C| = |Y\mathbf{C}_X(g)| = \frac{|Y||\mathbf{C}_X(g)|}{|Y \cap \mathbf{C}_X(g)|} = \frac{|Y||\mathbf{C}_X(g)|}{|\mathbf{C}_Y(g)|},$$

where the first equality holds by the definition of the group C. Then

$$|\mathbf{C}_X(g)| = |\mathbf{C}_Y(g)||\mathbf{C}_{X/Y}(g)| = \alpha(g)\beta(g),$$

and so $\alpha\beta$ is the permutation character of G on X, as claimed. Then

$$\sigma_X = \det(\alpha\beta) = \det(\alpha)^{\beta(1)} \det(\beta)^{\alpha(1)} = (\sigma_Y)^{|X/Y|}(\sigma_{X/Y})^{|Y|} = \sigma_Y \sigma_{X/Y},$$

where the last equality holds because $|X/Y|$ and $|Y|$ are odd and σ_Y and $\sigma_{X/Y}$ are sign characters. ∎

Theorem 2.32. *Let G be π-separable, where $2 \notin \pi$. Then the sign character $\nu_{(G)}$ of G is independent of the choice of the particular normal series used in its construction.*

Proof. Let $1 = N_0 \subseteq \cdots \subseteq N_r = G$ be the normal series used to construct $\nu_{(G)}$. If F is a trivial factor in this series, then σ_F is the principal character of G, and so we can delete repeats in the given series and assume that $N_i < N_{i+1}$ for $0 \le i < r$.

Next, we show that if we refine the given series by inserting a new normal subgroup M beween two of its terms, this does not change $\nu_{(G)}$. To see this, suppose we have $N_{i-1} < M < N_i$, where $M \lhd G$. If N_i/N_{i-1} is a π'-group, the insertion of M does not affect any of the π-factors in our series, and since the construction of $\nu_{(G)}$ involves only the π-factors, the insertion of M leaves $\nu_{(G)}$ unchanged.

The remaining possibility is that $F = N_i/N_{i-1}$ is a π-group, and in this case F has odd order since $2 \notin \pi$. Lemma 2.31 tells us that σ_F is the product of the permutation sign characters corresponding to the two new π-factors, N_i/M and M/N_{i-1}, and again, there is no effect on $\nu_{(G)}$.

We can now assume that the series used to define $\nu_{(G)}$ cannot be further refined, and thus it is a chief series for G. By the Jordan–Hölder theorem, all chief series for G have G-isomorphic factors, so every chief series yields the same character $\nu_{(G)}$. ■

We continue to assume that G is π-separable with $2 \notin \pi$. If $H \subseteq G$, we can now define the π-standard sign character by setting

$$\delta_{(G,H)} = \nu_{(H)}(\nu_{(G)})_H \,.$$

Note that $\delta_{(G,H)}$ really is a sign character of H, as promised.

The following establishes a few elementary properties of the π-standard sign character.

Lemma 2.33. *Assume that G is π-separable, where $2 \notin \pi$.*

(a) *If $Y \subseteq X \subseteq G$, then $\delta_{(G,Y)} = (\delta_{(G,X)})_Y \delta_{(X,Y)}$.*

(b) *If $N \lhd G$, then $\delta_{(G,N)} = 1_N$, the principal character.*

(c) *If $N \cap X = Y$, where $N \lhd G$, then $(\delta_{(G,X)})_Y = \delta_{(N,Y)}$.*

Proof. For (a), we compute that

$$(\delta_{(G,X)})_Y \delta_{(X,Y)} = (\nu_{(G)})_Y (\nu_{(X)})_Y (\nu_{(X)})_Y \nu_{(Y)} = (\nu_{(G)})_Y \nu_{(Y)} = \delta_{(G,Y)} \,,$$

where the penultimate equality holds because $(\nu_{(X)})^2$ is trivial.

For (b), we show that if $N \lhd G$, then $(\nu_{(G)})_N = \nu_{(N)}$, and thus

$$\delta_{(G,N)} = (\nu_{(G)})_N \nu_{(N)} = (\nu_{(N)})^2 = 1_N \,,$$

as wanted. To see this, consider a normal series for G with π and π'-factors, where N is one of the terms of the series. By definition, $\nu_{(G)}$ is the product of all of the permutation sign characters σ_F, where F is a π-factor of the series. Also, $\nu_{(N)}$ is the product of the restrictions to N of the sign characters σ_F for π-factors F lying below N. To prove our assertion, therefore, it suffices to observe that the restrictions to N of the permutation sign characters σ_F for factors F above N are trivial. This is clear, however, since if $N \subseteq L \subseteq K$ with $K, L \lhd G$, then N acts trivially on K/L.

Finally, for (c), two applications of (a) yield

$$(\delta_{(G,X)})_Y\delta_{(X,Y)} = \delta_{(G,Y)} = (\delta_{(G,N)})_Y\delta_{(N,Y)}\,.$$

Now (b) tells us that $\delta_{(G,N)}$ is principal, and since $Y \lhd X$, also $\delta_{(X,Y)}$ is principal. Then $(\delta_{(G,X)})_Y = \delta_{(N,Y)}$, as wanted. ∎

We need an elementary fact.

Lemma 2.34. *Let $H \subseteq G$, and suppose that $L \subseteq M$, where L and M are normal subgroups of G such that $MH = G$ and $M \cap H = L$. Then the right multiplication action of H on the set of right cosets of H in G and the conjugation action of H on the elements of M/L are permutation isomorphic.*

Proof. Each right coset of H in G has the form Hm for some element $m \in M$, and it is easy to check that $Hm \cap M = Lm$. It follows that intersection with M defines a bijection from the set of right cosets of H in G onto the set of cosets of L in M. To prove that the actions of H on these two sets agree, we show that if $h \in H$ and S is a right coset of H in G, then $Sh \cap M = (S \cap M)^h$. This is clear, however, because $S^h = h^{-1}Sh = Sh$ since S is closed under left multiplication by elements of H. ∎

Next, we compute $\delta_{(G,H)}$ in a special case.

Lemma 2.35. *Let G be π-separable with $2 \notin \pi$, and suppose that $H \subseteq G$ and that $|G : H|$ is a π-number. If there exist normal subgroups $L \subseteq M$ of G such that $MH = G$ and $M \cap H = L$, then $\delta_{(G,H)} = \det(((1_H)^G)_H)$. In particular, this is the case if H is maximal in G.*

Proof. Since $|M : L| = |G : H|$ is a π-number, we see that L and M are consecutive terms in some normal series $\mathcal{S}$ of G in which each factor is a π-group or a π'-group. By Theorem 2.32, we can assume that $\mathcal{S}$ is the normal series used to define $\nu_{(G)}$, and thus we can write $\nu_{(G)} = \tau\sigma_{M/L}$, where τ is a product of permutation sign characters corresponding to the π-factors in $\mathcal{S}$ above M and the π-factors in $\mathcal{S}$ below L.

The intersections of the terms of $\mathcal{S}$ with H form a normal series of H that we can use to compute $\nu_{(H)}$. The factors of this series are H-isomorphic to all of the factors of $\mathcal{S}$ other than M/L, and it follows that $\nu_{(H)}$ is the restriction τ_H. Then

$$\delta_{(G,H)} = (\nu_{(G)})_H\nu_{(H)} = (\sigma_{M/L})_H\tau_H\nu_{(H)} = (\sigma_{M/L})_H(\nu_{(H)})^2 = (\sigma_{M/L})_H\,.$$

Lemma 2.34 guarantees that $(\sigma_{M/L})_H$ is equal to the sign of the permutation character of the action of H on the right cosets of H in G, and hence $\delta_{(G,H)} = (\sigma_{M/L})_H = \det(((1_H)^G)_H)$, as wanted.

To complete the proof, we must show that if H is maximal in G, then there exist normal subgroups L and M as in the statement of the lemma. Let $L = \mathrm{core}_G(H)$, and note that $L < G$, so we can choose a chief factor M/L of G. Then $M \not\subseteq H$, and thus $HM = G$ by the maximality of H. We have

$$|G : H| = |HM : H| = |M : (M \cap H)|,$$

and this divides $|M : L|$. Since $|G : H|$ is a nontrivial π-number, we see that M/L is not a π'-group, and it follows that the chief factor M/L must be a π-group. Now M/L has odd order because $2 \notin \pi$, and hence M/L is solvable by the Feit–Thompson theorem. Since M/L is minimal normal in G/L, we see that M/L is abelian, and thus $M \cap H$ is normal in M. Since $M \cap H$ is clearly also normal in H and $HM = G$, we see that $M \cap H \lhd G$, and thus $M \cap H = L$, as wanted. ∎

We close our discussion of the π-standard sign character $\delta_{(G,H)}$ by showing that (as promised) it has the induction determinant property for appropriate subgroups H.

Corollary 2.36. *Let G be π-separable, where $2 \notin \pi$, and let $H \subseteq G$. Assume that $|G : H|$ is a π-number, and write $\delta = \delta_{(G,H)}$. Then δ has the induction determinant property for H in G. In particular, for all characters ψ of H, we see that $o(\psi^G)$ divides $o(\delta\psi)$ and $o((\delta\psi)^G)$ divides $o(\psi)$.*

Proof. Since $\delta_{(G,G)}$ is principal, there is nothing to prove if $H = G$, so we can assume that $H < G$, and we proceed by induction on $|G : H|$. If H is maximal in G, then $\delta_{(G,H)} = \det(((1_H)^G)_H)$ by Lemma 2.35, and we know by Lemma 2.30(b) that this character has the induction determinant property for H in G. We can thus assume that there exists a subgroup K such that $H < K < G$.

Let ψ be a character of H, so $\psi^G = (\psi^K)^G$. Since $|G : K| < |G : H|$, we can apply the inductive hypothesis to deduce that $o((\psi^K)^G)$ divides $o(\delta_{(G,K)}(\psi^K)) = o((\delta_{(G,K)})_H\psi)^K$. Also, since $|K : H| < |G : H|$, we can apply the inductive hypothesis again to conclude that $o((\delta_{(G,K)})_H\psi)^K$ divides $o(\delta_{(K,H)}(\delta_{(G,K)})_H\psi)$. The result now follows since $\delta_{(G,H)} = \delta_{(K,H)}(\delta_{(G,K)})_H$ by Lemma 2.33(a). ∎

Finally, we are ready to prove our result on induced characters whose irreducible constituents are π-special.

Proof of Theorem 2.29. We are given a character $\psi \in \mathrm{Irr}(H)$, where $H \subseteq G$, and we assume that every irreducible constituent of ψ^G is π-special. We prove by induction on $|G|$ that $|G : H|$ is a π-number and that $\lambda\psi$ is π-special, where the linear character λ of H is principal if $2 \in \pi$, and

$\lambda = \delta_{(G,H)}$ if $2 \notin \pi$. Note that the assertion that $\psi(1)$ is a π-number is immediate once we establish that $\lambda\psi$ is π-special.

We can assume that $G > 1$, and we let M be a maximal normal subgroup of G. Write $U = H \cap M$, and fix an irreducible constituent θ of ψ_U. We wish to apply the inductive hypothesis in M, so we argue that each irreducible constituent φ of θ^M is π-special. To see this, observe that φ is a constituent of $(\psi_U)^M = (\psi^{HM})_M$, and thus φ lies under some irreducible constituent ξ of ψ^{HM}.

Now ξ lies over ψ, so each irreducible constituent of ξ^G lies over ψ, and thus by hypothesis the irreducible constituents of ξ^G are π-special. Since these constituents lie over φ and $M \triangleleft G$, it follows that φ is π-special, as wanted. By the inductive hypothesis, therefore, $|M : U|$ is a π-number and $\mu\theta$ is π-special, where μ is principal if $2 \in \pi$ and $\mu = \delta_{(M,U)}$ if $2 \notin \pi$.

We argue now that $\lambda_U = \mu$. If $2 \in \pi$, this is clear since both λ and μ are principal; if $2 \notin \pi$, the assertion holds because $(\delta_{(G,H)})_U = \delta_{(M,U)}$ by Lemma 2.33(c). We conclude that $\lambda\psi$ lies over $\mu\theta$.

Suppose first that G/M is a π-group. Then $|G : U| = |G : M||M : U|$ is a π-number, and thus $|G : H|$ is a π-number, as required. Also, H/U is a π-group, and since $\mu\theta$ is π-special, it follows by Theorem 2.4(a) that $\lambda\psi$ is π-special, and we are done in this case.

We can now assume that G/M is a π'-group, and we fix an irreducible constituent χ of ψ^G. Then χ is π-special, and we write $\varphi = \chi_M$, so φ is irreducible. It follows by Theorem 2.4(b) that χ is the unique π-special character of G that lies over φ, and we deduce that χ is the unique common irreducible constituent of ψ^G and φ^G.

Let $\xi = \chi_{HM}$. Then $\xi_M = \varphi$, so ξ is irreducible. Also ξ lies over ψ, so every irreducible constituent of ξ^G lies over both ψ and φ, and hence ξ^G is a multiple of χ. Also, $[\xi^G, \chi] = [\xi, \chi_{HM}] = [\xi, \xi] = 1$, and it follows that $\xi^G = \chi$, and since $\chi(1) = \xi(1)$, we deduce that $HM = G$. Then $|G : H| = |M : U|$, and this completes the proof that $|G : H|$ is a π-number. We work now to show that $\lambda\psi$ is π-special.

We argue next that θ is invariant in H. To see this, let T be the stabilizer of θ in H, and let $\alpha \in \mathrm{Irr}(T|\theta)$ be the Clifford correspondent of ψ with respect to θ. Then $\alpha^G = (\alpha^H)^G = \psi^G$, and hence all irreducible constituents of α^G are π-special. By the part of the theorem that we have already established, $|G : T|$ is a π-number, and thus $|H : T|$ is a π-number. Since $|H : T|$ also divides $|H : U| = |G : M|$, which is a π'-number, we conclude that $|H : T| = 1$, and thus θ is invariant in H, as claimed. We can thus write $\psi_U = e\theta$, for some integer e.

We have seen that χ is the only irreducible character of G that is a constituent of both ψ^G and φ^G, and it follows that $[\psi^G, \varphi^G] = [\psi^G, \chi][\varphi^G, \chi]$. Also, since $[\varphi^G, \chi] = [\varphi, \chi_M] = [\varphi, \varphi] = 1$, we have

$$\begin{aligned}[\psi^G, \chi] = [\psi^G, \varphi^G] = [(\psi^G)_M, \varphi] = [(\psi_U)^M, \varphi] &= [(e\theta)^M, \varphi] \\ &= e[\theta^M, \chi_M] \\ &= e[\theta^G, \chi]\,.\end{aligned}$$

Also, ψ is a constituent of θ^H, so we can write $\theta^H = \psi + \Delta$, where Δ is either a character of H or is zero. Then $\theta^G = \psi^G + \Delta^G$, and thus

$$[\theta^G, \chi] \geq [\psi^G, \chi]\,.$$

Combining this with our previous equality, we get

$$[\psi^G, \chi] = e[\theta^G, \chi] \geq e[\psi^G, \chi]\,,$$

and since $[\psi^G, \chi] \neq 0$, we deduce that $e = 1$, and hence $\psi_U = \theta$. In particular, $\psi(1) = \theta(1)$, and this is a π-number since $\mu\theta$ is π-special.

Since $\lambda_U = \mu$, we see that $\lambda\psi$ extends $\mu\theta$. Also, $\mu\theta$ has a canonical extension $\eta \in \operatorname{Irr}(H)$, and we know by Theorem 2.4(b) that η is π-special, and thus it suffices to show that $\lambda\psi = \eta$. By the Gallagher correspondence, there exists a (necessarily linear) character $\beta \in \operatorname{Irr}(H/U)$ such that $\beta\lambda\psi = \eta$, and we complete the proof by showing that β is the principal character.

We argue that both $o(\psi^G)$ and $o((\lambda\eta)^G)$ are π-numbers. First, by hypothesis, ψ^G is a sum of π-special characters, and thus $\det(\psi^G)$ is the product of the determinants of these characters. Since the order of each factor is a π-number, it follows that the order of the product is a π-number, as wanted.

To compute $o((\lambda\eta)^G)$, we consider separately the cases where $2 \in \pi$ and $2 \notin \pi$. First, if $2 \in \pi$, then λ is principal, so $o((\lambda\eta)^G) = o(\eta^G)$, and this divides the π-number $2o(\eta)$ by Lemma 2.30(a). If $2 \notin \pi$, then $\lambda = \delta_{(G,H)}$ has the induction determinant property by Corollary 2.36, and thus $o((\lambda\eta)^G)$ divides $o(\eta)$, which is a π-number. In both cases, therefore, $o((\lambda\eta)^G)$ is a π-number, as claimed.

Now β has an extension $\gamma \in \operatorname{Irr}(G/M)$, and since $\beta\lambda\psi = \eta$, we have $\lambda\eta = \beta\psi$, and thus $(\lambda\eta)^G = (\beta\psi)^G = \gamma\psi^G$. Then

$$\det((\lambda\eta)^G) = \det(\gamma\psi^G) = \gamma^d \det(\psi^G)\,,$$

where $d = \psi^G(1)$. Since both $o((\lambda\eta)^G)$ and $o(\psi^G)$ are π-numbers, it follows that $o(\gamma^d)$ is also a π-number. Furthermore, $d = \psi^G(1) = |G : H|\psi(1)$ is a π-number, and we conclude that $o(\gamma)$ is a π-number. As γ is a linear character of the π'-group G/M, we deduce that γ is principal, and thus $\beta = \gamma_H$ is principal, as required. ∎

We return now to the situation where $H \subseteq G$ and $\psi^G = \chi$, where $\psi \in \mathrm{Irr}(H)$ and $\chi \in \mathrm{Irr}(G)$. By Theorem 2.29, we know that if χ is π-special, then ψ is "almost" π-special, or more precisely, $\lambda\psi$ is π-special, where λ is principal if $2 \in \pi$ and $\lambda = \delta_{(G,H)}$ if $2 \notin \pi$. This suggests that perhaps the converse is true. If we assume that $\lambda\psi$ is π-special, we ask if it is true that χ must be π-special. In this generality, the answer is clearly "no" because there is no reason to think that $\chi(1)$ is a π-number.

Suppose that, in addition to assuming that $\lambda\psi$ is π-special, we also assume that $\chi(1)$ is a π-number or, equivalently, that $|G : H|$ is a π-number. In general, this is still not sufficient to guarantee that χ is π-special, but by Corollary 2.28, this condition is sufficient if we assume that $|G|$ is odd. Actually, a weaker oddness condition suffices: it is enough to assume that $2 \notin \pi$. Proving that this is sufficient, however, seems to require additional machinery, and we defer the argument to Chapter 7 (see Theorem 7.24).

If we replace ordinary induction by what might be called "Clifford induction", then by a result of Gajendragadkar, no oddness assumption is needed. (We refer here to the induction in the Clifford correspondence.)

Theorem 2.37 (Gajendragadkar). *Let $\theta \in \mathrm{Irr}(N)$, where $N \lhd G$ and G is π-separable, and let T be the stabilizer of θ in G. If $\psi \in \mathrm{Irr}(T|\theta)$ and $\chi \in \mathrm{Irr}(G|\theta)$ with $\psi^G = \chi$, then χ is π-special if and only if $|G : T|$ is a π-number and $\lambda\psi$ is π-special, where the linear character λ of T is principal if $2 \in \pi$ and $\lambda = \delta_{(G,T)}$ if $2 \notin \pi$.*

Proof. If χ is π-special, then $|G : T|$ is a π-number and $\lambda\psi$ is π-special by Theorem 2.29. It thus suffices to prove the converse, so we assume that $|G : T|$ is a π-number and $\lambda\psi$ is π-special, and we show by induction on $|G|$ that χ is π-special.

Clearly, we can assume that $N < G$, so we can choose a maximal normal subgroup M of G with $N \subseteq M$, and we write $S = M \cap T$. Then $S \lhd T$, and S is the stabilizer of θ in M. Also, $|M : S| = |TM : T|$, and since this divides the π-number $|G : T|$, we see that $|M : S|$ is a π-number.

Let α be an irreducible constituent of ψ_S and note that α lies over θ since ψ_N is a multiple of θ. Then α^M is irreducible by the Clifford correspondence, and we write $\varphi = \alpha^M$. Now χ lies over α, so χ is a constituent of $\alpha^G = (\alpha^M)^G = \varphi^G$, and thus φ lies under χ.

We wish to apply the inductive hypothesis in the group M to conclude that φ is π-special. Since we know that $|M : S|$ is a π-number, it suffices to show that $\mu\alpha$ is π-special, where μ is principal if $2 \in \pi$ and $\mu = \delta_{(M,S)}$ if $2 \notin \pi$.

We argue that $\lambda_S = \mu$. If $2 \in \pi$, this is clear since both λ and μ are principal; if $2 \notin \pi$, the assertion holds because $(\delta_{(G,T)})_S = \delta_{(M,S)}$ by

Lemma 2.33(c). Then $\lambda\psi$ lies over $\mu\alpha$, and since $S \lhd T$ and $\lambda\psi$ is π-special by assumption, we conclude that $\mu\alpha$ is π-special, as wanted, and thus φ is π-special by the inductive hypothesis. If G/M is a π-group, then since χ lies over φ, it follows by Theorem 2.4(a) that χ is π-special, as required.

We can now assume that G/M is a π'-group. Since $\lambda\psi$ is π-special, $\psi(1)$ is a π-number, and thus $\chi(1) = |G : T|\psi(1)$ is a π-number, and it follows that χ_M is irreducible. Then χ is an extension of φ, and it suffices by Theorem 2.4(b) to show that χ is the canonical extension or, equivalently, that $o(\chi)$ is a π-number.

Now $\chi = \psi^G$, so $o(\chi)$ divides $2o(\psi)$ by Lemma 2.30(a). If $2 \in \pi$, this is a π-number because in that case, ψ is π-special. If $2 \notin \pi$, Corollary 2.36 guarantees that $o(\chi)$ divides $o(\delta_{(G,T)}\psi)$, and this is a π-number since $\delta_{(G,T)}\psi = \lambda\psi$ is π-special. ∎

Corollary 2.38. *Let $\theta \in \mathrm{Irr}(N)$, where $N \lhd G$ and G is π-separable. Let T be the stabilizer of θ in G, and let $\lambda \in \mathrm{Irr}(T)$, where λ is principal if $2 \in \pi$ and $\lambda = \delta_{(G,T)}$ if $2 \notin \pi$. If $|G : T|$ is a π-number, then the map $\psi \mapsto (\lambda\psi)^G$ is a bijection from the set of π-special characters in $\mathrm{Irr}(T|\theta)$ onto the set of π-special characters in $\mathrm{Irr}(G|\theta)$.*

Proof. Observe first that $N \subseteq \ker(\lambda)$. This is clear if $2 \in \pi$ since λ is principal in that case. If $2 \notin \pi$, then $\lambda = \delta_{(G,T)}$, and we have $\lambda_N\delta_{(T,N)} = \delta_{(G,N)}$ by Lemma 2.33(a). Since $\delta_{(T,N)}$ and $\delta_{(G,N)}$ are principal by Lemma 2.33(b), it follows that λ_N is principal, as claimed. We conclude that the map $\psi \mapsto \lambda\psi$ is a bijection from $\mathrm{Irr}(T|\theta)$ to itself, and thus by the Clifford correspondence, the map $\psi \mapsto (\lambda\psi)^G$ is a bijection from $\mathrm{Irr}(T|\theta)$ onto $\mathrm{Irr}(G|\theta)$.

It suffices now to observe that if $\psi \in \mathrm{Irr}(T|\theta)$, then ψ is π-special if and only if $(\lambda\psi)^G$ is π-special. Since $|G : T|$ is a π-number, we see by Theorem 2.37 that $(\lambda\psi)^G$ is π-special if and only if $\lambda(\lambda\psi)$ is π-special. The result follows because λ^2 is principal. ∎

Using the notation of Corollary 2.38, we know that the numbers of π-special characters in $\mathrm{Irr}(T|\theta)$ and $\mathrm{Irr}(G|\theta)$ are equal, and it is natural to ask when it is true that these sets actually contain a π-special character. Clearly, a necessary condition is that θ is π-special, and although it may not be obvious, this condition is also sufficient. The key to proving this is the following general result, which appears (in slightly different form) as Theorem 13.31 of [**CTFG**].

Theorem 2.39. *Let K act via automorphisms on a group G, and suppose that $N \lhd G$ is K-invariant. Assume that $|K|$ and $|G : N|$ are relatively prime, and let $\theta \in \mathrm{Irr}(N)$ be K-invariant. Then there exists a K-invariant character in $\mathrm{Irr}(G|\theta)$.*

We will not present the proof of Theorem 2.39 here, but we mention some of its nontrivial ingredients. First, if G/N is solvable, then working by induction on $|G : N|$, one can reduce to the case where G/N is abelian, and in that situation, the result follows via Glauberman's lemma (Lemma 1.4). Otherwise, it follows by the odd-order theorem that $|G : N|$ is even, and thus $|K|$ is odd, and hence K is solvable. Using character-triple isomorphisms, it is possible to reduce the problem to the case where $|K|$ and $|G|$ are relatively prime, and then the proof can be completed using the Glauberman correspondence, presented in Chapter 13 of [**CTFG**].

Assuming Theorem 2.39, we can prove the following.

Theorem 2.40. *Let $\theta \in \mathrm{Irr}(N)$, where $N \lhd G$ and G is π-separable. Assume that θ is π-special and that it lies in a G-orbit whose size is a π-number. Then there exists a π-special character in* $\mathrm{Irr}(G|\theta)$.

Proof. We proceed by induction on $|G : N|$. Since there is nothing to prove if $N = G$, we can assume that $N < G$, and we choose a chief factor M/N of G. The index in G of the stabilizer of θ is a π-number, so the stabilizer contains some Hall π'-subgroup K of G, and we will show that $\mathrm{Irr}(M|\theta)$ contains a K-invariant π-special character φ. Assuming this for the moment, we see that since K stabilizes φ, the size of the G-orbit of φ is a π-number, and hence by the inductive hypothesis, $\mathrm{Irr}(G|\varphi)$ contains a π-special character, and since this character lies over θ, we are done.

If M/N is a π'-group, then θ is invariant in M, so $\mathrm{Irr}(M|\theta)$ contains a unique π-special character φ by Theorem 2.4(b), and since θ is K-invariant and uniquely determines φ, it follows that φ is K-invariant, as wanted. In the remaining case, M/N is a π-group, and thus every member of $\mathrm{Irr}(M|\theta)$ is π-special. At least one of these characters is K-invariant by Theorem 2.39, and this completes the proof. ∎

Problems

2.1. Let $A \lhd G$ and $B \lhd G$, where G is π-separable and $G = AB$. Write $N = A \cap B$, and assume that A/N is a π-group and B/N is a π'-group. Suppose that $\alpha \in \mathrm{Irr}(N)$ is invariant in B and is π-special and that $\beta \in \mathrm{Irr}(N)$ is invariant in A and is π'-special. Show that every member of $\mathrm{Irr}(G|\alpha\beta)$ is π-factored.

Hint: Show that every member of $\mathrm{Irr}(A|\alpha)$ is the restriction of some π-special member of $\mathrm{Irr}(G)$.

2.2. Suppose that G is π-separable with $2 \notin \pi'$, and assume that $\chi \in \mathrm{Irr}(G)$ is π-factored. If χ is fixed by the magic field automorphism for π, show that χ is π-special.

2.3. Suppose $|G|$ is odd and that $\chi \in \text{Irr}(G)$. Let π be a set of primes, and let σ be the magic field automorphism for π. Show that the following are equivalent.

(1) $\chi^\sigma = \chi$.

(2) If $N \lhd G$, then every irreducible constituent of χ_N is fixed by σ.

(3) $\chi = \alpha^G$, where α is a π-special character of some subgroup $H \subseteq G$.

Hint: To prove that (2) implies (3), show that if χ is not quasi-primitive, then χ is induced from some σ-fixed character of a proper subgroup of G.

2.4. Suppose that G is p-solvable for some prime p. Show that G has a nonprincipal p-special character if and only if there exist normal subgroups $K \subseteq Z$ of G such that Z/K is a nontrivial central p-subgroup of G/K.

2.5. Let G be π-separable. Suppose that $G = NH$, where $N \lhd G$, and $|G : H|$ is a π'-number, and write $D = N \cap H$. Let $\theta \in \text{Irr}(N)$ be π-special, and let $\varphi = \theta_D$.

(a) Show that φ is irreducible and that restriction defines a bijection from $\text{Irr}(G|\theta)$ onto $\text{Irr}(H|\varphi)$.

(b) Let $\psi \in \text{Irr}(H|\varphi)$ be π-special, and let χ be the unique member of $\text{Irr}(G|\theta)$ such that $\chi_H = \psi$. Show that χ is π-special.

Hint: For (b), proceed by induction on $|G : N|$, and consider a maximal normal subgroup M of G with $N \subseteq M$.

2.6 (Navarro). Let $H \subseteq G$, where $|G|$ is odd, and suppose that $\psi \in \text{Irr}(H)$, where $\psi^G = \chi$ is irreducible and ψ is π-factored. Let $N \lhd G$, and assume that the irreducible constituents of χ_N are π-factored. Show that ψ^{NH} is π-factored.

Hints: Show that one can assume that $|NH : H|$ is a π'-number. Write $\psi = \alpha\beta$, where α is π-special and β is π'-special, and use Problem 2.5 to deduce that α has a π-special extension to NH.

2.7. Let $\chi \in \text{Irr}(G)$, where G is π-separable, and let H be a Hall π-subgroup of G. Show that $\chi = \alpha^G$ for some character $\alpha \in \text{Irr}(H)$ if and only if $|G|/\chi(1)$ is a π-number.

Hints: For "if", write $\chi = \beta^G$, where β is a primitive character of some subgroup J of G, and observe that $|J|/\beta(1)$ is a π-number. Deduce that J is a π-group.

2.8. Let $W \subseteq G$, where G is π-separable and $|G : W|$ is a π'-number. Let $\psi \in \text{Irr}(W)$ be π-special and quasi-primitive, and assume that ψ does not extend to any subgroup of G that properly contains W. Show that ψ^G is irreducible.

Hints: Proceed by induction on $|G|$. Let $N = \text{core}_G(W)$, let $\alpha \in \text{Irr}(N)$ lie under ψ, and let T be the stabilizer of α in G. If $T < G$, the inductive hypothesis applies, so assume that α is invariant in G. Let M/N be a chief factor of G, and observe that α has a canonical extension $\theta \in \text{Irr}(M)$. Let $D = M \cap W$, and write $\varphi = \theta_D$. Now apply Lemma 2.11(a) in the group MW.

Notes and References

The set $\mathcal{X}(G)$ of p'-special characters of a p-solvable group G appears in Definition 2.2 of [**36**]. After seeing Isaacs's paper, Gajendragadkar defined the more general set $\mathcal{X}_\pi(G)$ for a π-separable group G. His definition appeared in his PhD thesis in 1978, and it was published in his paper [**23**], which contains many of the key properties of these characters, including the results explicitly credited to him in this chapter.

Theorem 2.9, which is the fact that quasi-primitive characters are factored, appears in [**37**].

The Feit "conjecture" was posed as a question in [**17**]. It was proved for solvable groups in [**3**], and approximately simultaneously, a proof appeared in [**18**].

Theorem 2.23 is a result of Isaacs and Navarro [**52**].

B. Huppert's Theorem 2.18 appears as Satz 6 of [**32**].

Much of the material in this chapter on the "π-standard sign character" appears in [**23**], and in particular, Theorem 2.37 of this chapter is Theorem 5.10 of that paper. An expository presentation closer to the version here can be found in [**42**]. Theorem 2.29 appears as Theorem C of that paper.

Chapter 3

Partial Characters

3A

To motivate our discusion of π-partial characters of π-separable groups, we begin with a very brief description of R. Brauer's "modular character theory". Brauer's theory is concerned with group representations in prime characteristic, and although this is a vast subject, we discuss here only the tiny fraction of it that is directly relevant to the main topic of this chapter. Much more detail about Brauer's work can be found in Chapter 15 of [**CTFG**], and in many other texts, but none of that material will actually be used here.

Let G be an arbitrary finite group, and let K be an algebraically closed field. As we know, one can study the K-representations of G by looking at the corresponding trace functions: the K-characters. This approach works extremely well in the case where $K = \mathbb{C}$, the complex numbers, but if K has prime characteristic, the character associated with a K-representation gives much less information than we might want. (In prime characteristic, for example, the K-character does not even determine the degree of the representation; the value at the identity only tells us the degree modulo the characteristic.)

Assume that K has prime characteristic p. (The most interesting case is where p divides $|G|$, but we do not assume that.) With each K-representation of G, Brauer associated an object that has proved to be much more useful than the K-character. This object, now called a "Brauer character", differs from a K-character in two significant respects: it is a *complex-valued* function; and it is defined only on the subset G^0 of G consisting of the elements with order not divisible by p, the p'-elements of G. (These are also referred to as the p-**regular** elements of G, while the elements of G with order divisible by p are said to be p-**singular**.) We will not actually define Brauer characters here, but we mention that although the set of Brauer characters of G depends on the characteristic p, it does not depend on the specific algebraically closed field K. As we will explain, a π-partial character of a π-separable group can be viewed as a generalization of a Brauer character, where we replace the set p' of primes different from p by an arbitrary set π of primes.

Brauer characters enjoy many of the properties of ordinary characters. For example, they are class functions: they lie in the complex vector space $\mathrm{cf}(G^0)$ of functions that are constant on the conjugacy classes of G contained in G^0. (Recall that Brauer characters are undefined on elements not in G^0.) Also, if $H \subseteq G$ and φ is a Brauer character of G, then the restriction φ_H (which is defined on $H^0 = H \cap G^0$) is a Brauer character of H.

The set of Brauer characters of G is closed under addition, and in fact, the sum of two Brauer characters is the Brauer character associated with the direct sum of the two corresponding representations. We say that a Brauer character is **irreducible** if it cannot be written as a sum of two (nonzero) Brauer characters, and we write $\mathrm{IBr}(G)$ to denote the set of irreducible Brauer characters of the group G. (Usually, the prime p is understood from the context, and it is not mentioned explicitly.) As one might guess, the irreducible Brauer characters of G are exactly the Brauer characters associated to irreducible K-representations, and nonisomorphic irreducible K-representations yield distinct irreducible Brauer characters. A word of caution is appropriate here, however, because in general, nonisomorphic reducible K-representations can yield identical Brauer characters.

If φ is a Brauer character, then $\varphi(1)$ is the degree of each of the K-representations with which it is associated, and in particular $\varphi(1)$ is a positive integer. It follows that each Brauer character of G is either irreducible, or it can be decomposed as a finite sum of irreducible Brauer characters. In fact, the set $\mathrm{IBr}(G)$ is linearly independent, and it follows that the decomposition of a Brauer character as a sum of one or more irreducible Brauer characters is unique.

One of the most significant facts about Brauer characters is that they provide a link between the "modular" (characteristic p) representation theory and the ordinary, classical, complex character theory of G. (In fact, the theory of Brauer characters can be applied to prove theorems about ordinary characters, and this was Brauer's original motivation.) To see how the connection between Brauer characters and ordinary characters works, suppose that χ is a character (i.e., an ordinary, complex character) of G, and write χ^0 to denote the restriction of χ to the set G^0 of p'-elements of G.

Then χ^0 is necessarily a Brauer character, although it may not be irreducible even if χ is irreducible. If $\varphi = \chi^0$, we say that the ordinary character χ is a **lift** of the Brauer character φ. It is easy to see that if χ is a lift of an irreducible Brauer character φ, then χ must be irreducible, but usually, χ is not uniquely determined by φ.

In general, a Brauer character need not have a lift. If G is p-solvable, however, then by a theorem of P. Fong and R. Swan, every Brauer character of G actually does lift to an ordinary character. In particular, if p does not divide $|G|$, then G is automatically p-solvable, and thus every Brauer character of G lifts. Since in this case, $G^0 = G$, it follows that every Brauer character of G is an ordinary character, and we conclude that if $|G|$ is not divisible by p, then $\mathrm{IBr}(G) = \mathrm{Irr}(G)$. (We mention that it is not hard to prove this directly, without appealing to the Fong–Swan theorem. See, for example, Theorem 15.13 of [**CTFG**].)

Let $\chi \in \mathrm{Irr}(G)$, where G is an arbitrary finite group. Since χ^0 is a Brauer character, and every Brauer character is uniquely a sum of one or more irreducible Brauer characters, we can write

$$\chi^0 = \sum_{\varphi \in \mathrm{IBr}(G)} d_{\chi,\varphi}\varphi,$$

where the coefficients $d_{\chi,\varphi}$ are uniquely determined nonnegative integers called the **decomposition numbers**. For each choice of χ, we see that $\chi^0(1) \neq 0$, and thus at least one of the decomposition numbers $d_{\chi,\varphi}$ must be nonzero. It is much less clear, however, that if $\varphi \in \mathrm{IBr}(G)$ is arbitrary, then at least one decomposition number $d_{\chi,\varphi}$ is nonzero. (If G is p-solvable, then this is an immediate consequence of the Fong–Swan theorem, but it is true in general, even if G is not p-solvable.)

Brauer used the decomposition numbers to associate an ordinary character

$$\Phi_\varphi = \sum_{\chi \in \mathrm{Irr}(G)} d_{\chi,\varphi}\chi$$

with each irreducible Brauer character $\varphi \in \mathrm{IBr}(G)$. These characters Φ_φ are often referred to as the **projective indecomposable** characters of G (with respect to the prime p) and they have a number of interesting properties. For example, the value of a projective indecomposable character on an element of G with order divisible by p is always zero. The characters Φ_φ are distinct for distinct $\varphi \in \mathrm{IBr}(G)$, and they form a basis for the subspace of the space of class functions of G that vanish on the p-singular elements of G.

Given a prime p, the decomposition numbers for a group G can also be used to define what are called the "p-blocks" of G. This provides one of several possible entries into the rich subject known as "block theory", and although we will resist the temptation to pursue this here, it seems appropriate at least to present the relevant definition. One constructs a (bipartite, undirected) graph whose vertex set consists of all members of $\mathrm{Irr}(G)$ together with all members of $\mathrm{IBr}(G)$. Each edge of this graph joins an ordinary character χ and a Brauer character φ, and an edge is drawn precisely when the decomposition number $d_{\chi,\varphi}$ is nonzero. The p-blocks of G are the connected components of this graph, and in particular, each p-block contains some characters $\chi \in \mathrm{Irr}(G)$ and some Brauer characters $\varphi \in \mathrm{IBr}(G)$.

Although we have proved nothing thus far, we have stated enough to deduce the following result of Brauer, which he proved using ring-theoretic techniques.

Theorem 3.1. *Let G be finite, and fix a prime p. Then $\mathrm{IBr}(G)$ is a basis for the vector space $\mathrm{cf}(G^0)$, where G^0 is the set of p'-elements of G. In particular, $|\,\mathrm{IBr}(G)|$ is equal to the number of classes of p'-elements in G.*

Proof. First, observe that the dimension of $\mathrm{cf}(G^0)$ is equal to the number of classes of p'-elements of G, so the second assertion will follow once we establish that $\mathrm{IBr}(G)$ is a basis for $\mathrm{cf}(G^0)$. We know that $\mathrm{IBr}(G) \subseteq \mathrm{cf}(G^0)$, and we have said that the set $\mathrm{IBr}(G)$ is linearly independent, so it suffices to observe that $\mathrm{IBr}(G)$ spans $\mathrm{cf}(G^0)$.

For each class K of p'-elements of G, let α_K be the function defined on G such that $\alpha_K(x) = 1$ for $x \in K$ and $\alpha_K(x) = 0$, otherwise. Then α_K is a class function of G, so it lies in the linear span of $\mathrm{Irr}(G)$, and thus $(\alpha_K)^0$ is a linear combination of the functions χ^0 for $\chi \in \mathrm{Irr}(G)$. Since χ^0 is a sum of members of $\mathrm{IBr}(G)$, it follows that the linear span of $\mathrm{IBr}(G)$ contains all of the functions $(\alpha_K)^0$. These functions, however, form a basis for $\mathrm{cf}(G^0)$, and thus $\mathrm{cf}(G^0)$ is the linear span of $\mathrm{IBr}(G)$. ∎

An unpleasant fact that we have so far managed to avoid mentioning is that, in general, the Brauer characters of a group G are not quite uniquely determined; there is a choice that must be made in their construction. We will ignore this complication, however, because not only is it not as problematic as one might expect, but in fact, for p-solvable groups, the ambiguity disappears: for p-solvable groups, the set of Brauer characters is uniquely determined. To see this, it suffices to observe that by the Fong–Swan theorem, the Brauer characters for the prime p of a p-solvable group G are exactly the functions χ^0, where χ runs over the set of ordinary characters of G.

Finally, we offer as an application a result about the ordinary characters of p-solvable groups.

Theorem 3.2. *Suppose that G is p-solvable, and consider the set $P(G)$ of restrictions of characters χ of G to the set G^0 of p'-elements of G. Let $I(G)$ be the subset of $P(G)$ consisting of those members that cannot be written as a sum of two (nonzero) members of $P(G)$. Then $|I(G)|$ is the number of classes of p'-elements in G.*

Proof. By the Fong–Swan theorem, $P(G)$ is exactly the set of Brauer characters of G, and thus $I(G) = \mathrm{IBr}(G)$. The result now follows by Theorem 3.1. ∎

Observe that Theorem 3.2 does not refer to Brauer characters or to characteristic p representations. The statement of this result would make sense, therefore, if the set p' of primes different from p were replaced by an arbitrary set π of primes. (Of course, we would also want to replace the p-solvability hypothesis by the assumption that G is π-separable.) In fact, not only does this π-generalization of Theorem 3.2 make sense; it is actually true. Entirely different techniques will be needed for its proof, however.

3B

Let G be π-separable for some set π of primes. Generalizing the notation of the previous section, we write G^0 to denote the set of elements of G whose orders are π-numbers. (These are referred to as the π-elements of G, and we say that the classes of G in which they lie are the π-classes.) If χ is a character (or any complex-valued class function) of G, we write χ^0 to denote the restriction of χ to G^0, and we observe that $\chi^0 \in \mathrm{cf}(G^0)$, the complex vector space of class functions is defined on G^0.

A complex-valued function φ on G^0 is said to be a π-**partial character** of G if $\varphi = \chi^0$ for some character χ of G. Borrowing the notation from the previous section, therefore, we could say that the π-partial characters of G are the members of $\mathrm{cf}(G^0)$ that lift to ordinary characters of G.

We now present a few of the most basic properties of π-partial characters. First, since the set of characters of G is closed under addition, it should be clear that the set of π-partial characters of G is also closed under addition. A π-partial character φ of G is **irreducible** if it cannot be written as a sum of two π-partial characters, and we write $\mathrm{I}_\pi(G)$ to denote the set of irreducible π-partial characters of the π-separable group G. Since the value of a π-partial character at the identity is a positive integer, we see that every π-partial character can be written as a sum of one or more members of $\mathrm{I}_\pi(G)$.

Note that if $\varphi \in \mathrm{I}_\pi(G)$ and χ is a character such that $\chi^0 = \varphi$, then χ must be irreducible. Otherwise, we could write χ as a sum of two characters, and then $\varphi = \chi^0$ would be a sum of two π-partial characters, and this would contradict the assumed irreducibility of φ.

It is interesting to note that the set $\mathrm{I}_\pi(G)$ of irreducible π-partial characters of a π-separable group G can be constructed from the character table of G, with no additional input. (The analogous statement about the irreducible Brauer characters of an arbitrary finite group is false.) First, by a result of G. Higman (Theorem 8.21 of [**CTFG**]) it is possible to identify the columns of the character table of G that correspond to π-classes. It is then a finite problem to identify the irreducible characters χ such that χ^0 cannot be written as a sum of π-partial characters of the form ψ^0 for irreducible characters ψ different from χ. It should be clear that the functions χ^0 found in this way are exactly the members of $\mathrm{I}_\pi(G)$.

Next, we discuss restriction and induction of π-partial characters. If $H \subseteq G$ and φ is a π-partial character of G, then the restriction φ_H is defined on $H^0 = H \cap G^0$, and it is easy to see that φ_H is a π-partial character of H. (This is so because we can write $\varphi = \chi^0$ for some character χ, and we have $\varphi_H = (\chi^0)_H = (\chi_H)^0$.) Note that if it happens that G^0 is contained in H, then $H^0 = G^0$, and thus φ and φ_H are actually the same function. We do not identify them, however, because one is viewed as a π-partial character of G and the other as a π-partial character of H. In fact, in this situation, it can happen that φ is irreducible while φ_H is reducible. Another interesting case is where H is a π-subgroup of G. In this situation, $H^0 = H$, and thus φ_H is actually an ordinary character of H.

We can also induce π-partial characters. If $H \subseteq G$ and θ is a π-partial character of H, we can define θ^G using the usual formula for induced characters, but applied only to π-elements of G. We see that θ^G is a π-partial character of G because for characters ψ of H we have $(\psi^0)^H = (\psi^H)^0$.

Finally, we observe that if $\pi = p'$, the set of primes different from p, then by the Fong–Swan theorem, the π-partial characters of G are exactly the p-Brauer characters, and thus $\mathrm{I}_\pi(G) = \mathrm{IBr}_p(G)$. (The Fong–Swan theorem applies because we are assuming that G is π-separable, and thus in the case where $\pi = p'$, the group is p-solvable.)

We can now state the π-analogue of Theorem 3.2.

Theorem 3.3. *Let G be π-separable. Then $\mathrm{I}_\pi(G)$ is a basis for the space $\mathrm{cf}(G^0)$ of class functions on the set G^0 of π-elements of G. In particular, $|\,\mathrm{I}_\pi(G)|$ is equal to the number of π-classes of G.*

Of course, $\mathrm{I}_\pi(G) \subseteq \mathrm{cf}(G^0)$, and by an argument identical to that used in the second paragraph of the proof of Theorem 3.1, it is easy to see that the linear span of $\mathrm{I}_\pi(G)$ is the full vector space $\mathrm{cf}(G^0)$. The proof that set $\mathrm{I}_\pi(G)$ is linearly independent, however, is much less trivial, and in particular it depends on the assumption that G is π-separable. (This fails, for example, if G is the alternating group A_5 and $\pi = \{3, 5\}$.

We will deduce the linear independence of $\mathrm{I}_\pi(G)$ from Theorem 3.4, but first, we introduce a bit of additional notation. The π-part and π'-part of a positive integer n are, respectively, the unique positive π-number r and the unique positive π'-number s such that $n = rs$. We write $r = n_\pi$ and $s = n_{\pi'}$.

Theorem 3.4. *Let $\varphi \in \mathrm{I}_\pi(G)$, where G is π-separable, and let H be a Hall π-subgroup of G. If α is an irreducible constituent of smallest possible degree of the character φ_H, then the following hold.*

(a) $\alpha(1) = \varphi(1)_\pi$, *the π-part of $\varphi(1)$.*

(b) $[\varphi_H, \alpha] = 1$.

(c) $[\mu_H, \alpha] = 0$ *if $\mu \in \mathrm{I}_\pi(G)$ with $\mu \neq \varphi$.*

In the situation of Theorem 3.4, the irreducible constituents of minimum degree of φ_H are referred to as the **Fong characters** associated with φ. (This eponym is explained by the fact that it was P. Fong who first studied the Brauer-character theory of p-solvable groups, and it was he who discovered some of the remarkable properties of the minimum-degree constituents of the restriction of an irreducible Brauer character to a Hall p'-subgroup of one of these groups.) We stress that in general there may be a number of different Fong characters associated with a given irreducible π-partial character φ, but of course they all have the same degree.

Assuming Theorem 3.4, we can easily complete the proof of Theorem 3.3.

Proof of Theorem 3.3. We have already seen that $\mathrm{I}_\pi(G)$ spans $\mathrm{cf}(G^0)$, so what remains to prove is that the set $\mathrm{I}_\pi(G)$ is linearly independent. To accomplish this, it suffices to show that the restrictions φ_H are linearly independent for $\varphi \in \mathrm{I}_\pi(G)$, where H is a Hall π-subgroup of G. This is clear, however, because by Theorem 3.4(c), each of the characters φ_H has an irreducible constituent that does not appear as a constituent of any of the others. ∎

Of course, every π-partial character θ of G is a sum of irreducible π-partial characters, and by Theorem 3.3, we see that these summands are uniquely determined. By analogy with ordinary characters, we refer to the members of $\mathrm{I}_\pi(G)$ that occur as summands of θ as the **irreducible constituents** of θ.

For the remainder of this section we continue to discuss consequences of Theorem 3.4, but we defer the proof of this key result to section 3C.

Corollary 3.5. *Let H be a Hall π-subgroup of a π-separable group G, and suppose that $\theta \in \mathrm{cf}(G^0)$. Then θ is a π-partial character of G if and only if θ_H is a character of H.*

Proof. If θ is a π-partial character, then $\theta = \psi^0$ for some character ψ of G, and thus $\theta_H = (\psi^0)_H = \psi_H$ is a character of H. Conversely, suppose that θ_H is a character of H. Since $\theta \in \mathrm{cf}(G^0)$ by assumption and $\mathrm{I}_\pi(G)$ is a basis for this vector space, we can write

$$\theta = \sum_{\mu \in \mathrm{I}_\pi(G)} c_\mu \mu \,,$$

where the coefficients c_μ are complex numbers.

Now let $\varphi \in \mathrm{I}_\pi(G)$, and let α be a Fong character for φ. Then

$$[\theta_H, \alpha] = \sum_{\mu \in \mathrm{I}_\pi(G)} c_\mu [\mu_H, \alpha] = c_\varphi \,,$$

where the final equality holds because by Theorem 3.4 we have $[\varphi_H, \alpha] = 1$ and $[\mu_H, \alpha] = 0$ for $\mu \neq \varphi$. Since θ_H is a character by assumption, $[\theta_H, \alpha]$ is a nonnegative integer, and it follows that c_φ is a nonnegative integer. Then θ is a nonnegative-integer linear combination of the members of $\mathrm{I}_\pi(G)$, and since θ is nonzero, it is a π-partial character of G. ∎

Of course, we have not yet proved Theorem 3.4, so at this point, Corollary 3.5 (and the other results in this section) must be viewed as provisional. What we have definitely established, however, is that for any group G in which the conclusion of Theorem 3.4 is valid, the conclusion of Corollary 3.5 will also hold. We will use this fact in our inductive proof of Theorem 3.4.

We observe next that the Fong–Swan theorem is an immediate consequence of Corollary 3.5. (Recall that this theorem asserts that if G is p-solvable, then every Brauer character of G for the prime p lifts to an ordinary character of G.) Of course, we cannot actually present a proof of the Fong–Swan theorem here because we have not even defined Brauer characters. Nevertheless, the result follows via Corollary 3.5 if we assume two basic facts from Brauer's theory:

(1) The Brauer characters of G for the prime p lie in the set $\mathrm{cf}(G^0)$, where G^0 is the set of p'-elements of G.

(2) If φ is a Brauer character of G for the prime p and H is a p'-subgroup of G, then the restriction φ_H is an ordinary character of H.

To prove the Fong–Swan theorem now, we apply Corollary 3.5 with $\pi = p'$, and we deduce that every Brauer character is a π-partial character. This means that it lifts to an ordinary character of G, as required.

A striking application of Corollary 3.5 is the following.

Theorem 3.6. *Let H be a Hall π-subgroup of a π-separable group G, and let ψ be a character of H. Then ψ is the restriction to H of some character of G if and only if $\psi(x) = \psi(y)$ whenever $x, y \in H$ are conjugate in G.*

Proof. If $\psi = \chi_H$ for some character χ of G, then since χ is a class function, it is immediate that $\psi(x) = \psi(y)$ whenever $x, y \in H$ are G-conjugate. For the converse, we argue first that each π-class K of G meets H nontrivially. To see this, let $x \in K$ be arbitrary. Then by Theorem 1.1, the π-group $\langle x \rangle$ is contained in some Hall π-subgroup of G. Also by Theorem 1.1, all Hall π-subgroups of G are conjugate, so we can write $\langle x \rangle \subseteq H^t$ for some element $t \in G$. Then $x^{t^{-1}}$ is an element of $K \cap H$, and it follows that $K \cap H$ is nonempty, as claimed.

Assuming now that $\psi(x) = \psi(y)$ whenever $x, y \in H$ are G-conjugate, we see that for every π-class K, the character ψ is constant on $K \cap H$. We can thus define a function θ on G^0 so that the value of θ on every element of K is equal to the constant value of ψ on the nonempty set $K \cap H$. By construction then, $\theta \in \mathrm{cf}(G^0)$ and $\theta_H = \psi$, which is a character of H. We deduce by Corollary 3.5 that θ is a π-partial character of G, and thus $\theta = \chi^0$ for some character χ of G. Then $\chi_H = (\chi^0)_H = \theta_H = \psi$, as required. ∎

Theorem 3.6 can be viewed as a generalization of a result of Gallagher that asserts that if H is a normal Hall π-subgroup of a group G and $\psi \in \mathrm{Irr}(H)$ is G-invariant, then ψ extends to G. To see why this follows from Theorem 3.6, observe that in the case where H is normal in G, the condition that $\psi(x) = \psi(y)$ for G-conjugate elements $x, y \in H$ is equivalent to saying that ψ is invariant in G. Also, the assumption in Theorem 3.6 that G is π-separable follows from (but is much weaker than) Gallagher's hypothesis that G has a normal Hall π-subgroup. (Note that Theorem 1.5 is another generalization of this result of Gallagher.)

We have one additional observation concerning Theorem 3.6: it does not generalize to arbitrary finite groups, even if we take H to be a Sylow subgroup. For example, suppose $G = A_5$, the alternating group on five symbols. A Sylow 3-subgroup H of G has order 3, and we can take ψ to be the sum of the two nonprincipal linear characters of H. Then $\psi(x) = -1$ for all nonidentity elements $x \in H$, and thus $\psi(x) = \psi(y)$ whenever $x, y \in H$ are G-conjugate. It is easy to see, however, that ψ cannot be the restriction of a character of G. This is because the unique degree 2 character of G is the sum of two copies of the principal character, and this does not agree with ψ on H. (Note, however, that ψ is the restriction of a *difference* of characters of G. This is a general phenomenon, which is a consequence of Brauer's "characterization of characters" theorem.)

Suppose that H is a Hall π-subgroup of a π-separable group G, then even without constructing the set $\mathrm{I}_\pi(G)$, it is easy to determine if a character $\alpha \in \mathrm{Irr}(G)$ is a Fong character: the Fong characters of H are exactly those characters that occur as minimum-degree constituents of restrictions χ_H for $\chi \in \mathrm{Irr}(G)$. To see this, observe first that if α is a Fong character for $\varphi \in \mathrm{I}_\pi(G)$, we can write $\varphi = \chi^0$ for some character $\chi \in \mathrm{Irr}(G)$, and thus α is a minimum-degree constituent of χ_H. Conversely, suppose that α is a minimum-degree constituent of χ_H, where $\chi \in \mathrm{Irr}(G)$. Then χ^0 is a sum of irreducible π-partial characters and α must be a constituent of the restriction of one of them, say φ, to H. Since every irreducible constituent of φ_H is a constituent of χ_H, it follows that α is a minimum-degree constituent of φ_H, and thus, by definition, α is a Fong character.

Given two Fong characters α and β of a Hall π-subgroup H of G, it is natural to ask when it is true that α and β correspond to the same irreducible π-partial character φ of G. One answer is that if α and β are Fong characters corresponding to $\varphi \in \mathrm{I}_\pi(G)$, then $\alpha^G = \beta^G$, and in fact, $\alpha^G = \Phi_\varphi = \beta^G$, where Φ_φ is the analogue of Brauer's projective indecomposable character.

To explain this, we start with an arbitrary member $\chi \in \mathrm{Irr}(G)$. The π-partial character χ^0 is a sum of irreducible π-partial characters, so we can write

$$\chi^0 = \sum_{\varphi \in \mathrm{I}_\pi(G)} d_{\chi,\varphi}\varphi\,,$$

where the decomposition numbers $d_{\chi,\varphi}$ are nonnegative integers, and by Theorem 3.3, they are uniquely determined.

We can now define the characters

$$\Phi_\varphi = \sum_{\chi \in \mathrm{Irr}(G)} d_{\chi,\varphi}\chi\,,$$

and we observe that these characters are distinct for distinct $\varphi \in \mathrm{I}_\pi(G)$, and in fact, they are linearly independent. To see this, note that if we choose $\chi \in \mathrm{Irr}(G)$ with $\chi^0 = \varphi$, then $d_{\chi,\varphi} = 1$, so χ occurs as a constituent of Φ_φ. If $\mu \in \mathrm{I}_\pi(G)$ with $\mu \neq \varphi$, however, then $d_{\chi,\mu} = 0$, and so χ is not a constituent of Φ_μ.

Next, we collect some fairly elementary facts connecting the characters Φ_φ of G and the irreducible characters of a Hall π-subgroup H of G. In particular, we will see that as claimed, a character $\alpha \in \mathrm{Irr}(H)$ is a Fong character for $\varphi \in \mathrm{I}_\pi(G)$ if and only if $\alpha^G = \Phi_\varphi$.

Theorem 3.7. *Let H be a Hall π-subgroup of a π-separable group G. Then $\Phi_\varphi(1) = |G : H|\varphi(1)_\pi$ for all members $\varphi \in \mathrm{I}_\pi(G)$. Also, if $\alpha \in \mathrm{Irr}(H)$ is arbitrary, the following hold.*

(a) $\alpha^G = \sum_\mu [\mu_H, \alpha]\Phi_\mu$, *where the sum runs over* $\mu \in \mathrm{I}_\pi(G)$. *In particular,* α^G *is a nonnegative-integer linear combination of the characters* Φ_μ.

(b) $\alpha(1) = \sum_\mu [\mu_H, \alpha]\mu(1)_\pi$.

(c) α *is a Fong character associated with* $\varphi \in \mathrm{I}_\pi(G)$ *if and only if* $\alpha^G = \Phi_\varphi$.

(d) *If* α *is a Fong character associated with* $\varphi \in \mathrm{I}_\pi(G)$ *and* ψ *is any minimum-degree irreducible constituent of* α^G, *then* $\psi^0 = \varphi$.

Proof. We prove (a) first. For $\chi \in \mathrm{Irr}(G)$, we have

$$\chi_H = \sum_{\mu \in \mathrm{I}_\pi(G)} d_{\chi,\mu}\mu_H ,$$

and thus

$$\begin{aligned}\sum_{\mu \in \mathrm{I}_\pi(G)} [\mu_H, \alpha]\Phi_\mu = \sum_{\substack{\mu \in \mathrm{I}_\pi(G) \\ \chi \in \mathrm{Irr}(G)}} d_{\chi,\mu}[\mu_H, \alpha]\chi &= \sum_{\chi \in \mathrm{Irr}(G)} [\chi_H, \alpha]\chi \\ &= \sum_{\chi \in \mathrm{Irr}(G)} [\chi, \alpha^G]\chi \\ &= \alpha^G ,\end{aligned}$$

as wanted. The coefficient $[\mu_H, \alpha]$ is a nonnegative integer because μ_H is a character, and this completes the proof of (a).

Now given $\varphi \in \mathrm{I}_\pi(G)$, let α be a Fong character for φ, so by definition, $\alpha(1) = \varphi(1)_\pi$, and also $[\varphi_H, \alpha] = 1$ and $[\mu_H, \alpha] = 0$ for $\mu \in \mathrm{I}_\pi(G)$ with $\mu \neq \varphi$. Now (a) yields $\alpha^G = \Phi_\varphi$, proving the "only if" part of (c). We also have

$$\Phi_\varphi(1) = \alpha^G(1) = |G : H|\alpha(1) = |G : H|\varphi(1)_\pi ,$$

as required.

To complete the proof of (c), we assume that $\alpha^G = \Phi_\varphi$, where $\alpha \in \mathrm{Irr}(H)$, and we show that α is a Fong character for φ. First, to see that α is a constituent of φ_H, we observe that

$$\Phi_\varphi = \alpha^G = \sum_{\mu \in \mathrm{I}_\pi(G)} [\mu_H, \alpha]\Phi_\mu ,$$

where the second equality holds by (a). It follows from the linear independence of the characters Φ_μ that $[\varphi_H, \alpha] = 1$, and thus α is a constituent of φ_H, as claimed. To prove that α is a Fong character for φ, it now suffices to show that $\alpha(1) = \varphi(1)_\pi$. We have

$$|G : H|\alpha(1) = \alpha^G(1) = \Phi_\varphi(1) = |G : H|\varphi(1)_\pi ,$$

and thus $\alpha(1) = \varphi(1)_\pi$, as required.

Now (a) and the fact that $\Phi_\mu(1) = |G : H|\mu(1)_\pi$ yield

$$|G : H|\alpha(1) = \alpha^G(1) = \sum_\mu [\mu_H, \alpha]\Phi_\mu(1) = \sum_\mu [\mu_H, \alpha]|G : H|\mu(1)_\pi\,,$$

and (b) follows.

Finally, to prove (d), suppose that α is a Fong character for φ. Then $\alpha^G = \Phi_\varphi$ by (c), and it suffices to show that if ψ is a minimum-degree irreducible constituent of Φ_φ, then $\psi^0 = \varphi$.

If ψ is an arbitrary irreducible constituent of Φ_φ, we have $d_{\psi,\varphi} > 0$, and thus $\psi(1) \geq \varphi(1)$. Also, if equality holds here, we must have $\psi^0 = \varphi$. To complete the proof of (d), therefore, we must show that Φ_φ actually has an irreducible constituent χ with degree as small as $\varphi(1)$. To see this, let $\chi \in \mathrm{Irr}(G)$ with $\chi^0 = \varphi$. Then certainly, $\chi(1) = \varphi(1)$, and to see that χ is a constituent of Φ_φ, it suffices to observe that $d_{\chi,\varphi} > 0$. ∎

The formula $\Phi_\varphi(1) = |G : H|\varphi(1)_\pi$ in Theorem 3.7 does not hold in general for Brauer's projective indecomposable characters (with $\pi = p'$). It does hold for p-solvable groups, of course, and that case was proved by Fong.

We can now describe a reasonably efficient algorithm that can be used to construct $\mathrm{I}_\pi(G)$ in a π-separable group G. Given a Hall π-subgroup H of G, we can locate the Fong characters in the set $\mathrm{Irr}(H)$ because, as we have seen, they are exactly the smallest degree irreducible constituents of the restrictions χ_H for characters $\chi \in \mathrm{Irr}(G)$. Next, we construct the induced characters α^G for Fong characters α, and we discard repeats. Theorem 3.7(c) guarantees that at this point we have found the characters Φ_φ, and we choose one irreducible constituent of minimum degree from each of them. By Theorem 3.7(d), the restrictions to G^0 of the members of $\mathrm{Irr}(G)$ selected in this way are exactly the members of $\mathrm{I}_\pi(G)$, each obtained just once.

As we observed previously, the irreducible π-partial characters of a π-separable group G can also be constructed from the character table of G, without using additional information, and in particular, without having access to a Hall π-subgroup. It follows that the character table determines the decomposition numbers $d_{\chi,\varphi}$, and thus the characters Φ_φ are also determined by the character table.

If μ is the "principal" π-partial character $(1_G)^0$, then clearly, the principal character 1_H of the Hall π-subgroup H is the unique Fong character associated with μ, and it follows by Theorem 3.7(c) that $(1_H)^G = \Phi_\mu$. We conclude that for a π-separable group G with Hall π-subgroup H, the character $(1_H)^G$ is determined by the character table. The value of this character at a π-element $x \in G$ is equal to $m|\mathbf{C}_G(x)|/|H|$, where m is the number of G-conjugates of x that lie in H. It follows that for π-elements of π-separable groups, the quantity m can be determined from the character table. (It is not known if the corresponding statement is true for p-elements and Sylow p-subgroups for arbitrary finite groups.)

Since we mentioned that Brauer's projective indecomposable characters vanish on elements with orders divisible by p, we note that it is easy to prove the corresponding result in our setting.

Corollary 3.8. *Let G be π-separable. Then the characters Φ_φ for $\varphi \in \mathrm{I}_\pi(G)$ form a basis for the subspace $V \subseteq \mathrm{cf}(G)$ consisting of those class functions on G that vanish on the set $G - G^0$.*

Proof. By Theorem 3.7, the characters Φ_φ are induced from a Hall π-subgroup H, so they can have nonzero values only on elements of G that are conjugate to elements of H. All such elements lie in G^0, and it follows that the linearly independent characters Φ_φ all lie in the subspace V.

The number of characters Φ_φ is equal to $|\,\mathrm{I}_\pi(G)|$, and this is equal to the number of π-classes of G. The result now follows from the observation that the number of π-classes of G is the dimension of the space V, and thus the set of characters Φ_φ forms a basis for V. ∎

3C

In this section, we establish Theorem 3.4. In order to state and prove our first preliminary results, it is convenient to use the "bar convention" for factor groups, and so we offer a brief description of the relevant notation. If $M \triangleleft G$, we can write $G/M = \overline{G}$, and in this situation, we view the overbar as the name of the canonical homomorphism $G \to \overline{G}$. In particular, if $g \in G$, then $\overline{g}$ is the coset Mg, and if $H \subseteq G$, then $\overline{H}$ is the image of H in $\overline{G}$, and thus $\overline{H} = HM/M$. Note that M is the kernel of the overbar map, so if $H \cap M = 1$, the map $h \mapsto \overline{h}$ is actually an isomorphism from H onto $\overline{H}$. Also, observe that in general if $x, y \in G$ are conjugate, then $\overline{x}$ and $\overline{y}$ are conjugate in $\overline{G}$, and thus the map $g \mapsto \overline{g}$ carries conjugacy classes to conjugacy classes.

Lemma 3.9. *Let $M \triangleleft G$, and write $\overline{G} = G/M$. Then the map $x \mapsto \overline{x}$ carries the set of π-elements of G onto the set of π-elements of $\overline{G}$. Also, if M is a π'-group, this map induces a bijection from the set of π-classes of G onto the set of π-classes of $\overline{G}$.*

Proof. If $x \in G$, then $o(\overline{x})$ divides $o(x)$, so if x is a π-element, then $\overline{x}$ is also a π-element. To prove surjectivity, suppose that $\overline{g}$ is an arbitrary π-element of $\overline{G}$. Let H be the full preimage in G of $\langle\overline{g}\rangle$, and observe that $\overline{H} = \langle\overline{g}\rangle$ is a π-group. Since $\langle g\rangle$ is an abelian group, we can write $\langle g\rangle = UV$, where U is a π-group and V is a π'-group. Then $V \subseteq \langle g\rangle \subseteq H$, so $\overline{V}$ is a π'-subgroup of the π-group $\overline{H}$, and hence $\overline{V}$ is the trivial group. We have $\overline{g} \in \overline{U}\,\overline{V} = \overline{U}$, and thus $\overline{g} = \overline{u}$ for some element $u \in U$. Since u is a π-element, it follows that the map $x \mapsto \overline{x}$ carries the set of π-elements of G onto the set of π-elements of $\overline{G}$, as claimed.

Our map takes classes to classes, and we now know that it carries the set of π-classes of G onto the set of π-classes of $\overline{G}$. Assuming that M is a π'-group, it remains to show that the induced map on π-classes is injective. If $x, y \in G$ are π-elements, therefore, and $\overline{x}$ and $\overline{y}$ are conjugate in $\overline{G}$, we must show that x and y are conjugate in G.

Replacing y by a conjugate if necessary, we can assume that $\overline{x} = \overline{y}$, and we let H be the full preimage in G of the subgroup $\langle\overline{x}\rangle = \langle\overline{y}\rangle$. Then $\overline{H} = H/M$ is a π-group, and hence M is a normal Hall π'-subgroup of H. By the Schur–Zassenhaus theorem, therefore, we can write $H = MU$, where U is a Hall π-subgroup of H. By Theorem 1.1, it follows that the π-elements x and y are conjugate in H to elements of U, and it suffices to show that they are conjugate to the *same* element of U. To see that this is so, suppose $x^h = s \in U$ and $y^k = t \in U$, where $h, k \in H$.

Since $\overline{H}$ is abelian, we have $\overline{s} = \overline{x} = \overline{y} = \overline{t}$. Also, $U \cap M = 1$, so the overbar map from U into $\overline{G}$ is injective, and it follows that $s = t$, as wanted. ∎

If M is a π'-group, then by Lemma 3.9 the canonical homomorphism $G \to \overline{G} = G/M$, defines a bijection from the set of π-classes of G onto the set of π-classes of $\overline{G}$, and it follows that there is a natural bijection from the set of class functions on π-elements of G onto the set of class functions on π-elements of $\overline{G}$. The image $\overline{\varphi}$ of the class function φ is defined by the formula $\overline{\varphi}(\overline{x}) = \varphi(x)$ for π-elements x of G.

Theorem 3.10. *Let G be π-separable, and suppose that $M \triangleleft G$ is a π'-group such that the conclusion of Theorem 3.4 holds in $\overline{G} = G/M$. Then there is a bijection $\varphi \mapsto \overline{\varphi}$ from $\mathrm{I}_\pi(G)$ onto $\mathrm{I}_\pi(\overline{G})$, where $\overline{\varphi}(\overline{x}) = \varphi(x)$ for all π-elements $x \in G$. Also, the conclusion of Theorem 3.4 holds for G.*

Proof. We have a bijection $\varphi \mapsto \overline{\varphi}$ from the set of class functions on π-elements of G to the set of class functions on π-elements of $\overline{G}$, and we show first that if φ is a π-partial character of G, then $\overline{\varphi}$ is a π-partial character of $\overline{G}$.

Certainly, $\overline{\varphi}$ is a class function on $\overline{G}$, and since by hypothesis, the conclusion of Theorem 3.4 holds for $\overline{G}$, we know that Corollary 3.5 also holds. To prove that $\overline{\varphi}$ is a π-partial character of $\overline{G}$, therefore, it suffices to show that the restriction of $\overline{\varphi}$ to a Hall π-subgroup of $\overline{G}$ is actually a character.

Let H be a Hall π-subgroup of G. Then $\overline{H} = HM/M$ is a Hall π-subgroup of $\overline{G}$. Every element of $\overline{H}$ has the form $\overline{h}$ for some element $h \in H$, and it follows from the definition of $\overline{\varphi}$ that $\overline{\varphi}(\overline{h}) = \varphi(h)$. Since φ is a π-partial character of G, we know that φ_H is a character, and thus since the map $h \mapsto \overline{h}$ defines an isomorphism from H onto $\overline{H}$, it follows that $\overline{\varphi}_{\overline{H}}$ is a character. Then $\overline{\varphi}$ is a π-partial character of $\overline{G}$, as claimed.

Next, we observe that every π-partial character of $\overline{G}$ has the form $\overline{\varphi}$ for some π-partial character φ of G. To see this, let μ be a π-partial character of $\overline{G}$ and write $\mu = \chi^0$ for some character χ of $\overline{G} = G/M$. Viewing χ as a character of G, we let $\varphi = \chi^0$, so φ is a π-partial character of G. Then $\varphi(x) = \chi(x) = \chi(\overline{x}) = \mu(\overline{x})$, and thus $\mu = \overline{\varphi}$, as wanted.

We now have a bijection $\varphi \mapsto \overline{\varphi}$ from the set of π-partial characters of G onto the set of π-partial characters of $\overline{G}$, and since this map clearly respects addition, it defines a bijection from $\mathrm{I}_\pi(G)$ onto $\mathrm{I}_\pi(\overline{G})$. Also, we have seen that if H is a Hall π-subgroup of G, then the characters φ_H and $\overline{\varphi}_{\overline{H}}$ agree with each other via the isomorphism $H \cong \overline{H}$. Since by assumption, the restrictions of the irreducible π-partial characters of $\overline{G}$ to the Hall π-subgroup $\overline{H}$ behave according to the requirements of Theorem 3.4, the same is true about the restrictions of the irreducible π-partial characters of G to the Hall π-subgroup H. ■

To complete the proof of Theorem 3.4, we will invoke the theory of character-triple isomorphisms, which was discussed in Chapter 1. In particular, we recall Theorem 1.13, which asserts that if (G, N, θ) is a character triple, then there exists an isomorphic character triple (G^*, N^*, θ^*) such that $N^* \subseteq \mathbf{Z}(G^*)$.

Given isomorphic character triples (G, N, θ) and (G^*, N^*, θ^*) and a subgroup H with $N \subseteq H \subseteq G$, then as in Chapter 1 we will usually write H^* to denote the subgroup of G^* corresponding to H. In particular, therefore, $N^* \subseteq H^* \subseteq G^*$ and $H/N \cong H^*/N^*$. Also, if $\xi \in \mathrm{Irr}(H|\theta)$, we will write ξ^* to denote the corresponding character of H^* lying over θ^*.

We need a refinement of Theorem 1.13.

Lemma 3.11. *Let (G, N, θ) be a character triple, where N is a π-group. Then there exists an isomorphic character triple (G^*, N^*, θ^*) such that $N^* \subseteq \mathbf{Z}(G^*)$ and N^* is a π-group.*

Proof. By Theorem 1.13, there exists a character triple (H, M, β) isomorphic to (G, N, θ), where $M \subseteq \mathbf{Z}(H)$. We complete the proof by showing that (H, M, β) is isomorphic to a character triple (G^*, N^*, θ^*) with the required properties.

Since M is abelian, β is a linear character of M, and thus we can write $\beta = \mu\nu$, where $o(\mu)$ is a π-number and $o(\nu)$ is a π'-number. We propose to apply Theorem 1.10 to show that ν extends to H, so we must check for every prime p that ν extends to P, where P/M is some Sylow p-subgroup of H/M.

If $p \in \pi$, then P/M is a π-group, and since $\nu(1) = 1$ and $o(\nu)$ is a π'-number, Gallagher's theorem (Theorem 1.5) guarantees that ν extends to P, as required. We can thus assume that $p \notin \pi$.

Now $G/N \cong H/M$, and we let $Q/N \subseteq G/N$ correspond to P/M under this isomorphism. Then Q/N is a p-group, and hence it is a π'-group. Since N is a π-group by hypothesis, we see that both $\theta(1)$ and $o(\theta)$ are π-numbers, and thus by Gallagher's theorem, θ extends to Q. Now (G, N, θ) and (H, M, β) are isomorphic character triples and Q and P are corresponding intermediate subgroups, and since θ extends to Q, Lemma 1.12 guarantees that β has an extension $\lambda \in \mathrm{Irr}(P)$.

Recall that β is linear and $\beta = \mu\nu$, and thus $\beta^{o(\mu)} = \nu^{o(\mu)}$. Also, since $o(\mu)$ and $o(\nu)$ are relatively prime, there exists an integer m such that $o(\mu)m \equiv 1 \mod o(\nu)$. Then $\beta^{o(\mu)m} = \nu^{o(\mu)m} = \nu$, and thus $\lambda^{o(\mu)m}$ is an extension of ν to P, as wanted.

At this point, we have shown that the linear character ν extends to H, and we let $\gamma \in \mathrm{Irr}(H)$ be the complex conjugate of some extension of ν. Then $\beta\gamma_M = \mu$, and it follows that if X is an intermediate subgroup of (H, M, β) and ξ is an admissible character of X, then the restriction of $\xi\gamma_X$ to M is a multiple of μ. We see now that multiplication by restrictions of the linear character γ to intermediate subgroups of (H, M, β) defines a character-triple isomorphism from (H, M, β) to (H, M, μ). Finally, if $K = \ker(\mu)$, then $K \lhd H$ since $M \subseteq \mathbf{Z}(H)$, and hence (H, M, μ) is isomorphic to $(H/K, M/K, \mu)$. Since $|M/K| = o(\mu)$ is a π-number, we can take $(G^*, N^*, \theta^*) = (H/K, M/K, \mu)$. ∎

We shall also need the following.

Lemma 3.12. *Let (G, N, θ) and (G^*, N^*, θ^*) be isomorphic character triples, where N and N^* are π-groups and G is π-separable. The following then hold.*

(a) *G^* is π-separable.*

(b) *If $\chi \in \mathrm{Irr}(G|\theta)$ and $\chi^0 \in \mathrm{I}_\pi(G)$, then $(\chi^*)^0 \in \mathrm{I}_\pi(G^*)$.*

(c) *Let $\zeta, \chi \in \mathrm{Irr}(G|\theta)$. Then $\zeta^0 = \chi^0$ if and only if $(\zeta^*)^0 = (\chi^*)^0$.*

Proof. By hypothesis, G is π-separable, so G/N is π-separable, and since $G^*/N^* \cong G/N$, we see that G^*/N^* is π-separable. Also, N^* is a π-group, and it follows that G^* is π-separable, proving (a).

Let $\chi \in \mathrm{Irr}(G|\theta)$. To prove (b), it suffices to assume that the π-partial character $(\chi^*)^0$ is not irreducible and to show that χ^0 is not irreducible.

By assumption, $(\chi^*)^0$ can be written as a sum of two π-partial characters of G^*, and thus there exist characters ξ^* and η^* of G^* such that $\chi^*(y) = \xi^*(y) + \eta^*(y)$ for all π-elements $y \in G^*$. In particular, since N^* is a π-subgroup, $(\chi^*)_{N^*} = (\xi^*)_{N^*} + (\eta^*)_{N^*}$, and because $(\chi^*)_{N^*}$ is a multiple of θ^*, it follows that $(\xi^*)_{N^*}$ and $(\eta^*)_{N^*}$ are also multiples of θ^*. There exist, therefore, admissible characters ξ and η of G corresponding to ξ^* and η^*.

Suppose that $x \in G$ is an arbitrary π-element, and let $U = N\langle x \rangle$. Then U is an intermediate subgroup of (G, N, θ), and $|U^*/N^*| = |U/N|$, which is a π-number. It follows that U^* is a π-group, and hence $(\chi^*)_{U^*} = (\xi^*)_{U^*} + (\eta^*)_{U^*}$. The restriction property for character-triple isomorphisms yields

$$(\chi_U)^* = (\chi^*)_{U^*} = (\xi^*)_{U^*} + (\eta^*)_{U^*} = (\xi_U)^* + (\eta_U)^* = (\xi_U + \eta_U)^*,$$

and so by the injectivity of the $*$-map, we have $\chi_U = \xi_U + \eta_U$. Then $\chi(x) = \xi(x) + \eta(x)$, and since $x \in G$ was an arbitrary π-element, it follows that $\chi^0 = \xi^0 + \eta^0$. Then χ^0 is not irreducible, proving (b).

To prove the "if" part of (c), let $\zeta, \chi \in \mathrm{Irr}(G|\theta)$, and suppose that $(\zeta^*)^0 = (\chi^*)^0$. If $\zeta^0 \neq \chi^0$, there exists a π-element x of G such that $\zeta(x) \neq \chi(x)$, and thus $\zeta_U \neq \chi_U$, where $U = N\langle x \rangle$. As before, U^* is a π-group, and thus

$$(\zeta_U)^* = (\zeta^*)_{U^*} = (\chi^*)_{U^*} = (\chi_U)^*,$$

where the second equality holds because we are assuming that $(\zeta^*)^0 = (\chi^*)^0$. Then $\zeta_U = \chi_U$ by the injectivity of the $*$-map, and this is a contradiction. It follows that $\zeta^0 = \chi^0$, as required. The "only if" part of (c) follows by applying what we have already established to the character-triple isomorphism in the reverse direction, from (G^*, N^*, θ^*) to (G, N, θ). ■

We need one more preliminary result for our inductive proof of Theorem 3.4.

Lemma 3.13. *Let $S \subseteq G$, where G is π-separable and S is a π-subgroup, and assume that the conclusion of Theorem 3.4 holds for the group G. Let $\theta \in \mathrm{I}_\pi(G)$, and let η be an irreducible constituent of the character θ_S. Then*

$$\eta(1) \geq \frac{\theta(1)_\pi |S|}{|G|_\pi}.$$

Also, if equality holds, then $[\theta_S, \eta] = 1$ and θ is the unique member of $\mathrm{I}_\pi(G)$ whose restriction to S has η as a constituent.

Proof. Let H be a Hall π-subgroup of G containing S. Since η is an irreducible constituent of θ_S, there exists a character $\xi \in \mathrm{Irr}(H)$ lying over η

such that ξ is a constituent of θ_H. By Theorem 3.4, the minimum of the degrees of the irreducible constituents of θ_H is $\theta(1)_\pi$, and thus

$$\theta(1)_\pi \le \xi(1) \le \eta^H(1) = \eta(1)|H : S| = \frac{\eta(1)|G|_\pi}{|S|},$$

and the desired inequality follows.

If equality holds, then $\xi = \eta^H$ and $\xi(1) = \theta(1)_\pi$. Since ξ is a constituent of θ_H, the latter equality tells us that ξ is a Fong character for θ, and thus by Theorem 3.4 we have

$$1 = [\theta_H, \xi] = [\theta_H, \eta^H] = [\theta_S, \eta],$$

as required. If η is a constituent of ν_S where $\nu \in \mathrm{I}_\pi(G)$, we have

$$0 < [\nu_S, \eta] = [\nu_H, \eta^H] = [\nu_H, \xi],$$

and since ξ is a Fong character for θ, Theorem 3.4 guarantees that $\nu = \theta$. ∎

We are now ready to prove Theorem 3.4.

Proof of Theorem 3.4. Let $N = \mathbf{O}_\pi(G)$. If $N = G$, then G is a π-group, and in this case $\mathrm{I}_\pi(G) = \mathrm{Irr}(G)$, and the result is trivial. We can assume, therefore, that $N < G$, and we proceed by induction on $|G : N|$.

Let $\varphi \in \mathrm{I}_\pi(G)$, and let U be a Hall π-subgroup of G. Choose $\chi \in \mathrm{Irr}(G)$ such that $\chi^0 = \varphi$, and note that $\varphi_U = \chi_U$. We must show that the minimum of the degrees of the irreducible constituents of φ_U is exactly $\varphi(1)_\pi$ and that each of these minimum-degree constituents has multiplicity 1 in φ_U. In addition, we must show that if $\mu \in \mathrm{I}_\pi(G)$ is different from φ, then none of the minimum-degree constituents of φ_U is also a constituent of μ_U.

Let $\alpha \in \mathrm{Irr}(N)$ be a constituent of φ_N, and assume first that α is invariant in G. (Note that we are not assuming that α is nonprincipal.) Then (G, N, α) is a character triple, and we can apply Lemma 3.11 to produce an isomorphic character triple (G^*, N^*, α^*), where $N^* \subseteq \mathbf{Z}(G^*)$ and N^* is a π-group. Then G^* is π-separable, and we argue that G^* satisfies the conclusion of Theorem 3.4. To see this, observe that $\mathbf{O}_\pi(G^*/N^*)$ is isomorphic to $\mathbf{O}_\pi(G/N)$, which is trivial. Also, $|G^* : N^*| = |G : N| > 1$, and thus $\mathbf{O}_{\pi'}(G^*/N^*) > 1$, and we write $L^*/N^* = \mathbf{O}_{\pi'}(G^*/N^*)$. Since N^* is a central Hall π-subgroup of L^*, we can write $L^* = N^* \times K$, where $K = \mathbf{O}_{\pi'}(L^*)$, and thus $1 < K \triangleleft G^*$. Now $L^*/K \subseteq \mathbf{O}_\pi(G^*/K)$, and since $|(G^*/K) : (L^*/K)| = |G^* : L^*| < |G^* : N^*| = |G : N|$, it follows by the inductive hypothesis that the group G^*/K satisfies the conclusion of Theorem 3.4. By Theorem 3.10, therefore, G^* satisfies the conclusion of Theorem 3.4, as claimed.

Now α is a constituent of $\varphi_N = \chi_N$, so $\chi \in \mathrm{Irr}(G|\alpha)$, and thus $(\chi^*)^0 \in \mathrm{I}_\pi(G^*)$ by Lemma 3.12(b), where χ^* corresponds to χ via the character-triple isomorphism.

Since U/N is a Hall π-subgroup of G/N, it follows that U^*/N^* is a Hall π-subgroup of G^*/N^*, and thus U^* is a Hall π-subgroup of G^*. We have seen that the conclusion of Theorem 3.4 is valid for G^*, and we know that $(\chi^*)^0 \in \mathrm{I}_\pi(G^*)$, and it follows that all of the minimum-degree constituents of $(\chi^*)_{U^*}$ occur with multiplicity 1, and their common degree is $\chi^*(1)_\pi$.

By the restriction property, we know that the $*$-map carries the irreducible constituents of χ_U bijectively to the irreducible constituents of $(\chi^*)_{U^*}$, preserving multiplicities. Also, if ξ is an irreducible constituent of χ_U, then by Lemma 1.12 we have

$$\frac{\xi(1)}{\alpha(1)} = \frac{\xi^*(1)}{\alpha^*(1)} = \xi^*(1)\,.$$

Then $\xi(1) = \alpha(1)\xi^*(1)$, and thus the $*$-map carries the minimum-degree constituents of χ_U to the minimum-degree constituents of $(\chi^*)_{U^*}$. It follows that the minimum-degree constituents of χ_U have multiplicity 1 and that their common degree is $\alpha(1)\chi^*(1)_\pi$. A similar calculation using Lemma 1.12 yields $\chi(1) = \alpha(1)\chi^*(1)$, and since $\alpha(1)$ is a π-number, $\chi(1)_\pi = \alpha(1)\chi^*(1)_\pi$. We have seen that this is exactly the common degree of the minimum-degree constituents of χ_U, as required.

We show next that if $\mu \in \mathrm{I}_\pi(G)$ and $\mu \neq \varphi$, then none of the minimum-degree constituents of φ_U is a constituent of μ_U. Since every irreducible constituent of $\varphi_U = \chi_U$ lies over α, this is clear if α is not a constituent of μ_N, and thus we can assume that $\mu = \zeta^0$, for some character $\zeta \in \mathrm{Irr}(G|\alpha)$. It follows by Lemma 3.12(b) that $(\zeta^*)^0 \in \mathrm{I}_\pi(G^*)$, and since $\chi^0 = \varphi \neq \mu = \zeta^0$, Lemma 3.12(c) guarantees that $(\zeta^*)^0 \neq (\chi^*)^0$. Since the conclusion of Theorem 3.4 holds in the group G^*, it follows that none of the minimum-degree constituents of $(\chi^*)_{U^*}$ is a constituent of $(\zeta^*)_{U^*}$, and hence by the restriction property and the injectivity of the $*$-map, no minimum-degree constituent of $\varphi_U = \chi_U$ is a constituent of $\mu_U = \zeta_U$. This completes the proof in the case where α is invariant in G.

We can now assume that the stabilizer T of α in G is a proper subgroup, and we let $\psi \in \mathrm{Irr}(T|\alpha)$ be the Clifford correspondent of χ. Then $\psi^G = \chi$, and thus $(\psi^0)^G = \chi^0 = \varphi$. We deduce that $\psi^0 \in \mathrm{I}_\pi(T)$, and we write $\theta = \psi^0$. Then $\theta^G = \varphi$ and $\varphi(1) = |G:T|\theta(1)$.

Now let $\delta \in \mathrm{Irr}(U)$, and suppose that δ lies over some G-conjugate of α. (In particular, this will be the case if δ is a constituent of $\varphi_U = \chi_U$, but for now, we do not assume this.) Replacing α by a suitable conjugate, we can assume that δ lies over α, and we let $S = T \cap U$, so that S is the stabilizer of α in U. By the Clifford correspondence, therefore, we can write $\delta = \eta^U$ for some character $\eta \in \mathrm{Irr}(S|\alpha)$. We have

$$[\varphi_U, \delta] = [\chi_U, \delta] = [\chi_U, \eta^U] = [\chi, (\eta^U)^G] = [\chi, (\eta^T)^G]\,.$$

Every irreducible constituent of η^T lies in $\mathrm{Irr}(T|\alpha)$, and by the Clifford correspondence we know that ψ is the unique member of $\mathrm{Irr}(T|\alpha)$ that lies under χ. Since $\psi^G = \chi$, it follows that

$$[\chi, (\eta^T)^G] = [\psi, \eta^T],$$

and thus

$$[\varphi_U, \delta] = [\chi, (\eta^T)^G] = [\psi, \eta^T] = [\psi_S, \eta] = [\theta_S, \eta].$$

Suppose now that δ actually is an irreducible constituent of φ_U. We show in this case that $\delta(1) \geq \varphi(1)_\pi$. We have

$$[\theta_S, \eta] = [\varphi_U, \delta] > 0,$$

and thus η is a constituent of θ_S. We wish to apply Lemma 3.13 in the group T, and to justify this, we observe that by the inductive hypothesis, the theorem holds for the group T because $N \subseteq T$, and thus $|T : \mathbf{O}_\pi(T)| \leq |T : N| < |G : N|$. We conclude from Lemma 3.13 that

$$\eta(1) \geq \frac{\theta(1)_\pi |S|}{|T|_\pi},$$

and thus since $|U| = |G|_\pi$, we have

$$\begin{aligned}\delta(1) = |U : S|\eta(1) \geq \frac{|G|_\pi}{|S|}\frac{\theta(1)_\pi |S|}{|T|_\pi} = |G : T|_\pi \theta(1)_\pi = (|G : T|\theta(1))_\pi \\ = \varphi(1)_\pi,\end{aligned}$$

as wanted.

If equality holds here, then equality holds in Lemma 3.13, and hence by that result $[\theta_S, \eta] = 1$. Since $[\varphi_U, \delta] = [\theta_S, \eta]$, we conclude that if δ is an irreducible constituent of φ_U such that $\delta(1) = \varphi(1)_\pi$, then $[\varphi_U, \delta] = 1$.

Continuing to assume that $\delta(1) = \varphi(1)_\pi$, we conclude from Lemma 3.13 that θ is the unique member of $\mathrm{I}_\pi(T)$ whose restriction to S has η as a constituent, and we argue that φ is the unique member of $\mathrm{I}_\pi(G)$ whose restriction to U has δ as a constituent. To see this, suppose that δ is a constituent of μ_U, where $\mu \in \mathrm{I}_\pi(G)$. Then α is a constituent of μ_N, and reasoning as we did with φ, we can write $\mu = \nu^G$, where $\nu \in \mathrm{I}_\pi(T)$. Now replacing φ by μ and θ by ν, our earlier computation shows that

$$[\nu_S, \eta] = [\mu_U, \delta] > 0,$$

and it follows that $\nu = \theta$. Then $\mu = \theta^G = \varphi$, as claimed.

Finally, to complete the proof, we argue that φ_U actually does have an irreducible constituent δ with degree equal to the lower bound $\varphi(1)_\pi$. Let K be a Hall π-subgroup of T, and observe that some conjugate of K is contained in U. We can thus replace α by a conjugate and assume that $K \subseteq U$, so that in our previous notation, $S = T \cap U$ is a full Hall π-subgroup of T. By the inductive hypothesis applied in T, there exists an irreducible constituent η of θ_S such that $\eta(1) = \theta(1)_\pi$, and we let $\delta = \eta^U$, so that $\delta \in \mathrm{Irr}(U)$ by the Clifford correspondence. Then δ lies over α, and so by our earlier calculation, $[\varphi_U, \delta] = [\theta_S, \eta] > 0$, and thus δ is a constituent of φ_U. Also

$$\delta(1) = |U : S|\eta(1) = \frac{|G|_\pi}{|T|_\pi}\theta(1)_\pi = (|G : T|\theta(1))_\pi = \varphi(1)_\pi ,$$

as wanted. ∎

3D

We shall see in Chapter 5 that there is an intimate connection between the theory of π-special characters and the theory of π-partial characters. We offer a glimpse of that connection now by proving the following result.

Theorem 3.14. *Let G be π-separable. Then the map $\chi \mapsto \chi^0$ defines a bijection from the set of π-special characters of G onto the set of those irreducible π-partial characters of G whose degree is a π-number.*

Proof. Let H be a Hall π-subgroup of G. Suppose first that $\chi \in \mathrm{Irr}(G)$ is π-special, so Gajendragadkar's restriction theorem (Theorem 2.10) guarantees that $(\chi^0)_H = \chi_H$ is irreducible. Then χ^0 is irreducible, and since $\chi^0(1) = \chi(1)$ is a π-number, our map is well defined. To see that this map is injective, assume that ψ is π-special and $\psi^0 = \chi^0$. Then $\psi_H = \chi_H$, and it follows that $\psi = \chi$ because the restriction map in Theorem 2.10 is injective.

For surjectivity, let $\varphi \in \mathrm{I}_\pi(G)$, where $\varphi(1)$ is a π-number. We seek a π-special character χ of G such that $\chi^0 = \varphi$, and we observe that if G is a π-group, we can take $\chi = \varphi$, and there is nothing to prove. We can assume, therefore, that G is not a π-group, and in particular G is nontrivial, and we proceed by induction on $|G|$.

Let $\alpha \in \mathrm{Irr}(H)$ be a Fong character for φ. Then α is an irreducible constituent of φ_H, and $\alpha(1) = \varphi(1)_\pi = \varphi(1)$, and it follows that $\varphi_H = \alpha$, and in particular, φ_H is irreducible.

Let $N = \mathbf{O}^\pi(G)$ and $M = \mathbf{O}^{\pi'}(N)$. Now $N > 1$ because G is not a π-group, and thus $N > M$. Also, since G/N is a π-group, we have $NH = G$. Write $K = N \cap H$, so K is a Hall π-subgroup of N. Since N/M is a π'-group, we have $K \subseteq M$. Also, $|G : MH| = |N : M| > 1$, and thus $MH < G$.

Since φ_H is irreducible, it follows that φ_{MH} is irreducible. We can thus apply the inductive hypothesis in the group MH to deduce that there exists a π-special character $\xi \in \mathrm{Irr}(MH)$ such that $\xi^0 = \varphi_{MH}$, and so $\xi_H = \alpha$.

Now $M \triangleleft MH$ and $\xi \in \mathrm{Irr}(MH)$ is π-special, and thus the irreducible constituents of ξ_M are π-special, and we let ψ be one of these. Now K is a

Hall π-subgroup of M, and it follows by Theorem 2.10 that ψ_K is irreducible. Write $\psi_K = \beta$, and note that ξ lies over β, and thus $\alpha = \xi_H$ lies over β.

Theorem 2.10 guarantees that ψ is the unique π-special extension of β to M, and it follows that each element of G that normalizes K and stabilizes β must stabilize ψ. We will show that $\mathbf{N}_N(K)$ stabilizes β, and thus $\mathbf{N}_N(K)$ stabilizes ψ.

Assuming this for the moment, we observe that $N = M\mathbf{N}_N(K)$ by the Frattini argument, and thus ψ is invariant in N. Now N/M is a π'-group and ψ is π-special, and it follows that ψ has a canonical extension $\zeta \in \mathrm{Irr}(N)$, and we see that ζ is π-special and ζ lies over β.

Under the assumption that $\mathbf{N}_N(K)$ stabilizes β, we have found a π-special character ζ of N lying over β. Now α is a constituent of β^H, and it follows that α is a constituent of $(\zeta_K)^H = (\zeta^G)_H$. Some irreducible constituent χ of ζ^G, therefore, lies over α, and we see that χ is π-special because ζ is π-special and G/N is a π-group. Now χ_H is irreducible by Theorem 2.10, and thus $\chi_H = \alpha = \varphi_H$. We conclude that χ^0 and φ agree on elements of H, and so these π-partial characters agree on all π-elements of G. It follows that $\chi^0 = \varphi$, as wanted.

To complete the proof, it suffices to show that $\mathbf{N}_N(K)$ stabilizes β. To see this, let $L = \mathbf{N}_G(K)$, and note that $H \subseteq L$. Since $\varphi_H = \alpha$ is irreducible, we see that φ_L is an irreducible π-partial character of L, so there exists $\eta \in \mathrm{Irr}(L)$ such that $\eta^0 = \varphi_L$. Now η lies over α, and thus η lies over β. Writing T to denote the stabilizer of β in L, we see by Clifford's theorem that $|L : T|$ divides $\eta(1) = \varphi(1)$, and thus $|L : T|$ is a π-number. Observe now that $(N \cap L)/K$ is a normal Hall π'-subgroup of L/K. Then $(N \cap L)/K \subseteq T/K$, and hence $\mathbf{N}_N(K) = N \cap L \subseteq T$. Then $\mathbf{N}_N(K)$ stabilizes β, as required. ∎

Corollary 3.15. *Let H be a Hall π-subgroup of G, where G is π-separable, and let $\alpha \in \mathrm{Irr}(H)$. If α extends to G, then α has a unique π-special extension to G.*

Proof. Let $\psi \in \mathrm{Irr}(G)$ extend α. Then $(\psi^0)_H = \psi_H = \alpha$ is irreducible, so $\psi^0 \in \mathrm{I}_\pi(G)$. Since $\psi^0(1) = \alpha(1)$ is a π-number, it follows by Theorem 3.14 that there exists a π-special character $\chi \in \mathrm{Irr}(G)$ such that $\chi^0 = \psi^0$. Then $\chi_H = (\chi^0)_H = (\psi^0)_H = \alpha$, as wanted. Also, χ is the unique π-special extension of α by the injectivity in Gajendragadkar's restriction theorem (Theorem 2.10). ∎

Recall that if H is a Hall π-subgroup of a π-separable group G and $\alpha \in \mathrm{Irr}(H)$, then it follows by Theorem 3.6 that α extends to G if and only if $\alpha(x) = \alpha(y)$ whenever $x, y \in H$ are conjugate in G. By Corollary 3.15

and Theorem 2.10, therefore, the characters $\alpha \in \mathrm{Irr}(H)$ that satisfy this condition are exactly the restrictions to H of the π-special characters of G.

Next, as an application of Corollary 3.15, we present what seems to be a somewhat surprising result.

Theorem 3.16. *Let $\alpha \in \mathrm{Irr}(H)$, where H is a Hall π-subgroup of a π-separable group G. Then there is a unique largest subgroup W with the property that $H \subseteq W \subseteq G$ and α extends to W.*

Proof. If $H = G$, there is nothing to prove, so we can assume that $H < G$, and we proceed by induction on $|G|$. Let N be a maximal normal subgroup of G, and suppose first that G/N is a π'-group. Then $H \subseteq N$, and by the inductive hypothesis, there is a unique largest subgroup V such that $H \subseteq V \subseteq N$ and α extends to V. By Corollary 3.15, there exists a unique π-special character $\beta \in \mathrm{Irr}(V)$ such that β is an extension of α.

Let W be the stabilizer of β in $\mathbf{N}_G(V)$. Since V contains the Hall π-subgroup H of G, we see that W/V is a π'-group, and thus Theorem 2.4(b) guarantees that the π-special character β of V has an extension to W, and in particular α extends to W. To prove that W is the unique largest subgroup with this property, we assume that $H \subseteq K \subseteq G$ and that α extends to K, and we show that $K \subseteq W$.

Now $H \subseteq K \cap N \lhd K$, and thus $K = (N \cap K)\mathbf{N}_K(H)$ by the Frattini argument. Since α extends to $N \cap K$ and V is the unique largest subgroup of N to which α extends, we have $N \cap K \subseteq V \subseteq W$, and it remains to show that $\mathbf{N}_K(H) \subseteq W$.

Now $\mathbf{N}_K(H)$ stabilizes H and it leaves α invariant because α extends to K. Also, of course, $\mathbf{N}_K(H)$ stabilizes N, and since V is uniquely determined by H, α, and N, it follows that $\mathbf{N}_K(H)$ stabilizes V. Since α uniquely determines β, we see that $\mathbf{N}_K(H)$ stabilizes β, and thus $\mathbf{N}_K(H) \subseteq W$, as required.

We can now assume that G/N is not a π'-group, so it is a π-group, and thus $HN = G$ and $H \cap N$ is a Hall π-subgroup of N. Let $\varphi_1, \varphi_2, \ldots, \varphi_t$ be the distinct irreducible constituents of $\alpha_{H\cap N}$.

By the inductive hypothesis applied in the group N, there exist unique largest subgroups V_i of N with $H \cap N \subseteq V_i$, and such that φ_i extends to V_i. Now H stabilizes N and $H \cap N$, and by Clifford's theorem the characters φ_i are permuted by H. It follows that the subgroups V_i are permuted by H.

Now write $M = \bigcap V_i$ and $W = HM$, and observe that H normalizes M and thus W is a group and $M \lhd W$. Also, $H \cap M = H \cap N$ is a Hall π-subgroup of M and φ_1 extends to M. By Corollary 3.15, therefore, φ_1 has a unique π-special extension $\theta \in \mathrm{Irr}(M)$, and the uniqueness guarantees that θ

is stabilized by the stabilizer of φ_1 in H. We can now apply Lemma 2.11(a) to deduce that restriction defines a bijection from $\mathrm{Irr}(W|\theta)$ onto $\mathrm{Irr}(H|\varphi_1)$, and in particular α extends to W.

To complete the proof, we show that if α extends to K, where $H \subseteq K \subseteq G$, then $K \subseteq W$. We have $HN = G$, so by Dedekind's lemma, $K = H(K \cap N)$, and it suffices to show that $K \cap N \subseteq M$. By Corollary 3.15, we can choose a π-special extension β of α to K, and since $K \cap N \lhd K$, we see that the irreducible constituents of $\beta_{K\cap N}$ are π-special, and so by Theorem 2.10, they restrict irreducibly to the Hall π-subgroup $H \cap N$ of $K \cap N$. This shows that all irreducible constituents of $\alpha_{H\cap N} = \beta_{H\cap N}$ extend to $K \cap N$, and thus $K \cap N \subseteq V_i$ for all i. Then $K \cap N \subseteq M$, as required. ∎

Theorem 3.14 suggests the question of when it is true that the degree of *every* member of $\mathrm{I}_\pi(G)$ is a π-number.

Theorem 3.17. *Let H be a Hall π-subgroup of a π-separable group G. Then the following are equivalent.*

(1) *$\varphi(1)$ is a π-number for all π-partial characters $\varphi \in \mathrm{I}_\pi(G)$.*

(2) *Every character $\alpha \in \mathrm{Irr}(H)$ extends to G.*

(3) *G has a normal Hall π'-subgroup.*

Proof. Assuming (1), let $\alpha \in \mathrm{Irr}(H)$, and let $\psi \in \mathrm{Irr}(G)$ lie over α. Then α is a constituent of $\psi_H = (\psi^0)_H$, and thus α is a constituent of φ_H, where $\varphi \in \mathrm{I}_\pi(G)$ is some irreducible summand of ψ^0. (In other words, the decomposition number $d_{\psi,\varphi}$ is nonzero.) By assumption, $\varphi(1) = \varphi(1)_\pi$, which is the degree of a Fong constituent of φ_H, and thus φ_H is irreducible, and we deduce that $\varphi_H = \alpha$. Now $\varphi = \chi^0$ for some character $\chi \in \mathrm{Irr}(G)$, and thus $\chi_H = \varphi_H = \alpha$, proving (2).

To prove that (2) implies (3), we may assume that $|G| > 1$, and we proceed by induction on $|G|$. Let $M = \mathbf{O}_{\pi'}(G)$, and suppose first that $M > 1$. Then HM/M is a Hall π-subgroup of G/M, and we argue that every irreducible character α of HM/M extends to G/M. To see this, view α as a character of HM with kernel containing M, and observe that α_H is irreducible. Then α_H extends to G by assumption, and thus α_H has a π-special extension $\chi \in \mathrm{Irr}(G)$ by Corollary 3.15. Since χ is π-special and $M = \mathbf{O}_{\pi'}(G)$, it follows that $M \subseteq \ker(\chi)$.

If $h \in H$ and $m \in M$, we have $\chi(hm) = \chi(h) = \alpha(h) = \alpha(hm)$, and this shows that χ is an extension of α to G/M. We can apply the inductive hypothesis to conclude that G/M has a normal Hall π'-subgroup K/M. Then K is a normal Hall π'-subgroup of G, as required. We may assume, therefore, that $\mathbf{O}_{\pi'}(G) = 1$, and we prove (3) by showing that G is a π-group.

Now let L be a maximal normal subgroup of G, and let $U = L \cap H$. Then U is a Hall π-subgroup of L, and we argue that every character $\beta \in \mathrm{Irr}(U)$ extends to L. To see this, let $\alpha \in \mathrm{Irr}(H)$ lie over β. By assumption, α extends to G, and thus α has a π-special extension $\chi \in \mathrm{Irr}(G)$, and we see that χ lies over β. Choose $\psi \in \mathrm{Irr}(L)$ lying over β and under χ, and note that ψ is π-special because χ is π-special and $L \triangleleft G$. Since U is a Hall π-subgroup of H, we conclude that ψ_U is irreducible, and thus $\psi_U = \beta$, as wanted. By the inductive hypothesis, therefore, L has a normal Hall π'-subgroup. Since $\mathbf{O}_{\pi'}(L) \subseteq \mathbf{O}_{\pi'}(G) = 1$, however, we conclude that L is a π-group.

If G/L is a π'-group, then L is a normal Hall π-subgroup of G, and thus $L = H$. By assumption, every irreducible character of H extends to G, and hence every such character is G-invariant. By Lemma 2.8, a Hall π'-subgroup K of G centralizes L, and thus $K \triangleleft G$, and this is a contradiction since $\mathbf{O}_{\pi'}(G) = 1$. It follows that G/L is a π-group, and thus G is a π-group, as required, and this completes the proof that (2) implies (3).

Finally, to prove that (3) implies (1), let M be a normal Hall π'-subgroup of G and write $\overline{G} = G/M$. By Theorem 3.10, we see that for every π-partial character $\varphi \in \mathrm{I}_\pi(G)$, there is a π-partial character $\overline{\varphi} \in \mathrm{I}_\pi(\overline{G})$ such that $\varphi(1) = \overline{\varphi}(1)$. Since $\overline{G}$ is a π-group, $\overline{\varphi} \in \mathrm{Irr}(\overline{G})$, so its degree is a π-number, proving (1) ∎

It was conjectured by W. Willems that for an arbitrary finite group G and prime p, the inequality $\sum_\mu \mu(1)^2 \geq |G|_{p'}$ always holds, where μ runs over $\mathrm{IBr}_p(G)$. Although this conjecture is still unresolved, it is known to hold for p-solvable groups. More generally, replacing $\mathrm{IBr}_p(G)$ by $\mathrm{I}_\pi(G)$, we show that the corresponding inequality holds for π-separable groups.

Theorem 3.18. *Let G be π-separable. The following then hold.*

(a) $\sum_\mu \mu(1)\mu(1)_\pi = |G|_\pi$, *where μ runs over $\mathrm{I}_\pi(G)$.*

(b) $\sum_\mu \mu(1)^2 \geq |G|_\pi$ *with equality if and only if G has a normal Hall π'-subgroup.*

Proof. First, for $\chi \in \mathrm{Irr}(G)$, we have $\chi(1) = \sum_\mu d_{\chi\mu}\mu(1)$, and thus

$$\sum_{\mu \in \mathrm{I}_\pi(G)} \Phi_\mu(1)\mu(1) = \sum_{\substack{\mu \in \mathrm{I}_\pi(G) \\ \chi \in \mathrm{Irr}(G)}} d_{\chi,\mu}\chi(1)\mu(1) = \sum_{\chi \in \mathrm{Irr}(G)} \chi(1)^2 = |G|\,.$$

If H is a Hall π-subgroup of G, then $\Phi_\mu(1) = |G : H|\mu(1)_\pi$ by Theorem 3.7, and thus $|G| = \sum_\mu |G : H|\mu(1)\mu(1)_\pi$. We conclude that $|H| = \sum_\mu \mu(1)\mu(1)_\pi$, as required for (a).

The inequality in (b) is now clear, and we see that equality holds if and only if $\mu(1) = \mu(1)_\pi$ for all $\mu \in \mathrm{I}_\pi(G)$. By Theorem 3.17, this happens precisely when G has a normal Hall π'-subgroup. ■

We note that Theorem 3.18(a) also follows easily from the formula

$$\alpha(1) = \sum_{\mu \in \mathrm{I}_\pi(G)} [\mu_H, \alpha]\mu(1)_\pi$$

in Theorem 3.7(b), where $\alpha \in \mathrm{Irr}(H)$ and H is a Hall π-subgroup of G. To see this, observe that $\mu(1) = \sum_\alpha [\mu_H, \alpha]\alpha(1)$, and thus

$$\sum_{\mu \in \mathrm{I}_\pi(G)} \mu(1)\mu(1)_\pi = \sum_{\substack{\mu \in \mathrm{I}_\pi(G) \\ \alpha \in \mathrm{Irr}(H)}} \alpha(1)[\mu_H, \alpha]\mu(1)_\pi = \sum_{\alpha \in \mathrm{Irr}(H)} \alpha(1)^2 = |H|\,,$$

as wanted.

Next, we consider the field $\mathbb{Q}(\varphi)$ generated by the values of an irreducible π-partial character φ of a π-separable group G. Since every member of G^0 is conjugate to an element of a Hall π-subgroup H, it is clear that $\mathbb{Q}(\varphi) = \mathbb{Q}(\varphi_H) \subseteq \mathbb{Q}_{|H|}$. We can do better than this, however: $\mathbb{Q}(\varphi)$ is contained in the field $\mathbb{Q}(\alpha)$, where α is some irreducible character of H.

Theorem 3.19. *Let H be a Hall π-subgroup of a π-separable group G, and let $\varphi \in \mathrm{I}_\pi(G)$. If $\alpha \in \mathrm{Irr}(H)$ is a Fong character for φ, then $\mathbb{Q}(\varphi) \subseteq \mathbb{Q}(\alpha)$.*

Proof. By Galois theory, it suffices to show that every automorphism σ of the cyclotomic field $\mathbb{Q}_{|H|}$ that acts trivially on $\mathbb{Q}(\alpha)$ also acts trivially on $\mathbb{Q}(\varphi)$. Now α is a minimum-degree constituent of φ_H, so $\alpha = \alpha^\sigma$ is a minimum-degree constituent of $(\varphi_H)^\sigma = (\varphi^\sigma)_H$. It should be clear that φ^σ is a π-partial character of G, and it is irreducible since φ is irreducible. By Theorem 3.4(c), we conclude that $\varphi^\sigma = \varphi$, and it follows that σ acts trivially on $\mathbb{Q}(\varphi)$, as required. ■

We close this chapter with the observation that we have proved nothing about the restrictions of irreducible π-partial characters to normal subgroups. Analogues of Clifford's theorem and the Clifford correspondence actually do hold, and these results are quite useful for applications. Unfortunately, their proofs seem to require much more theory than we have so far accumulated. The necessary machinery will be presented in the next two chapters.

Problems

3.1. Let $\chi \in \mathrm{Irr}(G)$, where G is π-separable, and let H be a Hall π-subgroup of G. If α is a minimum-degree irreducible constituent of χ_H, show that α is a Fong character for some member $\varphi \in \mathrm{I}_\pi(G)$. Also, show that every Fong character for φ is a constituent of χ_H.

3.2. Let $K \subseteq G$, where G is π-separable.

(a) Show that there is a Hall π-subgroup H of G such that $K \cap H$ is a Hall π-subgroup of K.

(b) Let $\mu \in \mathrm{I}_\pi(K)$, and let $\beta \in \mathrm{Irr}(K \cap H)$ be a Fong character for μ. Write $\varphi = \mu^G$, and assume that $\varphi \in \mathrm{I}_\pi(G)$. Show that β^H is a Fong character for φ.

3.3. Let $H \subseteq G$ and $\alpha \in \mathrm{Irr}(H)$, and assume that $\alpha = \gamma^H$, where γ is a character of some subgroup $J \subseteq H$. Suppose $J^g \subseteq H$ for some element $g \in G$, and let γ^g be the character of J^g defined by the formula $\gamma^g(x^g) = \gamma(x)$ for elements $x \in J$. Finally, let $\beta = (\gamma^g)^H$.

(a) Show that $\alpha^G = \beta^G$.

(b) Now assume that G is π-separable and H is a Hall π-subgroup of H. Show by example, that β need not be irreducible.

(c) Now assume in (b) that α is a Fong character associated with some π-partial character $\varphi \in \mathrm{I}_\pi(G)$, show that β is also a Fong character for φ, and in particular β is irreducible.

Hint: For (b), take G to be the symmetric group S_4.

3.4. Let G be π-separable, and suppose that χ is a character of G. If χ vanishes on elements of G that are not π-elements, show that χ is a nonnegative integer linear combination of the characters Φ_φ for $\varphi \in \mathrm{I}_\pi(G)$. Deduce that for an arbitrary character χ, the product $\Phi_\mu\chi$ is a nonnegative integer linear combination of the Φ_φ.

3.5. Let G be π-separable, and suppose that $\varphi, \mu \in \mathrm{I}_\pi(G)$. Show that

$$\frac{1}{|G|}\sum_{x\in G^0} \Phi_\varphi(x)\overline{\mu(x)} = \delta_{\varphi,\mu}\,,$$

the Kronecker symbol.

3.6. Let G be π-separable, and let $\alpha \in \mathrm{Irr}(H)$, where H is a Hall π-subgroup of G and α is primitive. Let $H \subseteq W$, where W is maximal such that α extends to W, and let ψ be the unique π-special extension of α to W as in Corollary 3.15. Show that ψ^G is irreducible.

Hint: Use Problem 2.8.

3.7. Let $J \subseteq G$, where G is π-separable and $|G : J|$ is a π'-number. Suppose that $\varphi \in \mathrm{I}_\pi(G)$ and that φ_J is a multiple of $\mu \in \mathrm{I}_\pi(J)$. Show that $\varphi_J = \mu$ and that $(\Phi_\mu)^G = \Phi_\varphi$.

3.8. Suppose that G has a normal Hall π-subgroup. If $\chi \in \mathrm{Irr}(G)$, show that the π-partial character χ^0 is a multiple of some member of $\mathrm{I}_\pi(G)$.

3.9. Let G be π-separable and suppose that for every character $\chi \in \mathrm{Irr}(G)$, the π-partial character χ^0 is a multiple of some member of $\mathrm{I}_\pi(G)$. Show that G has a normal Hall π-subgroup.

Notes and References

One can approach R. Brauer's modular representation theory via several different routes, including ring theory, module theory, and character theory. For many readers of this book, the character point of view may be the most natural, and a good reference for this is G. Navarro's text [**62**].

In several papers, including [**19**] and [**20**], Fong investigated Brauer's modular theory as applied to solvable and p-solvable groups, and he discovered a number of significant results. Swan pointed out in Theorem 6 of [**70**] that Fong's work yields the result we now know as the Fong–Swan theorem.

Brauer's theorem that $|\,\mathrm{IBr}(G)|$ is equal to the number of p-regular classes of G appears as (3A) in [**6**]. The proof given in that paper is ring theoretic, but this result can also be obtained via character theory, as in Theorem 3.1.

The notation $\mathrm{I}_\pi(G)$ appears in [**45**]. This set was initially defined in [**40**], where it is denoted $\mathrm{I}^\pi(G)$, and where the associated Fong characters are discussed. The proofs in [**40**] are quite different from those in this chapter. The presentation in [**45**], which was inspired by [**68**], more closely parallels what we have done here.

The essential content of Corollary (2E) of [**20**] is that if G is p-solvable and H is a Hall p'-subgroup of G, then given $\varphi \in \mathrm{IBr}(G)$, there exists an irreducible character α of H such that $\alpha^G = \Phi_\varphi$. By Theorem 3.7(c), it follows that Fong's character α is exactly what we refer to as a Fong character for φ, and that is the reason we refer to the characters defined in Theorem 3.4 as "Fong characters".

Theorem 3.6 is Theorem B of [**41**]. Also, Corollary 3.15 is part of Proposition 2.1 of that paper and Theorem 3.16 appears there as Theorem A.

Willems conjecture appears in his paper [**73**].

The equivalence of (2) and (3) in Theorem 3.17 is a result of C.-H. Sah, appearing in [**69**].

Problem 3.7 appears as Lemma 3.1 of [**59**].

Chapter 4

The Nucleus and B_π-Characters

4A

Let G be a π-separable group. We saw in Theorem 3.14 that the map $\chi \mapsto \chi^0$ defines a bijection from the set $\mathcal{X}_\pi(G)$ of π-special characters of G onto the subset of $\mathrm{I}_\pi(G)$ consisting of those irreducible π-partial characters φ such that $\varphi(1)$ is a π-number. In this chapter we define a subset $\mathrm{B}_\pi(G)$ of $\mathrm{Irr}(G)$, where $\mathrm{B}_\pi(G)$ contains $\mathcal{X}_\pi(G)$, and such that the map $\chi \mapsto \chi^0$ is a bijection from $\mathrm{B}_\pi(G)$ onto the whole of $\mathrm{I}_\pi(G)$. In other words, $\mathrm{B}_\pi(G)$ is a canonical set of lifts for the irreducible π-partial characters of G.

There are actually several ways to construct specific sets of lifts for $\mathrm{I}_\pi(G)$, but the set $\mathrm{B}_\pi(G)$ that we define here satisfies a useful additional condition: it shares with $\mathcal{X}_\pi(G)$ the property that it respects restriction to normal subgroups. More precisely, if $\chi \in \mathrm{B}_\pi(G)$ and $N \triangleleft G$, then the irreducible constituents of χ_N lie in $\mathrm{B}_\pi(N)$. (This will enable us to use $\mathrm{B}_\pi(G)$ to deduce information about the restrictions of irreducible π-partial characters of G to normal subgroups.) Since we want $\mathrm{B}_\pi(G)$ to respect restriction to normal subgroups, it will also have to respect restriction to subnormal subgroups, and that explains why we must deal with subnormality in the following.

Recall that an irreducible character ψ of a π-separable group G is said to be "π-factored" if we can write $\psi = \alpha\beta$, where α is π-special and β is π'-special. By Theorem 2.9, all primitive characters of π-separable groups are π-factored, and since every irreducible character is induced from a primitive character of a (not necessarily proper) subgroup, it follows that every irreducible character of a π-separable group is induced from a π-factored character. In fact, if $\chi \in \mathrm{Irr}(G)$, there are usually many pairs (W, γ) such that $W \subseteq G$ and $\gamma \in \mathrm{Irr}(W)$, where γ is π-factored and $\gamma^G = \chi$.

We digress briefly to introduce some notation. We say that (W, γ) is a **subgroup pair** in G if $W \subseteq G$ and $\gamma \in \mathrm{Irr}(W)$, and the subgroup pair (W, γ) is π-**factored** if γ is π-factored. Observe that G acts by conjugation on the set of subgroup pairs in G, where by definition $(W, \gamma)^g = (W^g, \gamma^g)$, and $\gamma^g \in \mathrm{Irr}(W^g)$ is defined by the formula $\gamma^g(x^g) = \gamma(x)$ for $x \in W$. It should be clear that the set of those π-factored subgroup pairs (W, γ) in G such that γ^G is equal to a given character $\chi \in \mathrm{Irr}(G)$ is invariant under the conjugation action of G.

Before we present any of the details, we offer a brief outline of the somewhat complicated construction of the set $\mathrm{B}_\pi(G)$, where G is π-separable. First, we select for each character $\chi \in \mathrm{Irr}(G)$ a uniquely determined, unambiguously defined conjugacy class of π-factored subgroup pairs (W, γ) such that $\gamma^G = \chi$. Each subgroup pair in the selected class is said to be a **nucleus** for χ, and if (W, γ) is one of these nuclei, we refer to γ as a **nucleus character** for χ.

By definition, a nucleus character γ for χ is π-factored, and we note that this includes the possibility that γ is π-special. We define $\mathrm{B}_\pi(G)$ to be the set of characters $\chi \in \mathrm{Irr}(G)$ such that a nucleus character for χ is π-special, and we observe that since all nuclei for χ are conjugate, this definition is unambiguous. We shall see that if χ is π-factored, then (G, χ) is the unique nucleus for χ, and it follows that the π-factored characters in $\mathrm{B}_\pi(G)$ are exactly the π-special characters. In particular, $\mathcal{X}_\pi(G)$ is contained $\mathrm{B}_\pi(G)$.

We begin by studying the set $\mathcal{S}(G)$ of π-factored subgroup pairs (S, γ) such that $S \lhd\lhd G$. We are especially interested in the maximal members of the set $\mathcal{S}(G)$, so to make this precise, we define a partial order on the set of subgroup pairs for a group G by setting $(H, \mu) \leq (K, \nu)$ if $H \subseteq K$ and μ lies under ν. In this situation, we sometimes say that (H, μ) lies below (K, ν) and that (K, ν) lies above (H, μ). (It is easy to check that this really does define a partial order.) Our first result concerns the set $\mathcal{S}^*(G)$ of maximal members of $\mathcal{S}(G)$.

Theorem 4.1. *Let $\chi \in \mathrm{Irr}(G)$, where G is π-separable. The following then hold.*

(a) *There exists a unique conjugacy class of pairs $(S, \varphi) \in \mathcal{S}^*(G)$ such that $(S, \varphi) \leq (G, \chi)$.*

(b) *If $(S, \varphi) \in \mathcal{S}(G)$ is maximal with the property that $(S, \varphi) \leq (G, \chi)$, then $(S, \varphi) \in \mathcal{S}^*(G)$.*

Our proof of Theorem 4.1 proceeds via a series of lemmas, but we begin by reviewing a few facts about subnormal subgroups. First, recall that if

$S \lhd\lhd G$ and S is a π-group, then $S \subseteq \mathbf{O}_\pi(G)$. (This follows easily by induction on $|G|$ because if $S \subseteq M \lhd G$ with $M < G$, then $S \subseteq \mathbf{O}_\pi(M) \subseteq \mathbf{O}_\pi(G)$, where the first containment follows by the inductive hypothesis and the second holds because $\mathbf{O}_\pi(M)$ is characteristic in M and hence is normal in G.) Another easy observation is that if $S \lhd\lhd G$ and $S < G$, then there exists a subgroup $U \lhd\lhd G$ such that $S \lhd U$ and U/S is a simple group. (To see this, take $U \lhd\lhd G$ minimal such that $S < U$.) Much less trivial is the fact that if $S, T \lhd\lhd G$, then $\langle S, T\rangle \lhd\lhd G$. (This result of H. Wielandt appears as Theorem 2.5 of [**FGT**].)

Recall from Chapter 2 that if $N \lhd G$, where G is π-separable, and if $\alpha, \beta \in \operatorname{Irr}(G)$ are π-special and π'-special, respectively, then each irreducible constituent of $(\alpha\beta)_N$ is the product of an irreducible constituent of α_N with an irreducible constituent of β_N. In particular, if $\chi \in \operatorname{Irr}(G)$ is π-factored, then all irreducible constituents of χ_N are π-factored, and similarly for restrictions to subnormal subgroups. The following results show that under certain conditions, knowing that some irreducible constituent of χ_N is π-factored is sufficient to guarantee that χ is π-factored.

Lemma 4.2. *Let $N \lhd G$, where G is π-separable, and suppose that G/N is a π'-group. Let $\alpha, \beta \in \operatorname{Irr}(N)$ be π-special and π'-special, respectively, and assume that α is invariant in G. Then every member of* $\operatorname{Irr}(G|\alpha\beta)$ *is π-factored.*

Proof. Theorem 2.4 guarantees that α has a π-special extension $\hat{\alpha} \in \operatorname{Irr}(G)$, and also all irreducible constituents γ of β^G are π'-special. Now $(\alpha\beta)^G = (\hat{\alpha}_N\beta)^G = \hat{\alpha}\beta^G$, and this character is a sum of factored irreducible characters of the form $\hat{\alpha}\gamma$, where γ runs over the irreducible constituents of β^G. ∎

Next, we present a generalization of Lemma 4.2. (This result appears as Problem 2.1.)

Lemma 4.3. *Let $N \lhd G$, where G is π-separable, and suppose $G/N = A/N \times B/N$, where A/N is a π-group and B/N is a π'-group. Let $\alpha, \beta \in \operatorname{Irr}(N)$ be π-special and π'-special, respectively, and assume that α is invariant in B and β is invariant in A. Then every member of* $\operatorname{Irr}(G|\alpha\beta)$ *is π-factored.*

Proof. Let $\chi \in \operatorname{Irr}(G)$ lie over $\alpha\beta$, and let $\psi \in \operatorname{Irr}(B)$ lie under χ and over $\alpha\beta$. Since α is invariant in B and B/N is a π'-group, it follows by Lemma 4.2 that ψ is π-factored, and we let γ be the π'-special factor of ψ. Now $G/B \cong A/N$ is a π-group, so if we can show that γ is invariant in G, it will follow from Lemma 4.2 (with the roles of π and π' reversed) that every irreducible constituent of ψ^G is π-factored, and in particular that χ is π-factored, as required.

To show that γ is invariant in G, observe first that since $\alpha\beta$ is a constituent of ψ_N, it follows that β is a constituent of γ_N. Now β has a canonical extension $\hat{\beta} \in \mathrm{Irr}(A)$, and γ is an irreducible constituent of $\beta^B = (\hat{\beta}_N)^B = (\hat{\beta}^G)_B$. Then γ lies under some character $\delta \in \mathrm{Irr}(G)$ such that δ lies over $\hat{\beta}$. Since $G/A \cong B/N$ is a π'-group and $\hat{\beta}(1) = \beta(1)$ is a π'-number, it follows that $\delta(1)$ is a π'-number. Also, $G/B \cong A/N$ is a π-group, and thus δ_B is irreducible. It follows that $\delta_B = \gamma$, and hence γ is G-invariant, as required. ■

Lemma 4.4. *Let G be π-separable, and suppose that $G = \langle A, B\rangle$, where A and B are subnormal in G. Let $N \subseteq A \cap B$, and assume that $N \triangleleft G$ and that each of A/N and B/N is either a π-group or a π'-group. Let $\xi \in \mathrm{Irr}(A)$ and $\eta \in \mathrm{Irr}(B)$ be π-factored, and assume that η_N and ξ_N have a common irreducible constituent θ. Then every member of $\mathrm{Irr}(G|\theta)$ is π-factored.*

Proof. Since $N \triangleleft G$ and θ lies under the π-factored characters ξ and η, we see that θ is π-factored, and we write $\theta = \alpha\beta$, where $\alpha, \beta \in \mathrm{Irr}(N)$ are π-special and π'-special, respectively. We can assume without loss that B/N is a π'-group, and thus the restriction to N of the π-special factor of η is irreducible. It follows that this restriction is equal to α, and thus α is invariant in B. If A/N is also a π'-group, similar reasoning shows that α is invariant in A, and thus α is invariant in $G = \langle A, B\rangle$. In this case, both A/N and B/N are contained in $\mathbf{O}_{\pi'}(G/N)$, and thus G/N is a π'-group. By Lemma 4.2, therefore, all members of $\mathrm{Irr}(G|\theta)$ are π-factored.

We can now assume that A/N is a π-group. Then $A/N \subseteq \mathbf{O}_\pi(G/N)$ and $B/N \subseteq \mathbf{O}_{\pi'}(G/N)$, and thus A/N and B/N centralize each other. Also, $N = A \cap B$, and since $G = \langle A, B\rangle$, we have $G/N = A/N \times B/N$. We saw previously that α is invariant in B, and similar reasoning shows that β is invariant in A. The result thus follows by Lemma 4.3. ■

Corollary 4.5. *Let $N \triangleleft G$, where G is π-separable, and suppose that G/N is either a π-group or a π'-group. Let $\chi \in \mathrm{Irr}(G)$ be π-factored, and let θ be an irreducible constituent of χ_N. Then every member of $\mathrm{Irr}(G|\theta)$ is π-factored.*

Proof. Take $A = G$ and $B = N$ in Lemma 4.4, with $\xi = \chi$ and $\eta = \theta$. ■

Next, we need a few observations about the sets $\mathcal{S}(G)$ and $\mathcal{S}^*(G)$.

Lemma 4.6. *Let $(D, \delta) \in \mathcal{S}(G)$, where G is π-separable, and assume that $(D, \delta) \notin \mathcal{S}^*(G)$. Then there exists a pair $(T, \tau) \in \mathcal{S}(G)$ such that $(D, \delta) < (T, \tau)$ and $D \triangleleft T$, where T/D is a simple group.*

Proof. By assumption, (D, δ) is not maximal in $\mathcal{S}(G)$, so we can choose $(T, \tau) \in \mathcal{S}(G)$ minimal such that $(D, \delta) < (T, \tau)$. Then $D < T$ and $D \triangleleft\triangleleft T$, and thus there exists a subgroup U such that $D < U \triangleleft\triangleleft T$, where $D \triangleleft U$ and

U/D is simple. Choose $\mu \in \mathrm{Irr}(U)$ lying over δ and under τ, and observe that μ is π-factored, and thus $(U,\mu) \in \mathcal{S}(G)$. Then $(D,\delta) < (U,\mu) \leq (T,\tau)$, and so $(T,\tau) = (U,\mu)$ by the minimality of (T,τ). Then $D \lhd T$ and T/D is simple, as required. ∎

Lemma 4.7. *Let $(S,\varphi) \in \mathcal{S}^*(G)$, where G is π-separable, and let $M \lhd G$, where G/M is either a π-group or a π'-group. Write $D = S \cap M$, and let $\delta \in \mathrm{Irr}(D)$ lie under φ. Then $(D,\delta) \in \mathcal{S}^*(M)$.*

Proof. First, note that $D \lhd S$ and $S/D \cong SM/M$, so S/D is either a π-group or a π'-group. Also, since φ is π-factored and $D \lhd S$, we see that δ is π-factored, and thus $(D,\delta) \in \mathcal{S}(M)$.

Assume now that (D,δ) is not maximal in $\mathcal{S}(M)$. Then by Lemma 4.6, there exists a pair $(T,\tau) \in \mathcal{S}(M)$ such that $(D,\delta) < (T,\tau)$ and $D \lhd T$ with T/D simple, and in particular, T/D is either a π-group or a π'-group.

Let $H = \langle S,T\rangle$, and note that $H \lhd\lhd G$. Also, $D \lhd S$ and $D \lhd T$, and thus $D \lhd H$. Now $\varphi \in \mathrm{Irr}(S)$ and $\tau \in \mathrm{Irr}(T)$ are π-factored, and δ is a common irreducible constituent of φ_D and τ_D. By Lemma 4.4, therefore, every member of $\mathrm{Irr}(H|\delta)$ is π-factored, and in particular, if we choose $\theta \in \mathrm{Irr}(H)$ lying over φ, then θ is π-factored. Then $(S,\varphi) \leq (H,\theta) \in \mathcal{S}(G)$, and since (S,φ) is maximal in $\mathcal{S}(G)$, we deduce that $S = H$. Then $T \subseteq H = S$, so $T \subseteq S \cap M = D$, and this is a contradiction. ∎

Proof of Theorem 4.1. First, we observe that there do exist members $(S,\varphi) \in \mathcal{S}(G)$ that lie below (G,χ). (For example, we could take S to be the trivial group.) To prove (b), let (S,φ) be maximal among members of $\mathcal{S}(G)$ that lie below (G,χ), and suppose that $(S,\varphi) \notin \mathcal{S}^*(G)$. Then by Lemma 4.6, there exists a pair $(T,\tau) \in \mathcal{S}(G)$ such that $(S,\varphi) < (T,\tau)$, where $S \lhd T$ and T/S is simple. In particular, T/S is a π-group or a π'-group, and since τ is π-factored, it follows by Corollary 4.5 that every member of $\mathrm{Irr}(T|\varphi)$ is π-factored. Now χ lies over φ, and thus χ lies over some character $\mu \in \mathrm{Irr}(T)$, where μ lies over φ. Then $(T,\mu) \in \mathcal{S}(G)$ and $(S,\varphi) < (T,\mu) \leq (G,\chi)$. This contradicts the maximality of (S,φ), and we conclude that $(S,\varphi) \in \mathcal{S}^*(G)$, as desired. In particular, this proves (b), and it establishes existence in (a).

Now fix $(S,\varphi) \in \mathcal{S}^*(G)$ with $(S,\varphi) \leq (G,\chi)$. To prove uniqueness in (a), we show by induction on $|G|$ that if $(T,\theta) \in \mathcal{S}(G)$ and $(T,\theta) \leq (G,\chi)$, then there exists $g \in G$ such that $(T,\theta)^g \leq (S,\varphi)$. If $T = G$, then $(T,\theta) = (G,\chi)$, and this pair lies above (S,φ). Since $(S,\varphi) \in \mathcal{S}^*(G)$, however, it follows that $(T,\theta) = (S,\varphi)$, and there is nothing further to prove. We may assume, therefore, that $T < G$, and because $T \lhd\lhd G$, we can choose a maximal normal subgroup M of G with $T \subseteq M$, and we observe that G/M is either a π-group or a π'-group. Let $\psi \in \mathrm{Irr}(M)$ lie under χ and over θ.

Let $D = S \cap M$, and let $\delta \in \mathrm{Irr}(D)$ lie under φ. Then δ lies under χ, and so some irreducible constituent ψ^x of χ_M lies over δ, where $x \in G$. We have $(D, \delta) \le (M, \psi^x)$, and furthermore, $(D, \delta) \in \mathcal{S}^*(M)$ by Lemma 4.7. Also $(T, \theta)^x \le (M, \psi^x)$ and $(T, \theta)^x \in \mathcal{S}(M)$, so by the inductive hypothesis, $(T, \theta)^{xm} \le (D, \delta) \le (S, \varphi)$ for some element $m \in M$. We can thus take $g = xm$, and the proof is complete. ■

We can now deduce a strong form of Lemma 4.4. This will not be needed in what follows, however.

Corollary 4.8. *Let G be π-separable, and suppose $G = \langle A, B\rangle$, where A and B are subnormal in G. Let $\chi \in \mathrm{Irr}(G)$, and assume that at least one irreducible constituent of each of χ_A and χ_B is π-factored. Then χ is π-factored.*

Proof. By hypothesis, there exist pairs (A, ξ) and (B, η) in $\mathcal{S}(G)$ such that $(A, \xi) \le (G, \chi)$ and $(B, \eta) \le (G, \chi)$. We can assume that each of these pairs is maximal among members of $\mathcal{S}(G)$ lying below (G, χ), and thus by Theorem 4.1(b), they lie in $\mathcal{S}^*(G)$. By Theorem 4.1, these pairs are conjugate in G, and in particular A and B are conjugate in G. Since $G = \langle A, B\rangle$, we see that A cannot can be contained in a proper normal subgroup of G. We deduce from the subnormality of A that $A = G$, and thus $\chi = \xi$ is π-factored. ■

We introduce a little more notation now. If $\varphi \in \mathrm{Irr}(S)$, where $S \subseteq G$, we write $\mathbf{N}_G(S, \varphi)$ to denote the stabilizer in G of the pair (S, φ). Then $\mathbf{N}_G(S, \varphi)$ is exactly the stabilizer of φ in $\mathbf{N}_G(S)$, and thus in the case where $S \lhd G$, the Clifford correspondence implies that induction defines a bijection $\mathrm{Irr}(T|\varphi) \to \mathrm{Irr}(G|\varphi)$, where $T = \mathbf{N}_G(S, \varphi)$. We show next that if G is π-separable, there is a similar bijection if $(S, \varphi) \in \mathcal{S}^*(G)$, even without assuming that S is normal.

Theorem 4.9. *Let $(S, \varphi) \in \mathcal{S}^*(G)$, where G is π-separable, and let $T = \mathbf{N}_G(S, \varphi)$. Then induction defines a bijection $\mathrm{Irr}(T|\varphi) \to \mathrm{Irr}(G|\varphi)$.*

Proof. We will show that induction defines an injective map from $\mathrm{Irr}(T|\varphi)$ into $\mathrm{Irr}(G|\varphi)$. Once that is established, it will follow that this map is surjective. To see this, observe that if $\chi \in \mathrm{Irr}(G|\varphi)$ and $\psi \in \mathrm{Irr}(T|\varphi)$ lies under χ, then ψ^G is irreducible, and hence $\psi^G = \chi$.

We proceed by induction on $|G|$. If $S = G$, there is nothing to prove, so we can assume that $S < G$, and since S is subnormal, we can choose $M \lhd G$ such that $S \subseteq M < G$, and we observe that $(S, \varphi) \in \mathcal{S}^*(M)$. Suppose first that $MT < G$. By the inductive hypothesis, induction defines an injection from $\mathrm{Irr}(T|\varphi)$ into $\mathrm{Irr}(MT|\varphi)$, so it suffices to show that induction defines an injection from $\mathrm{Irr}(MT|\varphi)$ into $\mathrm{Irr}(G)$.

Let $\theta \in \mathrm{Irr}(M)$ lie over φ, and let I be the stabilizer of θ in G. We argue that if $g \in G$ is an element such that θ^g lies over φ, then $g \in MT$, and in particular $I \subseteq MT$. To see this, observe that $(S,\varphi) \leq (M,\theta^g)$, and also $(S,\varphi)^g \leq (M,\theta^g)$ since $(S,\varphi) \leq (M,\theta)$. Both (S,φ) and $(S,\varphi)^g$ lie in $\mathcal{S}^*(M)$, however, and it follows by Theorem 4.1 that there exists $m \in M$ such that $(S,\varphi)^{gm} = (S,\varphi)$. Then $gm \in T$, and hence $g \in MT$, as claimed.

Now let $\psi \in \mathrm{Irr}(MT|\varphi)$ and choose $\theta \in \mathrm{Irr}(M)$ lying below ψ and above φ. By the result of the previous paragraph, we have $I \subseteq MT$, where I is the stabilizer of θ in G, and thus by the Clifford correspondence there exists $\alpha \in \mathrm{Irr}(I|\theta)$ such that $\psi = \alpha^{MT}$. Then $\psi^G = \alpha^G$, and since $M \triangleleft G$, this character is irreducible by the Clifford correspondence in G. This shows that induction defines a map from $\mathrm{Irr}(MT|\varphi)$ to $\mathrm{Irr}(G)$, as wanted.

To prove injectivity, write $\chi = \psi^G$, where as above $\psi \in \mathrm{Irr}(MT|\theta)$ and θ lies over φ. Suppose that also $\eta^G = \chi$, where $\eta \in \mathrm{Irr}(MT|\varphi)$. Since ψ lies over θ, we see that χ lies over θ, and it follows that the irreducible constituents of η_M are conjugate to θ in G. Since η lies over φ, at least one of the constituents of η_M lies over φ, so we may assume that θ^g lies over φ and under η. As we have seen, this implies that $g \in MT$, and since η lies over θ^g, we see that η lies over θ. By the Clifford correspondence, therefore, there exists $\beta \in \mathrm{Irr}(I|\theta)$ such that $\eta = \beta^{MT}$, and thus $\beta^G = \eta^G = \psi^G = \alpha^G$. It follows by the injectivity of the Clifford correspondence in G that $\alpha = \beta$, and thus $\eta = \beta^{MT} = \alpha^{MT} = \psi$, as required.

We can now assume that $MT = G$. Write $D = T \cap M$, and observe that $D \triangleleft T$ and $D = \mathbf{N}_M(S,\varphi)$. Since $(S,\varphi) \in \mathcal{S}^*(M)$, it follows by the inductive hypothesis that induction defines an injection from $\mathrm{Irr}(D|\varphi)$ to $\mathrm{Irr}(M)$. Now let $\mu \in \mathrm{Irr}(D|\varphi)$, and write $\theta = \mu^M$. Then θ is irreducible, and μ is the unique member of $\mathrm{Irr}(D|\varphi)$ that induces θ. Since T stabilizes D and φ, it follows from the uniqueness of μ that the stabilizer of θ in T stabilizes μ, and thus Lemma 2.11(b) applies, and we conclude that induction defines an injection from $\mathrm{Irr}(T|\mu)$ to $\mathrm{Irr}(G|\theta)$.

Now let $\psi \in \mathrm{Irr}(T|\varphi)$. Then ψ lies over some character $\mu \in \mathrm{Irr}(D|\varphi)$, and thus by the result of the previous paragraph the character $\chi = \psi^G$ is irreducible and lies over δ^M. To complete the proof, we assume that also $\chi = \eta^G$, where $\eta \in \mathrm{Irr}(T|\varphi)$, and we show that $\psi = \eta$.

We have seen that if $\mu \in \mathrm{Irr}(D|\varphi)$ lies below ψ, then μ^M is irreducible and lies below $\psi^G = \chi$. Similarly, if $\nu \in \mathrm{Irr}(D|\varphi)$ lies below η, then ν^M is irreducible and lies below $\eta^G = \chi$. Since χ is irreducible, μ^M is conjugate to ν^M in G and because $G = MT$, we can write $\nu^M = (\mu^M)^t = (\mu^t)^M$ for some element $t \in T$. Both ν and μ^t lie in $\mathrm{Irr}(D|\varphi)$, and it follows that $\nu = \mu^t$

because the induction map $\mathrm{Irr}(D|\varphi) \to \mathrm{Irr}(M)$ is injective. Since η lies over ν, it also lies over μ, and thus $\eta = \psi$ by the injectivity of the induction map from $\mathrm{Irr}(D|\delta)$ to $\mathrm{Irr}(G)$. ∎

We need one more preliminary result.

Lemma 4.10. *Let G be π-separable, and suppose $(S, \varphi) \in \mathcal{S}^*(G)$. If $\mathbf{N}_G(S, \varphi) = G$, then $S = G$.*

Proof. If $S < G$, then there exists a subgroup $U \lhd\lhd G$ such that $S \lhd U$ and U/S is simple. In particular, $S < U$ and U/S is either a π-group or a π'-group. Also, φ is invariant in U, and since φ is π-factored, it follows by Lemma 4.2 that every member of $\mathrm{Irr}(U|\varphi)$ is π-factored, and thus if $\theta \in \mathrm{Irr}(U|\varphi)$, we have $(S, \varphi) < (U, \theta) \in \mathcal{S}(G)$. This is a contradiction because by hypothesis, $(S, \varphi) \in \mathcal{S}^*(G)$. ∎

We can now present the definition toward which we have been working. Given a character $\chi \in \mathrm{Irr}(G)$, where G is π-separable, a **nucleus** for χ is any subgroup pair (W, γ) constructed inductively as in the following paragraph. Also, if (W, γ) is a nucleus for χ, we say that γ is a **nucleus character** for χ.

If χ is π-factored, take $(W, \gamma) = (G, \chi)$. Assuming now that χ is not π-factored, choose $(S, \varphi) \in \mathcal{S}^*(G)$ such that $(S, \varphi) \le (G, \chi)$. (This is possible by Theorem 4.1.) Let $T = \mathbf{N}_G(S, \varphi)$, and observe that $T < G$, since otherwise, Lemma 4.10 would imply that $S = G$, and thus $\chi = \varphi$ is π-factored, which is not the case. By Theorem 4.9, there exists a unique character $\psi \in \mathrm{Irr}(T|\varphi)$ such that $\psi^G = \chi$, and since $T < G$, the inductive hypothesis guarantees that a nucleus for ψ has already been constructed. Let (W, γ) be a nucleus for ψ.

Since we describe certain characters as being "π-factored" and not just "factored", we admit to being inconsistent when we refer to (W, γ) simply as a "nucleus" and not as a "π-nucleus". Our notation does have one advantage, however: it stresses the symmetry between π and π'. Just as a character is π-factored if and only if it is π'-factored, our definition of a "nucleus" is unchanged if π and π' are interchanged.

Corollary 4.11. *Let $\chi \in \mathrm{Irr}(G)$, where G is π-separable. The following then hold.*

(a) *All nuclei for χ are conjugate in G.*

(b) *If (W, γ) is a nucleus for χ, then the nucleus character γ is π-factored, and $\gamma^G = \chi$.*

(c) *Let $(S, \varphi) \in \mathcal{S}^*(G)$ with $(S, \varphi) \le (G, \chi)$, and let $T = \mathbf{N}_G(S, \varphi)$. Then there exists a nucleus (W, γ) for χ such that $(S, \varphi) \le (W, \gamma)$*

and $W \subseteq T$. *Also,* (W, γ) *is a nucleus for the unique character* $\psi \in \operatorname{Irr}(T|\varphi)$ *such that* $\psi^G = \chi$.

Proof. If χ is π-factored, then (G, χ) is the unique nucleus for χ, and in this case, all three assertions are clear. We can assume, therefore, that χ is not π-factored, and we proceed by induction on $|G|$.

There are precisely two points in the construction of a nucleus for χ where it was necessary to make a choice. We choose $(S, \varphi) \in \mathcal{S}^*(G)$ lying below (G, χ), and we choose a nucleus for ψ, working in the group $T = \mathbf{N}_G(S, \varphi) < G$. By Theorem 4.1, all members of $\mathcal{S}^*(G)$ lying below (G, χ) are conjugate in G, and thus the ambiguity introduced by the choice of (S, φ) cannot affect the G-conjugacy class of the constructed nucleus. Also, by the inductive hypothesis applied in T, the choice of a nucleus for ψ is unambiguous up to conjugacy in T, and it follows that the final result of the construction is unambiguous up to conjugation in G. This proves (a).

Now let (S, φ) and (T, ψ) be as in the construction of a nucleus (W, γ) for χ. In particular, $(S, \varphi) \in \mathcal{S}^*(G)$ lies below (G, χ) and $T = \mathbf{N}_G(S, \varphi)$. Also, $\psi \in \operatorname{Irr}(T|\varphi)$ with $\psi^G = \chi$ and (W, γ) is a nucleus for ψ. By Lemma 4.10, we see that $T < G$, so by the inductive hypothesis (b) holds for the group T, and it follows that γ is π-factored and that $\gamma^T = \psi$. Then $\gamma^G = \psi^G = \chi$, proving (b).

For (c), observe that $(S, \varphi) \leq (T, \psi)$ and $(S, \varphi) \in \mathcal{S}(T)$. By Theorem 4.1(b), there exists $(U, \theta) \in \mathcal{S}^*(T)$ such that $(S, \varphi) \leq (U, \theta) \leq (T, \psi)$. Since (c) holds in T by the inductive hypothesis, it follows that there exists a nucleus (W, γ) for ψ such that $(U, \theta) \leq (W, \gamma)$, and thus $(S, \varphi) \leq (W, \gamma)$ and $W \subseteq T = \mathbf{N}_G(S, \varphi)$. As (W, γ) is also a nucleus for χ, this establishes (c). ∎

Suppose that $N \subseteq \ker(\chi)$, where $\chi \in \operatorname{Irr}(G)$ and G is π-separable, and let (W, γ) be a nucleus for χ. Since $\gamma^G = \chi$ by Corollary 4.11, it follows that $N \subseteq W$ and $N \subseteq \ker(\gamma)$. In fact, it is not hard to see that if we view χ as a character of G/N, then $(W/N, \gamma)$ is a nucleus for χ. We will not present a formal proof of this, but we mention that the key observation is that if $(S, \varphi) \in \mathcal{S}^*(G)$ and $(S, \varphi) \leq (G, \chi)$, then $N \subseteq S$ and $N \subseteq \ker(\varphi)$. This follows by the maximality of (S, φ) because χ lies over a π-factored character $\theta \in \operatorname{Irr}(SN)$, where $\theta_S = \varphi$ and $N \subseteq \ker(\theta)$.

4B

Given a π-separable group G, we define $\mathrm{B}_\pi(G)$ to be the set of those characters $\chi \in \operatorname{Irr}(G)$ such that a nucleus character for χ is π-special. (This is unambiguous because by Corollary 4.11(a) all nuclei for χ are conjugate in G.)

In order to clarify the connections between $\mathrm{B}_\pi(G)$ and the set $\mathcal{X}_\pi(G)$ of π-special characters of G, we offer the following.

Theorem 4.12. *Let $\chi \in \mathrm{Irr}(G)$, where G is π-separable. The following are equivalent.*

(1) *χ is π-special.*

(2) *$\chi \in \mathrm{B}_\pi(G)$ and $\chi(1)$ is a π-number.*

(3) *$\chi \in \mathrm{B}_\pi(G)$ and χ is π-factored.*

In particular, the principal character 1_G of an arbitrary π-separable group G is π-special, and hence it must lie in $\mathrm{B}_\pi(G)$. Of course, 1_G also lies in $\mathrm{B}_{\pi'}(G)$, and we note that in fact, $\mathrm{B}_\pi(G) \cap \mathrm{B}_{\pi'}(G) = \{1_G\}$. This is true because if $\chi \in \mathrm{B}_\pi(G) \cap \mathrm{B}_{\pi'}(G)$, then a nucleus character γ for χ must be both π-special and π'-special. (We are, of course, using the fact here that the definition of a nucleus is unchanged if π and π' are interchanged.) Then γ is principal, and since $\chi = \gamma^G$, it follows that χ is principal.

We need an easy preliminary result for our proof of Theorem 4.12.

Lemma 4.13. *Let $\chi \in \mathrm{B}_\pi(G)$, where G is π-separable, and let $U \lhd\lhd G$. Then every π-factored irreducible constituent θ of χ_U is π-special.*

Proof. We have $(U, \theta) \in \mathcal{S}(G)$ and $(U, \theta) \leq (G, \chi)$. By Theorem 4.1, there exists a pair $(S, \varphi) \in \mathcal{S}^*(G)$ such that $(U, \theta) \leq (S, \varphi) \leq (G, \chi)$, and by Corollary 4.11(c), there exists a nucleus (W, γ) for χ such that $(S, \varphi) \leq (W, \gamma)$. By assumption, $\chi \in \mathrm{B}_\pi(G)$, so the nucleus character γ is π-special, and hence θ is π-special because $U \lhd\lhd W$. ∎

Proof of Theorem 4.12. First, assume (1), and observe that since χ is π-special, $\chi(1)$ is a π-number. Also, χ is π-factored, and thus (G, χ) is a nucleus for χ. Then $\chi \in \mathrm{B}_\pi(G)$ because its nucleus character χ is π-special.

Now assume (2). To prove (3), it suffices to show that χ is π-factored, so we assume the contrary and derive a contradiction. Let $(S, \varphi) \in \mathcal{S}^*(G)$ with $(S, \varphi) \leq (G, \chi)$, and note that $S < G$ since χ is not π-factored. Also, φ is π-special by Lemma 4.13.

Let $U \lhd\lhd G$, where $S \lhd U$ and U/S is simple, and note that U/S is either a π-group or a π'-group. Now $U < S$, and hence by the maximality of (S, φ), no member of $\mathrm{Irr}(U|\varphi)$ can be π-factored, and in particular, $\mathrm{Irr}(U|\varphi)$ contains no π-special character. Since φ is π-special, we conclude by Theorem 2.4(a) that U/S cannot be a π-group. Then U/S is a π'-group, and it follows by Theorem 2.4(b) that φ is not invariant in U.

Let $T = \mathbf{N}_G(S, \varphi)$ and $N = \mathbf{N}_G(S)$, and observe that $T \subseteq N$. Also, $|G : T|$ is a π-number because by assumption $\chi(1)$ is a π-number, and we know that χ is induced from some character of T. Then $|(N/S) : (T/S)|$ is a π-number, and hence T/S contains every normal π'-subgroup of N/S. Since U/S is a subnormal π'-subgroup of N/S, it is contained in some normal π'-subgroup, and therefore $U/S \subseteq T/S$ and we have $U \subseteq T$. This is a contradiction, however, since φ is not invariant in U.

To see that (3) implies (1), observe that if χ is π-factored, then χ is its own nucleus character, and thus if χ lies in $\mathrm{B}_\pi(G)$, it must be π-special. ■

Note that if G is a π'-group, then $\mathrm{B}_\pi(G) = \{1_G\}$. To see this, observe that every member of $\mathrm{Irr}(G)$ is π'-special, and hence is π-factored, and so if $\chi \in \mathrm{B}_\pi(G)$, then χ is π-special by Theorem 4.12, and thus $\chi = 1_G$. At the other extreme, if G is a π-group, then every irreducible character of G is π-special and hence lies in $\mathrm{B}_\pi(G)$.

In fact, Corollary 4.23, asserts that G is a π'-group if and only if $\mathrm{B}_\pi(G) = \{1_G\}$, and G is a π-group if and only if $\mathrm{B}_\pi(G) = \mathrm{Irr}(G)$. In order to prove that result, however, we must first study the fairly deep connections between the sets $\mathrm{B}_\pi(G)$ and $\mathrm{B}_\pi(N)$ for normal subgroups N of G. Before we begin that investigation, however, we mention that the necessary and sufficient conditions of Corollary 4.23 can be viewed as trivial consequences of the fact (which we prove in Chapter 5) that $|\mathrm{B}_\pi(G)|$ is equal to the number of classes of π-elements of G. This observation, however, does not yield a quick proof of Corollary 4.23 because the results of Chapter 5 also depend on the normal-subgroup machinery that we are about to present.

Suppose now that $N \lhd G$, and assume that G/N is a π-group. Let $\chi \in \mathrm{Irr}(G)$ and $\theta \in \mathrm{Irr}(N)$, and suppose that χ lies over θ. We know that in this situation χ is π-special if and only if θ is π-special, and our next goal is to prove the analogous result for B_π-characters.

Theorem 4.14. *Let $N \lhd G$, where G/N is a π-group and G is π-separable, and let $\psi \in \mathrm{Irr}(N)$ and $\chi \in \mathrm{Irr}(G|\psi)$. Then $\chi \in \mathrm{B}_\pi(G)$ if and only if $\psi \in \mathrm{B}_\pi(N)$.*

We shall need several preliminary results, and we begin with a general character-theoretic fact.

Lemma 4.15. *Let A act via automorphisms on G, and suppose that $N \lhd G$ is A-invariant. Assume that the action of A on G/N is trivial and that A stabilizes a character $\delta \in \mathrm{Irr}(N)$. The following then hold.*

(a) *The derived subgroup A' of A acts trivially on $\mathrm{Irr}(G|\delta)$.*

(b) *If $|A|$ and $|G : N|$ are relatively prime, then A acts trivially on $\mathrm{Irr}(G|\delta)$.*

Proof. By the Clifford correspondence, every member of $\mathrm{Irr}(G|\delta)$ is induced from a member of $\mathrm{Irr}(T|\delta)$, where T is the stabilizer of δ in G. Also, T is A-invariant, and the action of A on T/N is trivial. It is no loss, therefore, to replace G by T, and so we can assume that δ is invariant in G.

Let $g \in G$, and write $C = N\langle g\rangle$, so C/N is cyclic. Also, C is A-invariant and A acts trivially on C/N. To show that an element $a \in A$ fixes a character $\chi \in \mathrm{Irr}(G|\delta)$, we must show that $\chi(g^a) = \chi(g)$ for every choice of g. Since χ_N is a multiple of δ, we see that χ_C is a sum of characters in $\mathrm{Irr}(C|\delta)$, so to prove that a fixes χ, it suffices to show that a fixes all members of $\mathrm{Irr}(C|\delta)$. It is thus no loss to replace G by C, and so we can assume that G/N is cyclic.

Corollary 1.14 guarantees that δ has an extension $\theta \in \mathrm{Irr}(G)$. By the Gallagher correspondence, therefore, every member of $\mathrm{Irr}(G|\delta)$ has the form $\lambda\theta$ for some (necessarily linear) character λ of G/N. Also, each of these characters λ is A-invariant because A acts trivially on G/N. To show that an element $a \in A$ fixes all members of $\mathrm{Irr}(G|\delta)$, therefore, it suffices to show that a fixes θ.

If $a \in A$, then θ^a lies in $\mathrm{Irr}(G|\delta)$, so by the Gallagher correspondence, $\theta^a = \lambda_a\theta$ for some uniquely determined linear character λ_a of G/N, where λ_a depends on a. This defines a map $f : A \to \mathrm{Irr}(G/N)$, where $f(a) = \lambda_a$, and we argue that f is actually a group homomorphism. To see this, let $x, y \in A$. Then

$$\lambda_{xy}\theta = \theta^{xy} = (\theta^x)^y = (\lambda_x\theta)^y = \lambda_x\theta^y = \lambda_x\lambda_y\theta\,,$$

where the penultimate equality holds because λ_x is A-invariant. It follows by the uniqueness in the Gallagher correspondence that

$$f(xy) = \lambda_{xy} = \lambda_x\lambda_y = f(x)f(y)\,,$$

and thus f is a homomorphism from A into $\mathrm{Irr}(G/N)$, as claimed.

Since the group $\mathrm{Irr}(G/N)$ is abelian, $f(A')$ is trivial, and thus A' fixes θ, proving (a). For (b), we are assuming that $|A|$ and $|G:N|$ are relatively prime, and it follows that $f(A)$ is trivial, and thus A fixes θ, as required. ∎

Lemma 4.16. *Let $(S,\varphi) \in \mathcal{S}^*(G)$, where G is π-separable. Suppose that $(D,\delta) \le (S,\varphi)$, where $D \lhd S$ and S/D is either a π-group or a π'-group, and let $J = \mathbf{N}_G(D,\delta)$. Then $J \subseteq \mathbf{N}_G(S)$.*

Proof. Let $X = \langle S^t \mid t \in J\rangle$, and note that J normalizes X. Also, $X \lhd\lhd G$ since each of the subgroups S^t is subnormal in G. Now $D \lhd S$ and J normalizes D, and thus $D = D^t \lhd S^t$ for all $t \in J$, and it follows that $D \lhd X$.

We can assume that S/D is a π'-group, and thus each of the groups S^t/D for $t \in J$ is a subnormal π'-subgroup of X/D. It follows that $S^t/D \subseteq \mathbf{O}_{\pi'}(X/D)$ for $t \in J$, and hence X/D is a π'-group.

Now φ is π-factored and $D \triangleleft S$, and thus δ is π-factored. Furthermore, the π-special factor α of δ is the restriction to D of the π-special factor of φ, and hence α is invariant in S. Also, J stabilizes α because J stabilizes φ, and thus α is invariant in S^t for all $t \in J$, and we conclude that α is invariant in X. Since X/D is a π'-group, we deduce by Lemma 4.2 that every member of $\mathrm{Irr}(X|\delta)$ is π-factored.

Now let $\theta \in \mathrm{Irr}(X|\varphi)$. Then θ lies over δ, so θ is π-factored, and hence $(S, \varphi) \leq (X, \theta) \in \mathcal{S}(G)$. Since $(S, \varphi) \in \mathcal{S}^*(G)$, we deduce that $S = X$, and thus J normalizes S, as required. ∎

Lemma 4.17. *Let $N \triangleleft G$, where G is π-separable and G/N is a π-group, and let $(S, \varphi) \in \mathcal{S}^*(G)$. Let $D = S \cap N$, and let $\delta \in \mathrm{Irr}(D)$ lie under φ. Write $T = \mathbf{N}_G(S, \varphi)$ and $J = \mathbf{N}_N(D.\delta)$. Then*

(a) *$T \cap J \triangleleft T$.*

(b) *$T/(T \cap J)$ is a π-group.*

(c) *$T \cap J \triangleleft J$.*

(d) *$J/(T \cap J)$ is a π-group.*

Proof. Note that $|S : D| = |S : S \cap N| = |SN : N|$ is a π-number, so we can apply Lemma 4.16 to deduce that $J \subseteq \mathbf{N}_G(S)$. Also, $[J, S] \subseteq [N, S] \subseteq N$, and thus $[J, S] \subseteq S \cap N = D \subseteq J$. It follows that $S \subseteq \mathbf{N}_G(J)$, and we argue that in fact, $T \subseteq \mathbf{N}_G(J)$. To see this, note first that T normalizes both S and N, and thus $T \subseteq \mathbf{N}_G(D)$. Now T stabilizes φ, so T permutes the set of irreducible constituents of φ_D. Since S acts transitively on this set, we can write $T = SI$, where I is the stabilizer of δ in T. Also, since I stabilizes (D, δ), it follows that I normalizes $\mathbf{N}_N(D, \delta) = J$, and thus $T = SI$ normalizes J, as claimed. Assertion (a) now follows.

Now $|T : I| = |SI : I| = |S : S \cap I|$, and this divides $|S : D|$ because $D \subseteq S \cap I$. It follows that $|T : I|$ divides $|G : N|$. Also, $T \cap J = I \cap N$, so $|I : T \cap J| = |I : I \cap N|$, and this divides $|G : N|$. It follows that $|T : T \cap J| = |T : I||I : T \cap J|$ divides $|G : N|^2$, and this proves (b).

Now recall that J normalizes S, so $[S, J] \subseteq S \cap N = D$, and thus the action of J on S/D is trivial. Also, J stabilizes δ, so we can apply Lemma 4.15(a) to conclude that J' stabilizes φ. Then $J' \subseteq T \cap J$ and hence $T \cap J \triangleleft J$, proving (c).

Finally, if H is a Hall π'-subgroup of J, then $|H|$ and $|S/D|$ are relatively prime, and thus H stabilizes φ by Lemma 4.15(b). It follows that $H \subseteq T \cap J$, and (d) follows. ∎

We can now establish Theorem 4.14.

Proof of Theorem 4.14. We have $N \lhd G$, where G/N is a π-group. Also $\chi \in \mathrm{Irr}(G)$ lies over $\psi \in \mathrm{Irr}(N)$, and we must show that $\chi \in \mathrm{B}_\pi(G)$ if and only if $\psi \in \mathrm{B}_\pi(N)$. We can assume that $N < G$, and we proceed by induction on $|G|$.

Choose $(S, \varphi) \in \mathcal{S}^*(G)$ with $(S, \varphi) \leq (G, \chi)$, and let $D = S \cap N$. Let $\delta \in \mathrm{Irr}(D)$ lie under φ, and observe that $(D, \delta) \in \mathcal{S}^*(N)$ by Lemma 4.7.

Let $T = \mathbf{N}_G(S, \varphi)$, and suppose first that $T < G$. Also, let $J = \mathbf{N}_N(D, \delta)$, and write $M = T \cap J$. By Theorem 4.9, there exists $\xi \in \mathrm{Irr}(T|\varphi)$ such that $\xi^G = \chi$, and we note that ξ lies over δ because it lies over φ. It follows that there exists a character $\mu \in \mathrm{Irr}(M)$ lying under ξ and over δ.

Now χ lies over ξ, so χ lies over μ. Also, $M \subseteq J$ so we can choose $\eta \in \mathrm{Irr}(J)$ lying over μ and under χ. Since $\eta \in \mathrm{Irr}(J|\delta)$ and $(D, \delta) \in \mathcal{S}^*(N)$, it follows by Theorem 4.9 that η^N is irreducible, and in fact η^N is an irreducible constituent of χ_N, and it follows that η^N has the form ψ^g, for some element $g \in G$. Since ψ lies in $\mathrm{B}_\pi(N)$ if and only if ψ^g lies in $\mathrm{B}_\pi(N)$, there is no loss if we replace ψ by ψ^G, so we can assume that $\eta^N = \psi$.

A nucleus character for ξ is a nucleus character for χ, so $\xi \in \mathrm{B}_\pi(T)$ if and only if $\chi \in \mathrm{B}_\pi(G)$, and similarly, $\eta \in \mathrm{B}_\pi(J)$ if and only if $\psi \in \mathrm{B}_\pi(N)$. It suffices, therefore, to show that $\xi \in \mathrm{B}_\pi(T)$ if and only if $\eta \in \mathrm{B}_\pi(J)$.

By Lemma 4.17 we see that $M \lhd T$ and $M \lhd J$, and also, both T/M and J/M are π-groups. Now μ lies under both ξ and η, and since $J \subseteq N < G$, the inductive hypothesis applied in J guarantees that $\eta \in \mathrm{B}_\pi(J)$ if and only if $\mu \in \mathrm{B}_\pi(M)$. Also, since we are assuming that $T < G$, it follows by the inductive hypothesis applied in T that $\xi \in \mathrm{B}_\pi(T)$ if and only $\mu \in \mathrm{B}_\pi(M)$, and so we are done in this case.

We can assume now that $T = G$, so $S = G$ by Lemma 4.10, and thus χ is π-factored, and it follows that ψ is also π-factored. By Theorem 4.12, we see that $\chi \in \mathrm{B}_\pi(G)$ if and only if χ is π-special, and $\psi \in \mathrm{B}_\pi(N)$ if and only if ψ is π-special. Since χ is π-special if and only if ψ is π-special, the proof is complete. ∎

The following is an easy consequence of Theorem 4.14.

Corollary 4.18. *Let $N \lhd G$, where G is π-separable and G/N is a π-group, and let $N \subseteq K \subseteq G$. Let $\chi \in \mathrm{Irr}(G)$ and $\xi \in \mathrm{Irr}(K)$, and suppose that χ lies over ξ. Then $\xi \in \mathrm{B}_\pi(K)$ if and only if $\chi \in \mathrm{B}_\pi(G)$.*

Proof. Let ψ be an irreducible constituent of ξ_N, and note that χ lies over ψ. The result follows via two appeals to Theorem 4.14: once to compare χ with ψ and once to compare ξ with ψ.

4C

Next, we consider the case where $N \lhd G$ and G/N is a π'-group. Let $\psi \in \mathrm{Irr}(N)$ and recall that Theorem 2.4(b) guarantees that if $\chi \in \mathrm{Irr}(G)$ is π-special and lies over ψ, then ψ is π-special. We will show in this section that the analogous conclusion holds for B_π-characters. Theorem 2.4(b) also asserts that if ψ is π-special and invariant in G, then there is a unique π-special character $\chi \in \mathrm{Irr}(G|\psi)$, and in this case, $\chi_N = \psi$. We will see that this too holds for B_π-characters, and that in fact, almost everything goes through even without assuming that ψ is invariant. (Of course, if ψ is not invariant, we cannot expect that $\chi_N = \psi$, but nevertheless, if $\chi \in \mathrm{B}_\pi(G)$, then each irreducible constituent of χ_N occurs with multiplicity 1.) ■

Theorem 4.19. *Let $N \lhd G$, where G is π-separable and G/N is a π'-group, and let $\psi \in \mathrm{Irr}(N)$.*

(a) *If $\chi \in \mathrm{B}_\pi(G)$ lies over ψ, then $\psi \in \mathrm{B}_\pi(N)$.*

(b) *If $\psi \in \mathrm{B}_\pi(N)$, then there exists a unique character $\chi \in \mathrm{B}_\pi(G)$ that lies over ψ. Also, $[\chi_N, \psi] = 1$.*

(c) *Suppose that ψ and χ are as in* (b). *Let (W, γ) be a nucleus for ψ, and write $V = \mathbf{N}_G(W, \gamma)$. Then V/W is a π'-group, and (V, β) is a nucleus for χ, where β is the canonical extension of γ to V.*

The most significant conclusions of Theorem 4.19 are (a) and (b), but occasionally it is useful to be able to control nuclei, and that is why we have included (c). We need the following preliminary result.

Lemma 4.20. *Let $N \lhd G$, where G is π-separable and G/N is a π'-group, and suppose that $(D, \delta) \in \mathcal{S}^*(N)$, where δ is π-special. Then there exists a unique pair $(S, \varphi) \in \mathcal{S}^*(G)$ such that $(D, \delta) \leq (S, \varphi)$ and φ is π-special. Also, if (S, φ) has these properties, then $S \cap N = D$.*

Proof. Choose a pair $(S, \varphi) \in \mathcal{S}(G)$ with $(D, \delta) \leq (S, \varphi)$ and such that $|S|$ is as large as possible. Since $D \lhd\lhd S$ and δ is π-special and lies under φ, it follows that δ lies under the π-special factor of φ. Replacing φ by its π-special factor, we can assume that φ is π-special. Also, $(S, \varphi) \in \mathcal{S}^*(G)$ by the maximality of $|S|$.

To establish the uniqueness of (S, φ), we suppose that also $(D, \delta) \leq (T, \theta) \in \mathcal{S}(G)$, where θ is π-special, and we show that $(T, \theta) \leq (S, \varphi)$. Now $D \subseteq T \cap N$, and we argue that equality holds here. Let $\alpha \in \mathrm{Irr}(T \cap N)$ lie under θ and over δ. Then $(D, \delta) \leq (T \cap N, \alpha) \in \mathcal{S}(N)$, and since $(D, \delta) \in \mathcal{S}^*(N)$ by hypothesis we conclude that $D = T \cap N$, as claimed. It follows that $D \lhd T$ and that T/D is isomorphic to a subgroup of G/N, and hence

T/D is a π'-group. Since we can apply this reasoning with S in place of T, we also have $D = S \cap N$, as required, and also $D \lhd S$ and S/D is a π'-group.

Since S/D and T/D are π'-groups and φ and θ are π-special, it follows that $\varphi_D = \delta$ and $\theta_D = \delta$, and thus δ is invariant in S and T. Let $H = \langle S, T \rangle$, and observe that $D \lhd H$ and δ is invariant in H. Also, S/D and T/D are subnormal π'-subgroups of H/D, and thus H/D is a π'-group, and it follows that δ has a π-special extension $\mu \in \mathrm{Irr}(H)$. Now $H \lhd\lhd G$, and thus $(D,\delta) \le (H,\mu) \in \mathcal{S}(G)$. By the choice of S, we have $H = S$, and thus $T \subseteq S$. Also, φ is the canonical extension of δ to S, and thus φ_T is the canonical extension of δ to T, and we deduce that $\theta = \varphi_T$, and hence $(T,\theta) \le (S,\varphi)$, as required. ∎

Proof of Theorem 4.19. The result is trivial if $N = G$, so we assume that $N < G$, and we proceed by induction on $|G|$. By Theorem 4.1, we can choose a pair $(D,\delta) \in \mathcal{S}^*(N)$ with $(D,\delta) \le (N,\psi)$. Of course, δ is π-factored, and we argue that we can assume that δ is actually π-special. Otherwise, it follows by Lemma 4.13 that $\psi \notin \mathrm{B}_\pi(N)$ and that no member of $\mathrm{Irr}(G|\psi)$ lies in $\mathrm{B}_\pi(G)$, and thus (a), (b), and (c) all hold vacuously.

By Lemma 4.20, there exists a unique pair $(S,\varphi) \in \mathcal{S}^*(G)$ such that $(D,\delta) \le (S,\varphi)$ and φ is π-special. Lemma 4.20 also guarantees that $D = S \cap N$, so S/D is a π'-group, and thus $\varphi_D = \delta$, so δ is S-invariant and φ is the unique π-special member of $\mathrm{Irr}(S|\delta)$. Note, however, that all members of $\mathrm{Irr}(S|\delta)$ are π-factored by Lemma 4.2.

Let $T = \mathbf{N}_G(S,\varphi)$, and suppose first that $T = G$. Then $S = G$ by Lemma 4.10, and thus $D = N$ and $\psi = \delta$ is π-special, and thus $\psi \in \mathrm{B}_\pi(N)$ and (a) holds. Also, since all members of $\mathrm{Irr}(G|\psi) = \mathrm{Irr}(S|\delta)$ are π-factored, it follows by Theorem 4.12 that a character lies in this set lies in $\mathrm{B}_\pi(G)$ if and only if it is π-special. There is a unique π-special character χ lying over ψ and $\chi_N = \psi$, and thus (b) follows. In this situation, (c) is clear since $(W,\gamma) = (N,\psi)$ and $(V,\beta) = (G,\chi)$, and thus (V,β) is a nucleus for χ. We may assume, therefore, that $T < G$.

Let $J = \mathbf{N}_N(D,\delta)$, and observe that since the pairs (S,φ) and (D,δ) uniquely determine one another, it follows that $T \cap N = J$, and thus $J \lhd T$ and T/J is a π'-group.

By Theorem 4.9, induction defines bijections $\mathrm{Irr}(T|\varphi) \to \mathrm{Irr}(G|\varphi)$ and $\mathrm{Irr}(J|\delta) \to \mathrm{Irr}(N|\delta)$, and in particular, there exists $\eta \in \mathrm{Irr}(J|\delta)$ such that $\eta^N = \psi$. Also, η and ψ share a nucleus, and thus $\eta \in \mathrm{B}_\pi(J)$ if and only if $\psi \in \mathrm{B}_\pi(N)$.

Next, we show that if $\xi \in \mathrm{Irr}(T)$ lies over η and under some member of $\mathrm{B}_\pi(G)$, then $\xi \in \mathrm{B}_\pi(T)$. Also, we shall see that in this situation ξ lies over φ, and thus ξ^G is irreducible. Now η lies over δ so ξ lies over δ, and hence

ξ lies over some character $\mu \in \operatorname{Irr}(S|\delta)$, and we recall that μ is π-factored. By assumption, there exists $\chi \in \mathrm{B}_\pi(G)$ lying over ξ, and hence χ lies over μ, and it follows by Lemma 4.13 that μ is π-special, and thus $\mu = \varphi$ and $\xi \in \operatorname{Irr}(T|\varphi)$. Then ξ^G is irreducible, and thus $\xi^G = \chi$ and ξ and χ share a nucleus. Since $\chi \in \mathrm{B}_\pi(G)$, it follows that $\xi \in \mathrm{B}_\pi(T)$, as claimed.

Suppose that some character $\chi \in \mathrm{B}_\pi(G)$ lies over ψ. We show that $\psi \in \mathrm{B}_\pi(N)$ and that χ is the unique member of $\mathrm{B}_\pi(G)$ lying over ψ. Now $0 < [\chi_N, \psi] = [\chi_N, \eta^N] = [\chi_J, \eta]$, and hence some irreducible constituent ξ of χ_T lies over η. By the result of the previous paragraph, $\xi \in \mathrm{B}_\pi(T)$ and $\xi^G = \chi$. Since $T < G$, we see by the inductive hypothesis that $\eta \in \mathrm{B}_\pi(J)$, and thus $\psi \in \mathrm{B}_\pi(N)$, as required. Also, the inductive hypothesis guarantees that ξ is the unique member of $\mathrm{B}_\pi(T)$ lying over η, and since $\chi = \xi^G$, the uniqueness of χ follows.

In the situation of the previous paragraph, we see that every irreducible constituent of χ_T that lies over η is a member of $\mathrm{B}_\pi(T)$, and thus ξ is the unique such constituent. It follows that

$$[\chi_N, \psi] = [\chi_J, \eta] = [\chi_T, \xi][\xi_J, \eta] = [\xi_J, \eta]\,,$$

where the final equality holds because $\xi^G = \chi$ and thus $[\chi_T, \xi] = 1$. By the inductive hypothesis, $[\xi_J, \eta] = 1$, and thus $[\chi_N, \psi] = 1$, as required.

To complete the proof of (b), we must show that if $\psi \in \mathrm{B}_\pi(N)$, then there exists a character $\chi \in \mathrm{B}_\pi(G)$ that lies over ψ. In this situation, $\eta \in \mathrm{B}_\pi(J)$, so by the inductive hypothesis there exists a character $\xi \in \mathrm{B}_\pi(T)$, such that ξ lies over η. Reasoning as before, we know that ξ lies over some character $\mu \in \operatorname{Irr}(S|\delta)$, where μ is π-factored, and it follows by Lemma 4.13 that μ is π-special, and thus $\mu = \varphi$. Since $\xi \in \operatorname{Irr}(T|\varphi)$, we see that ξ^G is irreducible, and we write $\xi^G = \chi$. Then χ and ξ share a nucleus, and thus $\chi \in \mathrm{B}_\pi(G)$. Also, since χ lies over ξ, it also lies over η, and thus $[\chi_N, \psi] = [\chi_J, \eta] > 0$, as required.

Finally, to prove (c), we continue with the previous notation. We can replace the nucleus (W, γ) for ψ with an N-conjugate, so we can assume that (W, γ) is also a nucleus for η. Recall that $V = \mathbf{N}_G(W, \gamma)$, and thus γ is invariant in $V \cap N$. Since $\gamma^N = \psi$ is irreducible, it follows that $V \cap N = W$. (To see this, write $\tau = \gamma^{V \cap N}$, so τ is irreducible. Since γ is invariant in $V \cap N$, it follows that $\tau_N = e\gamma$, for some integer e. Then

$$|(V \cap N) : W| = \tau(1)/\gamma(1) = e = [\tau_N, \gamma] = [\tau, \tau] = 1\,,$$

and thus $V \cap N = W$.) It follows that V/W is a π'-group, and thus the π-special character γ has a canonical extension $\hat{\gamma} \in \operatorname{Irr}(V)$. By the inductive hypothesis, $(V \cap T, \hat{\gamma}_T)$ is a nucleus for ξ. Then $(\hat{\gamma}_{(V \cap T)})^T = \xi$, and thus $\hat{\gamma}_{V \cap T}$ induces irreducibly to G. However, $\hat{\gamma}_{V \cap T}$ extends to $\hat{\gamma} \in \operatorname{Irr}(V)$, and it follows that $V \cap T = V$, and thus $(V, \hat{\gamma})$ is a nucleus for ξ, and hence it is also a nucleus for χ. ∎

4D

In Theorems 4.14 and 4.19, we considered normal subgroups with index a π-number or a π'-number. Combining appropriate parts of these results, we obtain consequences for an arbitrary normal subgroup, and in fact, more generally, for an arbitrary subnormal subgroup.

Corollary 4.21. *Let $S \lhd\lhd G$, where G is π-separable.*

(a) *If $\chi \in \mathrm{B}_\pi(G)$, then every irreducible constituent of χ_S lies in $\mathrm{B}_\pi(S)$.*

(b) *If $\psi \in \mathrm{B}_\pi(S)$, then some irreducible constituent of ψ^G lies in $\mathrm{B}_\pi(G)$.*

Proof. We can assume that $S < G$, and we proceed by induction on $|G|$. Since S is subnormal, there exists a maximal normal subgroup M of G such that $S \subseteq M$, and we know that G/M is a π-group or a π'-group. If $\chi \in \mathrm{B}_\pi(G)$, then by Theorem 4.14 or 4.19, the irreducible constituents of χ_M lie in $\mathrm{B}_\pi(M)$. Since each irreducible constituent ψ of χ_S lies under at least one of the irreducible constituents of χ_M, it follows by the inductive hypothesis applied in the group M that $\psi \in \mathrm{B}_\pi(S)$, and this proves (a).

Now let $\psi \in \mathrm{B}_\pi(S)$. By the inductive hypothesis, some character $\xi \in \mathrm{B}_\pi(M)$ lies over ψ. By Theorem 4.14 or 4.19, some member of $\mathrm{B}_\pi(G)$ lies over ξ, and (b) follows. ∎

Corollary 4.22. *Let G be π-separable. Then $\mathbf{O}_{\pi'}(G)$ is the intersection of the kernels of the members of $\mathrm{B}_\pi(G)$.*

Proof. Write $N = \mathbf{O}_{\pi'}(G)$, and let $\chi \in \mathrm{B}_\pi(G)$. If ψ is an irreducible constituent of χ_N, then $\psi \in \mathrm{B}_\pi(N)$ by Corollary 4.21. Also, ψ is π'-special because N is a π'-group, and thus ψ is (trivially) π-factored. Since ψ lies in $\mathrm{B}_\pi(N)$, it follows by Theorem 4.12 that ψ is π-special. We conclude now that ψ is principal because it is simultaneously π'-special and π-special, and thus $N \subseteq \ker(\psi)$.

Now let K be the intersection of the kernels of all members of $\mathrm{B}_\pi(G)$. Then $N \subseteq K$, and we assume that $N < K$. Choose a chief factor M/N of G with $M \subseteq K$, and observe that M/N is a π-group because $N = \mathbf{O}_{\pi'}(G)$.

Let $\theta \in \mathrm{Irr}(M/N)$ be nonprincipal, and view θ as a character of M. Then θ lies over the principal character 1_N, and since $1_N \in \mathrm{B}_\pi(N)$ and M/N is a π-group, it follows by Theorem 4.14 that $\theta \in \mathrm{B}_\pi(M)$. By Corollary 4.21, there exists $\chi \in \mathrm{B}_\pi(G)$ lying over θ. Now $M \subseteq K \subseteq \ker(\chi)$, this is a contradiction because χ lies over $\theta \in \mathrm{Irr}(M)$ and θ is not principal. It follows that $N = K$, as required. ∎

Corollary 4.23. *Let G be π-separable.*

(a) *G is a π'-group if and only if $\mathrm{B}_\pi(G) = \{1_G\}$.*

(b) *G is a π-group if and only if $\mathrm{B}_\pi(G) = \mathrm{Irr}(G)$.*

Proof. We have already seen that if G is a π-group then $\mathrm{B}_\pi(G) = \mathrm{Irr}(G)$ and that if G is a π'-group, then $\mathrm{B}_\pi(G) = \{1_G\}$. Conversely, If $\mathrm{B}_\pi(G) = \{1_G\}$, then by Corollary 4.22, we have $G = \mathbf{O}_{\pi'}(G)$, and thus G is a π'-group, proving (a).

If $\mathrm{B}_\pi(G) = \mathrm{Irr}(G)$, then $\mathrm{B}_{\pi'}(G) = \{1_G\}$ since we know that in general, $\mathrm{B}_\pi(G) \cap \mathrm{B}_{\pi'}(G) = \{1_G\}$. (See the discussion following the statement of Theorem 4.12 for a proof.) Then G is a π-group by assertion (a) with the roles of π and π' interchanged. ∎

Suppose that $N \lhd G$ and G/N is a π'-group, and let $\psi \in \mathrm{Irr}(N)$ be G-invariant. If ψ is π-special, we know that ψ has a canonical extension $\chi \in \mathrm{Irr}(G)$ and χ is the unique π-special character in $\mathrm{Irr}(G|\psi)$. Also, if $N \subseteq K \subseteq G$, it is clear that χ_K is the canonical extension of ψ to K, and thus χ_K is π-special. The following result, whose proof relies on conclusion (c) of Theorem 4.19, is an analogue of these facts for B_π-characters.

Corollary 4.24. *Let $N \lhd G$, where G is π-separable and G/N is a π'-group, and let $\psi \in \mathrm{B}_\pi(N)$ be G-invariant. Then there is a unique character $\chi \in \mathrm{B}_\pi(G)$ that extends ψ. Also, if $N \subseteq K \subseteq G$, then $\chi_K \in \mathrm{B}_\pi(K)$.*

Proof. By Theorem 4.19(b), there exists a unique character $\chi \in \mathrm{B}_\pi(G)$ that lies over ψ, and we must show that χ is an extension of ψ. By Theorem 4.19(b), again, we have $[\chi_N, \psi] = 1$, and since ψ is G-invariant, it follows that $\chi_N = \psi$, as wanted.

Now suppose that $N \subseteq K \subseteq G$. By the first part of the proof, ψ has an extension ξ to K with $\xi \in \mathrm{B}_\pi(K)$, and we must show that $\chi_K = \xi$.

Let (W, γ) be a nucleus for ψ, and let $V = \mathbf{N}_G(W, \gamma)$. By Theorem 4.19(c), we know that V/W is a π'-group, and since γ is π-special, it has a canonical extension $\beta \in \mathrm{Irr}(V)$. Also, (V, β) is a nucleus for χ, and thus $\beta^G = \chi$.

Now observe that $K \cap V = \mathbf{N}_K(W, \gamma)$ and $\beta_{K\cap V}$ is the canonical extension of γ to $K \cap V$. Then $(K \cap V, \beta_{K\cap V})$ is a nucleus for ξ by Theorem 4.19(c), and it follows that $(\beta_{K\cap V})^K = \xi$. Now χ lies over β and hence over $\beta_{K\cap V}$, so

$$0 < [\chi_{K\cap V}, \beta_{K\cap V}] = [\chi_K, (\beta_{K\cap V})^K] = [\chi_K, \xi],$$

and since χ_K is irreducible, it follows that $\chi_K = \xi$, as required. ∎

The fact that $\chi_K \in \mathrm{B}_\pi(K)$ in Corollary 4.24 is a special case of something more general, which we present next.

Theorem 4.25. *Let $N \lhd G$, where G is π-separable and G/N is a π'-group. Let $\chi \in \mathrm{B}_\pi(G)$, and suppose that $N \subseteq K \subseteq G$. Then every irreducible constituent of χ_K lies in $\mathrm{B}_\pi(K)$.*

We need the following.

Lemma 4.26. *Let $N \lhd G$, where G is π-separable and G/N is a π'-group. Let $\psi \in \mathrm{B}_\pi(N)$, and let T be the stabilizer of ψ in G. Then the unique member η of $\mathrm{B}_\pi(T)$ in $\mathrm{Irr}(T|\psi)$ and the unique member χ of $\mathrm{B}_\pi(G)$ in $\mathrm{Irr}(G|\psi)$ correspond under the Clifford correspondence.*

Proof. Let (W,γ) be a nucleus for ψ, and write $V = \mathbf{N}_G(W,\gamma)$. Since $\gamma^N = \psi$ and V stabilizes γ, it follows that V stabilizes ψ, and thus $V \subseteq T$ and $V = \mathbf{N}_T(W,\gamma)$. By Theorem 4.19(c), we know that γ has a canonical extension $\beta \in \mathrm{Irr}(V)$ and that (V,β) is a nucleus for the unique member $\chi \in \mathrm{B}_\pi(G)$ lying over ψ and also for the unique member $\eta \in \mathrm{B}_\pi(T)$ lying over ψ. Then $\beta^T = \eta$ and $\beta^G = \chi$, and thus $\eta^G = \beta^G = \chi$, and hence χ and η are Clifford correspondents with respect to ψ. ∎

Proof of Theorem 4.25. Let ζ be an irreducible constituent of χ_K, and let $\psi \in \mathrm{Irr}(N)$ lie under ζ. Let T be the stabilizer of ψ in G, write $S = K \cap T$, and let $\xi \in \mathrm{Irr}(S|\psi)$ be the Clifford correspondent of ζ. Now χ lies over ζ, so χ lies over ξ, and hence χ lies over some character $\eta \in \mathrm{Irr}(T)$ such that η lies over ξ. Since ξ lies over ψ, we see that $\eta \in \mathrm{Irr}(T|\psi)$, and thus η^G is irreducible by the Clifford correspondence. Then $\eta^G = \chi$, so η is the Clifford correspondent of χ with respect to ψ.

By hypothesis, $\chi \in \mathrm{B}_\pi(G)$, and thus $\psi \in \mathrm{B}_\pi(N)$. Also, Lemma 4.26 guarantees that the Clifford correspondent η of χ lies in $\mathrm{B}_\pi(T)$. Now ψ is invariant in T and $N \subseteq S \subseteq T$, so Corollary 4.23 guarantees that η_S is irreducible, and in fact, $\eta_S \in \mathrm{B}_\pi(S)$. Since η lies over $\xi \in \mathrm{Irr}(S)$, we see that $\eta_S = \xi$, and thus $\xi \in \mathrm{B}_\pi(S)$. It follows by Lemma 4.26 applied in the group K that $\zeta \in \mathrm{B}_\pi(K)$, as required ∎

Next, we present a theorem that is the key to our proof in Chapter 5 that the map $\chi \mapsto \chi^0$ is a bijection from $\mathrm{B}_\pi(G)$ onto $\mathrm{I}_\pi(G)$. To state this result, we recall that for a positive integer n, we write n_π to denote the π-part of n, which is the largest π-number dividing n.

Theorem 4.27. *Suppose that G is π-separable, and let $\alpha \in \mathrm{Irr}(H)$, where H is a Hall π-subgroup of G. Then*

$$\alpha(1) = \sum_{\chi \in \mathrm{B}_\pi(G)} [\alpha^G, \chi]\chi(1)_\pi \,.$$

Of course, the formula of Theorem 4.27 can be rewritten as

$$\alpha(1) = \sum_{\chi \in \mathrm{B}_\pi(G)} [\chi_H, \alpha]\chi(1)_\pi ,$$

and it is interesting to compare this with the formula of Theorem 3.7(b), which asserts that

$$\alpha(1) = \sum_{\mu \in \mathrm{I}_\pi(G)} [\mu_H, \alpha]\mu(1)_\pi .$$

If we knew that $\chi \mapsto \chi^0$ is a bijection from $\mathrm{B}_\pi(G)$ onto $\mathrm{I}_\pi(G)$, then Theorem 4.27 would be an immediate corollary of Theorem 3.7(b). We will go in the opposite direction, however, proving Theorem 4.27 and then using it to establish the bijection.

It is convenient for our proof of Theorem 4.27 to create (temporarily) an integer-valued function F defined on (not necessarily irreducible) characters Ξ of a π-separable group G. If $\chi \in \mathrm{Irr}(G)$, we set

$$F(\chi) = \begin{cases} \chi(1)_\pi & \text{if } \chi \in \mathrm{B}_\pi(G), \\ 0 & \text{otherwise}, \end{cases}$$

and we extend F additively, so that for an arbitrary character Ξ of G, we have

$$F(\Xi) = \sum_{\chi \in \mathrm{B}_\pi(G)} [\Xi, \chi]\chi(1)_\pi .$$

Our goal, therefore, is to show that if $\alpha \in \mathrm{Irr}(H)$, where H is a Hall π-subgroup of G, then $\alpha(1) = F(\alpha^G)$. We need the following two preliminary results.

Lemma 4.28. *Let $N \lhd G$, where G is π-separable and G/N is a π'-group. Then $F(\psi^G) = F(\psi)$ for all characters ψ of N.*

Proof. Since both F and character induction are additive maps, we may assume that ψ is irreducible. If $\psi \notin \mathrm{B}_\pi(N)$, then by Theorem 4.19(a), no irreducible constituent of ψ^G lies in $\mathrm{B}_\pi(G)$, and thus $F(\psi^G) = 0 = F(\psi)$, as required.

Suppose now that $\psi \in \mathrm{B}_\pi(N)$. By Theorem 4.19(b), there is a unique irreducible constituent χ of ψ^G such that $\chi \in \mathrm{B}_\pi(G)$. Also, $1 = [\chi_N, \psi] = [\chi, \psi^G]$, so χ occurs with multiplicity 1 in ψ^G, and it follows that $F(\psi^G) = F(\chi) = \chi(1)_\pi$. By Clifford's theorem, $\chi_N = \sum \psi_i$, where the characters ψ_i for $1 \le i \le t$ are the members of the G-orbit of ψ. Then $\chi(1) = t\psi(1)$, and t divides the π'-number $|G : N|$. Then

$$F(\psi^G) = \chi(1)_\pi = \psi(1)_\pi = F(\psi) ,$$

and the result follows. ∎

Lemma 4.29. *Let $N \lhd G$, where G is π-separable and G/N is a π-group. Then $F(\chi_N) = F(\chi)$ for all characters χ of G.*

Proof. Since both F and character restriction are additive maps, we can assume that χ is irreducible. If $\chi \notin \mathrm{B}_\pi(G)$, then by Theorem 4.14 no irreducible constituent of χ_N lies in $\mathrm{B}_\pi(N)$, and thus $F(\chi_N) = 0 = F(\chi)$, as required.

Suppose now that $\chi \in \mathrm{B}_\pi(G)$, and write $\chi_N\, e \sum \psi_i$, where e is a positive integer and the characters ψ_i for $1 \leq i \leq t$ lie in $\mathrm{B}_\pi(N)$ and form a G-orbit. Then $F(\chi_N) = et\, F(\psi)$, where $\psi = \psi_1$. Now $\chi(1) = et\, \psi(1)$, and e and t are π-numbers because they divide the π-number $|G : N|$. Then

$$F(\chi) = \chi(1)_\pi = et\, \psi(1)_\pi = et\, F(\psi) = F(\chi_N)\,,$$

and the result follows. ∎

Proof of Theorem 4.27. Recall that our goal is to show that $\alpha(1) = F(\alpha^G)$, where $\alpha \in \mathrm{Irr}(H)$ and H is a Hall π-subgroup of G. In fact, we proceed by induction on $|G|$ to prove this for all characters α of H.

We can assume that G is nontrivial, and we let N be a maximal normal subgroup of G. Suppose first that G/N is a π'-group. Then $H \subseteq N$, and it follows by the inductive hypothesis that $\alpha(1) = F(\alpha^N) = F(\alpha^G)$, where the second equality holds by Lemma 4.28 because $\alpha^G = (\alpha^N)^G$.

We can now assume that G/N is a π-group. Then $HN = G$, and we write $K = H \cap N$. Then K is a Hall π-subgroup of N, so we can apply the inductive hypothesis in the group N, and thus

$$\alpha(1) = \alpha_K(1) = F((\alpha_K)^N) = F((\alpha^G)_N) = F(\alpha^G)\,,$$

where the final equality holds by Lemma 4.29. ∎

4E

Before we proceed in Chapter 5 to establish and exploit the connection between $\mathrm{B}_\pi(G)$ and $\mathrm{I}_\pi(G)$, we discuss a few other results about nuclei and B_π-characters.

Recall from Theorem 3.16 that if $\alpha \in \mathrm{Irr}(H)$, where H is a Hall π-subgroup of a π-separable group G, then there exists a unique largest subgroup W with $H \subseteq W \subseteq G$ such that α has an extension $\beta \in \mathrm{Irr}(W)$. Of course, in this situation, β cannot be extended to any subgroup properly containing W, and this suggests a general question: When is it true that an irreducible character θ of some subgroup J has no proper extension? For example, a sufficient condition for this is that θ^G is irreducible. (To see this, it suffices to observe that if θ^G is irreducible and $J \subseteq K$, then θ^K is the unique irreducible character of K lying over θ, and if $J < K$, this character is definitely not an extension of θ.)

We return now to the situation where $\alpha \in \mathrm{Irr}(H)$ and H is a Hall π-subgroup of G and W is the largest subgroup containing H to which α extends. If $\beta \in \mathrm{Irr}(W)$ is an extension of α, the observation of the previous paragraph suggests that perhaps β^G must be irreducible. In fact, this is true in the special case where $H \triangleleft G$ because in that situation W is exactly the stabilizer in G of α, and thus β^G is irreducible by the Clifford correspondence.

Unfortunately, it is false in general that if β is a maximal extension of α, where $\alpha \in \mathrm{Irr}(H)$ and H is a Hall π-subgroup of a π-separable group G, then β^G is necessarily irreducible. An easy counterexample in a solvable group is to take G to be the symmetric group of degree 4, where $\pi = \{2\}$. A Hall π-subgroup H of G has order 8 and index 3, so H is a maximal subgroup of G. Now H has an irreducible character α of degree 2, and α does not extend to G, so in this case, $W = H$ and $\beta = \alpha$, and yet β^G is not irreducible.

If $\alpha \in \mathrm{Irr}(H)$ is quasi-primitive, however, and W is the largest subgroup to which α extends, then it actually is true that every extension of α to W induces irreducibly to G. (Recall that by definition, an irreducible character is quasi-primitive if its restriction to every normal subgroup is a multiple of an irreducible character.) The following theorem proves the irreducibility of the induced character in the case of the unique π-special extension of α to W. (Note that the existence and uniqueness of a π-special extension is guaranteed by Corollary 3.15.) The corresponding result for an arbitrary maximal extension of a quasi-primitive character will follow easily using results in Chapter 5.

Theorem 4.30. *Let $\alpha \in \mathrm{Irr}(H)$ be quasi-primitive, where H is a Hall π-subgroup of a π-separable group G, and let $H \subseteq W \subseteq G$, where W is the largest subgroup to which α extends. Let $\gamma \in \mathrm{Irr}(W)$ be the unique π-special extension of α to W, and write $\chi = \gamma^G$. Then χ is irreducible, and in fact, $\chi \in \mathrm{B}_\pi(G)$. Also, (W, γ) is a nucleus for χ.*

Proof. We proceed by induction on $|G|$. Let S be the subgroup generated by all subgroups of W that are subnormal in G. Then S is the unique largest subnormal subgroup of G contained in W, and hence $S \triangleleft W$, and in particular, H normalizes S.

Write $D = H \cap S$, and note that $D \triangleleft H$ and D is a Hall π-subgroup of S. It follows from the quasi-primitivity of α that α_D is a multiple of some character $\beta \in \mathrm{Irr}(D)$, and since γ is an extension of α, we see that γ_D is a multiple of β. Now γ lies over β, and thus there exists a character $\varphi \in \mathrm{Irr}(S)$

lying over β and under γ. Also, φ is π-special because γ is π-special and $S \lhd W$. Since D is a Hall π-subgroup of S, it follows by Theorem 2.10 that φ_D is irreducible, and thus $\varphi_D = \beta$, and in fact, φ is the unique π-special character of S lying over β.

Now $\varphi \in \mathrm{Irr}(S)$ and $S \lhd W$, and we argue that φ is invariant in W. Since D is a Hall π-subgroup of $S \lhd W$, the Frattini argument yields $W = S\mathbf{N}_W(D)$, so it suffices to show that $\mathbf{N}_W(D)$ stabilizes φ. It is certainly true that $\mathbf{N}_W(D)$ stabilizes γ (because γ is a character of W), and thus $\mathbf{N}_W(D)$ stabilizes the unique irreducible constituent β of γ_D. Since φ is the unique π-special character of S lying over β, we deduce that $\mathbf{N}_W(D)$ stabilizes φ, and thus W stabilizes φ, as claimed.

Now $(S, \varphi) \in \mathcal{S}(G)$, and we argue that in fact, $(S, \varphi) \in \mathcal{S}^*(G)$. Otherwise, by Lemma 4.6, there exists a pair $(U, \mu) \in \mathcal{S}(G)$ such that $(S, \varphi) \leq (U, \mu)$, where $S \lhd U$ and U/S is simple, and thus U/S is either a π-group or a π'-group. Suppose first that U/S is a π-group. Since $U \lhd\lhd G$ and H is a Hall π-subgroup of G, it follows that $U \cap H$ is a Hall π-subgroup of U, and thus $U = S(U \cap H) \subseteq W$. This is a contradiction, however, since $S < U$, and yet S is maximal among the subnormal subgroups of G contained in W. We conclude that U/S is a π'-group.

Now $\mu \in \mathrm{Irr}(U)$ is a π-factored character that lies over the π-special character φ, and thus the π-special factor of μ lies over φ. Since U/S is a π'-group, φ is the restriction of the π-special factor of μ to S, and in particular, φ is invariant in U.

Recall now that $H \subseteq W$ and W normalizes S and stabilizes φ, and thus for each element $h \in H$, we have $S = S^h \lhd U^h$ and $\varphi = \varphi^h$ is invariant in U^h. Let $R = \langle U^h \mid h \in H \rangle$. Then H normalizes R and $S \lhd R \lhd\lhd G$, and φ is invariant in R. Also, since each of the π'-groups U^h/S is subnormal in R/S, it follows that R/S is a π'-group. Since φ is π-special and invariant in R, we deduce that φ has a π-special extension $\theta \in \mathrm{Irr}(R)$, and since $\varphi_D = \beta$, we see that $\theta_D = \beta$. Now $|R : D| = |R : S||S : D|$ is a π'-number, and thus D is a Hall π-subgroup of R, and we have $D = R \cap H$. Also, θ is the unique π-special extension of β to R.

Since H stabilizes $\varphi_D = \beta$, it follows by the uniqueness of θ that H stabilizes θ, and thus we can apply Lemma 2.11(a) in the group HR. (Recall that H normalizes R, and thus $R \lhd HR$.) We deduce that the restriction defines a bijection from $\mathrm{Irr}(HR|\theta)$ onto $\mathrm{Irr}(H|\beta)$, and thus α extends to HR. Now W is the unique largest subgroup to which α extends, and thus $R \subseteq HR \subseteq W$. This is a contradiction, however, since $R \lhd\lhd G$ and $S < U \subseteq R$. This completes the proof that $(S, \varphi) \in \mathcal{S}^*(G)$, as claimed.

Now let $T = \mathbf{N}_G(S, \varphi)$, and observe that $W \subseteq T$ since we showed that φ is invariant in W. If $T = G$, then $S = G$ by Lemma 4.10, and thus $W = G$

and γ is a π-special character of G. Then $\gamma \in \mathrm{B}_\pi(G)$, and there is nothing further to prove in this case.

We can assume, therefore, that $T < G$. By the inductive hypothesis applied in the group T, we know that γ^T is irreducible and that (W, γ) is a nucleus for this character. Since γ^T lies over φ, it follows by Theorem 4.9 that $\chi = \gamma^G = (\gamma^T)^G$ is irreducible. Also, by definition, (W, γ) is a nucleus for χ. Finally, since the nucleus character γ for χ is π-special, we see that $\chi \in \mathrm{B}_\pi(G)$. ∎

4F

If χ is a π-special character of a π-separable group G, then we can associate with χ a certain family of irreducible characters of G, namely the characters $\chi\mu$, where $\mu \in \mathrm{Irr}(G)$ is π'-special. Of course, the irreducibility of these characters is a consequence of Gajendragadkar's factorization theorem, which also implies that the characters $\chi\mu$ are distinct for distinct π'-special characters μ and that the sets of characters associated in this way with distinct π-special characters χ are disjoint.

If instead of assuming that χ is π-special, we assume only the weaker condition that $\chi \in \mathrm{B}_\pi(G)$, it remains possible to associate with χ a certain family of irreducible characters of G. This is the essential content of the following result of G. Navarro.

Theorem 4.31 (Navarro). *Let $\chi \in \mathrm{B}_\pi(G)$, where G is π-separable, and let (W, γ) be a nucleus for χ. Then the map $\mu \mapsto (\gamma\mu)^G$ is an injection from the set of π'-special characters μ of W into* $\mathrm{Irr}(G)$.

We refer to the characters $(\gamma\mu)^G$ in Navarro's theorem as the **satellites** of χ, and we observe that since all nuclei for χ are conjugate in G, the set of satellites of χ is independent of the choice of the nucleus (W, γ). We shall also see that the sets of satellites of distinct members of $\mathrm{B}_\pi(G)$ are disjoint. (Note that in contrast with the astronomical use of the word "satellite", each member of $\mathrm{B}_\pi(G)$ is a satellite of itself.)

We begin with a somewhat technical lemma.

Lemma 4.32. *Let G be π-separable, and let $(S, \varphi) \in \mathcal{S}^*(G)$, where φ is π-special. Write $\mathcal{X}$ to denote the set of π-factored irreducible characters of S for which the π-special factor is φ.*

(a) *If $\alpha \in \mathcal{X}$, then $(S, \alpha) \in \mathcal{S}^*(G)$.*

(b) *Let $T = \mathbf{N}_G(S, \varphi)$, and write $\mathcal{Y}$ to denote the set of irreducible characters of T that lie over at least one member of $\mathcal{X}$. Then induction defines an injective map from $\mathcal{Y}$ into* $\mathrm{Irr}(G)$.

Proof. Suppose that $\alpha \in \mathcal{X}$. Then $(S,\alpha) \in \mathcal{S}(G)$, and to show that $(S,\alpha) \in \mathcal{S}^*(G)$, we suppose that $(S,\alpha) \le (V,\tau) \in \mathcal{S}(G)$, and we argue that $(S,\alpha) = (V,\tau)$. Let σ be the π-special factor of τ, and observe that σ lies over the π-special factor φ of α, and thus $(S,\varphi) \le (V,\sigma) \in \mathcal{S}(G)$. By hypothesis, however, $(S,\varphi) \in \mathcal{S}^*(G)$, and thus $(S,\varphi) = (V,\sigma)$, and in particular $S = V$. Then $(S,\alpha) = (V,\tau)$, as required.

Now let $\psi \in \mathcal{Y}$, and let $\alpha \in \mathcal{X}$ lie under ψ. Let $I = \mathbf{N}_G(S,\alpha)$, and observe that since the elements of I stabilize α, they must also stabilize the π-special factor φ of α, and thus $I \subseteq T$, and in fact, I is the stabilizer of α in T. By the Clifford correspondence, therefore, there is a unique character $\eta \in \mathrm{Irr}(I|\alpha)$ such that $\eta^T = \psi$. Since $(S,\alpha) \in \mathcal{S}^*(G)$ by (a), it follows by Theorem 4.9 that η^G is irreducible. Also, $\psi^G = (\eta^T)^G = \eta^G$, so ψ^G is irreducible, as required.

Now write $\chi = \psi^G$, and suppose that also $\chi = \zeta^G$, where $\zeta \in \mathcal{Y}$, and let $\beta \in \mathcal{X}$ lie under ζ. Then each of α and β lies under χ, and since (S,α) and (S,β) are members of $\mathcal{S}^*(G)$ by (a), it follows by Theorem 4.1 that $(S,\alpha)^g = (S,\beta)$ for some element $g \in G$. Then $\alpha^g = \beta$, and hence $\varphi^g = \varphi$ because φ is the π-special factor of each of α and β. Then $g \in T$, and hence the irreducible characters α and β of S are conjugate in T. Since $\zeta \in \mathrm{Irr}(T)$ lies over β, we conclude that ζ also lies over α, and we let $\xi \in \mathrm{Irr}(I)$ be the Clifford correspondent of ζ with respect to α. Then $\xi^T = \zeta$, so $\xi^G = \zeta^G = \chi = \psi^G = \eta^G$, and by the injectivity of the map in Theorem 4.9, it follows that $\xi = \eta$. Then $\psi = \eta^T = \xi^T = \zeta$, and thus our induction map is injective, as required. ∎

Proof of Theorem 4.31. We proceed by induction on $|G|$. If χ is π-special, then $W = G$ and $\gamma = \chi$, and the result follows via Gajendragadkar's factorization theorem. We can assume, therefore, that χ is not π-special.

By Corollary 4.11, we can choose a pair $(S,\varphi) \in \mathcal{S}^*(G)$ such that $(S,\varphi) \le (W,\gamma)$, and $W \subseteq T$, where $T = \mathbf{N}_G(S,\varphi)$. If $T = G$, we have $S = G$ by Lemma 4.10, and thus $\chi = \varphi$, which is not the case since we are assuming that χ is not π-special. We can suppose, therefore, that $T < G$.

By definition, (W,γ) is also a nucleus for some character in $\mathrm{B}_\pi(T)$. If $\mu \in \mathrm{Irr}(W)$ is π'-special, therefore, it follows by the inductive hypothesis that the character $\psi = (\gamma\mu)^T$ is irreducible, and since $(\gamma\mu)^G = \psi^G$, we must show that ψ^G is irreducible. Now $\gamma\mu$ is a π-factored character of W, and thus an irreducible constituent α of $(\gamma\mu)_S$ is π-factored, and γ lies over the π-special factor of α. Since γ_S is a multiple of φ, however, it follows that the π-special factor of α is φ, and thus α lies in the set $\mathcal{X}$ of Lemma 4.32. Since ψ lies over α, we see that ψ lies in the set $\mathcal{Y}$ of the lemma, and thus by Lemma 4.32(b), we conclude that ψ^G is irreducible, as required.

Now suppose that also $\nu \in \mathrm{Irr}(W)$ is π'-special and that $(\gamma\nu)^G = (\gamma\mu)^G$. Reasoning as we did with μ, we write $\zeta = (\gamma\nu)^T$, so ζ is irreducible and lies over some member β of the set $\mathcal{X}$. Since $\psi^G = \zeta^G$, the injectivity of the map in Lemma 4.32(b) guarantees that $\psi = \zeta$, and thus by the inductive hypothesis applied in the group T, we have $\mu = \nu$, as required. ∎

Next, we show that no character of G can be a satellite of more than one member of $\mathrm{B}_\pi(G)$. Observe first that if $\psi \in \mathrm{Irr}(G)$ is a satellite of $\chi \in \mathrm{B}_\pi(G)$, then as in the proof of Theorem 4.31, there exists $(S, \alpha) \in \mathcal{S}^*(G)$ lying below (G, ψ) and such that if φ is the π-special factor of α, then $(S, \varphi) \in \mathcal{S}^*(G)$ and (S, φ) lies below (G, χ). Furthermore, although the assertion of the previous sentence was that *there exists* a member of $\mathcal{S}^*(G)$ lying below (G, ψ) and having a certain property, actually, this property holds for *every* member of $\mathcal{S}^*(G)$ below (G, ψ). This is because by Theorem 4.1, all such pairs are conjugate in G.

Theorem 4.33. *Let G be π-separable, and suppose that $\psi \in \mathrm{Irr}(G)$. Then ψ is a satellite of at most one member of* $\mathrm{B}_\pi(G)$.

Proof. Let $(S, \alpha) \in \mathcal{S}^*(G)$, where $(S, \alpha) \leq (G, \psi)$, and let φ be the π-special factor of α. Then assuming that ψ is a satellite, it follows that $(S, \varphi) \in \mathcal{S}^*(G)$ and φ lies under every member of $\mathrm{B}_\pi(G)$ for which ψ is a satellite.

Let $T = \mathbf{N}_G(S, \varphi)$. If $T = G$, then $S = G$ by Lemma 4.10, and thus $\psi = \alpha$ and φ is the unique member of $\mathrm{B}_\pi(G)$ having ψ as a satellite. We can assume, therefore, that $T < G$, and we proceed by induction on $|G|$.

Suppose that ψ is a satellite of both χ and ζ in $\mathrm{B}_\pi(G)$. Then $(S, \varphi) \leq (G, \chi)$ and $(S, \varphi) \leq (G, \zeta)$, and by Corollary 4.11, there exist nuclei (W, γ) and (V, δ) for χ and ζ, respectively, such that $W, V \subseteq T$ and $(S, \varphi) \leq (W, \gamma)$ and $(S, \varphi) \leq (V, \delta)$. Also by Corollary 4.11, we see that (W, γ) and (V, δ) are nuclei for the irreducible characters γ^T and δ^T. Furthermore, since γ and δ are π-special, it follows by definition that γ^T and δ^T lie in $\mathrm{B}_\pi(T)$.

Since ψ is a satellite for each of χ and ζ, there exist π'-special characters μ of W and ν of V such that $(\gamma\mu)^G = \psi = (\delta\nu)^G$. Now $(\gamma\mu)^T$ and $(\delta\nu)^T$ are irreducible and they lie over members of the set $\mathcal{X}$ of Lemma 4.32. Since they both induce ψ, it follows by the injectivity of the induction map in that lemma that $(\gamma\mu)^T = (\delta\nu)^T$. This character is thus a satellite for each of the characters γ^T and δ^T. By the inductive hypothesis, we have $\gamma^T = \delta^T$, and it follows that $\chi = \gamma^G = \delta^G = \zeta$, as required. ∎

A natural question at this point is whether or not every member of $\mathrm{Irr}(G)$ is a satellite of some character $\chi \in \mathrm{B}_\pi(G)$. In general, the answer is "no" since in particular, the only way that a member $\psi \in \mathrm{B}_{\pi'}(G)$ can be a satellite of some character in $\mathrm{B}_\pi(G)$ is if ψ is π'-special. To see this, observe that if $\psi \in \mathrm{B}_{\pi'}(G)$ and $(S, \alpha) \in \mathcal{S}^*(G)$ lies below (G, ψ), then Lemma 4.13 (with π' in place of π) guarantees that α is π'-special. The π-special factor of α is thus the principal character 1_S. If ψ is a satellite of a member of $\mathrm{B}_\pi(G)$, then as we have seen, it would follow that $(S, 1_S) \in \mathcal{S}^*(G)$, and this implies that $S = G$, and thus $\psi = \alpha$ is π'-special.

What is true in this connection, however, is that every character $\chi \in \mathrm{Irr}(G)$ for which the degree is a π'-number is a satellite of some member of $\mathrm{B}_\pi(G)$. Unfortunately, the proof of this fact seems to require more machinery than we have developed so far.

4G

Next, we discuss the field $\mathbb{Q}(\chi)$ generated by the values of a character $\chi \in \mathrm{Irr}(G)$, where G is π-separable. Each irreducible character χ of G is induced from a nucleus character γ, so we have $\mathbb{Q}(\chi) \subseteq \mathbb{Q}(\gamma)$. If $\chi \in \mathrm{B}_\pi(G)$, then γ is π-special, and it follows by Corollary 2.13 that $\mathbb{Q}(\gamma)$ is contained in some cyclotomic field $\mathbb{Q}_m$, where m is a π-number, and thus $\mathbb{Q}(\chi)$ is contained in this field.

If $|G|$ is odd and $\chi \in \mathrm{Irr}(G)$, we will see that conversely, if $\mathbb{Q}(\chi)$ is contained in some cyclotomic field $\mathbb{Q}_m$, where m is a π-number, then $\chi \in \mathrm{B}_\pi(G)$. (This generalizes Theorem 2.26, which asserts that under these conditions and assuming in addition that $\chi(1)$ is a π-number, χ must be π-special.) As we shall see, the condition that $|G|$ is odd can be relaxed somewhat; it will suffice to assume only that $\chi(1)$ is odd and that a Hall π'-subgroup of G has odd order.

The key to the proof of Theorem 2.26 and to its generalization for B_π-characters is the magic field automorphism for a set π of primes and a group G. Recall that this is the unique element of the Galois group $\mathrm{Gal}(\mathbb{Q}_{|G|}/\mathbb{Q})$ that fixes all π-roots of unity in $\mathbb{Q}_{|G|}$ and maps all π'-roots of unity in this field to their complex conjugates.

We will occasionally use without explicit mention the fact that if $H \subseteq G$ and σ is the magic field automorphism for π and G, then the restriction of σ to the field $\mathbb{Q}_{|H|}$ is the magic field automorphism for π and H. Another useful observation is that σ^2 is the trivial field automorphism.

If χ is a character of G with values in a cyclotomic field $\mathbb{Q}_m$, where m is a π-number, then the values of χ lie in the subfield of $\mathbb{Q}_{|G|}$ fixed by the magic field automorphism σ, and thus $\chi^\sigma = \chi$. In particular, if $\chi \in \mathrm{B}_\pi(G)$, then $\chi^\sigma = \chi$.

Theorem 4.34. *Let $\chi \in \mathrm{Irr}(G)$, where G is π-separable, and assume that $\chi(1)$ is odd and that a Hall π'-subgroup of G has odd order. Then $\chi \in \mathrm{B}_\pi(G)$*

if and only if χ is fixed by the magic field automorphism for π and G. Also, if χ is real valued, then $\chi \in \mathrm{B}_\pi(G)$.

Note that if G has odd order, then the assumptions in Theorem 4.34 that $\chi(1)$ is odd and that a Hall π'-subgroup of G has odd order will automatically be satisfied.

Corollary 4.35. *Assume that $|G|$ is odd and that $\chi \in \mathrm{Irr}(G)$. If π is a set of primes, then the following are equivalent.*

(1) $\chi \in \mathrm{B}_\pi(G)$.

(2) *χ has values in some cyclotomic field $\mathbb{Q}_m$, where m is a π-number.*

(3) *χ is fixed by the magic field automorphism for π and G.*

Proof. First, note that since $|G|$ is odd, G is solvable, and hence it is π-separable. We have already observed that (1) implies (2) and that (2) implies (3). That (3) implies (1) is a consequence of Theorem 4.34, which applies because we are assuming that G has odd order. ■

Note that it follows from Corollary 4.35 that if $|G|$ is odd, then the members of $\mathrm{B}_\pi(G)$ can be identified in the character table of G. It is not clear if this is true for an arbitrary π-separable group, however.

To prove Theorem 4.34, we need a standard fact from elementary character theory: if $|G|$ is odd, then the principal character 1_G is the unique real-valued irreducible character of G. (This result of W. Burnside appears as Problem 3.16 in [**CTFG**], and also it is immediate from the Brauer permutation lemma, which appears as Theorem 6.32 of [**CTFG**].) An easy consequence is the following.

Lemma 4.36. *Let $\gamma \in \mathrm{Irr}(G)$, where G is π-separable, and assume that a Hall π'-subgroup of G has odd order. Suppose that γ is π-factored and assume that $\gamma^\sigma = \gamma$, where σ is either the magic field automorphism for π and G, or σ is complex conjugation. Then γ is π-special.*

Proof. Write $\gamma = \alpha\beta$, where α is π-special and β is π'-special. Now $\gamma^\sigma = \gamma$, so we have $\alpha^\sigma\beta^\sigma = \alpha\beta$, and since α^σ and β^σ are π-special and π'-special, respectively, it follows by the uniqueness of Gajendragadkar's factorization that $\beta^\sigma = \beta$.

Now let K be a Hall π'-subgroup of G. Then β_K is irreducible by Theorem 2.10, and we have $(\beta_K)^\sigma = \beta_K$. Also, the values of β_K are sums of $|K|$-roots of unity, and $|K|$ is a π'-number. In either case, if σ is the magic automorphism or σ is complex conjugation, it follows that $(\beta_K)^\sigma$ is the complex conjugate of β_K, and we deduce that β_K is real valued. Since by assumption, K has odd order, we have $\beta_K = 1_K$, and it follows by the injectivity of the restriction map in Theorem 2.10 that $\beta = 1_G$. Then $\gamma = \alpha$ is π-special, as required. ■

Proof of Theorem 4.34. We have already observed that if $\chi \in \mathrm{B}_\pi(G)$, then the values of χ lie in some cyclotomic field $\mathbb{Q}_m$, where m is a π-number, and thus $\chi^\sigma = \chi$, where σ is the magic field automorphism for π and G. We assume now that $\chi^\sigma = \chi$, where σ is either the magic field automorphism or complex conjugation, and we work to show that $\chi \in \mathrm{B}_\pi(G)$.

Let (W, γ) be a nucleus for χ. Then (W, γ^σ) is a nucleus for $\chi^\sigma = \chi$, and it follows by Corollary 4.11(a) that there exists an element $x \in G$ such that $(W, \gamma)^x = (W, \gamma^\sigma)$. Then $x \in \mathbf{N}_G(W)$ and $\gamma^x = \gamma^\sigma$.

Now $\gamma^G = \chi$ has odd degree by hypothesis, and thus $|G : W|$ is odd, and hence $|\mathbf{N}_G(W) : W|$ is odd. Writing $\mathcal{O}$ to denote the $\mathbf{N}_G(W)$-orbit of γ, it follows that $|\mathcal{O}|$ is odd. Also, $\gamma^\sigma = \gamma^x \in \mathcal{O}$, because $x \in \mathbf{N}_G(W)$, and since σ permutes the $\mathbf{N}_G(W)$-orbits in $\mathrm{Irr}(W)$, we conclude that σ stabilizes the set $\mathcal{O}$. Now $\sigma^2 = 1$, and $|\mathcal{O}|$ is odd, and thus σ fixes some member of $\mathcal{O}$. It follows that σ fixes every member of $\mathcal{O}$, and in particular, $\gamma^\sigma = \gamma$.

Since γ is π-factored and a Hall π'-subgroup of G has odd order, it follows by Lemma 4.36 that γ is π-special, and thus $\chi \in \mathrm{B}_\pi(G)$, as required. ■

Corollary 4.37. *Let $H \subseteq G$ and $N \triangleleft G$, where $|G|$ is odd, and let π be a set of primes.*

(a) *If $\chi \in \mathrm{B}_\pi(G)$ and χ_H is irreducible, then $\chi_H \in \mathrm{B}_\pi(H)$.*
(b) *If $\psi \in \mathrm{B}_\pi(H)$ and ψ^G is irreducible, then $\psi^G \in \mathrm{B}_\pi(G)$.*
(c) *Let $\chi \in \mathrm{Irr}(G)$ lie over $\varphi \in \mathrm{Irr}(N)$. Also, let T be the stabilizer of φ in G, and let $\psi \in \mathrm{Irr}(T)$ be the Clifford correspondent of χ with respect to φ. Then $\psi \in \mathrm{B}_\pi(T)$ if and only if $\chi \in \mathrm{B}_\pi(G)$.*

Proof. For (a), suppose that $\chi \in \mathrm{B}_\pi(G)$, and observe that by Corollary 4.35 we have $\mathbb{Q}(\chi) \subseteq \mathbb{Q}_m$, where m is some π-number. Since χ_H is irreducible and $\mathbb{Q}(\chi_H) \subseteq \mathbb{Q}(\chi) \subseteq \mathbb{Q}_m$, it follows by Corollary 4.35 that $\chi_H \in \mathrm{B}_\pi(H)$. Similarly for (b), we have $\mathbb{Q}(\psi^G) \subseteq \mathbb{Q}(\psi) \subseteq \mathbb{Q}_m$ for some π-number m, and thus $\psi^G \in \mathrm{B}_\pi(G)$.

For (c), observe that $\psi^G = \chi$, so if $\psi \in \mathrm{B}_\pi(T)$, then $\chi \in \mathrm{B}_\pi(G)$ by (b). Conversely, suppose that $\chi \in \mathrm{B}_\pi(G)$. Then by Corollary 4.21, we have $\varphi \in \mathrm{B}_\pi(N)$. By Corollary 4.35, there exists a π-number m such that both $\mathbb{Q}(\chi)$ and $\mathbb{Q}(\varphi)$ are contained in $\mathbb{Q}_m$. We argue that $\mathbb{Q}(\psi) \subseteq \mathbb{Q}_m$, and thus $\psi \in \mathrm{B}_\pi(T)$, as wanted. To see this, suppose τ lies in the Galois group $\mathrm{Gal}(\mathbb{Q}_m(\psi)/\mathbb{Q}_m)$. Then τ fixes χ and φ, and since these characters uniquely determine the Clifford correspondent ψ, it follows that τ fixes ψ, and thus the values of ψ lie in the field $\mathbb{Q}_m$, as wanted. ■

The converse of Corollary 4.37(b) is also true. If $|G|$ is odd and $\psi^G \in \mathrm{B}_\pi(G)$, where ψ is a character of some subgroup $H \subseteq G$, then $\psi \in \mathrm{B}_\pi(H)$. It appears, however, that the techniques available to us at this point are not sufficient to establish this result, and we defer the proof until Chapter 7.

The following two assertions can be viewed as strong forms of Corollary 4.37(a) and (b).

Corollary 4.38. *Let $H \subseteq G$, where $|G|$ is odd, and let π be a set of primes.*

(a) *If $\chi \in \mathrm{B}_\pi(G)$, then there exists $\psi \in \mathrm{B}_\pi(H)$ such that $[\chi_H, \psi]$ is odd, and in particular, ψ is a constituent of χ_H.*

(b) *If $\psi \in \mathrm{B}_\pi(H)$, then there exists $\chi \in \mathrm{B}_\pi(G)$ such that $[\psi^G, \chi]$ is odd, and in particular, χ is a constituent of ψ^G.*

Proof. Let σ be the magic field automorphism for π and G, so by Theorem 4.34, a character $\chi \in \mathrm{Irr}(G)$ lies in $\mathrm{B}_\pi(G)$ if and only if $\chi^\sigma = \chi$. Also, if $H \subseteq G$, then clearly the restriction of σ to $\mathbb{Q}_{|H|}$ is the magic field automorphism for π and H, and thus a character $\psi \in \mathrm{Irr}(H)$ lies in $\mathrm{B}_\pi(H)$ if and only if $\psi^\sigma = \psi$.

Now suppose that $\chi \in \mathrm{B}_\pi(G)$, so $\chi^\sigma = \chi$, and thus χ_H is fixed by σ. We can thus write $\chi_H = \sum a_i\Delta_i$, where the coefficients a_i are integers and the characters Δ_i are distinct, and each of them is the sum of the one or two irreducible characters in a $\langle\sigma\rangle$-orbit. Since $|G|$ is odd, it follows that $\chi(1) = \sum a_i\Delta_i(1)$ is odd, and thus for some subscript i, both a_i and $\Delta_i(1)$ are odd. Then Δ_i is a σ-fixed irreducible character of H, and so it lies in $\mathrm{B}_\pi(H)$. Also, $[\chi_H, \Delta_i] = a_i$ is odd.

Similarly, if $\psi \in \mathrm{B}_\pi(H)$, then $\psi^\sigma = \psi$, and so σ permutes the irreducible constituents of ψ^G. We can thus write $\psi^G = \sum b_i\Xi_i$, where the b_i are integers and the Ξ_i are distinct $\langle\sigma\rangle$-orbit sums of irreducible characters of G. Now $\psi^G(1) = |G : H|\psi(1)$, which is odd, so for some subscript i both b_i and $\Xi_i(1)$ are odd. In particular, Ξ_i is a member of $\mathrm{B}_\pi(G)$ and $[\psi^G, \Xi_i] = b_i$ is odd. ∎

Corollary 4.39. *Suppose that $|G|$ is odd, and let H be a Hall π'-subgroup of G. If $\chi \in \mathrm{B}_\pi(G)$, then $[\chi_H, 1_H]$ is odd.*

Proof. By Corollary 4.38(a), there exists $\psi \in \mathrm{B}_\pi(H)$ such that $[\chi_H, \psi]$ is odd. Since H is a π'-group, however, $\mathrm{B}_\pi(H) = \{1_H\}$, and thus $\psi = 1_H$. Then $[\chi_H, 1_H] = [\chi_H, \psi]$, and this is odd, as required. ∎

The converse of Corollary 4.39 is also true. If H is a Hall π-subgroup of an odd-order group G and $[\chi_H, 1_H]$ is odd for some character $\chi \in \mathrm{Irr}(G)$, then $\chi \in \mathrm{B}_\pi(G)$. Since $[\chi_H, 1_H] = [\chi, (1_H)^G]$, it follows that $\mathrm{B}_\pi(G)$ is exactly the set of those irreducible constituents of $(1_H)^G$ that occur with odd multiplicity. This result appears to lie considerably deeper, however, and we defer its proof to Chapter 8.

The following characterization of the set $\mathrm{B}_\pi(G)$ in the case where a Hall π'-subgroup of G has odd order is somewhat reminiscent of the definition of π-special.

Theorem 4.40. *Assume that a Hall π'-subgroup of a π-separable group G has odd order, and let σ be the magic field automorphism for π and G. Then a character $\chi \in \mathrm{Irr}(G)$ lies in $\mathrm{B}_\pi(G)$ if and only if the irreducible constituents of χ_S are σ-fixed for all subnormal subgroups S of G.*

Proof. If $\chi \in \mathrm{B}_\pi(G)$, then Corollary 4.21(a) guarantees that the irreducible constituents of χ_S lie in $\mathrm{B}_\pi(S)$ whenever $S \lhd\lhd G$. These constituents are thus σ-fixed, as required.

For the converse, assume that $\chi \in \mathrm{Irr}(G)$, and suppose that for all subnormal subgroups S of G, all irreducible constituents of χ_S are σ-fixed. In particular, χ is σ-fixed, so if $\chi(1)$ is odd, Theorem 4.34 guarantees that χ lies in $\mathrm{B}_\pi(G)$, as wanted. Since we are not assuming that $\chi(1)$ is odd, however, we follow a different route.

We can assume that G is nontrivial, and we proceed by induction on $|G|$. Let M be a maximal normal subgroup of G, and observe that every subnormal subgroup of M is subnormal in G. If ψ is an irreducible constituent of χ_M, therefore, the irreducible constituents of ψ_S are σ-fixed for all $S \lhd\lhd M$, and it follows by the inductive hypothesis that $\psi \in \mathrm{B}_\pi(M)$. If G/M is a π-group, then by Theorem 4.14, we have $\chi \in \mathrm{B}_\pi(G)$, as required, so we can assume that G/M is a π'-group.

Let T be the stabilizer of ψ in G, and let $\eta \in \mathrm{Irr}(T|\psi)$ be the Clifford correspondent of χ. Since $\psi \in \mathrm{B}_\pi(M)$ and G/M is a π'-group, it follows by Lemma 4.26 that to show that χ lies in $\mathrm{B}_\pi(G)$, it suffices to check that η lies in $\mathrm{B}_\pi(T)$.

Since η is uniquely determined by χ and ψ, and each of these characters is σ-fixed by hypothesis, it follows that η is σ-fixed. By Theorem 4.19(b), there exists a character $\xi \in \mathrm{B}_\pi(T)$ that lies over ψ, and furthermore, $[\xi_M, \psi] = 1$. Now ψ is invariant in T, so ξ_M is a multiple of ψ, and it follows that $\xi_M = \psi$. By the Gallagher correspondence, therefore, we can write $\eta = \xi\beta$ for some unique character $\beta \in \mathrm{Irr}(T/M)$.

We have

$$\xi\beta = \eta = \eta^\sigma = \xi^\sigma\beta^\sigma = \xi\beta^\sigma\,,$$

where the last equality holds because $\xi \in \mathrm{B}_\pi(T)$, and thus $\xi^\sigma = \xi$. It follows by the uniqueness of β that $\beta = \beta^\sigma$, and since β is a character of the π'-group T/M, we have $\beta = \beta^\sigma = \overline{\beta}$, and thus β is real valued.

By assumption, a Hall π'-subgroup of G has odd order, and it follows that $|T/M|$ is odd. Since β is a real irreducible character of T/M, we conclude that β is the principal character, and thus $\eta = \xi$. Thus η lies in $\mathrm{B}_\pi(T)$, as wanted. ∎

Problems

4.1. Let $\chi \in \mathrm{Irr}(G)$, where G is π-separable.

(a) Show that there exists a unique normal subgroup N of G maximal with the property that some irreducible constituent μ of χ_N is π-factored.

(b) Let N and μ be as in (a), and suppose that I is the stabilizer of μ in G. If $I = G$, show that $N = G$ and $\mu = \chi$.

(c) Continuing with the above notation, assume that $I < G$ and let ψ be the Clifford correspondent of χ with respect to μ. Now repeat this process with ψ in place of χ and I in place of G. Continue like this to obtain a sequence of pairs $(G, \chi) > (I, \psi) > \cdots$, and let (W, γ) be the final pair in this sequence. We say that the pair (W, γ) is a **normal nucleus** for χ. Show that (W, γ) is uniquely determined up to conjugacy in G, and show that γ is π-factored and that $\gamma^G = \chi$.

4.2. Let G be π-separable, and define $\mathrm{N}_\pi(G)$ to be the subset of $\mathrm{Irr}(G)$ consisting of those members such that a normal nucleus character γ for χ is π-special. If $|G|$ is odd, show that $\mathrm{N}_\pi(G) = \mathrm{B}_\pi(G)$.

4.3. Let G be π-separable, and suppose that $\chi \in \mathrm{N}_\pi(G)$.

(a) If $U \triangleleft G$ and χ_U has a π-factored irreducible constituent α, show that α is π-special.

(b) Now assume that $\chi(1)$ is a π-number. Show that χ is π-special.

Hint: For (b), let $U \triangleleft G$ be maximal such that χ_U has a π-special irreducible constituent θ. Show that $U = G$ by considering a chief factor V/U of G.

4.4. Let G be π-separable. Show that every member of $\mathrm{B}_\pi(G)$ is linear if and only if G has a normal Hall π'-subgroup K such that G/K is abelian.

4.5. Let $\chi \in \mathrm{Irr}(G)$, where G is π-separable, and suppose that $\chi(1)$ is a π'-number. Show that χ lies in $\mathrm{B}_\pi(G)$ if and only if $[\chi_K, 1_K] > 0$, where K is a Hall π'-subgroup of G.

4.6. Let $\chi \in \mathrm{Irr}(G)$, where G is π-separable, and assume that $|G|/\chi(1)$ is a π-number. Show that χ lies in $\mathrm{B}_\pi(G)$.

4.7. Suppose that $N \triangleleft G$, where G is π-separable. Show that $\mathrm{B}_\pi(G/N) = \mathrm{B}_\pi(G) \cap \mathrm{Irr}(G/N)$. (Here, as usual, we identify the characters of G/N with the characters of G whose kernels contain N.)

Hint: See the discussion at the end of Section 4A.

4.8. In this problem, we work by induction on $|G|$ to construct an alternative proof of the less trivial part of Corollary 4.23. Let N be minimal normal in a π-separable group G. If $\mathrm{B}_\pi(G) = \{1_G\}$, show that both G/N and N are π'-groups, and thus G is a π'-group. Also, if $\mathrm{B}_\pi(G) = \mathrm{Irr}(G)$, show that both G/N and N are π-groups, and thus G is a π-group.

Notes and References

In the special case where $\pi = p'$, a large part of the theory presented in this chapter appears in [**36**]. In particular, if G is p-solvable, the set denoted $\mathcal{Y}(G)$ in that paper is exactly $\mathrm{B}_{p'}(G)$.

The general definition of $\mathrm{B}_\pi(G)$ and many of the properties of this set appear in [**40**].

As was pointed out in the text, Theorem 4.27 is the key to the proof in Chapter 5 of the fact that the map $\chi \mapsto \chi^0$ is a bijection from $\mathrm{B}_\pi(G)$ onto $\mathrm{I}_\pi(G)$. Theorem 4.27 appears as Proposition 8.3 of [**40**], where it also plays a crucial role.

If α is a quasi-primitive character of a Hall π-subgroup of a π-separable group G, then by Theorem 4.30, there is an extension γ of α to a subgroup W such that γ^G is irreducible. This fact also appears as part of Theorem B of [**43**].

The construction in Theorem 4.31, in which satellite characters are defined, appears as Theorem C of [**61**].

The fact (which is part of Corollary 4.35) that for an odd-order group G, an irreducible character χ lies in $\mathrm{B}_\pi(G)$ if and only if χ is fixed by the magic field automorphism is Theorem 8.3 of [**47**].

Theorem 4.40 appears as Theorem 12.2 of [**38**].

The definition and construction of the normal nucleus in Problem 4.1 is due to G. Navarro, and it appears in Section 2 of [**63**].

In general, it is not true that the set $\mathrm{N}_\pi(G)$ of Problem 4.2 is always equal to $\mathrm{B}_\pi(G)$. An example due to J. P. Cossey can be found in [**11**].

The construction of the normal nucleus is much easier than the construction of the nucleus, so one might ask what is gained by using subnormal subgroups instead of normal subgroups. One answer is that it is not clear how to control restrictions to normal subgroups of characters in $\mathrm{N}_\pi(G)$. By the example in [**11**] the irreducible constituents of such a restriction need not be N_π-characters.

Problem 4.5 is due to P. Centella and G. Navarro. It appears as Theorem 2.2 of [**7**].

Chapter 5

$\mathbf{B}_\pi(G)$ and $\mathbf{I}_\pi(G)$

5A

Recall that a character χ of a π-separable group G is said to be a **lift** of a π-partial character φ of G if $\varphi = \chi^0$, where χ^0 is the restriction of χ to the set G^0 of π-elements of G. By definition, every π-partial character of G has at least one lift, and it should be clear that every lift of an irreducible π-partial character must be irreducible. Usually, there are several members of $\mathrm{Irr}(G)$ that lift any given member of $\mathrm{I}_\pi(G)$, but it is often convenient to be able to choose in advance a **canonical** set of lifts. What is wanted, in other words, is a subset of $\mathrm{Irr}(G)$ constructed in some unambiguous way, and having the property that each member $\varphi \in \mathrm{I}_\pi(G)$ has a unique lift in this subset. As promised, we begin this chapter with the theorem that the set $\mathrm{B}_\pi(G)$ is such a set of canonical lifts. Following that, we derive a number of applications of this fact.

Theorem 5.1. *Let G be π-separable. Then the map $\chi \mapsto \chi^0$ is a bijection from $\mathrm{B}_\pi(G)$ onto $\mathrm{I}_\pi(G)$.*

Proof. We can assume that G is nontrivial, and we proceed by induction on $|G|$. Given $\chi \in \mathrm{B}_\pi(G)$, let $\varphi \in \mathrm{I}_\pi(G)$ be an irreducible constituent of χ^0. In other words, φ is one of the uniquely determined summands when χ^0 is written as a sum of irreducible π-partial characters, and in particular, we have $\chi(1) \geq \varphi(1)$. We will show that equality holds here, and thus $\chi^0 = \varphi$ is irreducible.

Fix a Hall π-subgroup H of G, and let $\alpha \in \mathrm{Irr}(H)$ be a Fong character for φ. Then $\alpha(1) = \varphi(1)_\pi$, and by Theorem 4.27 we also have

$$\alpha(1) = \sum_{\zeta \in \mathrm{B}_\pi(G)} [\alpha^G, \zeta]\zeta(1)_\pi \,.$$

Since α is a constituent of φ_H, it is also a constituent of $(\chi^0)_H = \chi_H$, and thus $[\alpha^G, \chi] > 0$. It follows that the term in the above sum corresponding to $\zeta = \chi$ is nonzero. We deduce that $\varphi(1)_\pi = \alpha(1) \geq \chi(1)_\pi$, and if equality holds, then χ must be the unique member of $\mathrm{B}_\pi(G)$ that lies over α.

Now let N be a maximal normal subgroup of G, and let ψ be an irreducible constituent of χ_N. By Clifford's theorem, every irreducible constituent of χ_N is conjugate to ψ in G, and thus we can write $\chi(1) = a\psi(1)$ for some integer a. Then $a = \chi(1)/\psi(1)$, and it follows from Theorem 1.15 that a divides $|G : N|$.

By Corollary 4.21, each irreducible constituent of χ_N lies in $\mathrm{B}_\pi(N)$, so by the inductive hypothesis, $(\chi^0)_N = (\chi_N)^0$ is a sum of irreducible π-partial characters of N, each of which is conjugate to ψ^0 in G. Since φ is a constituent of χ^0, it follows that φ_N is a sum of some of these conjugates of ψ^0, and thus we have $\varphi(1) = b\psi(1)$ for some integer $b \leq a$.

Suppose that G/N is a π-group. Then a is a π-number, and

$$\chi(1)_\pi = a\psi(1)_\pi \geq b\psi(1)_\pi \geq b_\pi\psi(1)_\pi = \varphi(1)_\pi \,.$$

Since we saw previously that $\varphi(1)_\pi \geq \chi(1)_\pi$, we have equality here, and thus $b = a$. Then $\chi(1) = a\psi(1) = b\psi(1) = \varphi(1)$, and thus $\chi^0 = \varphi$ in this case.

Now assume that G/N is a π'-group. Since $\chi \in \mathrm{B}_\pi(G)$, it follows by Theorem 4.19 that each irreducible constituent of χ_N occurs with multiplicity 1, and thus χ_N is a sum of a distinct conjugates of ψ. By the injectivity of the map in the inductive hypothesis, therefore, the π-partial character $(\chi_N)^0$ is the sum of a distinct conjugates of ψ^0, and thus φ_N must be a sum of some of these. Since φ_N is G-invariant and its decomposition into irreducible π-partial characters of N is unique, it follows that the set of irreducible constituents of φ_N is G-invariant, and thus each of the a conjugates of ψ^0 occurs as a constituent of φ_N. This yields $\varphi(1) = a\psi(1) = \chi(1)$, so in this case too we have $\chi^0 = \varphi$.

We have now shown that $\chi \mapsto \chi^0$ defines a map from $\mathrm{B}_\pi(G)$ into $\mathrm{I}_\pi(G)$, and it remains to show that this map is a bijection. If as before, $\chi^0 = \varphi$ and α is a Fong character for φ, then $\chi(1)_\pi = \varphi(1)_\pi$, and as we observed, this implies that χ is the unique member of $\mathrm{B}_\pi(G)$ that lies over α. It follows that our map is injective.

To prove surjectivity, let $\varphi \in \mathrm{I}_\pi(G)$ be arbitrary, and let α be a Fong character for φ. By Theorem 4.27, there exists $\chi \in \mathrm{B}_\pi(G)$ lying over α, and we have proved that $\chi^0 \in \mathrm{I}_\pi(G)$. Now α is a constituent of $\chi_H = (\chi^0)_H$, and since α is a Fong character for φ, we know that χ^0 cannot be a member of $\mathrm{I}_\pi(G)$ that is different from φ, and thus $\chi^0 = \varphi$. ■

As an immediate consequence of Theorem 5.1, we can count the B_π-characters of a π-separable group G. We have $|\mathrm{B}_\pi(G)| = |\,\mathrm{I}_\pi(G)|$, and we know by Theorem 3.3 that $|\,\mathrm{I}_\pi(G)|$ is the number of π-classes in G. It follows that $|\mathrm{B}_\pi(G)|$ is also equal to the number of π-classes of G. An amusing corollary of this is the fact that the equality $\mathrm{Irr}(G) = \mathrm{B}_\pi(G) \cup \mathrm{B}_{\pi'}(G)$ holds if and only if every class of G is either a π-class or a π'-class. (We leave it to the reader to supply the proof.)

There are several applications of Theorem 5.1 to the theory of Brauer characters of p-solvable groups, and we devote the remainder of this section to a discussion of some of these. Recall that if G is p-solvable and we take $\pi = \{p\}'$, then G is π-separable, and in this situation (as we discussed in Chapter 3) the set $\mathrm{IBr}(G)$ of irreducible Brauer characters of G for the prime p is exactly the set $\mathrm{I}_\pi(G)$ of irreducible π-partial characters.

An easy consequence of Theorem 5.1 is a strong form of the Fong–Swan theorem. Recall that the Fong–Swan theorem asserts that if G is p-solvable and φ is an irreducible Brauer character for the prime p, then φ has a lift $\chi \in \mathrm{Irr}(G)$. Since all values of φ are values of χ at π-elements, we see that $\mathbb{Q}(\varphi) \subseteq \mathbb{Q}(\chi)$. Also, $\mathbb{Q}(\varphi)$ is contained in some cyclotomic field $\mathbb{Q}_m$, where m is a π-number or, equivalently, p does not divide m.

We can now show that in fact, it is always possible to choose a lift χ of φ such that $\mathbb{Q}(\chi) = \mathbb{Q}(\varphi)$, and thus $\mathbb{Q}(\chi)$ is contained in a field $\mathbb{Q}_m$, where p does not divide m. (A character with values contained in such a field $\mathbb{Q}_m$ is said to be **p-rational**.) We can now strengthen the Fong–Swan theorem to say that every irreducible Brauer character of a p-solvable group has a p-rational lift.

To prove this, let $\varphi \in \mathrm{IBr}(G)$. Then $\varphi \in \mathrm{I}_\pi(G)$, where $\pi = \{p\}'$, and so Theorem 5.1 guarantees that φ has a lift $\chi \in \mathrm{B}_\pi(G)$. We know that $\mathbb{Q}(\varphi) \subseteq \mathbb{Q}(\chi)$, and we argue that in fact, these fields are equal, and thus χ is a p-rational lift for φ, as wanted. To see this, let τ be an element of the Galois group $\mathrm{Gal}(\mathbb{Q}(\chi)/\mathbb{Q}(\varphi))$. Then $\chi^\tau \in \mathrm{B}_\pi(G)$ and

$$(\chi^\tau)^0 = (\chi^0)^\tau = \varphi^\tau = \varphi\,,$$

and thus $\chi^\tau = \chi$ by the injectivity of the map in Theorem 5.1. It follows that $\tau = 1$, and thus the Galois group $\mathrm{Gal}(\mathbb{Q}(\chi)/\mathbb{Q}(\varphi))$ is trivial. We conclude that $\mathbb{Q}(\varphi) = \mathbb{Q}(\chi)$, as claimed.

If $p \neq 2$, we can say more: if $\varphi \in \mathrm{IBr}(G)$, then there exists a *unique* p-rational lift of χ. As we will explain, this is a consequence of the following result.

Theorem 5.2. *Let G be π-separable, and assume that a Hall π'-subgroup of G has odd order. Suppose $\chi \in \operatorname{Irr}(G)$ and that $\chi^0 \in \mathrm{I}_\pi(G)$. If χ is fixed by the magic field automorphism for π and G, then $\chi \in \mathrm{B}_\pi(G)$.*

Note that Theorem 5.2 can be viewed as a variation on Theorem 4.34. In that earlier result, we assumed that $\chi(1)$ is odd, and here we are assuming instead that $\chi^0 \in \mathrm{I}_\pi(G)$. In both situations, we show that if χ is fixed by the magic field automorphism for π and G, then $\chi \in \mathrm{B}_\pi(G)$.

Recall that the magic field automorphism σ fixes roots of unity with π-order and inverts roots of unity with π'-order. In the case where $\pi = \{p\}'$, therefore, every p-rational character of G is fixed by σ. Also, if $p \neq 2$, then a Hall π'-subgroup (i.e., a Sylow p-subgroup) has odd order, and thus Theorem 5.2 applies, and we conclude that a p-rational lift of a member of $\operatorname{IBr}(G) = \mathrm{I}_\pi(G)$ must lie in $\mathrm{B}_\pi(G)$. If $\varphi \in \operatorname{IBr}(G)$, therefore, the unique lift of φ in $\mathrm{B}_\pi(G)$ is the only p-rational lift, and thus φ has a unique p-rational lift. Note, however, that if $p = 2$, then an irreducible Brauer character may have more than one p-rational lift. If G is an elementary abelian 2-group, for example, then every irreducible character is rational, and hence is 2-rational, and all of these characters are lifts of the principal Brauer character of G.

Proof of Theorem 5.2. Let σ be the magic field automorphism for π and G, and let (W, γ) be a nucleus for χ. We will show that $\gamma^\sigma = \gamma$, and since γ is π-factored and a Hall π'-subgroup of W has odd order, it will follow by Lemma 4.36 that γ is π-special, and hence χ lies in $\mathrm{B}_\pi(G)$, as wanted.

Since (W, γ^σ) is a nucleus for $\chi^\sigma = \chi$ and all nuclei for χ are conjugate, there exists an element $x \in G$ such that $(W, \gamma)^x = (W, \gamma^\sigma)$. It follows that $x \in \mathbf{N}_G(W)$ and $\gamma^x = \gamma^\sigma$. Write $H = W\langle x \rangle$, and observe that $W \lhd H$.

Let $\theta = \gamma^0$. The values of θ lie in the cyclotomic field $\mathbb{Q}_m$, where $m = |W|_\pi$, and thus θ is fixed by σ, and we have

$$\theta^x = (\gamma^x)^0 = (\gamma^\sigma)^0 = \theta^\sigma = \theta\,.$$

Also,

$$\theta^G = (\gamma^0)^G = (\gamma^G)^0 = \chi^0 \in \mathrm{I}_\pi(G)\,,$$

and since θ^G is irreducible, we see that θ and θ^H are also irreducible.

By Theorem 5.1, we can choose $\psi \in \mathrm{B}_\pi(W)$ with $\psi^0 = \theta$, and by the injectivity of the map in Theorem 5.1, we know that θ uniquely determines ψ. Since $\theta^x = \theta$, it follows that $\psi^x = \psi$, and hence ψ is invariant in H. Also, $(\psi^H)^0 = (\psi^0)^H = \theta^H$ is irreducible, and thus ψ^H is irreducible. Since ψ is invariant in H and ψ^H is irreducible, we deduce that $W = H$, and thus $x \in W$, and we have $\gamma^\sigma = \gamma^x = \gamma$, as wanted. ∎

Another consequence of Theorem 5.1 that has an application to the theory of Brauer characters of p-solvable groups is the following.

Corollary 5.3. *Let G be π-separable, and assume that a Hall π'-subgroup of G has odd order. Let $\chi \in \mathrm{Irr}(G)$ be real valued and have odd degree. Then $\chi \in \mathrm{B}_\pi(G)$, and thus $\chi^0 \in \mathrm{I}_\pi(G)$.*

This is relevant in the situation where G is p-solvable with $p \neq 2$. If we take $\pi = \{p\}'$, then a Hall π'-subgroup of G is a Sylow p-subgroup, which has odd order. By Corollary 5.3, therefore, a real-valued character χ of G having odd degree will be Brauer irreducible. (In other words, the restriction χ^0 of χ to the set of p'-elements of G is an irreducible Brauer character for the prime p.)

Proof of Corollary 5.3. By Theorem 4.34, we see that $\chi \in \mathrm{B}_\pi(G)$, and thus $\chi^0 \in \mathrm{I}_\pi(G)$ by Theorem 5.1. ■

We have just seen that if G is p-solvable and $p \neq 2$, then a real irreducible character of G having odd degree must be Brauer irreducible. It is natural to ask, therefore, what happens if $p = 2$. In fact, if $p = 2$, the conclusion of Corollary 5.3 would fail rather dramatically. A theorem of P. Fong asserts that if G is an arbitrary finite group and $p = 2$, then the only odd-degree real-valued irreducible Brauer character of G is the principal character. Thus if $\chi \in \mathrm{Irr}(G)$ has odd degree and is real valued, then χ cannot be Brauer irreducible for $p = 2$ unless $\chi(1) = 1$.

Fong's theorem in the case where G is solvable is a consequence of the following.

Corollary 5.4. *Let G be π-separable, where $2 \notin \pi$, and let $\varphi \in \mathrm{I}_\pi(G)$ be real and have odd degree. Then φ is principal.*

Proof. Let $\chi \in \mathrm{B}_\pi(G)$ with $\chi^0 = \varphi$, and observe that $\chi(1) = \varphi(1)$ is odd. Also, χ is real valued since, as we have seen, $\mathbb{Q}(\chi) = \mathbb{Q}(\varphi)$. By Corollary 5.3 with the roles of π and π' interchanged, we deduce that $\chi \in \mathrm{B}_{\pi'}(\mathrm{G})$, and thus since also $\chi \in \mathrm{B}_\pi(G)$, we have $\chi = 1_G$, and so φ is principal. (See the discussion following the statement of Theorem 4.12.) ■

Finally, we mention a theorem of B. Huppert, which asserts that if $\varphi \in \mathrm{IBr}(G)$, where G is p-solvable, then there exists a subgroup $W \subseteq G$ and a Brauer character $\theta \in \mathrm{IBr}(W)$ such that $\theta(1)$ is a p'-number and $\theta^G = \varphi$. More generally, we have the following.

Corollary 5.5. *Let G be π-separable, and suppose $\varphi \in \mathrm{I}_\pi(G)$. Then there exists an irreducible π-partial character θ of some subgroup of G such that $\theta(1)$ is a π-number and $\theta^G = \varphi$.*

Proof. By Theorem 5.1, there exists $\chi \in \mathrm{B}_\pi(G)$ such that $\chi^0 = \varphi$. If (W, γ) is a nucleus for χ, then $\gamma^G = \chi$, so if $\theta = \gamma^0$, we have $\theta^G = \varphi$, and the irreducibility of φ guarantees that θ is irreducible. Also, because $\chi \in \mathrm{B}_\pi(G)$, its nucleus character γ is π-special, and thus $\theta(1) = \gamma(1)$ is a π-number. ∎

5B

We continue to discuss applications of Theorem 5.1. First, we present a result that includes the promised stronger version of Theorem 4.30.

Theorem 5.6. *Let $\alpha \in \mathrm{Irr}(H)$ be quasi-primitive, where H is a Hall π-subgroup of a π-separable group G, and let $H \subseteq W \subseteq G$, where W is maximal such that α has an extension to W. If β is any extension of α to W, then β^G is irreducible. Also, α is a Fong character, and it is the smallest degree irreducible constituent of $(\beta^G)_H$.*

Proof. By Theorem 4.30, we know that if γ is the unique π-special extension of α to W, then $\gamma^G \in \mathrm{B}_\pi(G)$. By Theorem 5.1, we have $(\gamma^0)^G = (\gamma^G)^0 \in \mathrm{I}_\pi(G)$, and we write $(\gamma^0)^G = \varphi$. Now $\beta_H = \alpha = \gamma_H$, and since H is a Hall π-subgroup of W, it follows that β and γ agree on all π-elements of W. Then $\beta^0 = \gamma^0$, and we deduce that

$$(\beta^G)^0 = (\beta^0)^G = (\gamma^0)^G = \varphi .$$

Then β^G is a lift of the irreducible π-partial character φ, and it follows that β^G is irreducible.

Now

$$\varphi(1) = \beta^G(1) = |G : W|\beta(1) = |G : W|\alpha(1) ,$$

and since $|G : W|$ is a π'-number, we see that $\varphi(1)_\pi = \alpha(1)$. Also, α is a constituent of $(\beta^G)_H = \varphi_H$, and thus α is a Fong character for φ. It follows that α has minimum degree among irreducible constituents of $\varphi_H = (\beta^G)_H$. ∎

As in Theorem 5.6, let H be a Hall π-subgroup of a π-separable group G. We shall see in Theorem 5.13 that if $\alpha \in \mathrm{Irr}(H)$ is actually primitive (and not just quasi-primitive), then not only is α a Fong character for some member $\varphi \in \mathrm{I}_\pi(G)$, but in fact all Fong characters for φ are conjugate to α in $\mathbf{N}_G(H)$, and so all of them are primitive.

We remark that in general, it is not true that the set of irreducible characters of H that are Fong characters for a given member $\varphi \in \mathrm{I}_\pi(G)$ are conjugate in $\mathbf{N}_G(H)$. In fact, it can even happen that the fields of values of the Fong characters for φ are not all the same. An example of this phenomenon occurs for $\pi = \{2\}$ in the group constructed as the semidirect product of the symmetric group S_3 acting to permute the factors of the direct product of three cyclic groups of order 4.

We need some results connecting irreducible π-partial characters and normal subgroups. We begin with an analogue of Clifford's theorem that follows easily from Theorem 5.1.

Corollary 5.7. *Let $N \lhd G$, where G is π-separable, and let $\varphi \in \mathrm{I}_\pi(G)$. Then*

$$\varphi_N = e\sum_{i=1}^{t} \theta_i$$

for some integer e, where the θ_i are distinct members of $\mathrm{I}_\pi(N)$ that form an orbit for the conjugation action of G on $\mathrm{I}_\pi(N)$. Also, et divides $|G:N|$, and if $|G:N|$ is a π'-number, then $e = 1$.

Proof. By Theorem 5.1, there exists $\chi \in \mathrm{B}_\pi(G)$ such that $\chi^0 = \varphi$, and by Clifford's theorem, we can write

$$\chi_N = e\sum_{i=1}^{t} \psi_i\,,$$

where the characters ψ_i lie in $\mathrm{Irr}(N)$, and they are distinct and form a G-orbit. Now $\psi_i \in \mathrm{B}_\pi(N)$ by Corollary 4.21, so by Theorem 5.1, the π-partial characters $(\psi_i)^0$ are irreducible and distinct, and they form a G-orbit. Writing $\theta_i = (\psi_i)^0$, we have

$$\varphi_N = (\chi^0)_N = (\chi_N)^0 = e\sum_{i=1}^{t}(\psi_i)^0 = e\sum_{i=1}^{t}\theta_i\,,$$

as required. Also, $\chi(1) = et\psi_1(1)$, and thus $et = \chi(1)/\psi_1(1)$, and this divides $|G:N|$ by Theorem 1.15(a). Finally, if $|G:N|$ is a π'-number, then $e = 1$ by Theorem 4.19(b). ∎

It is convenient to introduce some notation and language for partial characters that is analogous to what we have been using for ordinary characters. If $H \subseteq G$ and $\theta \in \mathrm{I}_\pi(H)$, we will write $\mathrm{I}_\pi(G|\theta)$ to denote the set of members of $\mathrm{I}_\pi(G)$ whose restrictions to H have θ as an irreducible constituent. If $\varphi \in \mathrm{I}_\pi(G|\theta)$, we will sometimes say that θ lies under φ and that φ lies over θ.

Lemma 5.8. *Let $H \subseteq G$, where G is π-separable, and let $\theta \in \mathrm{I}_\pi(H)$. Then $\mathrm{I}_\pi(G|\theta)$ contains some irreducible constituent of θ^G. In particular, $\mathrm{I}_\pi(G|\theta)$ is nonempty, and if $\theta^G = \varphi$ is irreducible, then $\varphi \in \mathrm{I}_\pi(G|\theta)$.*

Proof. Let $\psi \in \mathrm{Irr}(H)$ with $\psi^0 = \theta$, and write $\chi = \psi^G$. Then $\chi^0 = (\psi^G)^0 = (\psi^0)^G = \theta^G$. Now χ lies over ψ, so χ^0 lies over $\psi^0 = \theta$, and thus some irreducible constituent of $\chi^0 = \theta^G$ lies over θ, as required. ∎

Some caution is appropriate here because unlike the situation for ordinary characters, it is *false* in general that $\mathrm{I}_\pi(G|\theta)$ is equal to the set of irreducible constituents of θ^G. In fact, there can exist irreducible constituents of θ^G that do not lie in $\mathrm{I}_\pi(G|\theta)$, and there can exist members of $\mathrm{I}_\pi(G|\theta)$ that are not constituents of θ^G.

The first of these two phenomena occurs, for example, whenever H is a nonnormal Hall π-subgroup of G and θ is the principal character 1_H, viewed as a π-partial character of H. Since 1_H is a Fong character for the principal π-partial character φ of G, we see that φ is the unique member of $\mathrm{I}_\pi(G|\theta)$. If it were true that all irreducible constituents of θ^G were members of $\mathrm{I}_\pi(G|\theta)$, therefore, it would follow that $((1_H)^G)^0$ is a multiple of the principal π-partial character φ and thus $(1_H)^G(x) = (1_H)^G(1)$ for all π-elements x of G. Then $H \subseteq \ker((1_H)^G)$, and since the reverse containment always holds, we have $H = \ker((1_H)^G) \triangleleft G$. This contradicts our assumption that H is not normal.

An example where some member of $\mathrm{I}_\pi(G|\theta)$ fails to be a constituent of θ^G occurs when G is the symmetric group S_4 and H is a point stabilizer in G, so $H \cong S_3$. Let $\pi = \{2\}$, and let θ be the principal π-partial character of H. Now G has two irreducible characters χ and ζ of degree 3, and one can check that χ^0 and ζ^0 are distinct members of $\mathrm{I}_\pi(G)$. If $\alpha \in \mathrm{Irr}(H)$ is the character of degree 2, then α is a constituent of both χ_H and ζ_H. It is easy to see that θ is an irreducible constituent of α^0, and thus it is a constituent of both $(\chi^0)_H$ and $(\zeta^0)_H$. Then χ^0 and ζ^0 lie in $\mathrm{I}_\pi(G|\theta)$, but only one of these π-partial characters is a constituent of θ^G.

Before we proceed to develop any more of the general theory, we digress to present an amusing application.

Theorem 5.9. *Let $\alpha \in \mathrm{Irr}(H)$, where H is a Hall π-subgroup of a π-separable group G, and assume that π is not the set of all prime numbers. Then α extends to G if and only if there exists a prime $p \notin \pi$ and a (possibly reducible) character χ of G such that α is the unique irreducible constituent of χ_H with multiplicity not divisible by p.*

Recall that we already know a necessary and sufficient condition for α to have an extension to G: that $\alpha(x) = \alpha(y)$ for every pair of elements $x, y \in H$ that are conjugate in G (see Theorem 3.6). What seems remarkable about Theorem 5.9 is that it provides a sufficient condition for extendibility that seems to say very little about the character α and nothing at all about its values. Also, it seems surprising that there is no need to assume that the character χ is irreducible or that the prime p is divisor of the group order.

Although the sufficiency of the extendibility condition in Theorem 5.9 may be surprising, its necessity is completely obvious. If we take χ to be an extension α, then α is the unique irreducible constituent of χ_H, and since α occurs with multiplicity 1 in χ_H, it is certainly the unique irreducible constituent with multiplicity not divisible by p, where p is an arbitrary prime. (We need the assumption that π is a proper subset of the set of all prime numbers in Theorem 5.9 so that it is possible to satisfy the requirement that $p \notin \pi$.)

The following lemma lies at the heart of the proof of Theorem 5.9.

Lemma 5.10. *Let H be a Hall π-subgroup of a π-separable group G, and let $\varphi \in \mathrm{I}_\pi(G)$ and $\alpha \in \mathrm{Irr}(H)$. Suppose for some prime $p \notin \pi$ that α is the unique irreducible constituent of φ_H with multiplicity not divisible by p. Then $\varphi_H = \alpha$.*

Proof. We can suppose that $H < G$, so G is not a π-group, and we proceed by induction on $|G|$. Let $N = \mathbf{O}^\pi(G)$ and $M = \mathbf{O}^{\pi'}(N)$, and write $K = MH$. Then $N > 1$, and since $N = \mathbf{O}^\pi(N)$, it follows that $M < N$. Also, $M = K \cap N$, and thus $K < G$.

A Fong constituent of φ_H occurs with multiplicity 1, and since by hypothesis, α is the only constituent of φ_H with multiplicity not divisible by p, it follows that α is a Fong character for φ. In particular, α is a constituent of minimum degree of φ_H. Now let $\nu \in \mathrm{I}_\pi(K)$, where ν lies under φ and over α. Then α has minimum degree among the irreducible constituents of ν_H, and hence α is a Fong character for ν. We conclude that $\varphi(1)_\pi = \alpha(1) = \nu(1)_\pi$.

The multiplicity of α as a constituent of φ_H is 1, and thus the multiplicity of ν as a constituent of φ_K is also 1. If μ is any constituent of φ_K different from ν and $\beta \in \mathrm{Irr}(H)$ is a Fong character for μ, then β lies under φ and $\beta \neq \alpha$, and thus by hypothesis, the multiplicity of β as a constituent of φ_H is divisible by p. Also, β lies under no member of $\mathrm{I}_\pi(K)$ other than μ, so the multiplicities of μ in φ_K and β in φ_H are equal, and thus the multiplicity of μ as a constituent of φ_K is divisible by p. It follows that we can write $\varphi_K = \nu + p\Delta$, where Δ is either zero or is a not necessarily irreducible π-partial character of K. We conclude that for each member $\theta \in \mathrm{I}_\pi(M)$, the multiplicities of θ as a constituent of φ_M and as a constituent of ν_M are congruent modulo p.

Since $M \triangleleft K$, it follows by Corollary 5.7 that the multiplicities of the irreducible constituents of ν_M are equal. Also, this common multiplicity divides the π-number $|K : M|$, and hence it is not divisible by p. Each irreducible constituent of ν_M, therefore, has multiplicity not divisible by p and it follows that its multiplicity as a constituent of φ_M is also not divisible by p.

At least one irreducible constituent of φ_M is a constituent of ν_M, and since the multiplicity of that constituent in φ_M is not divisible by p, it follows by another application of Corollary 5.7 that none of the irreducible constituents of φ_M has multiplicity divisible by p. Each irreducible constituent of φ_M thus occurs with multiplicity not divisible by p in ν_M, and in particular, it occurs with nonzero multiplicity in ν_M.

We see now that φ_M and ν_M have identical sets of irreducible constituents, and by Corollary 5.7, these are permuted transitively by both G and K. We deduce that $G = KT$, where T is the stabilizer in G of one of the irreducible constituents of ν_M, say θ. Now $|G : T| = |K : K \cap T|$, and this divides the π-number $|K : M|$. Since N/M is a normal Hall π'-subgroup of G/M, it follows that $N \subseteq T$, and thus θ is invariant in N.

Let $\psi \in \mathrm{I}_\pi(N)$ lie over θ and under φ. By Corollary 5.7, we can write $\psi_M = e\theta$ for some integer e, and in fact, $e = 1$ because $|N : M|$ is a π'-number, and thus $\psi(1) = \theta(1)$. Then $\varphi(1)/\theta(1) = \varphi(1)/\psi(1)$, and this integer divides the π-number $|G : N|$ by Corollary 5.7. It follows that $\varphi(1)_{\pi'} = \theta(1)_{\pi'}$. Similarly, $\nu(1)/\theta(1)$ divides the π-number $|K : M|$, and thus $\nu(1)_{\pi'} = \theta(1)_{\pi'}$, and we conclude that $\varphi(1)_{\pi'} = \nu(1)_{\pi'}$. Since we saw previously that $\varphi(1)_\pi = \nu(1)_\pi$, we deduce that $\varphi(1) = \nu(1)$, and since φ lies over ν, we have $\varphi_K = \nu$.

Now $\varphi_H = \nu_H$, and thus α is the unique irreducible constituent of ν_H with multiplicity divisible by p. By the inductive hypothesis, therefore, $\nu_H = \alpha$, and the result follows. ∎

Proof of Theorem 5.9. As we have seen, it suffices to show that if χ is a character of G and $\alpha \in \mathrm{Irr}(H)$ is the unique irreducible constituent of χ_H with multiplicity not divisible by p, then α extends to G.

Write $\chi^0 = \sum a_\mu \mu$, where μ runs over $\mathrm{I}_\pi(G)$ and the coefficients a_μ are nonnegative integers. For each member $\mu \in \mathrm{I}_\pi(G)$, choose a Fong character β_μ for μ. Since β_μ has multiplicity 1 as a constituent of μ_H and β_μ lies under no member of $\mathrm{I}_\pi(G)$ other than μ, it follows that the multiplicity of β_μ as a constituent of $\chi_H = (\chi^0)_H$ is exactly a_μ. By hypothesis, therefore, a_μ is divisible by p whenever $\beta_\mu \neq \alpha$.

Since α occurs with multiplicity not divisible by p in $\chi_H = (\chi^0)_H$, we see that χ^0 is not p times some π-partial character, and thus at least one of the coefficients a_φ is not divisible by p. It follows that $\beta_\varphi = \alpha$, and thus for all members $\mu \in \mathrm{I}_\pi(G)$ with $\mu \neq \varphi$, we see that a_μ is divisible by p.

Now $\alpha = \beta_\varphi$ occurs with multiplicity 1 as a constituent of φ_H, and we argue that if $\gamma \in \mathrm{Irr}(H)$ is different from α, then the multiplicity $[\varphi_H, \gamma]$ is divisible by p. To see this, observe that

$$[\chi_H, \gamma] = \sum_{\mu} a_\mu [\mu_H, \gamma] ,$$

and by hypothesis the left-hand side of this equation is divisible by p. Since a_μ is divisible by p for $\mu \neq \varphi$, it follows that $a_\varphi[\varphi_H, \gamma]$ is divisible by p. We know, however, that a_φ is not divisible by p, and hence $[\varphi_H, \gamma]$ is divisible by p, as claimed. By Lemma 5.10, therefore, $\varphi_H = \alpha$, and thus each lift for φ is an extension of α to G. ∎

Returning now to our study of the connections between the normal subgroup structure of a π-separable group and its set of irreducible π-partial characters, we present an analogue of the Clifford correspondence.

Theorem 5.11. *Suppose that $\theta \in \mathrm{I}_\pi(N)$, where $N \lhd G$ and G is π-separable, and let T be the stabilizer of θ in G. Then induction defines a bijection $\mathrm{I}_\pi(T|\theta) \to \mathrm{I}_\pi(G|\theta)$. Also, if $\alpha \in \mathrm{I}_\pi(T|\theta)$ and $\beta = \alpha^G$, then α is the unique irreducible constituent of β_T that lies in $\mathrm{I}_\pi(T|\theta)$.*

Proof. Given $\beta \in \mathrm{I}_\pi(G|\theta)$, our first goal is to show that β_T has a unique irreducible constituent α lying over θ, and that furthermore $\alpha^G = \beta$.

By Theorem 5.1, there exists a unique member $\psi \in \mathrm{B}_\pi(N)$ such that $\psi^0 = \theta$, and a unique member $\chi \in \mathrm{B}_\pi(G)$ such that $\chi^0 = \beta$. Also, the irreducible constituents of χ_N lie in $\mathrm{B}_\pi(N)$ by Corollary 4.21, and since θ is a constituent of β_N, it follows that ψ is a constituent of χ_N, and in fact, the multiplicity $e = [\chi_N, \psi]$ is exactly the multiplicity with which θ occurs as a constituent of β_N.

Since ψ and θ uniquely determine each other, the stabilizer T of θ in G is also the stabilizer of ψ, and since χ lies over ψ, it follows by the Clifford correspondence that $\chi = \eta^G$ for some character $\eta \in \mathrm{Irr}(T|\psi)$. Then $[\eta_N, \psi] = [\chi_N, \psi] = e$, and thus $\eta_N = e\psi$.

Now write $\alpha = \eta^0$. Then

$$\alpha^G = (\eta^0)^G = (\eta^G)^0 = \chi^0 = \beta ,$$

so α^G is irreducible, and thus $\alpha \in \mathrm{I}_\pi(T)$. Also, since χ lies over η, it follows that $\chi^0 = \beta$ lies over $\eta^0 = \alpha$. Furthermore,

$$\alpha_N = (\eta^0)_N = (\eta_N)^0 = e\psi^0 = e\theta ,$$

and in particular α lies over θ.

We have now shown that $\alpha \in \mathrm{I}_\pi(T|\theta)$ and that α lies under β, and we argue that α is unique with these properties. To see this, recall that the multiplicity of θ as a constituent of β_N is exactly e, and we have seen that this is also the multiplicity of θ in α_N. It follows that no irreducible constituent of β_N other than α can lie over θ, and the uniqueness follows.

We can now define a map $f : \mathrm{I}_\pi(G|\theta) \to \mathrm{I}_\pi(T|\theta)$ by setting $f(\beta) = \alpha$, where α is the unique irreducible constituent of β_T that lies over θ. We saw that if $f(\beta) = \alpha$, then $\alpha^G = \beta$, and thus f followed by induction to G is the identity map on $\mathrm{I}_\pi(G|\theta)$. In particular, f is injective.

To show that the map f is surjective, let $\alpha \in \mathrm{I}_\pi(T|\theta)$ be arbitrary. By Lemma 5.8, we see that α^G has an irreducible constituent β that lies over α. In particular, β lies over θ, and so $f(\beta)$ is defined, and we have $f(\beta) = \alpha$, as required.

We know now that the map f is invertible, and we recall that $f(\beta)^G = \beta$ for all $\beta \in \mathrm{I}_\pi(G|\theta)$. If $\alpha \in \mathrm{I}_\pi(T|\theta)$ and we write $\beta = f^{-1}(\alpha)$, we have $\alpha^G = f(\beta)^G = \beta = f^{-1}(\alpha)$, and thus induction to G is exactly the map $f^{-1} : \mathrm{I}_\pi(T|\theta) \to \mathrm{I}_\pi(G|\theta)$. In particular, this induction map is a bijection. ∎

In order to make effective use of Theorem 5.11, it is convenient to be able to produce a subgroup $N \lhd G$ such that we can guarantee that φ_N has irreducible constituents that are not G-invariant, where $\varphi \in \mathrm{I}_\pi(G)$. The following result can be used for this purpose.

Lemma 5.12. *Let G be π-separable, and suppose that $\varphi \in \mathrm{I}_\pi(G)$, where $\varphi(1)$ is not a π-number. Then there exists a characteristic subgroup N of G such that the irreducible constituents of φ_N are not G-invariant, and furthermore, the common degree of these constituents is a π-number.*

Proof. By Corollary 5.7, we know that for each normal subgroup N of G, the irreducible constituents of φ_N are conjugate in G, and hence they have equal degrees. Let $\mathcal{S}$ be the set of characteristic subgroups N of G such that the common degree of the irreducible constituents of φ_N is a π-number. The set $\mathcal{S}$ is nonempty since it contains the trivial subgroup, so we can choose a maximal member N of $\mathcal{S}$, and we let ν be an irreducible constituent of φ_N.

Now $\nu(1)$ is a π-number, so to complete the proof, it suffices to show that ν is not invariant in G. We assume, therefore, that ν is invariant in G, and we derive a contradiction.

By hypothesis, $\varphi(1)$ is not a π-number, so $\nu \neq \varphi$. Then $N < G$, and we can choose a minimal characteristic subgroup M/N of G/N, and we observe that M is characteristic in G and M/N is either a π-group or a π'-group.

Let $\mu \in \mathrm{I}_\pi(M)$ lie over ν and under φ, and observe that $\mu(1)$ is not a π-number by the maximality of N. It follows that $\mu(1)/\nu(1)$ is not a π-number.

Applying Corollary 5.7 for the normal subgroup $N$of M, we deduce that $\mu(1)/\nu(1) = et = e$, where the second equality holds since $t = 1$. (Recall that we are assuming that ν is G-invariant.) Then e is not a π-number, and since e divides $|M : N|$, it follows that $|M : N|$ is not a π-group. We conclude that $|M : N|$ is a π'-number, and it follows by Corollary 5.7 that $e = 1$. Since we have seen that e is not a π-number, this is the desired contradiction. ■

We mention that Theorem 5.11 and Lemma 5.12 can be used to provide an alternative proof of Corollary 5.5, which asserts that every member $\varphi \in \mathrm{I}_\pi(G)$ is induced from some π-partial character having π-degree. If $\varphi(1)$ is a π-number, there is nothing to prove. Otherwise, Lemma 5.12 guarantees that there exist $N \triangleleft G$ and $\theta \in \mathrm{I}_\pi(N)$, such that $\varphi \in \mathrm{I}_\pi(G|\theta)$ and the stabilizer T of θ in G is a proper subgroup. Then φ is induced from T by Theorem 5.11, and the result follows by induction on $|G|$, applying the inductive hypothesis in T.

5C

The following is the main result of this section.

Theorem 5.13. *Let H be a Hall π-subgroup of a π-separable group G, and suppose that $\alpha \in \mathrm{Irr}(H)$ is primitive. Then α is a Fong character for some π-partial character $\varphi \in \mathrm{I}_\pi(G)$, and if $\beta \in \mathrm{Irr}(H)$ is an arbitrary Fong character for φ, then β is conjugate to α in $\mathbf{N}_G(H)$, and in particular, β is primitive.*

In the situation of Theorem 5.13, all members of $\mathrm{Irr}(H)$ that are Fong characters for φ are primitive. Since all Hall π-subgroups of G are conjugate, we observe that the same must be true if H is replaced by any other Hall π-subgroup of G.

We begin with an easy observation.

Lemma 5.14. *Let H be a Hall π-subgroup of a π-separable group G, and let*

$$\mathcal{X} = \{\varphi \in \mathrm{I}_\pi(G) \mid \varphi(1) \text{ is a } \pi\text{-number}\}.$$

Then restriction defines a map from $\mathcal{X}$ into $\mathrm{Irr}(H)$. Also, if $\varphi \in \mathcal{X}$ and $\varphi_H = \alpha$, then φ is the unique member of $\mathrm{I}_\pi(G)$ lying over α.

Proof. Recall from Theorem 3.4 that if $\varphi \in \mathrm{I}_\pi(G)$ and $\alpha \in \mathrm{Irr}(H)$ is a Fong character for φ, then α is a constituent of φ_H and $\alpha(1) = \varphi(1)_\pi$. If $\varphi \in \mathcal{X}$, then by definition $\varphi(1)$ is π-number, and hence $\alpha(1) = \varphi(1)_\pi = \varphi(1)$. Then $\varphi_H = \alpha$, so we have a well-defined map. Also, Theorem 3.4 guarantees that α is not a constituent of μ_H for any member $\mu \in \mathrm{I}_\pi(G)$ different from φ. ■

Proof of Theorem 5.13. We proceed by induction on $|G|$. First, since α is primitive, it is certainly quasi-primitive, and hence Theorem 5.6 applies, and we conclude that α is a Fong character for some member $\varphi \in \mathrm{I}_\pi(G)$, and in particular α is a constituent of φ_H. If $\varphi(1)$ is a π-number, then φ_H is irreducible by Lemma 5.14, and thus $\varphi_H = \alpha$. In this case, α is the unique member of $\mathrm{Irr}(H)$ that is a Fong character for φ, and there is nothing further to prove.

We can now assume that $\varphi(1)$ is not a π-number, so by Lemma 5.12, there is a subgroup $N \lhd G$ such that the irreducible constituents of φ_N are not invariant in G, and their common degree is a π-number. Let $M = H \cap N$, and note that M is a full Hall π-subgroup of N because $N \lhd G$.

Let $\beta \in \mathrm{Irr}(H)$ be an arbitrary Fong character for φ, and let γ and δ be irreducible constituents of α_M and β_M, respectively. Then γ lies under φ, so there is some irreducible constituent σ of φ_N such that γ lies under σ. Since $\sigma(1)$ is a π-number, it follows by Lemma 5.14 that σ_M is irreducible, and thus $\sigma_M = \gamma$. Similarly, there is some irreducible constituent τ of φ_N such that $\tau_M = \delta$.

By the Frattini argument, $G = N\mathbf{N}_G(M)$, and since we know by Corollary 5.7 that σ and τ are conjugate in G, it follows that there exists an element $g \in \mathbf{N}_G(M)$ such that $\tau^g = \sigma$, and thus we have

$$\delta^g = (\tau_M)^g = (\tau^g)_M = \sigma_M = \gamma\,.$$

Now $M \lhd H$ and α is quasi-primitive, and since γ is a constituent of α_M, we see that γ is invariant in H.

Lemma 5.14 guarantees that σ is the unique member of $\mathrm{I}_\pi(N)$ lying over γ, and since H stabilizes γ, it follows that H is contained in the stabilizer T of σ in G. Similarly, the stabilizer U of γ in H^g is contained in T. Since U is a π-subgroup of T, it is contained in some Hall π-subgroup K of T, and we observe that K is a full Hall π-subgroup of G because T contains the Hall π-subgroup H of G.

Now β lies over δ, so β^g is an irreducible character of H^g that lies over $\delta^g = \gamma$, and thus by the Clifford correspondence there exists a character $\eta \in \mathrm{Irr}(U|\gamma)$ such that $\eta^{H^g} = \beta^g$.

Now φ lies over β, so φ also lies over β^g, and thus φ lies over η. It follows that there exists $\varepsilon \in \mathrm{Irr}(K)$ lying over η and under φ. Now K is a Hall π-subgroup of G, so the minimum of the degrees of the irreducible constituents of φ_K is the degree of a Fong character for φ, namely $\beta(1)$.

Since ε is an irreducible constituent of φ_K, we have

$$\varepsilon(1) \geq \beta(1) = \beta^g(1) = \eta(1)|H^g : U| = \eta(1)|K : U| = \eta^K(1) \geq \varepsilon(1)\,,$$

where the last inequality holds because ε lies over η, and hence is a constituent of η^K. We thus have equality, and it follows that $\varepsilon = \eta^K$, and thus $\varepsilon(1) = \beta(1)$.

Theorem 5.11 guarantees that there exists $\mu \in \mathrm{I}_\pi(T|\sigma)$ such that $\mu^G = \varphi$. We have $\varphi(1) = |G : T|\mu(1)$, and since $|G : T|$ is a π'-number because $H \subseteq T$, we deduce that $\varphi(1)_\pi = \mu(1)_\pi$. The degree of a Fong character for μ, therefore, is the same as the degree of a Fong character for φ, namely $\alpha(1) = \beta(1) = \varepsilon(1)$.

We argue next that α and ε actually are Fong characters for μ. Now $\alpha \in \mathrm{Irr}(H)$ and $\varepsilon \in \mathrm{Irr}(K)$, and both H and K are Hall π-subgroups of T. So to see that α and ε are Fong characters for μ, it suffices to show that α and ε lie under μ.

It follows by Theorem 5.11 that μ is the unique irreducible constituent of φ_T that lies over σ. Also, we have seen that σ is the unique member of $\mathrm{I}_\pi(N)$ that lies over γ, and it follows that μ is the unique irreducible constituent of φ_T that lies over γ. Since each of α and ε lies over γ and under φ, it follows that α and ε lie under μ, and we conclude that as claimed, α and ε are Fong characters for μ.

Now $K^t = H$ for some element $t \in T$, and thus $\varepsilon^t \in \mathrm{Irr}(H)$ is a Fong character for μ. Since $T < G$ (because σ is not invariant in G) the inductive hypothesis guarantees that α and ε^t are conjugate in $\mathbf{N}_T(H)$, so we can write $\varepsilon^{tx} = \alpha$ for some element $x \in \mathbf{N}_T(H)$, and since α is primitive, it follows that ε is primitive. Now $\varepsilon = \eta^K$, where η is a character of U, and thus the primitivity of ε implies that $K = U \subseteq H^g$, and thus $K = H^g$. (Note that this is the first place where we use the hypothesis that α is actually primitive and not just quasi-primitive.) We now have $H^{gtx} = K^{tx} = H^x = H$, and so $gtx \in \mathbf{N}_G(H)$. Also, since $K = U = H^g$, we have $\varepsilon = \eta = \beta^g$, so $\beta^{gtx} = \varepsilon^{tx} = \alpha$, and thus α and β are conjugate in $\mathbf{N}_G(H)$, as required. ∎

Next, we present a result that can sometimes be used to show that a Fong character is *not* primitive. Recall that Corollary 5.5 asserts that if $\varphi \in \mathrm{I}_\pi(G)$, then there exists a subgroup U and a π-partial character $\theta \in \mathrm{I}_\pi(U)$ such that $\theta(1)$ is a π-number and $\theta^G = \varphi$.

Lemma 5.15. *Let $\varphi \in \mathrm{I}_\pi(G)$, where G be π-separable. Suppose that $\varphi = \theta^G$, where $\theta \in \mathrm{I}_\pi(U)$ for some subgroup $U \subseteq G$, and assume that $\theta(1)$ is a π-number. Let K be a Hall π-subgroup of U, and let H be a Hall π-subgroup of G containing K. Then $(\theta_K)^H$ is irreducible, and it is a Fong character for φ.*

Proof. Observe that φ lies over θ by Lemma 5.8. Also θ_K is irreducible by Lemma 5.14, and since φ lies over θ_K, we see that there exists a character

$\alpha \in \operatorname{Irr}(H)$ lying under φ and over θ_K, and we note that α is a constituent of $(\theta_K)^H$. Then

$$\begin{aligned}\alpha(1) \le (\theta_K)^H(1) = \theta(1)|H:K| = \frac{\theta(1)|G|_\pi}{|U|_\pi} &= \theta(1)|G:U|_\pi \\ &= (\theta(1)|G:U|)_\pi \\ &= \theta^G(1)_\pi \\ &= \varphi(1)_\pi\,,\end{aligned}$$

where the third from last equality holds because $\theta(1)$ is a π-number. Since α lies under φ and the minimum of the degrees of the irreducible constituents of φ_H is $\varphi(1)_\pi$, it follows that equality holds in the above equation, and thus $\alpha = (\theta_K)^H$ is a Fong character for φ. ∎

As an application, we prove a result about M-groups. It had been conjectured that each Hall subgroup of an M-group is itself an M-group. If that were true, an immediate consequence would be that a primitive character of a Hall subgroup of an M-group must be linear. The conjecture is now known to be false, but nevertheless, it is true that all primitive characters of Hall subgroups of M-groups are linear.

Corollary 5.16. *Let α be a primitive character of a Hall π-subgroup of an M-group G. Then α is linear.*

Proof. By Theorem 5.13, we know that α is a Fong character for some π-partial character $\varphi \in \mathrm{I}_\pi(G)$ and that every Fong character for φ (for every Hall π-subgroup) is primitive. Let $\chi \in \operatorname{Irr}(G)$ be a lift of φ. Since G is an M-group, we can write $\chi = \lambda^G$, where λ is a linear character of some subgroup U. Let K be a Hall π-subgroup of U, and choose a Hall π-subgroup H of G containing K.

Writing $\theta = \lambda^0$, we see that $\theta^G = \chi^0 = \varphi$, and also, $\theta(1) = \lambda(1) = 1$ is a π-number, so Lemma 5.15 applies. It follows that $(\theta_K)^H$ is a Fong character for φ, and hence this character is primitive. Then $H = K$ and the Fong character $(\theta_K)^H = \theta_K$ is linear. Since the degrees of all Fong characters for φ are equal, it follows that α is linear.

5D

To motivate the results of this section, we begin with a bit of background from representation theory. Given a field F and a finite group G, an FG-**module** is an F-space V on which G acts via linear transformations. A nonzero FG-module V is said to be **simple** if it has no nonzero proper G-invariant subspace, and it is **indecomposable** if it cannot be written as a direct sum of two nonzero G-invariant subspaces. Of course, simple modules are guaranteed to be indecomposable, but in general, the converse of this statement is false. (In fact, a necessary and sufficient condition for every indecomposable FG-module to be simple is that the characteristic of F fails to divide $|G|$.)

Now assume that F has prime characteristic p. There is a construction due to J. A. Green that assigns to each indecomposable FG-module V a unique conjugacy class of p-subgroups of G, which are called the **vertices** for V. If φ is an irreducible Brauer character of G for the prime p, then φ corresponds to a unique (up to isomorphism) simple FG-module V, where F is algebraically closed, and since V is necessarily indecomposable, there is an associated conjugacy class of vertices, which we refer to as the vertices for φ. (It does not matter which algebraically closed field of characteristic p we work with; we get the same vertices.)

We will not give Green's general definition of the vertices for an indecomposable module here, but we mention the special case that will be most important for us. Given $\varphi \in \mathrm{IBr}(G)$, it may be that there exist some subgroup $U \subseteq G$ and some Brauer character $\theta \in \mathrm{IBr}(U)$, such that $\theta^G = \varphi$ and $\theta(1)$ is a p'-number. (In fact, if G is p-solvable, then by Corollary 5.5 and the discussion preceding it, U and θ definitely exist.) In this case, it follows from Green's definition that a Sylow p-subgroup of U is a vertex for φ. In particular, Green's theory implies that although U may not be unique up to conjugacy in G, the Sylow p-subgroups of U are uniquely determined up to G-conjugacy. In this section, we will prove this uniqueness for p-solvable groups, and in fact, we will establish a more general π-version of this result.

Recall that the irreducible Brauer characters for some prime p for a p-solvable group G are exactly the irreducible π-partial characters of G, where $\pi = \{p\}'$. Since Green's vertices are p-groups, they are π'-groups in this case.

Now suppose that G is π-separable for an arbitrary set π of primes, and let $\varphi \in \mathrm{I}_\pi(G)$. We define a **vertex** for φ to be any Hall π'-subgroup of a subgroup U of G such that there exists $\theta \in \mathrm{I}_\pi(U)$, where $\theta^G = \varphi$ and $\theta(1)$ is a π-number. We write $\mathrm{vtx}(\varphi)$ to denote the set of all vertices for φ, so $\mathrm{vtx}(\varphi)$ is a set of π'-subgroups.

Observe that the set $\mathrm{vtx}(\varphi)$ is never empty. This is because U and θ are guaranteed to exist by Corollary 5.5. Also, we mention that in the case where $\pi = \{p\}'$, so $\mathrm{I}_\pi(G) = \mathrm{IBr}(G)$, it can be shown that the members of $\mathrm{vtx}(\varphi)$ are exactly the vertices in the sense of Green for a simple module corresponding to $\varphi \in \mathrm{IBr}(G)$.

An observation that is sometimes useful is that if $H \subseteq G$ and $\alpha \in \mathrm{I}_\pi(H)$ with $\alpha^G = \varphi$, then $\mathrm{vtx}(\alpha) \subseteq \mathrm{vtx}(\varphi)$. This is clear since every vertex of α is a Hall π'-subgroup of some subgroup $U \subseteq H$ such that there exists $\theta \in \mathrm{I}_\pi(U)$, where $\theta(1)$ is a π-number and $\theta^H = \alpha$. In this situation, $\theta^G = \varphi$, and so the Hall π'-subgroups of U are also vertices for φ.

Our principal result concerning vertices is the following. ■

Theorem 5.17. *Let G be π-separable, and let $\varphi \in \mathrm{I}_\pi(G)$. Then $\mathrm{vtx}(\varphi)$ is a single conjugacy class of π'-subgroups of G.*

We already know that $\mathrm{vtx}(\varphi)$ is a nonempty set of π'-subgroups of G, and that it is invariant under conjugation by elements of G. The essential content of Theorem 5.17, therefore, is that all vertices for a member $\varphi \in \mathrm{I}_\pi(G)$ are conjugate in G. Although it will take some work to prove this, it is easy to see that all vertices for φ have the same order, and we prove this next.

Corollary 5.18. *Let G be π-separable and let $\varphi \in \mathrm{I}_\pi(G)$. If $Q \in \mathrm{vtx}(\varphi)$, then $|Q| = |G|_{\pi'}/\varphi(1)_{\pi'}$.*

Proof. Let $Q \in \mathrm{vtx}(\varphi)$. By definition, Q is a Hall π'-subgroup of some subgroup $U \subseteq G$ such that there exists $\theta \in \mathrm{I}_\pi(U)$, where $\theta(1)$ is a π-number and $\theta^G = \varphi$. We have $\varphi(1) = \theta^G(1) = |G : U|\theta(1)$, and since $\theta(1)$ is a π-number, this yields

$$\varphi(1)_{\pi'} = |G : U|_{\pi'} = \frac{|G|_{\pi'}}{|U|_{\pi'}} = \frac{|G|_{\pi'}}{|Q|}\,,$$

and the result follows. ■

It follows that a vertex for a π-partial character $\varphi \in \mathrm{I}_\pi(G)$ is a full Hall π'-subgroup of G if and only if $\varphi(1)_{\pi'} = 1$, or equivalently, $\varphi(1)$ is a π-number. Recall from Theorem 3.14 that the members φ of $\mathrm{I}_\pi(G)$ with this property are exactly the partial characters χ^0, where $\chi \in \mathrm{Irr}(G)$ is π-special.

We begin work now toward a proof of Theorem 5.17 with an easy lemma.

Lemma 5.19. *Let $U \subseteq G$, where G is π-separable and $|G : U|$ is a π'-number, and suppose that $\varphi \in \mathrm{I}_\pi(G)$ lies over $\theta \in \mathrm{I}_\pi(U)$. Then*

(a) $\theta(1)_\pi \geq \varphi(1)_\pi$.

(b) *If equality holds in* (a), *then φ is the unique member of $\mathrm{I}_\pi(G)$ lying over θ.*

Proof. Let H be a Hall π-subgroup of U. Since $|G : U|$ is a π'-number, H is also a Hall π-subgroup of G, and thus each irreducible constituent of φ_H has degree at least the degree of a Fong character for φ, which is $\varphi(1)_\pi$. If $\alpha \in \mathrm{Irr}(H)$ is a Fong character for θ, then α lies under θ and hence also under φ, and we have $\theta(1)_\pi = \alpha(1) \geq \varphi(1)_\pi$, proving (a). If equality holds, then α is a Fong character for φ, and thus no member of $\mathrm{I}_\pi(G)$ other than φ can lie over α. It follows in this situation that no member of $\mathrm{I}_\pi(G)$ other than φ can lie over θ. ■

Our next two preliminary results lie somewhat deeper.

Lemma 5.20. *Let G be π-separable, and suppose that $G = NU$, where $N \lhd G$ and $|G : U|$ is a π'-number. Write $D = N \cap U$, let $\varphi \in \mathrm{I}_\pi(N)$ and $\theta \in \mathrm{I}_\pi(D)$, and suppose that $\varphi_D = \theta$. Then restriction defines a bijection from $\mathrm{I}_\pi(G|\varphi)$ onto $\mathrm{I}_\pi(U|\theta)$.*

Lemma 5.20 says more than we actually need; we will use only the fact that the restriction of a member $\mathrm{I}_\pi(G|\varphi)$ to U is irreducible. Once we have established that, however, the fact that we have a bijection follows with little extra work.

Proof of Lemma 5.20. If $N = G$, then $D = U$ and the result is clear, so we can assume that $N < G$. We prove by induction on $|G : N|$ that restriction defines a map from $\mathrm{I}_\pi(G|\varphi)$ to $\mathrm{I}_\pi(U|\theta)$. (We will show afterward that this map is a bijection.)

Suppose first that there exists a subgroup $M < G$ with $N < M \lhd G$. Write $V = M \cap U$, and observe that the inductive hypothesis with M in place of G and V in place of U guarantees that restriction maps $\mathrm{I}_\pi(M|\varphi)$ to $\mathrm{I}_\pi(V|\theta)$. If $\mu \in \mathrm{I}_\pi(M|\varphi)$ and $\nu = \mu_V$, it follows that ν is irreducible and lies over θ. We can thus apply the inductive hypothesis again, this time with M in place of N and μ and ν in place of φ and θ, and we deduce that restriction defines a map from $\mathrm{I}_\pi(G|\mu)$ to $\mathrm{I}_\pi(U|\nu)$.

Let $\alpha \in \mathrm{I}_\pi(G|\varphi)$. Since α lies over φ, it lies over some member $\mu \in \mathrm{I}_\pi(M|\varphi)$, and we write $\nu = \mu_V$. As we have seen, ν is irreducible and lies over θ, and thus by the result of the previous paragraph, $\alpha_U \in \mathrm{I}_\pi(U|\nu)$. In particular, α_U is irreducible, and since it lies over ν, it lies over θ, as required.

We can now assume that N is a maximal normal subgroup in G, and thus $|G : N|$ is either a π-number or a π'-number. Let T be the stabilizer of φ in G, and note that $T \cap U$ stabilizes $\varphi_D = \theta$. We argue that in fact, $T \cap U$ is the full stabilizer S of θ in U. To see this, observe that $|N : D| = |G : U|$ is a π'-number and that $\varphi(1) = \theta(1)$, and thus by Lemma 5.19(b), it follows that no member of $\mathrm{I}_\pi(N)$ other than φ can lie over θ. Then S stabilizes φ, and hence $S \subseteq T \cap U$, and we have equality, as claimed. Write $t = |G : T|$, and observe that $G = TU$, so $t = |TU : T| = |U : T \cap U|$, and thus both the G-orbit of φ and the U-orbit of θ have size t.

Let $\alpha \in \mathrm{I}_\pi(G|\varphi)$, so α lies over θ, and we choose an irreducible constituent β of α_U such that β lies over θ. By Corollary 5.7, we can write $\alpha(1) = etd$, where $d = \varphi(1) = \theta(1)$ and e is the multiplicity of φ as a constituent of α_N, and similarly, $\beta(1) = ftd$, where f is the multiplicity of θ as a constituent of β_D. If $|G : N| = |U : D|$ is a π'-number, then $e = 1 = f$ by Corollary 5.7, and thus $\alpha(1) = \beta(1)$, and we have $\alpha_U = \beta$. Otherwise,

$|G : N| = |U : D|$ is a π-number, and thus e and f are π-numbers by Corollary 5.7. Now $\alpha(1) \geq \beta(1)$ so $e \geq f$. Also, $\alpha(1)_\pi \leq \beta(1)_\pi$ by Lemma 5.19(a), and it follows that $e = e_\pi \leq f_\pi = f$, so in this case too, we have $e = f$, and thus $\alpha_U = \beta$.

We have now shown that the restriction defines a map from $\mathrm{I}_\pi(G|\varphi)$ to $\mathrm{I}_\pi(U|\theta)$. To see that this map is injective, let $\alpha \in \mathrm{I}_\pi(G|\varphi)$ and write $\beta = \alpha_U$. Then $\alpha(1) = \beta(1)$, and it follows by Lemma 5.19(b) that α is the unique member of $\mathrm{I}_\pi(G)$ lying over β. We conclude that our map is injective, as claimed.

Finally, to prove surjectivity, let $\beta \in \mathrm{I}_\pi(U|\theta)$ be arbitrary. We know by Lemma 5.8 that $\mathrm{I}_\pi(G|\beta)$ cannot be empty, so let α be a member of this set. Then α lies over θ, so it lies over some member of $\mathrm{I}_\pi(N)$ lying over θ. We saw, however, that φ is the unique member of $\mathrm{I}_\pi(N|\theta)$, and hence $\alpha \in \mathrm{I}_\pi(G|\varphi)$. Then as we have shown, α_U is irreducible, and it follows that $\alpha_U = \beta$. ∎

Lemma 5.21. *Let $N \triangleleft G$ and $U \subseteq G$, where G is π-separable, and let $\alpha \in \mathrm{I}_\pi(U)$. Suppose that α^G is irreducible and that the common degree of the irreducible constituents of $(\alpha^G)_N$ is a π-number. Then $|NU : U|$ is a π-number.*

Proof. The result is trivial if $N = 1$, so we can assume that $N > 1$, and we proceed by induction on $|N|$. Let N/M be a chief factor of G, and note that $U \subseteq MU \subseteq NU$.

Let $\beta = \alpha^{MU}$ and $\gamma = \beta^{NU}$, and note that $\gamma^G = \alpha^G$ is irreducible, and thus γ is irreducible. Also, γ lies under α^G by Lemma 5.8. Let $\varphi \in \mathrm{I}_\pi(N)$ lie under γ. Then φ lies under α^G, so by hypothesis, $\varphi(1)$ is a π-number. If μ is an irreducible constituent of φ_M, it follows by Corollary 5.7 that $\mu(1)$ divides $\varphi(1)$, and thus $\mu(1)$ is a π-number, and this is the common degree of the irreducible constituents of $(\alpha^G)_M$. By the inductive hypothesis, therefore, $|MU : U|$ is a π-number, and it suffices to show that $|NU : MU|$ is also a π-number.

Let $D = N \cap MU$. Then $|NU : MU| = |N : D|$, and this divides $|N : M|$, which is either a π-number or a π'-number. We can thus assume that $|N : M|$ is a π'-number, and we will complete the proof in that case by showing that $|NU : MU| = 1$.

Now $\varphi(1)$ is a π-number, so $\varphi(1)/\mu(1)$ is a π-number, and by Corollary 5.7, it divides the π'-number $|N : M|$. Then $\varphi(1)/\mu(1) = 1$, and thus $\varphi_M = \mu$ is irreducible, and we conclude that φ_D is irreducible. Also, since $|NU : MU| = |N : D|$ is a π'-number, we can apply Lemma 5.20 with NU in place of G and MU in place of U. Since $\gamma \in \mathrm{I}_\pi(NU|\varphi)$, we conclude that γ_{MU} is irreducible. Recall, however, that $\gamma = \beta^{NU}$

and thus β lies under γ by Lemma 5.8. It follows that $\gamma_{MU} = \beta$. Then $\beta(1) = \gamma(1) = \beta^G(1) = |NU : MU|\beta(1)$, and it follows that $|NU : MU| = 1$, as wanted. ■

Proof of Theorem 5.17. We proceed by induction on $|G|$. If $\varphi(1)$ is a π-number, then by Corollary 5.18, every member of $\mathrm{vtx}(\varphi)$ is a Hall π'-subgroup of G, and hence all vertices for φ are conjugate in G. We can assume, therefore, that $\varphi(1)$ is not a π-number, and thus by Lemma 5.12, there exists $N \triangleleft G$ such that the irreducible constituents of φ_N are not G-invariant, and their common degree is a π-number. Let θ be one of the irreducible constituents of φ_N, and let T be the stabilizer of θ in G. By Theorem 5.11, there exists a unique member $\eta \in \mathrm{I}_\pi(T|\theta)$ such that $\eta^G = \varphi$. Since $T < G$, the inductive hypothesis implies that all members of $\mathrm{vtx}(\eta)$ are conjugate in T, and we will complete the proof by showing that every member of $\mathrm{vtx}(\varphi)$ is G-conjugate to some member of $\mathrm{vtx}(\eta)$.

Let $Q \in \mathrm{vtx}(\varphi)$. By definition, Q is a Hall π'-subgroup of some subgroup $U \subseteq G$ such that there exists $\alpha \in \mathrm{I}_\pi(U)$, where $\alpha(1)$ is a π-number and $\alpha^G = \varphi$. By Lemma 5.21, we see that $|NU : U|$ is a π-number, and thus Q is a Hall π'-subgroup of NU. Also, $(\alpha^{NU})^G = \varphi$, and $\alpha^{NU}(1) = \alpha(1)|NU : U|$ is a π-number. We can thus replace U by NU and α by α^{NU}, so it is no loss to assume that $U \supseteq N$.

Now $\varphi = \alpha^G$ lies over α by Lemma 5.8, and thus the irreducible constituents of α_N lie under φ. By Corollary 5.7, each of these constituents is conjugate to θ in G, so if we replace U by an appropriate conjugate, we can assume that α lies over θ. (Of course, by changing U in this way, we must replace Q by a conjugate. But since our goal is to determine Q up to conjugacy, there is no loss in doing this.)

Now $\alpha \in \mathrm{I}_\pi(U|\theta)$ so by Theorem 5.11, there exists $\beta \in \mathrm{I}_\pi((U \cap T)|\theta)$ such that $\beta^U = \alpha$. Since $\alpha(1) = \beta(1)|U : U \cap T|$ is a π-number, it follows that $\beta(1)$ is a π-number and that $|U : U \cap T|$ is also a π-number. Now $(\beta^T)^G = \beta^G = (\beta^U)^G = \alpha^G = \varphi$, and thus β^T is irreducible. Lemma 5.8 guarantees that β^T lies over β, and hence over θ, and thus $\beta^T \in \mathrm{I}_\pi(T|\theta)$. Since η is the unique member of $\mathrm{I}_\pi(T|\theta)$ that induces φ, it follows that $\beta^T = \eta$, and thus a Hall π'-subgroup P of $U \cap T$ lies in $\mathrm{vtx}(\eta)$. We saw, however, that $|U : U \cap T|$ is a π-number, and thus P is a full Hall π'-subgroup of U, and thus Q is conjugate to $P \in \mathrm{vtx}(\eta)$, as claimed. ■

5E

In this section we prove the following remarkable result.

Theorem 5.22. *Let G be π-separable, and let $Q \in \mathrm{vtx}(\varphi)$, where $\varphi \in \mathrm{I}_\pi(G)$. Then there exists a Hall π'-subgroup S of G and an element $n \in \mathbf{N}_G(Q)$ such that $Q = S \cap S^n$.*

Our proof of Theorem 5.22 depends on the following fact, whose proof relies on Theorem 5.27 below. Although we will derive Theorem 5.23 from Theorem 5.27, we shall not be able to give a complete proof of Theorem 5.27 here.

Theorem 5.23. *Let A act via automorphisms on N, and suppose that $|A|$ and $|N|$ are relatively prime. Then there is a bijection f from $\mathrm{Irr}(N)$ onto the set of conjugacy classes of N such that for all $\theta \in \mathrm{Irr}(N)$, the stabilizers in A of θ and of $f(\theta)$ are identical.*

Actually, what we need is the following corollary, which is somewhat weaker than Theorem 5.23.

Corollary 5.24. *Let A act via automorphisms on N, and suppose that $|A|$ and $|N|$ are relatively prime. Then for each character $\theta \in \mathrm{Irr}(N)$, there exists an element $n \in N$ such that $\mathbf{C}_A(n)$ is the stabilizer of θ in A.*

Since Corollary 5.24 relies ultimately on Theorem 5.27, which we will not prove here, it would be pleasant if we could find some alternative proof of the main result of this section that is independent of Corollary 5.24. It seems extremely unlikely, however, that there is a proof of Theorem 5.22 that avoids an appeal to Corollary 5.24 or some similar result. This is because the conclusion of Corollary 5.24 is actually a consequence of Theorem 5.22, and we digress briefly to explain this.

We need an easy lemma, which will also be used in the proof of Theorem 5.22.

Lemma 5.25. *Let $G = NA$, where $N \lhd G$ and $A \cap N = 1$. If $n \in N$, then $\mathbf{C}_A(n) = A \cap A^n$.*

Proof. We have $\mathbf{C}_A(n) = (\mathbf{C}_A(n))^n \subseteq A^n$, and thus $\mathbf{C}_A(n) \subseteq A \cap A^n$. To prove the reverse containment, let $b \in A \cap A^n$. Then $b^n \in A^n$, so the commutator $[b, n] = b^{-1}b^n$ lies in A^n. Since $n \in N \lhd G$, we also have $[b, n] \in N$, so $[b, n] \in A^n \cap N = (A \cap N)^n = 1$, and thus $b \in \mathbf{C}_A(n)$, as required. ∎

To see that the conclusion of Corollary 5.24 is a consequence of Theorem 5.22, suppose that A acts on N, where $|A|$ and $|N|$ are relatively prime, and let π be the set of prime divisors of $|N|$. Let $G = NA$ be the semidirect product, and observe that G is π-separable. Let $\theta \in \mathrm{Irr}(N)$, and let B be the stabilizer of θ in A. We will use Theorem 5.22 to produce an element $n \in N$ such that $B = \mathbf{C}_A(n)$.

The stabilizer of θ in G is NB, and thus θ has a π-special extension $\psi \in \operatorname{Irr}(NB)$. Write $\chi = \psi^G$ and $\varphi = \chi^0$, and observe that $\chi \in \mathrm{B}_\pi(G)$ by Lemma 4.26, and thus $\varphi \in \mathrm{I}_\pi(G)$. Then $B \in \operatorname{vtx}(\varphi)$, and since we are assuming Theorem 5.22, it follows that B is the intersection of two Hall π'-subgroups of G. We can thus write $B = A^u \cap A^v$, where $u, v \in N$.

Now $B \subseteq A \cap A^u = \mathbf{C}_A(u)$ by Lemma 5.25, so u centralizes B. Writing $n = vu^{-1}$, we have

$$B = B^{u^{-1}} = (A^u \cap A^v)^{u^{-1}} = A \cap A^{vu^{-1}} = A \cap A^n = \mathbf{C}_A(n),$$

as wanted, where the final equality follows by Lemma 5.25.

To derive Corollary 5.24 from Theorem 5.23, we will use a fairly standard application of the Glauberman lemma (Lemma 1.4).

Theorem 5.26. *Let A act via automorphisms on N, where $|A|$ and $|N|$ are relatively prime, and write $C = \mathbf{C}_N(A)$. Then the map $K \mapsto K \cap C$ is a bijection from the set of A-invariant conjugacy classes of G onto the set of all conjugacy classes of C.*

Proof. If K is an A-invariant conjugacy class of N, then A acts on K, and of course, N acts on K via conjugation. It is routine to check that these two actions are compatible with the given action of A on N, so we can apply Lemma 1.4(a) to deduce that A fixes some element of K, so $K \cap C$ is nonempty. Now $K \cap C$ is clearly invariant under conjugation by elements of C, and Lemma 1.4(b) guarantees that all elements of $K \cap C$ are conjugate in C, and thus $K \cap C$ is a full conjugacy class of C.

We know now that $K \mapsto K \cap C$ is a map from the set of A-invariant classes of N into the set of classes of C. To see that this map is injective, note that if K and L are A-invariant classes of N such that $K \cap C = L \cap C$, then $K \cap L$ contains the nonempty set $K \cap C$. It follows that $K \cap L$ is nonempty, so $K = L$, as required.

For surjectivity, let X be an arbitrary class of C, and note that X is contained in a unique class K of N. Since $X \subseteq C$, we see that X is A-invariant, and thus K is A-invariant because X uniquely determines K. Now $K \cap C$ contains X, and since $K \cap C$ is a class of C, it follows that $K \cap C = X$. ∎

Proof of Corollary 5.24. Given $\theta \in \operatorname{Irr}(N)$, let B be the stabilizer of θ in A, and recall that we want to find an element $n \in N$ such that $B = \mathbf{C}_A(n)$. By Theorem 5.23, there exists a class K of N such that B is the stabilizer of K in A. It follows by Theorem 5.26 that $K \cap \mathbf{C}_N(B)$ is nonempty, so choose an element $n \in K \cap \mathbf{C}_N(B)$. Then $B \subseteq \mathbf{C}_A(n)$, and to prove that we have equality, observe that since $\mathbf{C}_A(n)$ fixes n, it leaves the class K of n invariant, and thus $\mathbf{C}_A(n) \subseteq B$. ∎

As we have indicated, the following fundamental character-counting result is an essential ingredient of the proof of Theorem 5.23.

Theorem 5.27. *Let A act via automorphisms on N, where $|A|$ and $|N|$ are relatively prime, and write $C = \mathbf{C}_N(A)$. Then the number of A-invariant irreducible characters of N is equal to $|\operatorname{Irr}(C)|$.*

If A is solvable, Theorem 5.27 follows from the Glauberman correspondence, which is discussed in Chapter 6 and constructed in Chapter 13 of [**CTFG**]. If A is nonsolvable, then by the odd-order theorem, $|A|$ is even, so $|N|$ is odd, and in particular, N is solvable. In this case, where N is solvable of odd order, the result follows from the Isaacs correspondence (Theorem 8.11).

Actually, much more is true than is stated in Theorem 5.27. Not only do the set of A-invariant irreducible characters of G and the set of all irreducible characters of the fixed-point subgroup C have equal cardinalities, but also, there is an unambiguously defined bijective correspondence connecting these two sets. The constructions of the specific character correspondences in the Glauberman case (where A is solvable) and in the Isaacs case (where $|N|$ is odd) are completely different, but amazingly, in situations where both maps are defined, the two maps are identical. (This is a result of T. R. Wolf.)

Assuming Theorem 5.27 now, we obtain the following corollary.

Corollary 5.28. *Let A act via automorphisms on N, and suppose that $|A|$ and $|N|$ are relatively prime. Then the numbers of A-invariant conjugacy classes and A-invariant irreducible characters of N are equal.*

Proof. By Theorem 5.27, the number of A-invariant irreducible characters of N is equal to the number of irreducible characters of $C = \mathbf{C}_N(A)$, and of course, that is equal to the number of conjugacy classes of C. Finally, by Theorem 5.26, the number of classes of C is equal to the number of A-invariant classes of N, and the result follows. ∎

Using the following general observation, it is very easy to derive Theorem 5.23 from Corollary 5.28. This result also appears as Lemma 13.23 of [**CTFG**].

Lemma 5.29. *Let A be a finite group that acts on two finite sets $\mathcal{X}$ and $\mathcal{Y}$, and assume that for every subgroup B of A, the numbers of B-fixed points in $\mathcal{X}$ and in $\mathcal{Y}$ are equal. Then there exists a bijection $f : \mathcal{X} \to \mathcal{Y}$ such for all* *points $x \in \mathcal{X}$ and all elements $a \in A$, we have $f(x^a) = f(x)^a$. In particular, the stabilizers in A of x and $f(x)$ are identical.*

Proof. Since $\mathcal{X}$ and $\mathcal{Y}$ are the sets of fixed points of the trivial subgroup in the two actions, it follows by hypothesis that $|\mathcal{X}| = |\mathcal{Y}|$, so if $|\mathcal{X}| = 1$, then $|\mathcal{Y}| = 1$, and the result is obvious. We can thus assume that $|\mathcal{X}| > 1$, and we proceed by induction on $|\mathcal{X}|$.

Choose an A-orbit $\mathcal{U}$ of minimal size in $\mathcal{X}$, and let B be the stabilizer of some point $u \in \mathcal{U}$. Then $|A : B| = |\mathcal{U}|$, so by the minimality of $|\mathcal{U}|$, no subgroup of A properly containing B can stabilize a point in $\mathcal{X}$. By hypothesis, B stabilizes some point $v \in \mathcal{Y}$ and no subgroup of A properly containing B stabilizes any point of $\mathcal{Y}$. It follows that B is the full stabilizer of v in A, and thus if $\mathcal{V}$ is the A-orbit of v in $\mathcal{Y}$, we have $|\mathcal{V}| = |A : B| = |\mathcal{U}|$.

Since u and v have a common stabilizer in A, we see that for $s, t \in A$, the element st^{-1} stabilizes u if and only if it stabilizes v. It follows that $u^s = u^t$ if and only if $v^s = v^t$, and since A is transitive on $\mathcal{U}$ and on $\mathcal{V}$, we can define a bijection $g : \mathcal{U} \to \mathcal{V}$ by setting $g(u^s) = v^s$.

Next, we show that if $x \in \mathcal{U}$ and $a \in A$, then $g(x^a) = g(x)^a$, and it follows that a fixes x if and only if a fixes $g(x)$. To see this, write $x = u^s$, where $s \in A$, and note that $g(x) = v^s$. Then $g(x^a) = g(u^{sa}) = v^{sa} = g(x)^a$, as wanted.

Since $|\mathcal{X}| = |\mathcal{Y}|$ and $|\mathcal{U}| = |\mathcal{V}|$, it follows that if $\mathcal{U} = \mathcal{X}$, then $\mathcal{V} = \mathcal{Y}$. We can take f to be the map g in this case, and there is nothing further to prove. We can thus assume that $\mathcal{X} - \mathcal{U}$ and $\mathcal{Y} - \mathcal{V}$ are nonempty, and we see that A acts on each of these sets.

For each subgroup $C \subseteq A$, we argue that C fixes equal numbers of points in $\mathcal{X} - \mathcal{U}$ and in $\mathcal{Y} - \mathcal{V}$. By hypothesis, C fixes equal numbers of points in $\mathcal{X}$ and $\mathcal{Y}$, so it suffices to show that C fixes equal numbers of points in $\mathcal{U}$ and $\mathcal{V}$. This is clear, however, because we have seen that for each point $x \in \mathcal{U}$, the stabilizers of x and $g(x)$ in A are identical, and thus C fixes x if and only if C fixes $g(x)$. By the inductive hypothesis, therefore, there exists a bijection h from $\mathcal{X} - \mathcal{U}$ onto $\mathcal{Y} - \mathcal{V}$ such that $h(x^a) = h(x)^a$ for all points $x \in \mathcal{X} - \mathcal{U}$ and elements $a \in A$. The map $f : \mathcal{X} \to \mathcal{Y}$ defined to be g on $\mathcal{U}$ and h on $\mathcal{X} - \mathcal{U}$ is thus the required bijection. ∎

Proof of Theorem 5.23. The group A acts on the set $\operatorname{Irr}(N)$ and on the set of classes of N, and for each subgroup B of A, it follows by Corollary 5.28 that B fixes equal numbers of points in the two sets. The result now follows by Lemma 5.29. ∎

We need two more preliminary results for our proof of Theorem 5.22.

Lemma 5.30. *Let $N \triangleleft G$, where G is π-separable and N is a π'-subgroup. If $U \subseteq G$ and $\theta \in \mathrm{I}_\pi(U)$, where θ^G is irreducible, then $N \subseteq U$.*

Proof. Since $(\theta^{NU})^G = \theta^G$ is irreducible, it follows that θ^{NU} is irreducible. Since N is a π'-group, we see that the principal π-partial character of N is the unique member of $\mathrm{I}_\pi(N)$, and thus the irreducible constituents of $(\theta^{NU})_N$ all have degree 1. By Lemma 5.21, therefore, $|NU : U|$ is a π-number, and since $|NU : U|$ divides the π'-number $|N|$, we have $|NU : U| = 1$, so $N \subseteq U$. ∎

Before stating our final lemma, we review a bit of basic character theory. Suppose that $N \subseteq H \subseteq G$, where $N \triangleleft G$. Given a character ψ of H with $N \subseteq \ker(\psi)$, it is customary to identify ψ with a character of $\overline{H} = H/N$, and similarly, a character χ of G with $N \subseteq \ker(\chi)$ is usually identified with a character of $\overline{G} = G/N$. Nevertheless, it is convenient to introduce some temporary notation to distinguish these two points of view, and so we write $\overline{\psi}$ and $\overline{\chi}$ to denote the characters ψ and χ when they are viewed as characters of $\overline{H}$ or $\overline{G}$, respectively.

Now suppose $\chi = \psi^G$, where ψ is a character of H and $N \subseteq \ker(\psi)$. Since $N \triangleleft G$, it is easy to see that $N \subseteq \ker(\chi)$, and that (in the notation of the previous paragraph) $\overline{\psi}^{\overline{G}} = \overline{\chi}$. We use this observation to prove our next result.

Lemma 5.31. *Let $N \triangleleft G$, where G is π-separable and N is a π'-subgroup, and let $\varphi \in \mathrm{I}_\pi(G)$ and $Q \in \mathrm{vtx}(\varphi)$. Then $N \subseteq Q$ and Q/N is a vertex for some member of $\mathrm{I}_\pi(G/N)$.*

Proof. By definition, Q is a Hall π'-subgroup of some subgroup $U \subseteq G$, where there exists $\theta \in \mathrm{I}_\pi(U)$ such that $\theta^G = \varphi$ and $\theta(1)$ is a π-number. We have $N \subseteq U$ by Lemma 5.30, and thus $N \subseteq Q$.

Writing $\overline{G} = G/N$, we recall that Theorem 3.10 guarantees the existence of a π-partial character $\overline{\varphi} \in \mathrm{I}_\pi(\overline{G})$ such that $\overline{\varphi}(\overline{x}) = \varphi(x)$ for all π-elements $x \in G$. We complete the proof by showing that $\overline{Q} = Q/N$ is a vertex for $\overline{\varphi}$.

We can apply Theorem 3.10 in the group U to produce the π-partial character $\overline{\theta} \in \mathrm{I}_\pi(\overline{U})$ such that $\overline{\theta}(\overline{x}) = \theta(x)$ for all π-elements $x \in U$. In particular, $\overline{\theta}(\overline{1}) = \theta(1)$, so the degree of $\overline{\theta}$ is a π-number. Since $\overline{Q}$ is a Hall π'-subgroup of $\overline{U}$, it suffices to show that $(\overline{\theta})^{\overline{G}} = \overline{\varphi}$ in order prove that $\overline{Q} \in \mathrm{vtx}(\overline{\varphi})$.

One way to compute $(\overline{\theta})^{\overline{G}}$ is to consider a lift $\overline{\psi} \in \mathrm{Irr}(\overline{U})$ for $\overline{\theta}$. If $u \in U$ is a π-element, then $\psi(u) = \overline{\psi}(\overline{u}) = \overline{\theta}(\overline{u}) = \theta(u)$, so $\psi^0 = \theta$, and thus ψ is a lift of θ. Now write $\chi = \psi^G$, and note that $N \subseteq \ker(\chi)$ since $N \triangleleft G$ and

$N \subseteq \ker(\psi)$. Now $\overline{\psi}^{\overline{G}} = \overline{\chi}$, as in the discussion preceding the statement of the lemma.

As x runs over the π-elements of G, we know that $\overline{x}$ runs over all π-elements of $\overline{G}$, and we have

$$\overline{\theta}^{\overline{G}}(\overline{x}) = \overline{\psi}^{\overline{G}}(\overline{x}) = \overline{\chi}(\overline{x}) = \chi(x) = \psi^G(x) = \theta^G(x) = \varphi(x) = \overline{\varphi}(\overline{x}),$$

where the first equality holds because $\overline{\psi}$ is a lift of $\overline{\theta}$ and the fifth holds because ψ is a lift of θ. Then $(\overline{\theta})^{\overline{G}} = \overline{\varphi}$, and the proof is complete. ∎

We are now ready to present the proof of Theorem 5.22. In order to make our inductive argument work, however, we state and prove a somewhat stronger result.

Theorem 5.32. *Let $M \triangleleft G$, where G is π-separable and M is a π-group. Let $\varphi \in \mathrm{I}_\pi(G)$ and $Q \in \mathrm{vtx}(\varphi)$, and assume that φ_M is a multiple of an irreducible character of M. Then there exists an element $n \in \mathbf{N}_G(MQ)$ and a Hall π'-subgroup S of G such that $MQ = MS \cap (MS)^n$.*

Of course, we can recover Theorem 5.22 from Theorem 5.32 by taking M to be the trivial subgroup of G. It should be noted, however, that Theorem 5.32 is not merely a restatement of Theorem 5.22, and in the general case, it is not clear that Theorem 5.32 is a consequence of Theorem 5.22.

Proof of Theorem 5.32. Observe that what we want to prove can be restated as the assertion that MQ/M has the form $H \cap H^x$, where H is some Hall π'-subgroup of G/M and $x \in G/M$ normalizes MQ/M. If $M = G$, this is obvious, so we can assume that $M < G$, and we proceed by induction on the index $|G : M|$.

Since $Q \in \mathrm{vtx}(\varphi)$, we see that Q is a Hall π'-subgroup of some subgroup $U \subseteq G$, where there exists $\mu \in \mathrm{I}_\pi(U)$ such that $\mu^G = \varphi$ and $\mu(1)$ is a π-number. Now $|MU : U|$ is a π-number since M is a π-group, and thus Q is a Hall π'-subgroup of MU. Also, if $\nu = \mu^{MU}$, then $\nu^G = \varphi$ and $\nu(1) = \mu(1)|MU : U|$, and this is a π-number. We can thus replace U by MU and μ by ν, so it is no loss to assume that $M \subseteq U$.

Let $\alpha \in \mathrm{Irr}(M)$ be the unique irreducible constituent of φ_M. Then α is invariant in G, and thus (G, M, α) is a character triple. By Lemma 3.11, there exists an isomorphic character triple (G^*, M^*, α^*), where $M^* \subseteq \mathbf{Z}(G^*)$ and M^* is a π-group, and we observe that G^* is π-separable. Furthermore, since $M \subseteq U$, there is a subgroup U^* of G^* corresponding to U, and we have $M^* \subseteq U^*$.

Now write $W = MQ$, and let W^* be the corresponding subgroup of G^*. Then $M \subseteq W \subseteq U$, and hence $M^* \subseteq W^* \subseteq U^*$ and $|W^* : M^*| = |W : M|$, which is a π'-number, and $|U^* : W^*| = |U : W|$, which is a π-number. It follows that we can write $W^* = M^*P$, where P is a Hall π'-subgroup of U^*.

Let $\psi \in \mathrm{Irr}(U)$ with $\psi^0 = \mu$, and write $\chi = \psi^G$. Then

$$\chi^0 = (\psi^G)^0 = (\psi^0)^G = \mu^G = \varphi\,,$$

and thus χ is a lift of φ, and in particular $\chi \in \mathrm{Irr}(G)$. Also, $\chi_M = \varphi_M$, which is a multiple of α, and thus there is a character $\chi^* \in \mathrm{Irr}(G^*)$ corresponding to χ, where χ^* lies over α^*. Furthermore, if $\psi^* \in \mathrm{Irr}(U^*)$ corresponds to ψ, we have $\chi^* = (\psi^*)^{G^*}$.

Since $\chi^0 = \varphi \in \mathrm{I}_\pi(G)$, it follows by Lemma 3.12 that $(\chi^*)^0 \in \mathrm{I}_\pi(G^*)$, and we write $\varphi^* = (\chi^*)^0$. Also, writing $\mu^* = (\psi^*)^0$, we see that

$$(\mu^*)^{G^*} = ((\psi^*)^0)^{G^*} = ((\psi^*)^{G^*})^0 = (\chi^*)^0 = \varphi^*\,.$$

Also,

$$\mu^*(1) = \psi^*(1) = \frac{\psi^*(1)}{\alpha^*(1)} = \frac{\psi(1)}{\alpha(1)} = \frac{\mu(1)}{\alpha(1)}\,,$$

and this is a π-number. It follows that the Hall π'-subgroup P of U^* is a vertex for φ^*.

We argue now that it suffices to prove the theorem for the group G^* with respect to the normal π-subgroup M^* and the π-partial character φ^*. To see why this is sufficient, suppose we can show that $W^*/M^* = M^*P/M^*$ has the form $J \cap J^y$, where J is some Hall π'-subgroup of G^*/M^* and $y \in G^*/M^*$ normalizes W^*/M^*. Since $G^*/M^* \cong G/M$, it follows that $W/M = H \cap H^x$, where H is a Hall π'-subgroup of G/M and $x \in G/M$ normalizes W/M, and this is exactly what is required. Since $|G : M| = |G^* : M^*|$, it follows that we can replace G, M, and φ with G^*, M^*, and φ^*, and so we can assume that $M \subseteq \mathbf{Z}(G)$.

Now suppose that $N \lhd G$, where N is a π'-group, and write $\overline{G} = G/N$. By Lemma 5.31, we know that $N \subseteq Q$ and that $\overline{Q}$ is a vertex for some member $\theta \in \mathrm{I}_\pi(\overline{G})$. Also, $\overline{M}$ is a central π-subgroup of $\overline{G}$, and thus $\theta_{\overline{M}}$ is a multiple of an irreducible character of $\overline{M}$. If $N > 1$, then

$$|\overline{G} : \overline{M}| = |G : MN| < |G : M|\,,$$

and hence the inductive hypothesis guarantees that $\overline{M}\,\overline{Q} = \overline{M}\,\overline{S} \cap (\overline{M}\,\overline{S})^{\overline{n}}$, where S is some Hall π'-subgroup of G and $\overline{n}$ normalizes $\overline{M}\,\overline{Q}$. Since N is contained in MQ and MS, we conclude that $MQ = MS \cap (MS)^n$ and that n normalizes MQ, and we are done in this case. We may suppose, therefore, that $\mathbf{O}_{\pi'}(G) = 1$.

Recall that $M < G$, so we can choose a minimal normal subgroup K/M of G/M. If K/M is a π'-group, then since M is central in K, we see that $\mathbf{O}_{\pi'}(K) > 1$, and this is a contradiction since $\mathbf{O}_{\pi'}(K) \subseteq \mathbf{O}_{\pi'}(G) = 1$. It follows that K/M is a π-group, and thus K is a π-group.

Let β be an irreducible constituent of φ_K, and let T be the stabilizer of β in G. By Theorem 5.11, there exists a member $\eta \in \mathrm{I}_\pi(T)$ lying over β and such that $\eta^G = \varphi$. Also, since every π-partial character that induces η also induces φ, it follows that $\mathrm{vtx}(\eta) \subseteq \mathrm{vtx}(\varphi)$. Replacing Q by a conjugate, if necessary, therefore, we can assume that Q is a vertex for η.

Now $|T : K| \leq |G : K| < |G : M|$, and since K is a normal π-subgroup of T and η_K is a multiple of β, the inductive hypothesis applies in the group T, and we have

$$KQ = KH \cap (KH)^n ,$$

where H is some Hall π'-subgroup of T and $n \in T$ normalizes KQ.

Since Q is contained in some Hall π'-subgroup of KH, we can assume without loss that $Q \subseteq H$. Also, $n \in \mathbf{N}_T(KQ) = K\mathbf{N}_T(Q)$ by the Frattini argument, and thus we can replace n by an element of $\mathbf{N}_T(Q)$ without changing $(KH)^n$, and hence we can assume that n normalizes Q.

Now let S be a Hall π'-subgroup of G containing H, so $H = S \cap T$, and since $K \subseteq T$, we have $KS \cap T = K(S \cap T) = KH$ by Dedekind's lemma. Then $(KH)^n = (KS \cap T)^n = (KS)^n \cap T$, where the second equality holds because $n \in T$. It follows that

$$KS \cap (KS)^n \cap T = KH \cap (KH)^n = KQ .$$

Now $S \cap T$ is the stabilizer of β in S, and it follows by Corollary 5.24 that there exists an element $c \in K$ such that $S \cap T = \mathbf{C}_S(c)$, and thus $Q \subseteq H = S \cap T = \mathbf{C}_S(c)$, so c centralizes Q. Next, observe that $S^n \cap KS$ is a π'-subgroup of KS, and since S is a Hall π'-subgroup of KS, it follows that $S^n \cap KS \subseteq S^k$ for some element $k \in K$.

We have $Q = Q^n \subseteq S^n$ and also $Q \subseteq KS$, so $Q \subseteq S^n \cap KS \subseteq S^k$. Then $Q \subseteq S \cap S^k$, and it follows by Lemma 5.25 that Q centralizes k, and thus Q centralizes ck, and we have $Q \subseteq S^{ck}$. It follows that $Q \subseteq S^n \cap S^{ck}$, and we argue that in fact $Q = S^n \cap S^{ck}$. To see this, observe first that $S \cap S^c = \mathbf{C}_S(c) = S \cap T$ by Lemma 5.25, and thus

$$S^n \cap S^{ck} = S^n \cap KS \cap S^{ck} \subseteq S^k \cap S^{ck} = (S \cap S^c)^k \subseteq T^k = T ,$$

where the first equality holds because $ck \in K$, and thus $S^{ck} \subseteq KS$. We now have

$$Q \subseteq S^n \cap S^{ck} = S^n \cap S^{ck} \cap T \subseteq (KS)^n \cap KS \cap T = KQ ,$$

so $S^n \cap S^{ck}$ is a π'-subgroup of KQ containing Q, and thus $S^n \cap S^{ck} = Q$, as claimed. Now let $r = ckn^{-1}$, so $r \in \mathbf{N}_G(Q)$ and $Q = S \cap S^r$.

Finally, recall that M is a central π-subgroup of G, so S is the unique Hall π'-subgroup of MS, and similarly, S^r is the unique Hall π'-subgroup of MS^r. It follows that $S \cap S^r = Q$ is a full Hall π' subgroup of $MS \cap MS^r$, and since M is a Hall π-subgroup of this intersection, we have $MS \cap MS^r = MQ$, as required. Since r normalizes MQ, the proof is complete. ■

5F

In this section, we apply some of the theory that we have developed to prove a striking result for a π-separable group G in which a Hall π-subgroup H is **self-normalizing** in G. (This means that $H = \mathbf{N}_G(H)$.) We begin by introducing a bit of notation that is convenient for discussing irreducible character degrees. Given a set π of primes, we write

$$\mathrm{Irr}_\pi(G) = \{\chi \in \mathrm{Irr}(G) \mid \chi(1) \text{ is a } \pi \text{ number}\}.$$

Theorem 5.33. *Let H be a self-normalizing Hall π-subgroup of a π-separable group G. Then*

(a) *If $\chi \in \mathrm{Irr}_{\pi'}(G)$, then χ_H has a unique linear constituent λ, and we have $[\chi_H, \lambda] = 1$.*

(b) *If $\lambda \in \mathrm{Irr}(H)$ is linear, then λ^G has a unique constituent χ that lies in $\mathrm{Irr}_{\pi'}(G)$.*

(c) *The map $\chi \mapsto \lambda$, where χ and λ are as in* (a) *and* (b) *is a bijection from $\mathrm{Irr}_{\pi'}(G)$ onto the set of linear characters of H.*

(d) *If χ and λ are as in* (a) *and* (b)*, then there exists a unique pair (W, γ) where $H \subseteq W \subseteq G$ and $\gamma \in \mathrm{Irr}(W)$, and such that $\gamma_H = \lambda$ and $\gamma^G = \chi$.*

Note that if π consists of just a single prime p, the existence of the linear character λ in Theorem 5.33(a) is obvious, as is the existence of the character $\chi \in \mathrm{Irr}_{p'}(G)$ in Theorem 5.33(b). For general prime sets π, however, these existence statements are nontrivial, and perhaps they are somewhat surprising.

To put Theorem 5.33 into context, we mention that when $\pi = \{p\}$, the result is actually a strong form of a special case of the McKay conjecture. This conjecture asserts that if G is an arbitrary finite group and p is an arbitrary prime, then $|\mathrm{Irr}_{p'}(G)| = |\mathrm{Irr}_{p'}(N)|$, where $N = \mathbf{N}_G(P)$ and $P \in \mathrm{Syl}_p(G)$. In the case where P is self-normalizing in G, we have $N = P$, so $\mathrm{Irr}_{p'}(N)$ is the set of linear characters of P, and by the conjecture, we expect that the number of such linear characters should equal $|\mathrm{Irr}_{p'}(G)|$.

If we assume that G is p-solvable, then Theorem 5.33 applies, and not only does it guarantee that $|\mathrm{Irr}_{p'}(G)|$ is equal to the number linear characters of P, but also the theorem provides a natural bijection from $\mathrm{Irr}_{p'}(G)$ onto the set of linear characters of P. In this situation, furthermore, we see by Theorem 5.33(a) that if $\chi \in \mathrm{Irr}_{p'}(G)$, then $\chi(1) \equiv 1 \bmod p$. (We mention that this fact is consistent with the Isaacs–Navarro conjecture, which is a strengthened version of the McKay conjecture.)

Although for p-solvable groups G in which a Sylow p-subgroup is self-normalizing, Theorem 5.33 yields a natural bijection from $\mathrm{Irr}_{p'}(G)$ onto $\mathrm{Irr}_{p'}(N)$, it is not true for p-solvable groups in general that the McKay equality $|\,\mathrm{Irr}_{p'}(G)| = |\,\mathrm{Irr}_{p'}(N)|$ is always realized by a natural bijection. For example, if $G = GL(2,3)$, which is solvable of order 48, then G has exactly eight irreducible characters, with degrees $1, 1, 2, 2, 2, 3, 3, 4$. Taking $p = 3$, therefore, we see that $|\,\mathrm{Irr}_{p'}(G)| = 6$. The degrees of the irreducible characters of the normalizer N of a Sylow 3-subgroup of G are $1, 1, 1, 1, 2, 2$, so $|\,\mathrm{Irr}_{p'}(N)| = 6$, as expected. There cannot be a natural bijection in this case, however, because two of the degree 2 characters of G are complex conjugates, but all six characters of N are real valued.

The status of the McKay conjecture as of this writing is that although it remains unproved in general, it is known to hold in a large number of cases. In particular, the McKay conjecture is true (for the prime p) whenever G is p-solvable, and we present a proof of this in Chapter 6.

We need a few preliminary results for our proof of Theorem 5.33.

Lemma 5.34. *Let H be a self-normalizing Hall π-subgroup of a π-separable group G, and let $K \triangleleft G$. Then HK/K is a self-normalizing Hall π-subgroup of G/K.*

Proof. It is clear that HK/K is a Hall π-subgroup of G/K, and we write $N/K = \mathbf{N}_{G/K}(HK/K)$. Then $HK \triangleleft N$, so by the Frattini argument, we have $N = \mathbf{N}_N(H)(HK) = HK$, where the second equality holds since $\mathbf{N}_N(H) = H$. Then $N/K = HK/K$, and thus HK/K is self-normalizing in G/K, as required. ∎

Note that in the foregoing proof, the only use that was made of the assumption that G is π-separable was to allow us to apply the Frattini argument, which works because of the conjugacy of Hall π-subgroups. If π consists of just a single prime, the conjugacy follows by the Sylow theorem, so in that case, the π-separability hypothesis is unnecessary in Lemma 5.34.

The proof of the following lemma uses the fact that if a group H acts on a group M, where $|H|$ and $|M|$ are relatively prime, and if some non-principal irreducible character θ of M is H-invariant, then $\mathbf{C}_M(H)$ must be nontrivial. This is a consequence of Theorem 5.27, but since we have not actually provided a proof of that result, we recall that in the case where H is solvable, the fact that we need is an easy consequence of the Glauberman correspondence.

If H is nonsolvable, then H has even order, so M has odd order, and hence M is solvable. In that case, we do not need the full strength of the Isaacs correspondence to conclude that $\mathbf{C}_M(H) > 1$. There is a fairly easy proof of this, and we offer a quick sketch of the argument, which proceeds by induction on $|M|$. If θ is linear, then $[M, H] < M$, and since $M = [M, H]\mathbf{C}_M(H)$ by Lemma 4.28 of [**FGT**], we have $\mathbf{C}_M(H) > 1$, as required. Otherwise, θ is nonlinear, so the irreducible constituents of the restriction of θ to the derived subgroup M' are nonprincipal,. At least one of these constituents is H-invariant by Theorem 13.27 of [**CTFG**], and the result follows by the inductive hypothesis applied to the group M'.

Lemma 5.35. *Let G be π-separable, and suppose that a Hall π-subgroup H of G is self-normalizing. Then the principal character 1_G is the unique π'-special character of G.*

Proof. Let $\chi \in \mathrm{Irr}(G)$ be π'-special, and let $K = \ker(\chi)$. We can view χ as an irreducible character of G/K, and as such, it is π'-special. By Lemma 5.34, we can replace G by G/K, and thus we can assume that χ is faithful. If G is the trivial group, there is nothing to prove, so we can assume that G is nontrivial, and we work to obtain a contradiction.

Let M be minimal normal in G, and let θ be an irreducible constituent of χ_M. Since χ is faithful and π'-special, θ is nonprincipal and π'-special, and thus M cannot be a π-group. It follows that M is a π'-group.

Let T be the stabilizer of θ in G, and note that $|G : T|$ divides $\chi(1)$, and thus $|G : T|$ is a π'-number. Then T contains some Hall π-subgroup of G, and replacing θ by a conjugate if necessary, we can assume that $H \subseteq T$, and thus H stabilizes θ. Since θ is nonprincipal, it follows by Theorem 5.27 (or by the discussion preceding the statement of the lemma) that $\mathbf{C}_M(H) > 1$. This is a contradiction since $\mathbf{C}_M(H) \subseteq M \cap \mathbf{N}_G(H) = M \cap H = 1$. ∎

Theorem 5.36. *Let $\chi \in \mathrm{Irr}_{\pi'}(G)$, where G is π-separable, and a Hall π-subgroup of G is self-normalizing. Then χ is monomial and lies in $\mathrm{B}_\pi(G)$.*

Proof. Let (W, γ) be a nucleus for χ. Then $\gamma^G = \chi$ and γ is π-factored, and we have $\chi(1) = |G : W|\gamma(1)$. It follows that both $|G : W|$ and $\gamma(1)$ are π'-numbers, and in particular, W contains a full Hall π-subgroup H of G. Also, H is self-normalizing in W, so by Lemma 5.35, the principal character of W is the unique π'-special character of W. Since γ is π-factored, it follows that γ is π-special, and hence by definition, $\chi \in \mathrm{B}_\pi(G)$. Since $\gamma(1)$ divides $\chi(1)$, it is a π'-number, but also, since γ is π-special, $\gamma(1)$ is a π-number, and thus γ is linear, and hence χ is monomial. ∎

Proof of Theorem 5.33. For (a), let $\chi \in \mathrm{Irr}_{\pi'}(G)$. Then $\chi \in \mathrm{B}_\pi(G)$ by Theorem 5.36, and thus $\chi^0 \in \mathrm{I}_\pi(G)$ by Theorem 5.1. Write $\varphi = \chi^0$, and let $\lambda \in \mathrm{Irr}(H)$ be a Fong character for φ. Then $[\chi_H, \lambda] = [\varphi_H, \lambda] = 1$ and $\lambda(1) = \varphi(1)_\pi = \chi(1)_\pi = 1$ by Theorem 3.4, where the last equality holds since by assumption, $\chi(1)$ is a π'-number.

Since λ is linear, it is primitive, so we can apply Theorem 5.13 to conclude that every Fong character for φ in $\mathrm{Irr}(H)$ is conjugate to λ in $\mathbf{N}_G(H) = H$. It follows that λ is the unique member of $\mathrm{Irr}(H)$ that is a Fong character for φ. If μ is any linear constituent of χ_H, then μ is a constituent of smallest degree of φ_H, and thus μ is a Fong character for φ. Then $\mu = \lambda$, and thus λ is the unique linear constituent of χ_H. This completes the proof of (a).

Now let λ be an arbitrary linear character of H, and suppose $H \subseteq W \subseteq G$, where W is maximal such that λ has some extension $\gamma \in \mathrm{Irr}(W)$. Since λ is quasi-primitive, we see by Theorem 5.6 that γ^G is irreducible, and we write $\chi = \gamma^G$, so $\chi(1) = |G : W|\gamma(1) = |G : W|$. Since $H \subseteq W$, it follows that $\chi \in \mathrm{Irr}_{\pi'}(G)$, and of course, χ is a constituent of λ^G.

Now $\chi \in \mathrm{B}_\pi(G)$ by Theorem 5.36, so $\varphi = \chi^0$ is irreducible by Theorem 5.1. Also, since λ is a constituent of smallest degree of φ_H, it follows that λ is a Fong character for φ, and thus Theorem 3.4(c) guarantees that φ is the unique member of $\mathrm{I}_\pi(G)$ that lies over λ. By the injectivity of the map in Theorem 5.1, we see now that χ is the unique member of $\mathrm{B}_\pi(G)$ that lies over λ. It follows by Theorem 5.36 that χ is the unique member of $\mathrm{Irr}_{\pi'}(G)$ that lies over λ, and this proves (b). Assertion (c) is now immediate since the map $\chi \mapsto \lambda$ is well defined by (a); it is surjective by the existence in (b), and it is injective by the uniqueness in (b).

Now let χ and λ be as in (a) and (b). As before, suppose $H \subseteq W \subseteq G$, where W is maximal such that λ extends to W, and let γ be an extension of λ to W. Also, recall that $\gamma^G = \chi$. Now let (U, δ) be any pair such that $H \subseteq U \subseteq G$ and λ extends to $\delta \in \mathrm{Irr}(U)$, and where $\delta^G = \chi$. By Theorem 3.16, there is a unique largest subgroup containing H to which λ extends, and thus $U \subseteq W$. Since γ and δ are linear, we have $|G : W| = \chi(1) = |G : U|$, and it follows that $U = W$, as required. Now, γ and δ are constituents of χ_W, and $\gamma_H = \lambda = \delta_H$. Since $[\chi_H, \lambda] = 1$, we deduce that $\gamma = \delta$, and this completes the proof of (d). ■

We can do more within this circle of ideas. For example, the following is a converse to Lemma 5.35.

Lemma 5.37. *Let G be π-separable, and suppose that the principal character 1_G is the unique π'-special character of G. Then a Hall π-subgroup of G is self-normalizing.*

Proof. We can assume that G is nontrivial, and we proceed by induction on $|G|$. Let N be minimal normal in G, and recall that if $N \subseteq \ker(\chi)$, then χ is π'-special if and only if when χ is viewed as a character of G/N, it is π'-special. It follows that we can apply the inductive hypothesis to G/N, and thus a Hall π-subgroup H/N of G/N is self-normalizing in G/N, and it follows that H is self-normalizing in G. If N is a π-group, then H is a Hall π-subgroup of G, and there is nothing further to prove.

We can assume now that N is a π'-group, and thus $H = NK$, where K is a Hall π-subgroup of G. Now write $M = \mathbf{N}_G(K)$ and observe that $M \subseteq \mathbf{N}_G(H) = H = NK$. Then $M = (N \cap M)K$ by Dedekind's lemma. Also, $[(N \cap M), K] \subseteq N \cap K = 1$, so $N \cap M = \mathbf{C}_N(K)$, and so to prove that $M = K$, it suffices to show that $\mathbf{C}_N(K) = 1$. Otherwise, it follows by Theorem 5.27 that there exists a nonprincipal K-invariant character $\theta \in \mathrm{Irr}(N)$. Now θ is π'-special since N is a π'-group, and so by Theorem 2.40, there exists a π'-special character $\chi \in \mathrm{Irr}(G|\theta)$. By hypothesis, $\chi = 1_G$, and thus $\theta = 1_N$, and this is a contradiction. ∎

It is interesting to ask which properties of a group can be determined by examining its character table. Since the normal subgroups of G are "visible" in the character table, the table certainly determines whether or not G is π-separable, and we show next that in fact, the character table of a π-separable group G determines whether or not a Hall π-subgroup of G is self-normalizing. By Lemmas 5.35 and 5.37, we know that a Hall π-subgroup of G is self-normalizing if and only if the principal character of G is the unique π'-special character of G.

Although in general it is not clear that the π'-special characters of a π-separable group can be identified in the character table, the following result shows that it is possible to determine whether or not the principal character of G is the unique π'-special character, and thus the character table determines whether or not a Hall π-subgroup of G is self-normalizing.

Theorem 5.38. *Let G be π-separable. Then the following are equivalent.*

(1) *G has a self-normalizing Hall π-subgroup.*

(2) *1_G is the unique π'-special character of G.*

(3) *The principal π'-partial character is the unique member of $\mathrm{I}_{\pi'}(G)$ having π'-degree.*

By the discussion in Chapter 3, section B, it is easy to construct the set of irreducible π'-partial characters of a π-separable group G from its character table. It follows that the third of the equivalent conditions in Theorem 5.38 can be checked by examining the character table of G, and hence, as claimed, the character table determines whether or not a Hall π-subgroup of G is self-normalizing.

Proof of Theorem 5.38. As we have observed, the equivalence of (1) and (2) follows via Lemmas 5.35 and 5.37. Now assume (2), and suppose that $\varphi \in \mathrm{I}_{\pi'}(G)$, where $\varphi(1)$ is a π'-number. By Theorem 3.14 (with π' in place of π), there exists a π'-special character χ of G such that $\chi^0 = \varphi$, and since $\chi = 1_G$ by (2), it follows that φ is the principal π'-partial character of G, proving (3). Assuming (3) now, it follows by Theorem 3.14 that G has a unique π'-special character, and thus (2) holds. ■

5G

In this section, we prove a theorem that concerns neither $\mathrm{B}_\pi(G)$ nor $\mathrm{I}_\pi(G)$, so it does not properly belong in this chapter. We have chosen to present it here, however, because it is another application of Theorem 5.27.

As motivation, we recall that by Corollary 4.22, the intersection of the kernels of the members of $\mathrm{B}_\pi(G)$ is exactly $\mathbf{O}_{\pi'}(G)$ (assuming, of course, that G is π-separable). We have not, however, discussed the analogous question of giving a group-theoretic description of the intersection of the kernels of the π-special characters of G. (Of course, all π-special characters lie in $\mathrm{B}_\pi(G)$, so the intersection of their kernels contains $\mathbf{O}_{\pi'}(G)$, but in general, this interesection can be much larger.)

Theorem 5.39. *Let $N = \mathbf{N}_G(H)$, where H is a Hall π'-subgroup of a π-separable group G. Then the intersection of the kernels of the π-special characters of G is the unique largest normal subgroup K of G such that $K \cap N$ is a π'-group.*

Of course, it is not obvious that there is a unique largest normal subgroup of G whose intersection with the normalizer of a Hall π'-subgroup of G is a π'-group, but of course it is a consequence of Theorem 5.39 that such a normal subgroup actually does exist if G is π-separable, and since it is unique, this normal subgroup must be characteristic in G.

An interesting case of Theorem 5.39 is where the Hall π'-subgroup H is self-normalizing in G. In that situation, N is a π'-group, and thus the largest normal subgroup K of G such that $K \cap N$ is a π'-group is the whole group G. It follows from the theorem that in this case, the principal character of G is the unique π-special character, and we observe that this is consistent with Theorem 5.38 with the roles of π and π' interchanged.

Proof of Theorem 5.39. Given $L \triangleleft G$, we prove that L is contained in the kernel of every π-special character of G if and only if $L \cap N$ is a π'-group. Since there is clearly a unique largest normal subgroup K of G contained in the kernel of every π-special character, it will follow that K is also the unique largest normal subgroup of G such that $K \cap N$ is a π'-group.

Suppose first that $L \cap N$ is a π'-group, and let χ be a π-special character of G. We show by induction on $|L|$ that L is contained in $\ker(\chi)$. We can assume that $L > 1$, so we can choose a subgroup $M \triangleleft G$ such that L/M is a chief factor of G. Since $M \cap N$ is a π'-group, it follows by the inductive hypothesis that $M \subseteq \ker(\chi)$, and hence we can view χ as a π-special character of G/M. (See the discussion preceding the statement of Corollary 2.1.) Seeking a contradiction, we can assume that $L \not\subseteq \ker(\chi)$, so L/M is not contained in the kernel of χ viewed as a character of G/M. Since χ is π-special, L/M is not a π'-group, and it follows that L/M is a π-group.

Let θ be an irreducible constituent of χ_L. Then θ is nonprincipal, and we can view θ as a character of L/M. Since $\chi(1)$ is a π-number, it follows that the stabilizer of θ in G contains a full Hall π'-subgroup of G, and replacing θ by a conjugate if necessary, we can assume that H stabilizes θ. Now H is a π'-group and H stabilizes the nonprincipal character θ of the π-group L/M, so by Theorem 5.27, the fixed-point subgroup $C/M = \mathbf{C}_{L/M}(H)$ must be nontrivial.

Now $[H, C] \subseteq M$, so $C \subseteq \mathbf{N}_G(MH)$, and since by the Frattini argument, $\mathbf{N}_G(MH) = M\mathbf{N}_G(H) = MN$, we have $C \subseteq MN$. By Dedekind's lemma, $C = M(C \cap N) \subseteq M(L \cap N)$, and thus the nontrivial π-number $|C : M|$ divides $|L \cap N|$. This is the desired contradiction since $L \cap N$ was assumed to be a π'-group.

Conversely now, we show by induction on $|L|$ that if $L \triangleleft G$ and L is contained in the kernel of every π-special character of G, then $L \cap N$ is a π'-group. We can certainly assume that $L > 1$, so as before, we can choose $M \triangleleft G$ such that L/M is a chief factor of G. Then M is contained in the kernel of every π-special character of G, so by the inductive hypothesis, $M \cap N$ is a π'-group, and it suffices to show that $|(L \cap N) : (M \cap N)|$ is a π'-number. Assuming now that this is not the case, we complete the proof by deriving a contradiction.

Write $C = M(L \cap N)$, so $C \subseteq L$ and

$$|C : M| = |M(L \cap N) : M| = |(L \cap N) : (M \cap N)|\,,$$

and since by assumption, this is not a π'-number, it follows that C/M is not a π'-group. Now $C/M \subseteq L/M$, and thus L/M is not a π'-group, and therefore, L/M is a π-group and C/M is a nontrivial subgroup.

Now $L \triangleleft G$ and $H \triangleleft N$, and thus $[L \cap N, H] \subseteq L \cap H \subseteq M$, where the second containment follows since L/M is a π-group. Then H acts trivially on $M(L \cap N)/M = C/M$, and since C/M is nontrivial, Theorem 5.27 guarantees the existence of a nonprincipal H-invariant character $\theta \in \mathrm{Irr}(L/M)$. Also, since θ is π-special when viewed as a character of the π-group L/M, it is also π-special as a character of L. Since H stabilizes θ, the size of the G-orbit of θ is a π-number, and thus by Theorem 2.40, there exists a π-special character χ of G lying over θ. Since θ is not principal, it follows that L is not contained in $\ker(\chi)$, and this is a contradiction. ■

As we indicated, it is a consequence of Theorem 5.39 that if G is π-separable and $N = \mathbf{N}_G(H)$, where H is a Hall π'-subgroup of G, then there is a unique largest normal subgroup K of G such that $K \cap N$ is a π'-group. It is interesting to ask if the assumption that G is π-separable is really necessary here. Since in general, Hall subgroups need not exist, we specialize to Sylow subgroups, and we have the following result.

Theorem 5.40. *Let $N = \mathbf{N}_G(P)$, where P is a Sylow p-subgroup of an arbitrary finite group G. Then there is a unique largest normal subgroup K of G such that $K \cap N$ is a p-group.*

To prove Theorem 5.40, it suffices to show that if K and L are normal subgroups of G such that $K \cap N$ and $L \cap N$ are p-groups, then $KL \cap N$ is also a p-group. In fact, more is true. We will see that $KL \cap N = (K \cap N)(L \cap N)$, and thus $KL \cap N$ is a product of p-subgroups, and hence it is a p-group, as required. As the following result shows, something more general is true.

Theorem 5.41. *Let $P \in \mathrm{Syl}_p(G)$ and $K, L \subseteq G$, and assume that KP, LP, and KL are subgroups. Then*

$$\mathbf{N}_{KL}(P) = \mathbf{N}_K(P)\mathbf{N}_L(P)\,.$$

Note that we are not assuming that K and L are normal in G in Theorem 5.41. Although this result has nothing to do with character theory, we have decided to include the elementary proof anyway.

Proof of Theorem 5.41. It is obvious that $\mathbf{N}_K(P)\mathbf{N}_L(P) \subseteq \mathbf{N}_{KL}(P)$, so it suffices to prove the reverse containment. We establish first the special case where $P \subseteq K \cap L$. If $x \in \mathbf{N}_{KL}(P)$, we can write $x = kl^{-1}$, where $k \in K$ and $l \in L$. Since $P^x = P$, it follows that $P^k = P^l$, and this subgroup (with two names) is contained in both K and L. By the Sylow conjugacy theorem applied in the group $K \cap L$, we can write $P^k = P^l = P^d$, for some element $d \in K \cap L$. Then $kd^{-1} \in \mathbf{N}_K(P)$ and $dl^{-1} \in \mathbf{N}_L(P)$, and thus $x = kl^{-1} = (kd^{-1})(dl^{-1}) \in \mathbf{N}_K(P)\mathbf{N}_L(P)$, as required.

For the general case, observe that $(KP)(LP) = KLP = LKP = (LP)(KP)$, so $(KP)(LP)$ is a subgroup. Applying the result of the previous paragraph with KP in place of K and LP in place of L, we deduce that

$$\mathbf{N}_{KL}(P) \subseteq \mathbf{N}_{(KP)(LP)}(P) \subseteq \mathbf{N}_{KP}(P)\mathbf{N}_{LP}(P)\,.$$

By Dedekind's lemma, $\mathbf{N}_{KP}(P) = \mathbf{N}_K(P)P$ and $\mathbf{N}_{LP}(P) = P\mathbf{N}_L(P)$, and thus

$$\mathbf{N}_{KL}(P) \subseteq \mathbf{N}_K(P)P\mathbf{N}_L(P)\,.$$

If $x \in \mathbf{N}_{KL}(P)$, therefore, we can write $x = kul$, where $k \in \mathbf{N}_K(P)$, $u \in P$, and $l \in \mathbf{N}_L(P)$. Now x, k, and l all lie in KL, and it follows that $u \in KL \cap P$, and thus,

$$\mathbf{N}_{KL}(P) \subseteq \mathbf{N}_K(P)(KL \cap P)\mathbf{N}_L(P)\,.$$

It suffices now to show that $KL \cap P \subseteq (K \cap P)(L \cap P)$. Since the containment in the opposite direction is clear, we see that if we can show that $|KL \cap P| = |(K \cap P)(L \cap P)|$, it will follow that $KL \cap P = (K \cap P)(L \cap P)$, and the proof will be complete.

Observe that $|K : K \cap P| = |KP : P|$ is not divisible by p, and thus $|K|_p = |K \cap P|$, and similarly, $|L|_p = |L \cap P|$. Since $|K \cap L|_p \geq |K \cap L \cap P|$ and $|KL|_p \geq |KL \cap P|$, we have

$$\begin{aligned} |KL|_p = \frac{|K|_p|L|_p}{|K \cap L|_p} \leq \frac{|K \cap P||L \cap P|}{|K \cap L \cap P|} &= |(K \cap P)(L \cap P)| \\ &\leq |KL \cap P| \\ &\leq |KL|_p\,. \end{aligned}$$

Equality thus holds throughout, and therefore $|KL \cap P| = |(K \cap P)(L \cap P)|$, as wanted. ■

As we observed, Theorem 5.40 follows from Theorem 5.41. We mention that the proof of Theorem 5.40 did not really require that P is a Sylow subgroup of G; it would be sufficient, for example, to assume that G is π-separable and that P is a Hall π'-subgroup. In particular, our argument can be used to give a character-free proof of the existence of the normal subgroup of Theorem 5.39.

Problems

5.1. Let G be π-separable. Show that $\mathrm{Irr}(G) = \mathrm{B}_\pi(G) \cup \mathrm{B}_{\pi'}(G)$ if and only if there is no element $g \in G$ such that the order of g is divisible by a prime in π and also by a prime in π'.

5.2. Let G be π-separable, and let $\mathrm{N}_\pi(G)$ be as in Problem 4.2. Show that $\mathrm{N}_\pi(G)$ is a set of lifts for $\mathrm{I}_\pi(G)$.

Hints: Let $M \lhd G$ be maximal such that the common degree of the irreducible constituents of φ_M is a π-number, and let μ be one of these constituents. Let T be the stabilizer of μ, and let $\eta \in \mathrm{I}_\pi(T|\mu)$ induce φ. If $T < G$, let $\psi \in \mathrm{N}_\pi(T)$ lift η and show that ψ^G lies in $\mathrm{N}_\pi(G)$.

5.3. Let G be π-separable.

(a) Let X be a matrix with rows indexed by $\mathrm{B}_\pi(G)$ and columns indexed by a set of representatives for the π-classes of G, and where the entry corresponding to χ and x is $\chi(x)$. Show that X is invertible.

(b) Show that G has equal numbers of real-valued members of $\mathrm{B}_\pi(G)$ and real π-classes.

5.4. Let G be π-separable, and let N be the intersection of the kernels of those members $\chi \in \mathrm{B}_\pi(G)$ such that $\chi(1)$ is a π'-number. Show that N is contained in the kernel of every linear character of a Hall π-subgroup of G.

5.5. Let $N \lhd G$, where G is π-separable, and suppose that $|G : N|$ is a π'-number. If $\mu \in \mathrm{I}_\pi(N)$ is G-invariant, show that $\mathrm{I}_\pi(G|\mu)$ has a single member φ, and show that $\varphi_N = \mu$.

5.6. Let $N \lhd G$, where G is π-separable, and let $\varphi \in \mathrm{I}_\pi(G)$ and $\theta \in \mathrm{I}_\pi(N)$.

(a) If $|G : N|$ is a π-number, show that the multiplicity of θ as a constituent of φ_N is equal to the multiplicity of φ as a constituent of θ^G.

(b) Show that the equality in (a) can fail if the assumption on the index $|G : N|$ is dropped.

Notes and References

In [**40**] it is proved that the set χ^0 for $\chi \in \mathrm{B}_\pi(G)$ forms a basis for the space of class functions on π-elements of π-separable groups. It is also proved in that paper that for every character ψ of G, the π-partial character ψ^0 is a nonnegative integer linear combination of members of this basis. These facts taken together are essentially Theorem 5.1 of this chapter.

Theorem 5.2 is essentially Theorem 2.3 of [**40**].

The theorem of P. Fong mentioned in connection with Corollary 5.4 of this chapter is essentially Lemma 1 of [**21**].

The theorem of B. Huppert that we generalized in Corollary 5.5 appears as Satz 7 of [**32**].

Theorem 5.6 on quasi-primitive characters of Hall π-subgroups appears as Theorem B of [**43**], and its companion result, Theorem 5.13 on primitive characters of Hall π-subgroups is Theorem 6.1 of that paper.

Theorem 5.9 is Theorem 9.1 of [**47**].

Examples showing that Hall subgroups of M-groups need not be M-groups were found by H. Fukushima in [**22**]. Corollary 5.16 of this chapter, which asserts that primitive characters of Hall subgroups of M-groups must be linear, appears as Theorem C of [**43**].

The definition of a vertex for a member of $\mathrm{I}_\pi(G)$ and the fact that all such vertices are conjugate appear in [**49**]. That paper also contains Theorem 5.22, which asserts that each vertex is an intersection of two Hall π-subgroups.

The theorem of T. R. Wolf asserting that the Glauberman and Isaacs correspondences agree when both are defined appears in [**75**].

The case of Theorem 5.33 where π consists a single prime appears as Theorem F of [**64**].

The Isaacs–Navarro conjecture, which we mentioned in connection with Theorem 5.33, appears in [**51**]. To state it, suppose that $N = \mathbf{N}_G(P)$, where $P \in \mathrm{Syl}_p(G)$ and G is an arbitrary finite group. The McKay conjecture proposes that G and N have equal numbers of irreducible characters having p'-degree. The Isaacs–Navarro conjecture, which is a strong form of the McKay conjecture, asserts that if k is any integer not divisible by p, then G and N have equal numbers of irreducible characters with degrees that are congruent modulo p to $\pm k$.

For more information about the McKay conjecture and its history, see the notes in Chapter 6.

Theorems 5.35 and 5.37 form a special case of the general result that the number of π'-special characters of a π-separable group G is equal to the number of classes of the group $\mathbf{N}_G(H)/H$, where H is a Hall π-subgroup of G. This theorem appears as Corollary 1.16 of [**76**].

Theorem 2.2 of [**53**] is the case where $\pi = \{p\}'$ of Theorem 5.39 of this chapter, and Corollary 2.1 of that paper is Theorem 5.41.

Part 2

Character Counts and Correspondences

Chapter 6

The Okuyama–Wajima Argument

6A

There are a number of interesting theorems and unresolved conjectures that have a common general form: the number of irreducible characters of a group G that satisfy a certain condition is equal to the number of irreducible characters of some other group H that satisfy a certain related condition. Sometimes, stronger forms of these character-counting assertions are true, and not only are the cardinalities of the two given sets of characters equal, but in addition there exists a bijection connecting the sets that is unambiguously defined and is independent of whatever choices may have been made in its construction. Such a map is called a **canonical** bijection.

We offer three specific examples. First, consider the following familiar situation. Given $\theta \in \mathrm{Irr}(N)$, where $N \triangleleft G$, let T be the stabilizer in G of θ. Then the Clifford correspondence is a bijection from the set of irreducible characters of T that lie over θ onto the set of irreducible characters of G that lie over θ, and furthermore, this map is canonically defined by character induction: $\eta \mapsto \eta^G$.

Next we consider the Glauberman correspondence, which lies much deeper. It is defined and studied in Chapter 13 of [**CTFG**], but since it plays a prominent role in what follows, we will summarize some of its properties here. Let S be a solvable group that acts via automorphisms on a group G, where $|S|$ and $|G|$ are relatively prime, and let $C = \mathbf{C}_G(S)$, the fixed-point subgroup of the action. The Glauberman correspondence is a bijection from the set $\mathrm{Irr}_S(G)$ of S-invariant irreducible characters of G onto the set $\mathrm{Irr}(C)$ of all irreducible characters of C.

Unfortunately, it is somewhat awkward to define the Glauberman map because it appears to be necessary to make some arbitrary choices in its construction. Nevertheless, one can prove that the map is independent of these choices, and thus it is canonical. If S is a p-group, however, there is an easy description of the Glauberman correspondence, and in this case, at least, it is trivial to see that the map is canonical. The description of the map when S is a p-group appears as part of Theorem 13.14 of [**CTFG**], and since we will need it later, we state it formally here.

Theorem 6.1. *Let S be a p-group that acts via automorphisms on a p'-group G, and let $C = \mathbf{C}_G(S)$. If $\chi \in \mathrm{Irr}_S(G)$, then there is a unique irreducible character $\psi \in \mathrm{Irr}(C)$ such that $[\chi_C, \psi] \not\equiv 0 \bmod p$, and in fact, $[\chi_C, \psi] \equiv \pm 1 \bmod p$. In this situation, ψ is the image of χ under the Glauberman correspondence $\mathrm{Irr}_S(G) \to \mathrm{Irr}(C)$.*

Note that an obvious consequence of Theorem 6.1 is that if S is the trivial group, then the associated Glauberman correspondence is the identity map.

We will also need the following *transitivity* property.

Theorem 6.2. *Let S act via automorphisms on a group G, where $|G|$ and $|S|$ are relatively prime and S is solvable. Write $C = \mathbf{C}_G(S)$, and let $f : \mathrm{Irr}_S(G) \to \mathrm{Irr}(C)$ be the associated Glauberman map. Let $T \lhd S$ and $B = \mathbf{C}_G(T)$. Then S/T acts on B, and $C = \mathbf{C}_B(S/T)$. Let $g : \mathrm{Irr}_T(G) \to \mathrm{Irr}(B)$ and $h : \mathrm{Irr}_{S/T}(B) \to \mathrm{Irr}(C)$ be the Glauberman maps associated with the actions of T on G and S/T on B. Then $f(\chi) = h(g(\chi))$ for all $\chi \in \mathrm{Irr}_S(G)$.*

This is essentially Corollary 13.9 of [**CTFG**], and although we will not reproduce the proof here, we should check that everything makes sense. First, since $T \subseteq S$, it is clear that $B \supseteq C$. Also, $T \lhd S$, so S stabilizes $\mathbf{C}_G(T) = B$, and hence S acts on B. Since T acts trivially on B, the action of S on B induces an action of S/T, and the fixed-point subgroup of S/T in B is the fixed-point subgroup of S, which is C. If $\chi \in \mathrm{Irr}_S(G)$, then certainly, χ is T-fixed, and thus $g(\chi)$ is a well-defined member of $\mathrm{Irr}(B)$. Also, $g(\chi)$ is S-invariant because χ is S-invariant and χ uniquely determines $g(\chi)$ since the map g is canonical. Then $g(\chi)$ is S/T-invariant, so it has a Glauberman correspondent $h(g(\chi))$ lying in $\mathrm{Irr}(C)$, and the theorem asserts that this is exactly the Glauberman correspondent $f(\chi)$ of χ with respect to the action of S on G.

One can use Theorems 6.1 and 6.2 to give a recursive description of the Glauberman map f. If S is trivial, then f is the identity map, so we can suppose that S is nontrivial, and assume that we already know how to construct the Glauberman correspondence for acting groups that are smaller than S. Since S is solvable, we can choose a nontrivial normal p-subgroup T of S for some prime p, and we let $B = \mathbf{C}_G(T)$. By Theorem 6.1, the correspondence $g : \mathrm{Irr}_T(G) \to \mathrm{Irr}(B)$ of Theorem 6.2 is easy to describe: $g(\chi)$ is the unique irreducible constituent of χ_B having p'-multiplicity. Also, by the inductive hypothesis, we know how to compute the map h of Theorem 6.2, and we know that the map f is simply g followed by h.

Of course, it is not obvious that the map constructed by this recursive procedure is canonical; it could conceivably depend on the choice of the subgroup T. By Theorems 6.1 and 6.2, however, the map that we constructed is the Glauberman map, which is known to be canonical. The map defined by our procedure, therefore, is independent of the choice of T.

It should be clear from our recursive description of the Glauberman correspondence that if S acts trivially on G, then the associated Glauberman map is the identity. In the situation of Theorem 6.2, therefore, we see that if T acts trivially on G, then $B = G$ and g is the identity map, and thus $f = h$. In other words, if T acts trivially, then the Glauberman maps associated with the actions of S and S/T are identical. (We will use this unsurprising observation later.)

Corollary 6.3. *Let S act via automorphisms on a group G, where $|G|$ and $|S|$ are relatively prime and S is solvable, and write $C = \mathbf{C}_G(S)$. Let $\chi \in \mathrm{Irr}_S(G)$, and let $\psi \in \mathrm{Irr}(C)$ be the image of χ under the Glauberman map $\mathrm{Irr}_S(G) \to \mathrm{Irr}(C)$. Then ψ is a constituent of χ_C.*

Proof. If $S = 1$, then $C = G$ and $\psi = \chi$, and the result is clear. We can assume, therefore, that S is nontrivial, and we proceed by induction on $|S|$. Let T be a minimal normal subgroup of S, so T is a p-group for some prime p, and let $B = \mathbf{C}_G(T)$. As in Theorem 6.2, let $g : \mathrm{Irr}_T(G) \to \mathrm{Irr}(B)$ and $h : \mathrm{Irr}_{S/T}(B) \to C$ be the relevant Glauberman maps. If $\beta = g(\chi)$, then χ lies over β by Theorem 6.1. Also, $\psi = h(\beta)$ by Theorem 6.2, and since $|S/T| < |S|$, the inductive hypothesis guarantees that β lies over ψ. It follows that χ lies over ψ, as required. ∎

For our final example, we discuss the so-called (and still unproved) McKay conjecture, which was mentioned in Chapter 5, and to which we return in Chapter 8. Recall that this conjecture asserts that if G is an arbitrary finite group and p is an arbitrary prime, and if $N = \mathbf{N}_G(P)$, where $P \in \mathrm{Syl}_p(G)$, then $|\,\mathrm{Irr}_{p'}(G)| = |\,\mathrm{Irr}_{p'}(N)|$, where $\mathrm{Irr}_{p'}(X)$ denotes the set of p'-degree irreducible characters of a group X. (J. McKay noticed that this appears to be true for $p = 2$ for simple groups G, but he did not actually state the conjecture in its most general form: for all primes and all finite groups.) The McKay conjecture has been proved for many large families of groups, including p-solvable groups, where p is the relevant prime. (One of the major goals of this chapter is to present a proof for p-solvable groups using an argument of T. Okuyama and M. Wajima.) The McKay conjecture is also known to be true for all primes p for symmetric groups, for many simple groups and for all groups with the property that every composition factor has an abelian Sylow 2-subgroup. Recently, the McKay conjecture has been proved for $p = 2$ for all finite groups G.

For groups of odd order, it is known that there is a canonical bijection that realizes the McKay conjecture, but in general, no such bijection can exist. To see why this is so, consider the group $G = GL(2,3)$ of order 48, and take $p = 3$. One can check that $\mathrm{Irr}(G)$ contains exactly six characters with $3'$-degree: four of these are real valued, and the other two (which have degree 2) are complex conjugates of each other. If $N = \mathbf{N}_G(P)$, where $P \in \mathrm{Syl}_3(G)$, then every member of $\mathrm{Irr}(N)$ has $3'$-degree, and as the McKay conjecture predicts, there are six such characters. All of the irreducible characters of N are real valued, however, and this precludes the possibility that there is a canonical bijection because such a map would have to map the complex conjugate of a character to the complex conjugate of its image.

Since even for p-solvable groups, we cannot hope to prove the McKay conjecture by constructing some natural bijection, we must rely on other techniques for counting characters, and in particular, the Gallagher character count of Theorem 1.19 will be invaluable.

6B

A suitable generalization of the technique used by Okuyama and Wajima in their proof of the McKay conjecture for p-solvable groups can also be applied to other character-counting problems. We present this generalized Okuyama–Wajima argument in this chapter, and we use it to prove a π-analogue of the p-solvable case of the McKay conjecture. We also obtain a strong form of the p-solvable case of another unresolved problem: the Alperin weight conjecture.

We begin by describing the somewhat complicated setup for the (generalized) Okuyama–Wajima argument. Let $K \lhd G$ and $Q \subseteq G$, where $|K|$ and $|Q|$ are relatively prime and Q is solvable, and assume that $KQ \lhd G$. Also, let $\theta \in \mathrm{Irr}(K)$ be invariant in G. Under these assumptions, we will say that (G, K, Q, θ) is an **OW-configuration.** (We mention that in the original Okuyama–Wajima situation, Q is a Sylow subgroup of G.)

Given an OW-configuration (G, K, Q, θ), we define a character triple (N, L, φ), which we call the **associated character triple**. Here $N = \mathbf{N}_G(Q)$ and $L = N \cap K$, so $L \lhd N$, as required. To define φ, we observe first that $L = \mathbf{C}_K(Q)$. To see this, note that

$$\mathbf{C}_K(Q) \subseteq K \cap \mathbf{N}_G(Q) = K \cap N = L\,.$$

To obtain the reverse containment, observe that $[L,Q] \subseteq K \cap Q = 1$ because $L \subseteq K \lhd G$ and $L \subseteq N = \mathbf{N}_G(Q)$. Then L centralizes Q, and we have $L = \mathbf{C}_K(Q)$, as claimed. Now Q is solvable and acts on K, and by assumption, $|K|$ and $|Q|$ are relatively prime, and since $L = \mathbf{C}_K(Q)$, the Glauberman map $\mathrm{Irr}_Q(K) \to \mathrm{Irr}(L)$ is defined. Also, we are assuming that θ is G-invariant, so $\theta \in \mathrm{Irr}_Q(K)$, and we let $\varphi \in \mathrm{Irr}(L)$ be the Glauberman correspondent of θ. Finally, to verify that (N, L, φ) is a character triple, we must show that φ is invariant in N. This follows because φ is uniquely determined by θ, and θ is G-invariant.

If (N, L, φ) is the associated character triple for the OW-configuration (G, K, Q, θ), we argue that $G = NK$. Observe that Q is a complement in KQ for the normal Hall subgroup K. By the conjugacy part of the Schur–Zassenhaus theorem, all complements for K in KQ are conjugate in KQ, and since $KQ \lhd G$, the Frattini argument applies, and we have $G = (KQ)N = KN$, as claimed. It follows that there is a natural isomorphism from N/L onto G/K, where the coset $Ln \in N/L$ maps to the coset $Kn \in G/K$. The corresponding bijection $B \mapsto C$ from the set of subgroups B intermediate between K and G onto the set of subgroups C intermediate between L and N is defined by $C = B \cap N$, and we observe that in this situation $B/K \cong C/L$.

Recall from the Gallagher character count (Theorem 1.19) that if $K \lhd G$ and $\theta \in \mathrm{Irr}(K)$ is G-invariant, then $|\,\mathrm{Irr}(G|\theta)|$ is equal to the number of θ-good classes of G/K. When this is combined with the following generalized Okuyama–Wajima theorem, the result is a powerful character counting tool.

Theorem 6.4 (Okuyama and Wajima). *Let (G, K, Q, θ) be an OW-configuration with associated character triple (N, L, φ). Then the coset Ln is φ-good in N/L if and only if the coset Kn is θ-good in G/K. In particular, the numbers of φ-good classes of N/L and θ-good classes of G/K are equal.*

An immediate consequence is the following.

Corollary 6.5. *Assuming the hypotheses and notation of Theorem 6.4, we have $|\,\mathrm{Irr}(G|\theta)| = |\,\mathrm{Irr}(N|\varphi)|$.*

Proof. By Theorem 6.4, the numbers of θ-good classes in G/K and φ-good classes in N/L are equal. By Theorem 1.19, these are equal to $|\,\mathrm{Irr}(G|\theta)|$ and $|\,\mathrm{Irr}(N|\varphi)|$, respectively. ∎

If (G, K, Q, θ) is an OW-configuration, then clearly (G, K, θ) is a character triple. Also, we have seen that if (N, L, φ) is the associated character triple, then $G/K \cong N/L$ and $|\,\mathrm{Irr}(G|\theta)| = |\,\mathrm{Irr}(N|\varphi)|$. This suggests that perhaps (G, K, θ) and (N, L, φ) are isomorphic triples, and in fact, this is true, at least in the case where Q is p-group. Unfortunately, however, this theorem of E. C. Dade does not appear to be accessible at the level of this book.

We need several preliminary results for our proof of Theorem 6.4.

Lemma 6.6. *Let $N \triangleleft G$, where G/N is abelian, and let $\theta \in \mathrm{Irr}(N)$ be G-invariant. Then $|\mathrm{Irr}(G|\theta)| \le |G : N|$, and equality holds if and only if every member of $\mathrm{Irr}(G|\theta)$ is an extension of θ.*

Proof. Let $n = |\mathrm{Irr}(G|\theta)|$. For each character $\chi \in \mathrm{Irr}(G|\theta)$, we can write $\chi_N = a_\chi \theta$ for some positive integer a_χ. It follows by Frobenius reciprocity that $\theta^G = \sum a_\chi \chi$, where χ runs over the n members of $\mathrm{Irr}(G|\theta)$. Also,

$$|G : N|\theta(1) = \theta^G(1) = \sum_{\chi \in \mathrm{Irr}(G|\theta)} a_\chi \chi(1) = \sum_{\chi \in \mathrm{Irr}(G|\theta)} (a_\chi)^2 \theta(1)\,,$$

and thus

$$|G : N| = \sum_{\chi \in \mathrm{Irr}(G)} (a_\chi)^2 \ge n\,,$$

as required. Also, equality holds if and only if all $a_\chi = 1$ or, equivalently, $\chi_N = \theta$ for all $\chi \in \mathrm{Irr}(G|\theta)$. ∎

It is interesting to observe that in the situation of Lemma 6.6, the degrees of all members of $\mathrm{Irr}(G|\theta)$ are equal. In fact, more is true: the group of linear characters of G/N acts transitively by multiplication on $\mathrm{Irr}(G|\theta)$. This fact will not be needed until Chapter 7, where it appears as part of Lemma 7.1.

Theorem 6.7. *Suppose that (G, K, Q, θ) is an OW-configuration with associated character triple (N, L, φ), and assume that Q is a p-group. Let $B \triangleleft G$, where $KQ \subseteq B$ and B/K is abelian, and let $C = B \cap N$. Then θ has a G-invariant extension to B if and only if φ has an N-invariant extension to C.*

Proof. Let X/K be the Hall p'-subgroup of the abelian group B/K, and write $Y = X \cap N$, so Y/L is the Hall p'-subgroup of C/L. Then X is a p'-group that is acted on by the p-group Q, and it is easy to check that $Y = \mathbf{C}_X(Q)$. Also $Y/L \cong X/K$.

Suppose first that θ has a G-invariant extension to B. Then θ has a G-invariant extension $\chi \in \mathrm{Irr}(X)$, and since $\chi_K = \theta$, it follows by the Gallagher correspondence that

$$\mathrm{Irr}(X|\theta) = \{\lambda\chi \mid \lambda \in \mathrm{Irr}(X/K)\}\,,$$

where distinct characters λ yield distinct members of $\mathrm{Irr}(X|\theta)$. Since B/K is abelian and $Q \subseteq B$, all of the characters $\lambda \in \mathrm{Irr}(X/K)$ are linear and

Q-invariant. Since χ is G-invariant, it is Q-invariant, and it follows that $\mathrm{Irr}(X|\theta)$ consists of $|X:K|$ extensions of θ, all of which are Q-invariant.

We show next that the Glauberman map $\mathrm{Irr}_Q(X) \to \mathrm{Irr}(Y)$ carries $\mathrm{Irr}(X|\theta)$ into $\mathrm{Irr}(Y|\varphi)$. If $\xi \in \mathrm{Irr}(X|\theta)$, then ξ extends θ, and thus $\xi_L = \theta_L$. By assumption, Q is a p-group and φ is the Glauberman correspondent of θ, so $[\theta_L, \varphi]$ is not divisible by p by Theorem 6.1. Then $[\xi_L, \varphi]$ is not divisible by p, and it follows that there is some irreducible constituent η of ξ_Y such that η lies over φ and the multiplicity of η as a constituent of ξ_Y is not divisible by p. Since $Y = \mathbf{C}_X(Q)$, we conclude by another application of Theorem 6.1 that η is the image of ξ under the Glauberman map $\mathrm{Irr}_Q(X) \to \mathrm{Irr}(Y)$, and thus this map carries $\mathrm{Irr}(X|\theta)$ into $\mathrm{Irr}(Y|\varphi)$, as claimed.

It follows that $|\mathrm{Irr}(Y|\varphi)| \geq |\mathrm{Irr}(X|\theta)| = |X:K| = |Y:L|$, and since Y/L is abelian, we deduce from Lemma 6.6 that every member of $\mathrm{Irr}(Y|\varphi)$ is an extension of φ.

Recall now that $\chi \in \mathrm{Irr}(X|\theta)$, so if $\psi \in \mathrm{Irr}(Y)$ is the image of χ under the Glauberman map, we deduce that ψ extends φ. Now χ is G-invariant, and since χ uniquely determines ψ, we conclude that ψ is N-invariant, and in particular, ψ is C-invariant. Now Y is a normal Hall subgroup of C, so Gallagher's theorem (Theorem 1.5) tells us that ψ has a unique canonical extension to C. Since ψ is N-invariant, its canonical extension to C is the required N-invariant extension of φ.

Conversely now, suppose that φ has an N-invariant extension to C. Then φ has an N-invariant extension $\psi \in \mathrm{Irr}(Y)$, and we have

$$\mathrm{Irr}(Y|\varphi) = \{\lambda\psi \mid \lambda \in \mathrm{Irr}(Y/L)\}\,,$$

where distinct characters λ yield distinct members of $\mathrm{Irr}(Y|\varphi)$. Since Y/L is abelian, all of the characters $\lambda \in \mathrm{Irr}(Y/L)$ are linear, and thus every member of $\mathrm{Irr}(Y|\varphi)$ is an extension of φ, and we have $|\mathrm{Irr}(Y|\varphi)| = |Y:L|$.

We show next that the set $\mathrm{Irr}(Y|\varphi)$ is contained in the image under the Glauberman map $\mathrm{Irr}_Q(X) \to \mathrm{Irr}(Y)$ of the set $\mathrm{Irr}_Q(X|\theta)$ of Q-invariant members of $\mathrm{Irr}(X|\theta)$. If $\eta \in \mathrm{Irr}(Y|\varphi)$, then

$$[(\eta^X)_K, \theta] = [(\eta_L)^K, \theta] = [\varphi^K, \theta] = [\varphi, \theta_L]\,,$$

and this number is not divisible by p.

Now η and θ are Q-invariant, and Q normalizes X, and it follows that Q permutes the set of irreducible constituents of η^X that lie over θ. The orbits of this action have p-power size, and as ξ runs over one of these orbits, the multiplicities $[\xi_K, \theta]$ are equal. It follows that the contribution of each nontrivial Q-orbit to the multiplicity of θ in $(\eta^X)_K$ is divisible by p. Since the total multiplicity is not divisible by p, there must exist a Q-invariant character $\xi \in \mathrm{Irr}(X)$ lying over θ and such that $[\eta^X, \xi]$ is not divisible by p.

Then $[\xi_Y, \eta]$ is not divisible by p, and it follows by Theorem 6.1 that η is the image of ξ under the Glauberman map, and thus $\mathrm{Irr}(Y|\varphi)$ is contained in the image of $\mathrm{Irr}_Q(X|\theta)$, as claimed.

We now have

$$|\,\mathrm{Irr}(X|\theta)| \geq |\,\mathrm{Irr}_Q(X|\theta)| \geq |\,\mathrm{Irr}(Y|\varphi)| = |Y : L| = |X : K|\,,$$

so by Lemma 6.6 equality holds throughout, and thus every member of $\mathrm{Irr}(X|\theta)$ is a Q-invariant extension of θ. Also, there is a unique extension χ of θ to X that is carried by the Glauberman map to the N-invariant character ψ, and thus χ is N-invariant. Since $NK = G$, it follows that χ is G-invariant, and thus the canonical extension of χ to B is the desired G-invariant extension of θ to B. ∎

Lemma 6.8. *Let $K \subseteq M \lhd G$, where $K \lhd G$ and $|K|$ and $|M : K|$ are relatively prime, and suppose $K \subseteq A \subseteq G$. Also, let $\theta \in \mathrm{Irr}(K)$ be invariant in G. Then the following are equivalent.*

(1) *The canonical extension of θ to M is extendible to MA.*

(2) *θ is extendible to MA.*

(3) *θ is extendible to A.*

Proof. It is obvious that (1) implies (2) and that (2) implies (3). To show that (3) implies (1), let $\psi \in \mathrm{Irr}(A)$ be an extension of θ. Write $N = A \cap M$, and let α be the canonical extension of θ to N. We work first to show that α has an extension to A.

Since α is an extension of θ to N and $\psi_N \in \mathrm{Irr}(N|\theta)$, we can write $\psi_N = \lambda\alpha$ for some character $\lambda \in \mathrm{Irr}(N/K)$, and we see that λ is linear because $\psi(1) = \theta(1) = \alpha(1)$. Writing $\mu = \det(\psi)$, we have

$$\mu_N = \det(\psi_N) = \det(\lambda\alpha) = \lambda^{\alpha(1)}\det(\alpha) = \lambda^{\theta(1)}\det(\alpha)\,.$$

Also, $o(\alpha) = o(\theta)$, since α is the canonical extension of θ to N. Then $\det(\alpha)^{o(\theta)}$ is trivial, and thus

$$(\mu_N)^{o(\theta)} = \lambda^{\theta(1)o(\theta)}\,.$$

Now $\theta(1)$ and $o(\theta)$ divide $|K|$, and since $o(\lambda)$ divides $|N : K|$, which divides $|M : K|$, it follows that $\theta(1)o(\theta)$ and $o(\lambda)$ are relatively prime, and hence there exists an integer t such that $\theta(1)o(\theta)t \equiv 1 \bmod o(\lambda)$. Then

$$(\mu^{o(\theta)t})_N = (\mu_N)^{to(\theta)} = \lambda^{\theta(1)o(\theta)t} = \lambda\,,$$

and writing $\nu = \mu^{o(\theta)t}$, we see that $\nu^{-1}\psi$ is an extension of α to A.

Changing notation now, we can assume that $\psi_N = \alpha$, and we let β be the canonical extension of θ to M, so $\beta_N = \alpha$. We have

$$[(\psi^{MA})_M, \beta] = [(\psi_N)^M, \beta] = [\psi_N, \beta_N] = [\alpha, \alpha] = 1\,,$$

and hence for some irreducible constituent χ of ψ^{MA}, we have $[\chi_M, \beta] = 1$. By assumption θ is G-invariant, and since $M \lhd G$ and θ uniquely determines its canonical extension $\beta \in \mathrm{Irr}(M)$, it follows that β is G-invariant. Then χ_M is a multiple of β, and we conclude that $\chi_M = \beta$. This establishes (1), as required. ■

Next, we recall a general representation-theoretic fact.

Lemma 6.9. *Let V be an abelian group, and suppose that an abelian group A acts faithfully and irreducibly on V. (In other words, no proper, nontrivial subgroup of V is A-invariant, and no nonidentity element of A acts trivially on V.) Then A is cyclic.*

Proof. Let R be the subring of the endomorphism ring of V consisting of those endomorphisms that commute with the action of A. By Schur's lemma, every nonzero element of R is invertible, and thus R is a division ring. If $a \in A$, then since A is abelian, it follows that the automorphism σ_a of V induced by a commutes with the action of A, and hence $\sigma_a \in R$. The map $a \mapsto \sigma_a$, therefore, is a homomorphism from A into the multiplicative group $R^\times$ of R. Since the action of A on V is faithful, this map is injective, and thus A is isomorphic to some subgroup $B \subseteq R^\times$. Now B is abelian, so the subdivision ring F of R generated by B is commutative. Then F is a field and B is a finite subgroup of the multiplicative group of F, and this implies that B is cyclic. Since $A \cong B$, we deduce that A is cyclic. ■

Our next result lies at the heart of the proof of Theorem 6.4.

Theorem 6.10. *Let (G, K, Q, θ) be an OW-configuration with associated character triple (N, L, φ), and suppose $L \subseteq A \subseteq N$, where A/L is abelian. Then θ extends to KA if and only if φ extends to A.*

Proof. If Q acts trivially on K, then $N = G$ and $\theta = \varphi$. Also $A = KA$ in this case, so there is nothing to prove. We can thus assume that Q acts nontrivially on K, and in particular, $Q > 1$ and $K > 1$. We proceed by double induction, first on $|Q|$ and then on $|K|$.

Let P be the kernel of the action of Q on K. Then $P \lhd Q$ and P is normalized by K, and thus $P \lhd KQ$. In fact, since $|K|$ and $|Q|$ are relatively prime, it is easy to see that $P = \mathbf{O}_\pi(KQ)$, where π is the set of prime divisors of $|Q|$. Since $KQ \lhd G$, it follows that $P \lhd G$.

Now $KP = K \times P$, so θ has a unique extension $\alpha \in \mathrm{Irr}(KP)$ such that $P \subseteq \ker(\alpha)$. Because θ is G-invariant and α is unique, it follows that α is invariant in G, and thus $(G/P, KP/P, Q/P, \alpha)$ is an OW-configuration. We argue that the associated character triple is $(N/P, LP/P, \beta)$, where β is the unique extension of φ to LP such that $P \subseteq \ker(\beta)$.

It is clear that $N/P = \mathbf{N}_{G/P}(Q/P)$, and since $N \cap KP = (N \cap K)P = LP$, we see that $(KP/P) \cap (N/P) = LP/P$, and it remains to show that β is the Glauberman correspondent of α relative to the action of Q/P on KP/P. To see this, observe first that Q/P acts on K since P acts trivially. Also, $L = \mathbf{C}_K(Q/P)$, and thus the Glauberman maps corresponding to the actions of Q and Q/P on K are identical, and therefore θ maps to φ under the Glauberman map associated with the action of Q/P on K. (This follows via the remark following Theorem 6.2.)

Now K is naturally isomorphic to KP/P, and the actions of Q/P on KP/P and on K respect this isomorphism. Also, the natural map from $\mathrm{Irr}(K)$ onto $\mathrm{Irr}(KP/P)$ carries θ to α. Similarly, L is naturally isomorphic to LP/P, and the natural map from $\mathrm{Irr}(L)$ onto $\mathrm{Irr}(LP/P)$ carries φ to β. Because the Glauberman map corresponding to the action of Q/P is canonical and it maps θ to φ, it follows that it maps α to β, and this shows that $(N/P, LP/P, \beta)$ is the associated character triple of $(G/P, KP/P, Q/P, \alpha)$, as claimed.

Since $P \subseteq \ker(\alpha)$, we have $P \subseteq \ker(\det(\alpha))$, and it follows that α is the canonical extension of θ to KP, and similarly, β is the canonical extension of φ to LP. We can now apply Lemma 6.8 with KP in the role of M and KA in the role of A, and we deduce that θ extends to KA if and only if α extends to $(KP)(KA) = KPA$. We can also apply Lemma 6.8 in the group N, with L in the role of K and LP in the role of M, and we see that φ extends to A if and only if β extends to $(LP)A = PA$. It suffices, therefore, to show that α extends to KPA if and only if β extends to PA.

Now $LP/P \subseteq PA/P \subseteq N/P$ and $(PA/P)/(PL/P)$ is abelian since it is a homomorphic image of A/L. If $P > 1$, then $|Q/P| < |Q|$, so we can apply the inductive hypothesis (on $|Q|$) to the OW-configuration $(G/P, KP/P, Q/P, \alpha)$, and we conclude that α extends to KPA if and only if β extends to PA, and the result follows in this case. We can thus assume that $P = 1$, so Q acts faithfully on K.

It should be clear that we can assume that $G = KQA$, and thus $N = LQA = QA$. Recall that Q is nontrivial and $Q \triangleleft N$, so we can choose a minimal normal subgroup V of N with $V \subseteq Q$, and since Q is solvable, it follows that V is an elementary abelian p-group for some prime p. Observe that VK is normalized by N, and since $NK = G$, it follows that $VK \triangleleft G$, and thus (G, K, V, θ) is an OW-configuration, and we let (M, H, γ) be its associated character triple, so $M = \mathbf{N}_G(V)$ and $H = M \cap K$, and we have $N \subseteq M \subseteq G$, and $L \subseteq H \subseteq K$. Also, $HQ \triangleleft M$ since $HQ = KQ \cap M$, and thus (M, H, Q, γ) is an OW-configuration.

Now $\gamma \in \mathrm{Irr}(H)$ is the image of θ under the Glauberman map corresponding to the action of V, and by Theorem 6.2, we see that φ is the image of γ under the Glauberman map with respect to the action of Q/V on H, or equivalently, with respect to the action of Q on H. It follows that (N, L, φ) is the character triple associated with (M, H, Q, γ). Note also that $H < K$ since $V > 1$ and Q acts faithfully on K, and it follows by the inductive hypothesis (on $|K|$) that φ extends to A if and only if γ extends to HA. Furthermore, HA/H is abelian, so if $V < Q$, the inductive hypothesis (on $|Q|$) guarantees that γ extends to HA if and only if θ extends to KA.

To complete the proof now, we can assume that $V = Q$, so Q is a minimal normal subgroup of N, and since Q is solvable, it is an elementary abelian p-group for some prime p. Let $C = \mathbf{C}_G(Q)$, so $LQ \subseteq C \triangleleft N$. We argue that N/C is cyclic and C/L is abelian.

First, $CA = N$ since $QA = N$, and thus $N/C = CA/C \cong A/(A \cap C)$, and since this is a homomorphic image of the abelian group A/L, it follows that N/C is abelian. Also, N/C acts faithfully and irreducibly on the abelian group Q, and hence by Lemma 6.9, we see that N/C is cyclic.

To see that C/L is abelian, observe that $Q \subseteq C \subseteq N = QA$, so $C = Q(C \cap A)$ by Dedekind's lemma. Then

$$\frac{C}{L} = \frac{LQ}{L}\frac{(C \cap A)L}{L},$$

and we see that the first factor is central in C/L because Q is central in C. The second factor is abelian since by hypothesis A/L is abelian, and it follows that C/L is abelian, as claimed.

Let $B = KC$, so $B \triangleleft G$ and $B/K \cong C/L$ is abelian, and also $C = B \cap N$. We can thus apply Theorem 6.7 to conclude that θ has a G-invariant extension to B if and only if φ has an N-invariant extension to C. Now $G/B \cong N/C$ is cyclic, and it follows by Corollary 1.14 that θ has a G-invariant extension to B if and only if θ extends to G. Similarly, since N/C is cyclic, φ has an N-invariant extension to C if and only if φ extends to N. We deduce that θ extends to G if and only if φ extends to N.

Because $G = KQA$, we see by Lemma 6.8 with KQ in the role of M and KA in the role of A that θ extends to KA if and only if it extends to G. Similarly, since $N = QA$, another application of Lemma 6.8 shows that φ extends to A if and only if it extends to N, and the result follows. ∎

It is now easy to complete the proof of the Okuyama–Wajima theorem.

Proof of Theorem 6.4. First, note that since $NK = G$, every element of G/K is a coset of the form Kn, for some element $n \in N$, and we are required to prove that Kn is a θ-good element of G/K if and only if Ln is a φ-good element of N/L.

By definition, Kn is θ-good if and only if θ extends to $\langle K, n, x\rangle$ for every element $x \in G$ such that $\langle K, n, x\rangle/K$ is abelian. The coset Kx contains an element $y \in N$, and thus $\langle K, n, x\rangle = \langle K, n, y\rangle$. Also, the natural isomorphism from N/L to G/K carries $\langle L, n, y\rangle/L$ to $\langle K, n, y\rangle/K$, so $\langle K, n, y\rangle/K$ is abelian if and only if $\langle L, n, y\rangle/L$ is abelian. We see, therefore, that Kn is θ-good if and only if θ extends to $\langle K, n, y\rangle$ whenever $y \in N$ and $\langle L, n, y\rangle/L$ is abelian. Also Ln is φ-good if and only if φ extends to $\langle L, n, y\rangle$ whenever $y \in N$ and $\langle L, N, y\rangle/L$ is abelian. Now write $A = \langle L, n, y\rangle$, so $L \subseteq A \subseteq N$, and $\langle K, n, y\rangle = KA$. We see now that it suffices to show that if $L \subseteq A \subseteq N$ where A/L is abelian, then θ extends to KA if and only if φ extends to A. This is exactly what Theorem 6.10 tells us. ∎

6C

Given a set π of primes, we recall the notation

$$\mathrm{Irr}_\pi(G) = \{\chi \in \mathrm{Irr}(G) \mid \chi(1) \text{ is a } \pi\text{-number}\}\,.$$

In this section, we use Theorem 6.4 and some ideas of T. R. Wolf to prove the following π-analogue of the p-solvable case of the McKay conjecture.

Theorem 6.11. *Let U be a Hall π-subgroup of a π-separable group G, and assume that U is solvable. Writing $N = \mathbf{N}_G(U)$, we have $|\mathrm{Irr}_{\pi'}(G)| = |\mathrm{Irr}_{\pi'}(N)|$.*

Of course, if π consists of a single prime p in Theorem 6.11, then U is a Sylow p-subgroup of G and U is automatically solvable. In this case, therefore, the assertion of the theorem is exactly the McKay conjecture for p-solvable groups.

In order to prove Theorem 6.11 inductively, we generalize it as follows.

Theorem 6.12. *Let $B \lhd G$, where G is π-separable, and suppose that U/B is a solvable Hall π-subgroup of G/B. Write $N = \mathbf{N}_G(U)$, and let $\beta \in \mathrm{Irr}(B)$ be U-invariant. Then $|\mathrm{Irr}_{\pi'}(G|\beta)| = |\mathrm{Irr}_{\pi'}(N|\beta)|$.*

In particular, if B is trivial, then β is the trivial character and U is a solvable Hall π-subgroup of G. So this case of Theorem 6.12 is exactly Theorem 6.11.

We begin with a result concerning good elements.

Lemma 6.13. *Let $K \subseteq N \lhd G$, where $K \lhd G$ is a π'-subgroup and N/K is a π-group, and suppose that the coset Kg is a π'-element of G/K. Also, let $\theta \in \mathrm{Irr}(K)$ be G-invariant, and let $\hat\theta$ be the canonical extension of θ to N. Then Kg is θ-good in G/K if and only if Ng is $\hat\theta$-good in G/N.*

Proof. First, suppose that Ng is $\hat{\theta}$-good in G/N. To show that Kg is θ-good, we consider an element $x \in G$ such that $\langle K, g, x\rangle/K$ is abelian. Then $[g, x] \in K \subseteq N$, so $\langle N, g, x\rangle/N$ is abelian, and thus $\hat{\theta}$ extends to $\langle N, g, x\rangle$. It follows that θ extends to $\langle N, g, x\rangle$, so θ extends to $\langle K, g, x\rangle$, and we conclude that Kg is θ-good.

Conversely, assume that Kg is θ-good, and let $x \in G$, where $\langle N, g, x\rangle/N$ is abelian. We must show that $\hat{\theta}$ extends to $\langle N, g, x\rangle$, and for this purpose, we can assume that $\langle N, g, x\rangle = G$. Now G/N is abelian, and we let M/N be the normal Hall π-subgroup of G/N. Then M/K is a normal Hall π-subgroup of G/K, and G/M is abelian, and thus $\langle M, g\rangle \triangleleft G$. Also, $G/\langle M, g\rangle$ is cyclic since it is generated by the image of the element x.

By the Schur–Zassenhaus theorem, we can choose a Hall π'-subgroup A/K of G/K containing the π'-element Kg. Then $G/M \cong A/K$, and thus A/K is abelian. Also, $\langle M, g\rangle \cap A = \langle K, g\rangle$, and thus $A/\langle K, g\rangle \cong G/\langle M, g\rangle$, which is cyclic, so we can write $A = \langle K, g, y\rangle$ for some element $y \in A$. By assumption, Kg is θ-good, and since A/K is abelian, it follows that θ extends to A. We can now apply Lemma 6.8 to conclude that the canonical extension α of θ to M extends to $MA = G$. Since α is an extension of $\hat{\theta}$, it follows that $\hat{\theta}$ extends to G, as required. ∎

Corollary 6.14. *Let (G, K, Q, θ) be an OW-configuration, where Q is a Hall π-subgroup of G, and let (N, L, φ) be the associated character triple. Write $\hat{\theta}$ and $\hat{\varphi}$ to denote the canonical extensions of θ and φ to KQ and LQ, respectively. Then $|\operatorname{Irr}(G|\hat{\theta})| = |\operatorname{Irr}(N|\hat{\varphi})|$.*

Proof. By Theorem 1.19, it suffices to show that the numbers of $\hat{\theta}$-good classes of G/KQ and $\hat{\varphi}$-good classes of N/LQ are equal. Now N/LQ is a π'-group, and it follows that every element of N/LQ is a coset of the form $(LQ)n$, where $n \in N$ is a π'-element. (One way to see this is to observe that $N = (LQ)H$, where H is a Hall π'-subgroup of N.) By Lemma 6.13, we see that $(LQ)n$ is $\hat{\varphi}$-good if and only if Ln is φ-good, and by Theorem 6.4 that happens if and only if Kn is θ-good. Finally, by Lemma 6.13 once more, Kn is θ-good if and only if $(KQ)n$ is $\hat{\theta}$-good, and this completes the proof. ∎

Theorem 6.15. *Let (G, K, Q, θ) be an OW-configuration, where Q is a Hall π-subgroup of G, and let (N, L, φ) be the associated character triple. Then $|\operatorname{Irr}_{\pi'}(G|\theta)| = |\operatorname{Irr}_{\pi'}(N|\varphi)|$.*

Proof. If $\chi \in \operatorname{Irr}_{\pi'}(G|\theta)$, then since $KQ \triangleleft G$, it is clear that the irreducible constituents of χ_{KQ} lie in $\operatorname{Irr}_{\pi'}(KQ|\theta)$. Conversely, we argue that if $\psi \in \operatorname{Irr}_{\pi'}(KQ|\theta)$, then every irreducible constituent χ of ψ^G lies in $\operatorname{Irr}_{\pi'}(G|\theta)$. Certainly, χ lies over θ, and to see that its degree is a π'-number, it suffices to observe that $\chi(1)/\psi(1)$ divides the π'-number $|G : KQ|$.

Let $\hat{\theta} \in \mathrm{Irr}(KQ)$ be the canonical extension of θ, and observe that by the Gallagher correspondence, each member $\psi \in \mathrm{Irr}(KQ|\theta)$ has the form $\psi = \lambda\hat{\theta}$ for some unique character $\lambda \in \mathrm{Irr}(KQ/K)$. Also, by the definition of an OW-configuration, $|K|$ and $|Q|$ are relatively prime, and thus K is a π'-group. We conclude that $\hat{\theta}(1) = \theta(1)$ is a π'-number, and since KQ/K is a π-group, it follows that

$$\mathrm{Irr}_{\pi'}(KQ|\theta) = \{\lambda\hat{\theta} \mid \lambda \in \mathrm{Irr}(KQ/K) \text{ is linear}\}\,.$$

Now $\hat{\theta}$ is G-invariant, so the stabilizer G_λ of a linear character $\lambda \in \mathrm{Irr}(KQ/K)$ stabilizes $\lambda\hat{\theta}$. Furthermore, $\lambda\hat{\theta}$ uniquely determines λ, and it follows that G_λ is the full stabilizer of $\lambda\hat{\theta}$ in G. Also, as λ runs over a G-orbit of linear characters of $\mathrm{Irr}(KQ/K)$, we see that $\lambda\hat{\theta}$ runs over a G-orbit in $\mathrm{Irr}_{\pi'}(KQ|\theta)$, and all G-orbits in $\mathrm{Irr}_{\pi'}(KQ|\theta)$ arise in this way.

Choose a set $\mathcal{X}$ of representatives for the G-orbits in the set of linear characters of KQ/K, and observe that the characters $\lambda\hat{\theta}$ for $\lambda \in \mathcal{X}$ form a set of representatives for the G-orbits in $\mathrm{Irr}_{\pi'}(KQ|\theta)$. Each member of $\mathrm{Irr}_{\pi'}(G|\theta)$ lies over exactly one such orbit, and since G_λ is the stabilizer of $\lambda\hat{\theta}$, it follows by the Clifford correspondence that

$$|\,\mathrm{Irr}_{\pi'}(G|\theta)| = \sum_{\lambda\in\mathcal{X}} |\,\mathrm{Irr}(G_\lambda|\lambda\hat{\theta})|\,.$$

Next, observe that KQ/K is a normal Hall subgroup of G_λ/K, so λ has an extension $\nu \in \mathrm{Irr}(G_\lambda)$. The map $\xi \mapsto \nu\xi$ is a bijection from $\mathrm{Irr}(G_\lambda|\hat{\theta})$ onto $\mathrm{Irr}(G_\lambda|\lambda\hat{\theta})$, and it follows that $|\,\mathrm{Irr}(G_\lambda|\lambda\hat{\theta})| = |\,\mathrm{Irr}(G_\lambda|\hat{\theta})|$, and thus

$$|\,\mathrm{Irr}_{\pi'}(G|\theta)| = \sum_{\lambda\in\mathcal{X}} |\,\mathrm{Irr}(G_\lambda|\hat{\theta})|\,.$$

Restriction defines a natural bijection from the set of linear characters of KQ/K onto the set of linear characters of LQ/L, and since $KN = G$, we see that the image of the set $\mathcal{X}$ under this map is a set $\mathcal{Y}$ of representatives for the N-orbits of linear characters of LQ/L.

Reasoning as we did previously, we conclude that

$$|\,\mathrm{Irr}_{\pi'}(N|\varphi)| = \sum_{\mu\in\mathcal{Y}} |\,\mathrm{Irr}(N_\mu|\hat{\varphi})|\,,$$

where N_μ is the stabilizer of μ in N. To complete the proof, therefore, it suffices to show that if $\lambda \in \mathcal{X}$ corresponds to $\mu \in \mathcal{Y}$, then

$$|\,\mathrm{Irr}(G_\lambda|\hat{\theta})| = |\,\mathrm{Irr}(N_\mu|\hat{\varphi})|\,.$$

Since $N_\mu = G_\lambda \cap N$, we see that the associated character triple of the OW-configuration $(G_\lambda, K, Q, \theta)$ is (N_μ, L, φ), so the desired equality follows by Corollary 6.14. ∎

We need two more preliminary results for our proof of Theorem 6.12.

Lemma 6.16. *Let $V \lhd G$, and suppose that $\lambda \in \mathrm{Irr}(V)$ is linear and invariant in G. Assume that the order $o(\lambda)$ is a π-number, and that some character $\chi \in \mathrm{Irr}_{\pi'}(G)$ lies over λ. Then λ is extendible to G.*

Proof. Let $m = \chi(1)$, so $\chi_V = m\lambda$. Writing $\mu = \det(\chi)$, we have

$$\mu_V = \det(\chi_V) = \det(m\lambda) = \lambda^m .$$

Now $o(\lambda)$ is a π-number and m is a π'-number, and thus there exists an integer t such that $mt \equiv 1 \bmod o(\lambda)$. Then $(\mu^t)_V = \lambda^{mt} = \lambda$, and thus μ^{mt} is an extension of λ to G, as required. ∎

Lemma 6.17. *Let Q be a Hall π-subgroup of G, where G is π-separable, and write $N = \mathbf{N}_G(Q)$. Let $M \lhd G$, and let $\mathcal{X}$ be a set of representatives for the orbits of the action of N on the set of Q-invariant members of $\mathrm{Irr}_{\pi'}(M)$. Then each character $\chi \in \mathrm{Irr}_{\pi'}(G)$ lies over a unique member of $\mathcal{X}$.*

Proof. Let $\chi \in \mathrm{Irr}_{\pi'}(G)$, and observe that the irreducible constituents of χ_M lie in $\mathrm{Irr}_{\pi'}(M)$, and they form a G-orbit with size dividing the π'-number $\chi(1)$. The index in G of the stabilizer of each member of this orbit is thus a π'-number, and since G is π-separable, each of these stabilizers contains a conjugate of the Hall π-subgroup Q. It follows that some irreducible constituent of χ_M is Q-invariant, and thus the set of irreducible constituents of χ_M contains a full N-orbit of Q-invariant members of $\mathrm{Irr}_{\pi'}(M)$. We conclude that χ lies over some member $\theta \in \mathcal{X}$.

Suppose that χ also lies over $\varphi \in \mathcal{X}$, so $\varphi = \theta^g$ for some element $g \in G$. Then both Q and Q^g are contained in the stabilizer T of φ in G. Since Q and Q^g are Hall π-subgroups of the π-separable group T, they are conjugate in T, and thus $Q^{gt} = Q$ for some element $t \in T$. Then $gt \in N$, and we have $\theta^{gt} = \varphi^t = \varphi$, so θ and φ lie in the same N-orbit. Since both of these characters lie in the set $\mathcal{X}$ of N-orbit representatives, we conclude that $\theta = \varphi$. ∎

Proof of Theorem 6.12. Recall that we have $B \lhd G$ and $N = \mathbf{N}_G(U)$, where U/B is a solvable Hall π-subgroup of G/B. Also, $\beta \in \mathrm{Irr}(B)$ is U-invariant, and we must show that $|\mathrm{Irr}_{\pi'}(G|\beta)| = |\mathrm{Irr}_{\pi'}(N|\beta)|$.

The result is trivial if $B = G$, so we can assume that $B < G$, and we proceed by induction on $|G : B|$. Also, we can assume that $\beta(1)$ is π'-number since otherwise

$$|\mathrm{Irr}_{\pi'}(G|\beta)| = 0 = |\mathrm{Irr}_{\pi'}(N|\beta)| .$$

Let T be the stabilizer of β in G, and note that $U \subseteq T$. Write $S = T \cap N$, so $S = \mathbf{N}_T(U)$ and S is the stabilizer of β in N. Also, $U \subseteq S \subseteq T$, so $|G : T|$

and $|N : S|$ are π'-numbers, and thus

$$|\operatorname{Irr}_{\pi'}(G|\beta)| = |\operatorname{Irr}_{\pi'}(T|\beta)| \qquad \text{and} \qquad |\operatorname{Irr}_{\pi'}(N|\beta)| = |\operatorname{Irr}_{\pi'}(S|\beta)|$$

by the Clifford correspondence. If $T < G$, then $|\operatorname{Irr}_{\pi'}(T|\beta)| = |\operatorname{Irr}_{\pi'}(S|\beta)|$ by the inductive hypothesis, and the result follows. We can thus assume that β is G-invariant.

Let (G^*, B^*, β^*) be a character triple isomorphic to (G, B, β), where $B^* \subseteq \mathbf{Z}(G^*)$. As usual, we write X^* to denote the subgroup intermediate between B^* and G^* corresponding to a given subgroup X intermediate between B and G.

Since $G/B \cong G^*/B^*$, it follows that U^*/B^* is a solvable Hall π-subgroup of G^*/B^*, and $N^* = \mathbf{N}_{G^*}(U^*)$. Also, there are bijections from $\operatorname{Irr}(G|\beta)$ onto $\operatorname{Irr}(G^*|\beta^*)$ and from $\operatorname{Irr}(N|\beta)$ onto $\operatorname{Irr}(N^*|\beta^*)$, and since $\beta(1)$ and $\beta^*(1)$ are π'-numbers, Lemma 1.12 guarantees that these maps define bijections on the sets of π'-degree characters in each case. It suffices, therefore, to show that $|\operatorname{Irr}_{\pi'}(G^*)| = |\operatorname{Irr}_{\pi'}(N^*)|$, and since $|G : B| = |G^* : B^*|$, we can replace (G, B, β) by (G^*, B^*, β^*), and so we can assume that $B \subseteq \mathbf{Z}(G)$, and in particular, β is linear.

Let V be the Hall π-subgroup of B, and write $\lambda = \beta_V$. Then λ is linear and invariant in G, and every member of $\operatorname{Irr}_{\pi'}(G|\beta)$ and every member of $\operatorname{Irr}_{\pi'}(N|\beta)$ lies over λ. If

$$|\operatorname{Irr}_{\pi'}(G|\lambda)| = 0 = |\operatorname{Irr}_{\pi'}(N|\lambda)|\,,$$

then

$$|\operatorname{Irr}_{\pi'}(G|\beta)| = 0 = |\operatorname{Irr}_{\pi'}(N|\beta)|\,,$$

and there is nothing further to prove. We may thus assume that at least one of the sets $\operatorname{Irr}_{\pi'}(G|\lambda)$ or $\operatorname{Irr}_{\pi'}(N|\lambda)$ is nonempty, and we argue that in either case, λ extends to G.

If $\operatorname{Irr}_{\pi'}(G|\lambda)$ is nonempty, then λ extends to G by Lemma 6.16. We can assume, therefore, that $\operatorname{Irr}_{\pi'}(N|\lambda)$ is nonempty, and hence by Lemma 6.16 again, λ extends to N. To show that λ extends to G, it suffices by Theorem 1.10 to show for each prime number p that λ extends to P, where P/V is some Sylow p-subgroup of G/V. If $p \in \pi$, we can choose $P \subseteq U$, and thus λ extends to P because $P \subseteq N$, and we know that λ extends to N. If $p \notin \pi$, then V is a Hall π-subgroup of P, and thus λ extends to P by Theorem 1.5. It follows that λ extends to G in this case also.

Let μ be an extension of λ to G, and write $\gamma = (\mu^{-1})_B\beta$, so $\gamma \in \operatorname{Irr}(B)$ and $V \subseteq \ker(\gamma)$. The map $\chi \mapsto \mu^{-1}\chi$ is a bijection from $\operatorname{Irr}_{\pi'}(G|\beta)$ onto $\operatorname{Irr}_{\pi'}(G|\gamma)$, and similarly, multiplication by $(\mu^{-1})_N$ defines a bijection from $\operatorname{Irr}_{\pi'}(N|\beta)$ onto $\operatorname{Irr}_{\pi'}(N|\gamma)$, and hence it suffices to show that $|\operatorname{Irr}_{\pi'}(G|\gamma)| = |\operatorname{Irr}_{\pi'}(N|\gamma)|$. Now V is contained in the kernels of all of the relevant characters, and since $|(G/V):(B/V)| = |G:B|$, we can replace G by G/V and β by γ. We can thus assume that $V = 1$, and hence B is a π'-subgroup. It follows that $U = B \times Q$, where Q is a Hall π-subgroup of G. Also, $N = \mathbf{N}_G(Q)$.

Now let K/B be a minimal normal subgroup of G/B, and let $\mathcal{X}$ be a set of representatives for the N-orbits of Q-invariant members of $\mathrm{Irr}_{\pi'}(K)$. By Lemma 6.17, each character χ in $\mathrm{Irr}_{\pi'}(G|\beta)$ lies over a unique member $\theta \in \mathcal{X}$, and since χ_B is a multiple of β, we see that θ lies over β.

Let $M = KN$, and observe that $M = \mathbf{N}_G(KQ)$ by the Frattini argument. By Lemma 6.17 again, every member of $\mathrm{Irr}_{\pi'}(M|\beta)$ lies over a unique member $\theta \in \mathcal{X}$, and reasoning as before, we see that θ lies over β.

Now $|G:K| < |G:B|$, so the inductive hypothesis (with K in place of B and KQ in place of U) guarantees that for each member $\theta \in \mathcal{X}$, we have $|\,\mathrm{Irr}_{\pi'}(G|\theta)| = |\,\mathrm{Irr}_{\pi'}(M|\theta)|$. Summing over those characters $\theta \in \mathcal{X}$ that lie over β, we obtain $|\,\mathrm{Irr}_{\pi}(G|\beta)| = |\,\mathrm{Irr}_{\pi'}(M|\beta)|$. If $M < G$, then $|M:B| < |G:B|$, and it follows by the inductive hypothesis (with M in place of G) that $|\,\mathrm{Irr}_{\pi'}(M|\beta)| = |\,\mathrm{Irr}_{\pi'}(N|\beta)|$, and the result follows in this case.

We can now assume that $M = G$, and thus $KQ \lhd G$ and KQ/K is a normal Hall π-subgroup of G/K. If K/B is a π-group, then $K \subseteq U$, so $U = BQ = KQ$, and thus $N = G$, and there is nothing to prove.

In the remaining case, K/B is a π'-group, and thus K is a π'-group. Write $L = K \cap N$, so $L = \mathbf{C}_K(Q)$. Recalling that the members of $\mathcal{X}$ are Q-invariant, we let $\mathcal{Y}$ be the image of $\mathcal{X}$ under the Glauberman map $\mathrm{Irr}_Q(K) \to \mathrm{Irr}(L)$. Because this map is canonical, it respects the actions of $N = \mathbf{N}_G(Q)$ on $\mathrm{Irr}_Q(K)$ and on L, and it follows that $\mathcal{Y}$ is a set of representatives for the N-orbits on $\mathrm{Irr}(L)$.

We argue that if $\theta \in \mathcal{X}$ and $\varphi \in \mathcal{Y}$ correspond under the Glauberman map, then θ lies over β if and only if φ lies over β. To see why this is so, observe that φ is a constituent of θ_L by Corollary 6.3, so if φ lies over β, then certainly θ lies over β. Conversely, if θ lies over β, then since B is central, θ_B is a multiple of β, and thus φ lies over β.

By Lemma 6.17 once more, we see that every member of $\mathrm{Irr}_{\pi'}(N|\beta)$ lies over a unique member $\varphi \in \mathcal{Y}$, where φ lies over β. It suffices, therefore, to show that $|\,\mathrm{Irr}_{\pi'}(G|\theta)| = |\,\mathrm{Irr}_{\pi'}(N|\varphi)|$ for every pair of corresponding characters $\theta \in \mathcal{X}$ and $\varphi \in \mathcal{Y}$.

Suppose that $\theta \in \mathcal{X}$ and $\varphi \in \mathcal{Y}$ correspond. Let X be the stabilizer of θ in G, and write $Y = X \cap N$, so Y is the stabilizer of φ in N because θ and φ uniquely determine each other. Now $Q \subseteq X$ and $N = \mathbf{N}_G(Q)$, so $Y = \mathbf{N}_X(Q)$. Also, $|G:X|$ and $|N:Y|$ are π'-numbers since X and Y

contain Q, and it follows by the Clifford correspondence that

$$|\operatorname{Irr}_{\pi'}(G|\theta)| = |\operatorname{Irr}_{\pi'}(X|\theta)| \qquad \text{and} \qquad |\operatorname{Irr}_{\pi'}(N|\varphi)| = |\operatorname{Irr}_{\pi'}(Y|\varphi)|\,.$$

It suffices, therefore, to prove that the right-hand sides of these two equations are equal.

We observe now that (X, K, Q, θ) is an OW-configuration because $KQ \lhd X$ and θ is invariant in X. Also, $Y = \mathbf{N}_X(Q)$ and $L = \mathbf{C}_K(Q)$, and since φ is the image of θ under the Glauberman map from $\operatorname{Irr}_Q(K)$ to $\operatorname{Irr}(L)$, it follows that (Y, L, φ) is the associated character triple. By Theorem 6.15, therefore, we have $|\operatorname{Irr}_{\pi'}(X|\theta)| = |\operatorname{Irr}_{\pi'}(Y|\varphi)|$, and the proof is now complete. ∎

6D

In this section, we prove an analogue of the Gallagher character count (Theorem 1.19) for π-partial characters. Given $N \lhd G$ and a G-invariant character $\theta \in \operatorname{Irr}(N)$, we recall that $|\operatorname{Irr}(G|\theta)|$ is equal to the number of θ-good classes of G/N. To state and prove the corresponding theorem for π-partial characters, we must assume, of course, that G is π-separable, and in addition, we suppose that N is a π-subgroup. Then $\theta \in \mathrm{I}_\pi(N)$, so it is meaningful to consider the set $\mathrm{I}_\pi(G|\theta)$ of members of $\mathrm{I}_\pi(G)$ lying over θ. Since $|\mathrm{I}_\pi(G)|$ is equal to the number of π-classes of G, it is natural to guess that $|\mathrm{I}_\pi(G|\theta)|$ is equal to the number of π-classes of G/N that are θ-good, and in fact, this is correct.

Theorem 6.18. *Let $N \lhd G$, where G is π-separable and N is a π-subgroup, and let $\theta \in \operatorname{Irr}(N)$ be G-invariant. Then $|\mathrm{I}_\pi(G|\theta)|$ is the number of θ-good classes of π-elements of G/N.*

We mention that $|\mathrm{I}_\pi(G|\theta)| = |\mathrm{B}_\pi(G|\theta)|$, so Theorem 6.18 also counts the members of the set $\mathrm{B}_\pi(G)$ that lie over θ. This follows because by Theorem 5.1 the map $\chi \mapsto \chi^0$ is a bijection from $\mathrm{B}_\pi(G)$ onto $\mathrm{I}_\pi(G)$, and it is immediate that $\chi \in \mathrm{B}_\pi(G)$ lies over θ if and only if χ^0 lies over θ.

We begin work toward a proof of Theorem 6.18 with a general result.

Lemma 6.19. *Let $N \lhd G$, and suppose that $\theta \in \operatorname{Irr}(N)$ is G-invariant. Let $\chi \in \operatorname{Irr}(G|\theta)$, and assume that the coset Ng is not θ-good in G/N. Then $\chi(g) = 0$.*

Proof. Write $M = \langle N, g\rangle$. Since Ng is not θ-good, there exists $x \in G$ such that $\langle M, x\rangle/N$ is abelian, but θ is not extendible to $\langle M, x\rangle$. Now θ is invariant in M and M/N is cyclic, so Corollary 1.14 guarantees that θ has an extension $\sigma \in \operatorname{Irr}(M)$. Also, $\langle M, x\rangle/M$ is cyclic, but σ does not extend to $\langle M, x\rangle$, so we can conclude that σ is not invariant in $\langle M, x\rangle$, and hence $\sigma^x \neq \sigma$.

Since $(\sigma^x)_N = \theta^x = \theta = \sigma_N$, it follows by the Gallagher correspondence that $\sigma^x = \lambda\sigma$ for some linear character $\lambda \in \mathrm{Irr}(M/N)$. Also, since $\sigma^x \neq \sigma$, we see that λ is not the principal character, and hence $\lambda(g) \neq 1$.

Now let τ be an arbitrary irreducible constituent of χ_M. By the Gallagher correspondence, $\tau = \mu\sigma$ for some character $\mu \in \mathrm{Irr}(M/N)$, and thus

$$\tau^x = (\mu\sigma)^x = \mu^x\sigma^x = \mu^x\lambda\sigma = \mu\lambda\sigma = \lambda\tau\,,$$

where the penultimate equality holds because $\langle M, x\rangle/N$ is abelian, and thus $\mu^x = \mu$.

Since conjugation by x multiplies each irreducible constituent of χ_M by λ, we have $\chi_M = (\chi_M)^x = \lambda\chi_M$, where the first equality holds since χ is a class function. It follows that $\chi(g) = \lambda(g)\chi(g)$, and since $\lambda(g) \neq 1$, we deduce that $\chi(g) = 0$, as required. ∎

Theorem 6.20. *Let $\theta \in \mathrm{Irr}(N)$, where $N \subseteq \mathbf{Z}(G)$, and choose a subset $R \subseteq G$ such that the cosets Nr for $r \in R$ are distinct and form a set of representatives for the θ-good classes of G/N. Let M be the matrix with rows indexed by the set $\mathrm{Irr}(G|\theta)$ and columns indexed by the set R, and where the entry in position (χ, r) in the matrix M is $\chi(r)$. Then M is nonsingular.*

Proof. First, observe that M is a square matrix since $|R|$ is the number of θ-good classes of G/N, and by Theorem 1.19 this is equal to $|\,\mathrm{Irr}(G|\theta)|$. It suffices, therefore, to show that the rows of M are linearly independent. To see this, suppose that for each character $\chi \in \mathrm{Irr}(G|\theta)$, there is a complex number a_χ such that

$$\sum_{\chi\in\mathrm{Irr}(G|\theta)} a_\chi\chi(x) = 0$$

for all $x \in R$. We argue that in fact, this sum vanishes for all $x \in G$.

First, if Nx is not θ-good, then by Lemma 6.19, we have $\chi(x) = 0$ for all $\chi \in \mathrm{Irr}(G|\theta)$, and thus the sum vanishes at x, as wanted. We may now assume that Nx is θ-good, and thus Nx is conjugate in G/N to Nr for some element $r \in R$. Since we can replace x by a G-conjugate, we can assume that $Nx = Nr$, and thus $x = zr$ for some element $z \in N \subseteq \mathbf{Z}(G)$.

Let $\mathcal{X}$ be a representation affording χ. Then $\mathcal{X}(z)$ is the scalar matrix $\theta(z)I$, where I is an identity matrix, and thus $\mathcal{X}(x) = \mathcal{X}(zr) = \theta(z)\mathcal{X}(r)$. It follows that $\chi(x) = \theta(z)\chi(r)$ for all $\chi \in \mathrm{Irr}(G|\theta)$, and we have

$$\sum_{\chi\in\mathrm{Irr}(G|\theta)} a_\chi\chi(x) = \theta(z)\sum_{\chi\in\mathrm{Irr}(G|\theta)} a_\chi\chi(r) = 0\,,$$

as claimed. Since the set $\mathrm{Irr}(G|\theta)$ is linearly independent, we see that $a_\chi = 0$ for all $\chi \in \mathrm{Irr}(G|\theta)$, and thus the rows of M are linearly independent, as required. ∎

Next, we obtain some information about π-partial characters in the context of a character triple isomorphism.

Lemma 6.21. *Let (G, N, θ) and (G^*, N^*, θ^*) be isomorphic character triples, where G and G^* are π-separable and N and N^* are π-groups. Then there exists a unique bijection $\varphi \mapsto \varphi^*$ from $\mathrm{I}_\pi(G|\theta)$ onto $\mathrm{I}_\pi(G^*|\theta^*)$ such that if χ is a lift of φ in $\mathrm{Irr}(G|\theta)$ and $\chi^* \in \mathrm{Irr}(G^*|\theta^*)$ corresponds to χ under the character triple isomorphism, then χ^* is a lift of φ^*. In particular, $|\,\mathrm{I}_\pi(G|\theta)| = |\,\mathrm{I}_\pi(G^*|\theta^*)|$.*

Proof. Given $\varphi \in \mathrm{I}_\pi(G|\theta)$, choose $\chi \in \mathrm{Irr}(G)$ such that $\chi^0 = \varphi$, and note that $\chi_N = \varphi_N$, which is a multiple of θ, and thus $\chi \in \mathrm{Irr}(G|\theta)$. By Lemma 3.12(b), it follows that $(\chi^*)^0 \in \mathrm{I}_\pi(G^*)$, where $\chi^* \in \mathrm{Irr}(G^*|\theta^*)$ corresponds to χ. Write $\varphi^* = (\chi^*)^0$, and note that $\varphi^* \in \mathrm{I}_\pi(G^*|\theta^*)$. Also, φ^* is independent of the choice of the lift χ by Lemma 3.12(c), and thus $\varphi \mapsto \varphi^*$ is the unique map from $\mathrm{I}_\pi(G|\theta)$ to $\mathrm{I}_\pi(G^*|\theta^*)$ that satisfies the condition of the lemma.

Suppose now that also $\mu \mapsto \varphi^*$, where $\mu \in \mathrm{I}_\pi(G|\theta)$, and let $\zeta \in \mathrm{Irr}(G|\theta)$ with $\zeta^0 = \mu$. Then $(\zeta^*)^0 = \varphi^* = (\chi^*)^0$, and it follows by Lemma 3.12(c) that $\varphi = \chi^0 = \zeta^0 = \mu$, and thus our map is injective. We deduce that $|\,\mathrm{I}_\pi(G|\theta)| \le |\,\mathrm{I}_\pi(G^*|\theta^*)|$. Equality follows by symmetry, and hence our map is a bijection. ∎

Proof of Theorem 6.18. By Lemma 3.11, there exists a character triple (G^*, N^*, θ^*) isomorphic to (G, N, θ), where N^* is a central π-subgroup of G^*, and G^* is π-separable. It follows by Lemma 1.22 and Lemma 6.21 that replacing (G, N, θ) by (G^*, N^*, θ) will not change either the number of θ-good π-classes of G/N or the number of members of $\mathrm{I}_\pi(G|\theta)$. It is thus no loss to assume that $N \subseteq \mathbf{Z}(G)$.

Consider the nonsingular matrix M of Theorem 6.20, with rows indexed by the set $\mathrm{Irr}(G|\theta)$ and columns indexed by a subset R of G with the property that the cosets Nr for $r \in R$ form a set of distinct representatives for the θ-good conjugacy classes of G/N. Let R_0 be the set of π-elements in R, and note that since N is a π-group, the cosets Nr for $r \in R_0$ form a set of representatives for the θ-good π-classes of G/N. In particular, $|R_0|$ is the number of such classes.

Now let M_0 be the matrix formed by the columns of M that correspond to members of the subset R_0 of R. Since M is nonsingular, the columns of M_0 are linearly independent, and thus the rank of M_0 is equal to $|R_0|$.

For each member $\varphi \in \mathrm{I}_\pi(G|\theta)$ choose a lift $\chi \in \mathrm{Irr}(G)$, and observe that $\chi_N = \varphi_N$, and thus $\chi \in \mathrm{Irr}(G|\theta)$. Let X be the subset of $\mathrm{Irr}(G|\theta)$ selected in this way, and observe that $|X| = |\,\mathrm{I}_\pi(G|\theta)|$. We argue that the set of rows of M_0 corresponding to characters χ in X forms a basis for the row space of M_0. Once this is established, it will follow that $|X|$ is equal to the rank $|R_0|$ of M, and since $|R_0|$ is the number of θ-good π-classes of G/N, the proof will be complete.

Since R_0 consists of π-elements, the entries in the row of the matrix M_0 corresponding to an arbitrary character $\psi \in \mathrm{Irr}(G|\theta)$ have the form $\psi(r) = \psi^0(r)$ for $r \in R_0$. Now ψ^0 is a sum of members of $\mathrm{I}_\pi(G)$, and since ψ_N is a multiple of θ, it follows that the members of $\mathrm{I}_\pi(G)$ that occur as summands of ψ^0 all lie over θ. We conclude that ψ^0 is a sum of members of $\mathrm{I}_\pi(G|\theta)$, and thus the row of M_0 corresponding to ψ is a sum of rows indexed by members of the set X. It follows that these rows span the row space of M_0.

Finally, to show that the rows of M_0 corresponding to members of X are linearly independent, we mimic the proof of Theorem 6.20. Assume that

$$\sum_{\chi \in X} a_\chi \chi(x) = 0$$

for all $x \in R_0$, where the coefficients a_χ are complex numbers. We argue that in fact, this sum vanishes for all π-elements $x \in G$. This is clear if Nx is not θ-good because in that case, Lemma 6.19 guarantees that $\chi(x) = 0$ for all $\chi \in X$. If x is a π-element and Nx is θ-good, we can replace x by a conjugate and assume that $Nx = Nr$ for some element $r \in R_0$. Then $x = zr$ for some element $z \in N \subseteq \mathbf{Z}(G)$, and we have $\chi(x) = \theta(z)\chi(r)$ for all $\chi \in X$. It follows that the above sum vanishes at x, as required.

We have now shown that $\sum a_\chi \chi^0$ is identically zero on the full set G^0 of π-elements of G. Since the π-partial characters χ^0 for $\chi \in X$ are distinct and irreducible, they are linearly independent by Theorem 3.3, and thus $a_\chi = 0$ for all $\chi \in X$. It follows that the rows of M_0 corresponding to members of X are linearly independent, and the proof is complete. ∎

The following is the analogue of Corollary 6.5 for π-partial characters.

Corollary 6.22. *Let (G, K, Q, θ) be an OW-configuration with associated character triple (N, L, φ). Assume that G is π-separable, that K is a π-group, and that Q is a π'-group. Then $|\,\mathrm{I}_\pi(G|\theta)| = |\,\mathrm{I}_\pi(N|\varphi)|$.*

Proof. By Theorem 6.4, the natural isomorphism from N/L to G/K defines a bijection from the set of φ-good classes of N/L onto the set of θ-good classes of G/K. Of course, this isomorphism also carries the set of π-classes of N/L onto the set of π-classes of G/K, and so the number of φ-good

π-classes of N/L is equal to the number of θ-good π-classes of G/K. The result now follows by Theorem 6.18.

6E

Fix a prime p, and let P be a p-subgroup of an arbitrary finite group G. Write $N = \mathbf{N}_G(P)$, and let $\tau \in \mathrm{Irr}(N/P)$. Following J. Alperin, we say that the pair (P, τ) is a **p-weight** of G if τ has p-defect zero. (This means that the degree $\tau(1)$ is divisible by the full p-part of $|N/P|$.) For example, if P is a Sylow p-subgroup of G, then p does not divide $|N/P|$, and hence (P, τ) is a p-weight for every member $\tau \in \mathrm{Irr}(N/P)$.

Observe that G acts by conjugation on its set of p-weights. The Alperin weight conjecture (often abbreviated AWC) asserts that the number of conjugacy classes of p-weights of an arbitrary finite group G is equal to $|\,\mathrm{IBr}(G)|$, the number of irreducible Brauer characters of G for the prime p. (Recall that $|\,\mathrm{IBr}(G)|$ is equal to the number of p-regular conjugacy classes of G. An alternative formulation of the AWC that does not mention Brauer characters, therefore, is that the number of classes of p-weights of G is equal to the number of p-regular conjugacy classes of G.)

Although the AWC remains unproved in general, it is known to be valid in many cases, and in particular it holds for p-solvable groups for the prime p. In this section, we prove a strong form of the AWC for p-solvable groups, and we also obtain a π-analogue for π-separable groups G having nilpotent Hall π-subgroups.

First, we generalize the notion of a p-weight in the obvious way. We say that a pair (Q, τ) is a **π-weight** of G if Q is a π-subgroup and $\tau \in \mathrm{Irr}(N/Q)$, where $N = \mathbf{N}_G(Q)$ and τ has π-defect zero. (This means, of course, that $\tau(1)$ is divisible by $|N/Q|_\pi$.)

Recall that for a p-solvable group, the set $\mathrm{IBr}(G)$ of irreducible Brauer characters (for the prime p) is exactly $\mathrm{I}_{p'}(G)$, so by analogy, we expect the number of classes of π-weights in a π-separable group to be equal to $|\mathrm{I}_{\pi'}(G)|$, and we prove this under the additional assumption that a Hall π-subgroup of G is nilpotent. (Note that in the original Alperin situation, where $\pi = \{p\}$, we get the nilpotence for free, and consequently, our result implies that the AWC holds for p-solvable groups.)

Although there is no fundamental difference between π-partial characters and π'-partial characters, it has been our custom to focus on the former. For the sake of consistency with our earlier results, therefore, we switch notation now, interchanging π and π', and we work toward a proof that if G is π-separable and a Hall π'-subgroup of G is nilpotent, then the number of classes of π'-weights of G is equal to $|\,\mathrm{I}_\pi(G)|$. (Note that since π-separability and π'-separability are equivalent, our change of notation does not require us to speak of "π'-separable groups" in place of "π-separable groups".)

We continue to assume that G is π-separable, and we also assume that a Hall π'-subgroup of G is nilpotent. The π-version of the AWC suggests the question of whether or not there exists a canonical bijection from $\mathrm{I}_\pi(G)$ onto the set of classes of π'-weights. If such a bijection exists, then by taking the first components of the class of weights corresponding to a member $\varphi \in \mathrm{I}_\pi(G)$, we could associate a unique conjugacy class of π'-subgroups to each irreducible π-partial character φ.

We already know of just such a natural assignment of a conjugacy class of π'-subgroups of G to each member $\varphi \in \mathrm{I}_\pi(G)$: the vertices of φ. We might guess, therefore, that the vertices of φ are exactly the first components of the class of weights corresponding to φ. In fact, there does not always exist a canonical bijection from $\mathrm{I}_\pi(G)$ onto the set of π'-weights of G, so we cannot properly speak of the weights corresponding to φ, but nevertheless, there is a connection between π'-weights and vertices in π-separable groups. We will show the counts are as we might expect: for each π'-subgroup Q of G, the number of members $\varphi \in \mathrm{I}_\pi(G)$ with vertex Q is equal to the number of classes of π'-weights with first component conjugate to Q.

Actually, there is no need to speak of conjugacy classes of weights or of π'-subgroups in this context. This is because two conjugate weights that have the same first component must be equal. (To see this, observe that if $(Q,\tau)^g = (Q,\sigma)$, then $g \in \mathbf{N}_G(Q)$, and since τ is a character of $\mathbf{N}_G(Q)$, we have $\tau = \tau^g = \sigma$.) Our result relating vertices and weights can thus be stated as follows. ∎

Theorem 6.23. *Let G be π-separable, and assume that a Hall π'-subgroup of G is nilpotent. Then for each π'-subgroup Q of G, the number of π'-weights whose first component is Q is equal to the number of members of $\mathrm{I}_\pi(G)$ for which Q is a vertex.*

As an immediate consequence, we obtain the promised π-analogue of the AWC.

Corollary 6.24. *Let G be π-separable, and assume that a Hall π'-subgroup of G is nilpotent. Then the number of conjugacy classes of π'-weights in G is equal to $|\mathrm{I}_\pi(G)|$.*

Proof. Let $\mathcal{X}$ be a set of representatives for the conjugacy classes of π'-subgroups of G. Since conjugate weights having the same first component are equal, we can count the total number of classes of π'-weights of G by summing the numbers of π'-weights having first component Q as Q runs over $\mathcal{X}$. Similarly, we can compute $|\mathrm{I}_\pi(G)|$ by summing the number of members

$\varphi \in \mathrm{I}_\pi(G)$ having Q as a vertex, where Q runs over $\mathcal{X}$. The result now follows by Theorem 6.23. ■

In the original setting of the AWC, where $\pi' = \{p\}$, examples show that if G is not assumed to be p-solvable, then the conclusion of Theorem 6.23 can fail. (Of course, since G is not p-solvable in this situation, we cannot use the definition of a vertex given in Chapter 5, and we must revert to Green's original definition.) For example, taking $p = 2$ and $G = GL(3, 2)$, one can compute that a Sylow 2-subgroup of G is a vertex for three of the four members of $\mathrm{IBr}(G)$, but it is a first component of only one 2-weight.

The condition that a Hall π'-subgroup of G is nilpotent is necessary for Corollary 6.24 (and thus also for Theorem 6.23). To see this, consider, for example, the case where G is itself a π'-group. We argue that if the conclusion of Corollary 6.24 holds, then in fact, G must be nilpotent. To see this, suppose that Q is a self-normalizing subgroup of G. (In other words, $\mathbf{N}_G(Q) = Q$.) If τ is the principal character of the trivial group $\mathbf{N}_G(Q)/Q$, then (Q, τ) is a π'-weight, and so there are at least as many classes of π'-weights as there are classes of self-normalizing subgroups of G. Since we are assuming that G is a π'-group, however, we have $|\,\mathrm{I}_\pi(G)| = 1$, so the conclusion of Corollary 6.24 implies that G has at most one conjugacy class of self-normalizing subgroups, and we see that this must be the one-member class $\{G\}$. It follows that no proper subgroup of G is self-normalizing, and, as is well known, this implies that G is nilpotent.

Although we must assume that a Hall π'-subgroup of G is nilpotent if we expect the conclusion of Theorem 6.23 to hold for all π'-subgroups Q of G, a weaker condition suffices if we are interested only in the case where Q is a full Hall π'-subgroup. In that case, it suffices to assume that Q is solvable, so we have the following slight generalization of Theorem 6.23.

Theorem 6.25. *Let Q be a π'-subgroup of a π-separable group G, and assume either that a Hall π'-subgroup of G is nilpotent or else that Q is a full Hall π'-subgroup of G and Q is solvable. Then the number of π'-weights in G having first component Q is equal to the number of members of $\mathrm{I}_\pi(G)$ for which Q is a vertex.*

Although it may seem that Theorem 6.25 is not an especially significant generalization of Theorem 6.23, we have included it because this result has an interesting consequence, and its proof requires no extra work.

Corollary 6.26. *Let G be π-separable, and suppose that a Hall π'-subgroup Q of G is solvable. Then the number of π-special irreducible characters of G is equal to $|\,\mathrm{Irr}(N/Q)|$, where $N = \mathbf{N}_G(Q)$.*

Proof. By Theorem 3.14, the number of π-special characters of G is equal to the number of members $\varphi \in \mathrm{I}_\pi(G)$ such that $\varphi(1)$ is a π-number. Also, since Q is a full Hall π'-subgroup of G, it follows by Corollary 5.18 that $\varphi(1)$ is a π-number if and only if φ has vertex Q. By Theorem 6.25, the number of members of $\mathrm{I}_\pi(G)$ having vertex Q is equal to the number of π'-weights with first component Q, and thus it suffices to show that the number of such weights is $|\operatorname{Irr}(N/Q)|$. This is clear, however, since N/Q is a π-group, and thus every irreducible character of N/Q has π'-defect zero. ∎

We begin work now toward a proof of Theorem 6.25.

Lemma 6.27. *Let $Q \triangleleft N$, where N is π-separable, and let $\tau \in \operatorname{Irr}(N/Q)$, where τ has π'-defect zero. Then viewing τ as a character of N, we have $\tau \in \mathrm{B}_\pi(N)$.*

Proof. If $Q = N$, then $\tau = 1_N$, the principal character, and thus $\tau \in \mathrm{B}_\pi(N)$, as required. We can thus suppose that $Q < N$, and we proceed by induction on $|N : Q|$. Let M be a maximal normal subgroup of N containing Q, and let σ be an irreducible constituent of τ_M. Observe that $Q \subseteq \ker(\sigma)$, so we can view $\sigma \in \operatorname{Irr}(M/Q)$.

Suppose first that N/M is a π-group. Then $\tau(1)/\sigma(1)$ is a π-number, and thus

$$\sigma(1)_{\pi'} = \tau(1)_{\pi'} = |N : Q|_{\pi'} = |M : Q|_{\pi'}\,,$$

where the second equality holds because τ has π'-defect zero as a character of N/Q. Then σ has π'-defect zero when viewed as a character of M/Q, so by the inductive hypothesis, $\sigma \in \mathrm{B}_\pi(M)$. Since N/M is a π-group, it follows by Theorem 4.14 that $\tau \in \mathrm{B}_\pi(N)$, as required.

In the remaining case, N/M is a π'-group. Now $\sigma(1)$ divides $|M : Q|$ since $\sigma \in \operatorname{Irr}(M/Q)$. Also, $\tau(1)$ divides $|N : M|\sigma(1)$, and thus

$$|N : Q|_{\pi'} = \tau(1)_{\pi'} \leq |N : M|\sigma(1)_{\pi'} \leq |N : M||M : Q|_{\pi'} = |N : Q|_{\pi'}\,.$$

Equality thus holds throughout, and we deduce that $\sigma(1)_{\pi'} = |M : Q|_{\pi'}$, and thus σ has π'-defect zero as a character of M/Q. By the inductive hypothesis we have $\sigma \in \mathrm{B}_\pi(M)$ in this case too.

We also see from the above equality that $\tau(1)_{\pi'} = |N : M|\sigma(1)_{\pi'}$. Since N/M is a π'-group, we have $\tau(1)_\pi = \sigma(1)_\pi$, and thus $\tau(1) = |N : M|\sigma(1)$. Since τ lies over σ, it follows that $\tau = \sigma^N$. By Theorem 4.19(b) some character of N lying over σ lies in $\mathrm{B}_\pi(N)$, and we conclude that $\tau \in \mathrm{B}_\pi(N)$. ∎

Lemma 6.28. *Let Q be a π'-subgroup of a π-separable group G, and write $N = \mathbf{N}_G(Q)$. Then the number of π'-weights of G with first component Q is equal to the number of members of $\mathrm{I}_\pi(N)$ having vertex Q.*

Proof. We construct a bijection from the set $\mathcal{S}$ of π'-weights of G with first component Q onto the set of members of $\mathrm{I}_\pi(N)$ having Q as a vertex. If $(Q, \tau) \in \mathcal{S}$, then by definition τ is a π'-defect-zero character of N/Q, and thus $\tau \in \mathrm{B}_\pi(N)$ by Lemma 6.27. It follows by Theorem 5.1 that $\tau^0 \in \mathrm{I}_\pi(N)$ and that τ is uniquely determined by τ^0, and thus $(Q, \tau) \mapsto \tau^0$ is an injective map from $\mathcal{S}$ into $\mathrm{I}_\pi(N)$.

We show next that if $(Q, \tau) \in \mathcal{S}$, then Q is a vertex for τ^0. Since $Q \lhd N$, it follows by Lemma 5.31 that Q is contained in a vertex P for τ^0. By Corollary 5.18, we have

$$|P| = \frac{|N|_{\pi'}}{\tau^0(1)_{\pi'}} = \frac{|N|_{\pi'}}{|N : Q|_{\pi'}} = |Q|_{\pi'} = |Q|,$$

where the second equality holds since τ has π'-defect zero as a character of N/Q. We conclude that $P = Q$, and thus Q is a vertex for τ^0, as required.

Finally, we must show that if $\varphi \in \mathrm{I}_\pi(N)$ has vertex Q, then $\varphi = \tau^0$ for some π'-defect-zero character $\tau \in \mathrm{Irr}(N/Q)$. We know by Theorem 5.1 that $\varphi = \tau^0$ for some member $\tau \in \mathrm{B}_\pi(N)$. Also, $Q \subseteq \mathbf{O}_{\pi'}(N) \subseteq \ker(\tau)$, where the second containment holds by Corollary 4.22. Then τ can be viewed as a character of N/Q, and we must show that, as such, it has π'-defect zero. By assumption, Q is a vertex for φ, so by Corollary 5.18, we have

$$|Q| = \frac{|N|_{\pi'}}{\varphi(1)_{\pi'}}.$$

Then

$$\tau(1)_{\pi'} = \varphi(1)_{\pi'} = \frac{|N|_{\pi'}}{|Q|} = |N/Q|_{\pi'},$$

and thus τ is a π'-defect-zero character of N/Q, as required. ∎

We can now reformulate Theorem 6.25 in a more convenient form, without mentioning weights. Assuming the hypotheses of the theorem, it suffices by Lemma 6.28 to show that the numbers of members of $\mathrm{I}_\pi(G)$ and of $\mathrm{I}_\pi(N)$ that have vertex Q are equal, where $N = \mathbf{N}_G(Q)$. (Note that if Q is a full Hall π'-subgroup of G, then a member of $\mathrm{I}_\pi(G)$ or $\mathrm{I}_\pi(N)$ has vertex Q if and only if its degree is a π-number. In this case therefore, the desired equality is analogous to the McKay situation, except that we are counting partial characters instead of ordinary characters.)

In order to prove our theorem inductively, we generalize it in the spirit of Theorem 6.12. We introduce a π-subgroup $B \lhd G$ and a G-invariant character $\beta \in \mathrm{Irr}(B)$, and we consider only partial characters lying over β. For notational convenience we will write $n(G, Q, \beta)$ to denote the number of members of $\mathrm{I}_\pi(G|\beta)$ that have Q as a vertex.

Theorem 6.29. *Let Q be a π'-subgroup of a π-separable group G, and assume either that a Hall π'-subgroup of G is nilpotent or else that Q is a full Hall π'-subgroup of G and Q is solvable. Let $B \lhd G$ be a π-subgroup, and let $\beta \in \mathrm{Irr}(B)$ be G-invariant. Then writing $N = \mathbf{N}_G(BQ)$, we have $n(G, Q, \beta) = n(N, Q, \beta)$.*

If B is the trivial group (and β is the trivial character) in Theorem 6.29, then $N = \mathbf{N}_G(Q)$. In this case, the assertion of the theorem is that the numbers of members of $\mathrm{I}_\pi(G)$ and $\mathrm{I}_\pi(N)$ for which Q is a vertex are equal, and as we have observed, this will establish Theorem 6.25.

We devote the remainder of this chapter to a proof of Theorem 6.29, and we begin with a few fairly easy results concerning vertices.

Lemma 6.30. *Let $K \lhd G$, where G is π-separable and K is a π-subgroup, and suppose $Q \subseteq G$ is a π'-subgroup such that $KQ \lhd G$. Let $\theta \in \mathrm{Irr}(K)$ be G-invariant, and let $\varphi \in \mathrm{I}_\pi(G|\theta)$. Then Q is contained as a normal subgroup in some vertex of φ.*

Proof. By Theorem 5.1, there exists $\chi \in \mathrm{B}_\pi(G)$ such that $\chi^0 = \varphi$, and we observe that $\chi_K = \varphi_K$, so χ lies over θ. Since $|KQ : K|$ is a π'-number, θ has a canonical extension $\hat{\theta} \in \mathrm{Irr}(KQ)$, and $\hat{\theta}$ is π-special. It follows that $\hat{\theta}$ lies in $\mathrm{B}_\pi(KQ)$, and in fact Theorem 4.19(b) guarantees that $\hat{\theta}$ is the unique member of $\mathrm{B}_\pi(KQ)$ lying over θ. Since $\chi \in \mathrm{B}_\pi(G)$ and $KQ \lhd G$, it follows by Corollary 4.21 that the irreducible constituents of χ_{KQ} lie in $\mathrm{B}_\pi(KQ)$, and we conclude that $\hat{\theta}$ is a constituent of χ_{KQ}.

Now $\hat{\theta}$ is π-factored, and hence in the notation of Chapter 4, we have $(KQ, \hat{\theta}) \in \mathcal{S}(G)$ and $(KQ, \hat{\theta}) \le (G, \chi)$. By Theorem 4.1(b), there exists a pair $(S, \alpha) \in \mathcal{S}^*(G)$ such that $(KQ, \hat{\theta}) \le (S, \alpha) \le (G, \chi)$, and by Corollary 4.11(c) there exists a nucleus (W, γ) for χ such that $(S, \alpha) \le (W, \gamma)$.

Since γ is a nucleus character for $\chi \in \mathrm{B}_\pi(G)$, it follows that γ is π-special, and in particular $\gamma(1)$ is a π-number. Also, $\gamma^G = \chi$, so $(\gamma^0)^G = \chi^0 = \varphi$, and since $\gamma^0(1)$ is a π-number, we see that the Hall π'-subgroups of W are vertices for φ. Now $Q \subseteq W$, so Q is contained in some Hall π'-subgroup P of W, and since Q is a full Hall π'-subgroup of KQ, we have $Q = P \cap KQ \lhd P$. Then P is a vertex for φ containing Q as a normal subgroup, as wanted. ∎

Lemma 6.31. *Let $K \lhd G$, where G is π-separable and K is a π-subgroup. Also, let $Q \subseteq G$ be a π'-subgroup, and let $M \lhd G$, where $M \subseteq Q$. Let $\theta \in \mathrm{Irr}(K)$ be G-invariant, and let $\hat{\theta}$ be the extension of θ to KM such that $M \subseteq \ker(\hat{\theta})$. Then writing $\overline{G} = G/M$ and viewing $\hat{\theta}$ as a character of $\overline{K}$, we have $n(G, Q, \theta) = n(\overline{G}, \overline{Q}, \hat{\theta})$.*

Proof. Since K is a π-subgroup and M is a π'-subgroup, we see that $KM = K \times M$, and thus θ has a unique extension $\hat{\theta} \in \mathrm{Irr}(KM)$ such that $M \subseteq \ker(\hat{\theta})$. Then $\hat{\theta}$ can be viewed as a character of $KM/M = \overline{K}$, as claimed. The number $n(\overline{G}, \overline{Q}, \hat{\theta})$ is thus defined.

By Theorem 3.10, there is a bijection $\varphi \mapsto \overline{\varphi}$ from $\mathrm{I}_\pi(G)$ onto $\mathrm{I}_\pi(\overline{G})$, where $\overline{\varphi}(\overline{x}) = \varphi(x)$ for all π-elements $x \in G$. We argue that φ lies over θ if and only if $\overline{\varphi}$ lies over $\hat{\theta}$, where we view $\hat{\theta}$ as a character of $\overline{K}$. To see this, let $x \in K$ and note that $\theta(x) = \hat{\theta}(x) = \hat{\theta}(\overline{x})$. Assuming that φ lies over θ, we have $\varphi_K = e\theta$ for some integer e. Then

$$\overline{\varphi}(\overline{x}) = \varphi(x) = e\theta(x) = e\hat{\theta}(\overline{x}),$$

and thus $\overline{\varphi}_{\overline{K}} = e\hat{\theta}$, as wanted. Conversely, if $\overline{\varphi}$ lies over $\hat{\theta}$, then $\overline{\varphi}_{\overline{K}} = e\hat{\theta}$ for some integer e, and we have

$$\varphi(x) = \overline{\varphi}(\overline{x}) = e\hat{\theta}(\overline{x}) = e\theta(x),$$

and thus $\varphi_K = e\theta$.

It suffices now to show that Q is a vertex for φ if and only if $\overline{Q}$ is a vertex for $\overline{\varphi}$. Assume first that Q is a vertex for φ. Then by definition, Q is a Hall π'-subgroup of some subgroup $U \subseteq G$, and there exists $\mu \in \mathrm{I}_\pi(U)$ such that $\varphi = \mu^G$ and $\mu(1)$ is a π-number. Now $M \subseteq Q \subseteq U$, so by Theorem 3.10 there exists $\overline{\mu} \in \mathrm{I}_\pi(\overline{U})$ such that $\overline{\mu}(\overline{x}) = \mu(x)$ for π-elements $x \in U$. It is routine to check that

$$(\overline{\mu})^{\overline{G}}(\overline{x}) = \mu^G(x) = \varphi(x) = \overline{\varphi}(\overline{x})$$

for π-elements $x \in G$, and thus $\overline{\mu}^{\overline{G}} = \overline{\varphi}$. Also, $\overline{\mu}(1) = \mu(1)$ is a π-number, and hence the Hall π'-subgroups of $\overline{U}$ are vertices for $\overline{\varphi}$. In particular, $\overline{Q}$ is a vertex for $\overline{\varphi}$, as wanted.

Conversely, assume that $\overline{Q}$ is a vertex for $\overline{\varphi}$. Then $\overline{Q}$ is a Hall π'-subgroup of some subgroup $\overline{U} \subseteq \overline{G}$, where we can assume that $M \subseteq U$, and thus Q is a Hall π'-subgroup of U. By Theorem 3.10, every member of $\mathrm{I}_\pi(\overline{U})$ has the form $\overline{\mu}$, for some member $\mu \in \mathrm{I}_\pi(U)$, so there exists $\mu \in \mathrm{I}_\pi(U)$ such that $\overline{\mu}^{\overline{G}} = \overline{\varphi}$, where $\overline{\mu}(1)$ is a π-number. Reasoning as before, we see that $\mu^G = \varphi$ and $\mu(1)$ is a π-number, and thus Q is a vertex for φ. ∎

Next, we present a refinement of Lemma 6.21 that respects vertices.

Lemma 6.32. *Let (G, N, θ) and (G^*, N^*, θ^*) be isomorphic character triples, where G and G^* are π-separable and N and N^* are π-groups. Also, let $Q \subseteq G$ and $Q^* \subseteq G^*$ be π'-groups, and assume that the intermediate subgroups NQ and N^*Q^* correspond via the character-triple isomorphism. Then $n(G, Q, \theta) = n(G^*, Q^*, \theta^*)$.*

Proof. By Lemma 6.21, there is a bijection $\varphi \mapsto \varphi^*$ from $\mathrm{I}_\pi(G|\theta)$ onto $\mathrm{I}_\pi(G^*|\theta^*)$ such that if $\chi \in \mathrm{Irr}(G|\theta)$ and $\chi^0 = \varphi$, then $(\chi^*)^0 = \varphi^*$, where $\chi^* \in \mathrm{Irr}(G^*|\theta^*)$ corresponds to χ via the character-triple isomorphism. We will show that if Q is a vertex for φ, then Q^* is a vertex for φ^*, and consequently, $n(G, Q, \theta) \le n(G^*, Q^*, \theta^*)$. The result will then follow by symmetry.

Suppose that Q is a vertex for $\varphi \in \mathrm{I}_\pi(G|\theta)$. By definition, Q is a Hall π'-subgroup of some subgroup $U \subseteq G$, where there exists $\mu \in \mathrm{I}_\pi(U)$ such that $\mu^G = \varphi$ and $\mu(1)$ is a π-number. By assumption, N is a π-group, so we can replace U by NU and μ by μ^{NU}, and thus we can assume that $N \subseteq U$. Now $\varphi = \mu^G$, so it follows by Lemma 5.8 that φ lies over μ, and since φ_N is a multiple of θ, we conclude that μ lies over θ .

Let $\psi \in \mathrm{Irr}(U)$ with $\psi^0 = \mu$, and note that $\psi_N = \mu_N$, so ψ lies over θ. Write $\chi = \psi^G$, and observe that $\chi^0 = (\psi^G)^0 = (\psi^0)^G = \mu^G = \varphi$, and thus χ is irreducible and lies over θ. By Lemma 6.21, we have $\varphi^* = (\chi^*)^0$, where $\chi^* \in \mathrm{Irr}(G^*|\theta^*)$ corresponds to χ.

Recall that $\chi = \psi^G$, so by the properties of character-triple isomorphisms, $\chi^* = (\psi^*)^{G^*}$, where $\psi^* \in \mathrm{Irr}(U^*|\theta^*)$ corresponds to ψ. Writing $\mu^* = (\psi^*)^0$, we see that

$$(\mu^*)^{G^*} = ((\psi^*)^0)^{G^*} = ((\psi^*)^{G^*})^0 = (\chi^*)^0 = \varphi^* .$$

Now $\mu(1) = \psi(1)$ is a π-number, and by Lemma 1.12 we have $\psi(1)/\theta(1) = \psi^*(1)/\theta^*(1)$, so $\psi^*(1)$ is also a π-number. Then $\mu^*(1)$ is a π-number, and since $(\mu^*)^{G^*} = \varphi^*$, the Hall π'-subgroups of U^* are vertices for φ^*.

Now NQ and N^*Q^* are corresponding subgroups, so $|U^* : N^*Q^*| = |U : NQ|$, and this is a π-number because Q is a Hall π'-subgroup of U. It follows that the Hall π'-subgroup Q^* of N^*Q^* is a Hall π'-subgroup of U^*, and thus Q^* is a vertex for φ^*, as required. ∎

Our next result is an analogue of Lemma 6.17. Before we state it, however, we recall that if G is π-separable and $M \lhd G$, then Theorem 5.11 guarantees that for each member $\sigma \in \mathrm{I}_\pi(M)$, induction defines a bijection from $\mathrm{I}_\pi(T|\sigma)$ onto $\mathrm{I}_\pi(G|\sigma)$, where T is the stabilizer of σ in G. Since this is analogous to the Clifford correspondence for ordinary characters, we will also refer to this map as the Clifford correspondence.

Shifting the point of view somewhat, it follows that if we fix $\varphi \in \mathrm{I}_\pi(G)$, then for each irreducible constituent σ of φ_M, there is a unique member $\alpha \in \mathrm{I}_\pi(T|\sigma)$ such that $\alpha^G = \varphi$, where T is the stabilizer of σ. We refer to α as the Clifford correspondent of φ with respect to σ, and we observe that all vertices for α are vertices for φ. Note also that if $x \in G$ and $\alpha \in \mathrm{I}_\pi(T)$ is the Clifford correspondent of φ with respect to σ, then $\alpha^x \in \mathrm{I}_\pi(T^x)$ is the Clifford correspondent of φ with respect to σ^x. Furthermore, if Q is a vertex for α, then Q^x is a vertex for α^x.

Lemma 6.33. *Let $Q \subseteq G$, where G is π-separable and Q is a π'-subgroup, and write $N = \mathbf{N}_G(Q)$. Let $M \lhd G$, and let $\mathcal{X}$ be a set of representatives for the orbits of the action of N on the set of Q-invariant members of $\mathrm{I}_\pi(M)$. Then given $\varphi \in \mathrm{I}_\pi(G)$ with vertex Q, there exists a unique member $\sigma \in \mathcal{X}$ such that φ lies over σ and Q is a vertex for the Clifford correspondent of φ with respect to σ.*

Proof. Suppose that Q is a vertex for $\varphi \in \mathrm{I}_\pi(G)$, and let $\sigma \in \mathrm{Irr}(M)$ be an irreducible constituent of φ_M. Let T be the stabilizer of σ in G, and suppose $\alpha \in \mathrm{I}_\pi(T)$ is the Clifford correspondent of φ with resect to σ. If P is a vertex for α, then P is also a vertex for φ, and thus $P^x = Q$ for some element $x \in G$. Replacing σ by σ^x and α by α^x, we can assume that Q is a vertex for α, and in particular $Q \subseteq T$. Then σ is Q-invariant, so there exists a member of $\mathcal{X}$ in the N-orbit containing σ. There exists $n \in N$, therefore, such that $\sigma^n \in \mathcal{X}$. Then φ lies over σ^n, and the Clifford correspondent of φ with respect to σ^n is α^n. Also, $Q = Q^n$ is a vertex for α^n, as wanted.

For uniqueness now, suppose that φ lies over $\sigma \in \mathcal{X}$ and $\tau \in \mathcal{X}$, and suppose that Q is a vertex for the Clifford correspondent α of φ with respect to σ and also for the Clifford correspondent β of φ with respect to τ.

Since both σ and τ lie under φ, we can write $\sigma = \tau^g$ for some element $g \in G$. Then $\alpha = \beta^g$, and thus Q^g is a vertex for α. By assumption, Q is also a vertex for α, and thus Q and Q^g are conjugate in the domain of α, which is the stabilizer T of σ. We can thus write $Q^{gt} = Q$, for some element t stabilizing σ, and thus $gt \in N$ and $\tau^{gt} = \sigma^t = \sigma$. Then σ and τ lie in the same N-orbit, and since by assumption, σ and τ lie in $\mathcal{X}$, we deduce that $\sigma = \tau$. ∎

Our proof of Theorem 6.29 will proceed by induction on the index $|G : B|$. It will thus be convenient to give this quantity a name, and we call it the **relevant index**. The key to the inductive argument is the following.

Theorem 6.34. *Let (G, K, Q, θ) be an OW-configuration with associated character triple (N, L, φ), and suppose that K is a π-group and Q is a π'-group. Assume either that a Hall π'-subgroup of G is nilpotent or else that Q is a full Hall π'-subgroup of G. Assume also that Theorem 6.29 is valid whenever the relevant index r satisfies $r \le |G : K|$. Then $n(G, Q, \theta) = n(N, Q, \varphi)$.*

Proof. Note that $N = \mathbf{N}_G(Q)$, and suppose that $Q \subseteq P \subseteq N$, where P is a π'-subgroup. We claim that if $g \in G$ and $Q \subseteq P^g \subseteq N$, then $g \in N$, and in particular $\mathbf{N}_G(P) \subseteq N$. To see this, observe that since Q is a full Hall π'-subgroup of KQ and $Q \subseteq KQ \cap P$, we have $Q = KQ \cap P$, and similarly $Q = KQ \cap P^g$. Since $KQ \lhd G$, we have

$$Q^g = (KQ \cap P)^g = KQ \cap P^g = Q\,,$$

and thus $g \in \mathbf{N}_G(Q) = N$, as claimed.

Now let $\mathcal{P}$ be a set of representatives for the conjugacy classes of π'-subgroups of N that contain Q. Then no two distinct members of $\mathcal{P}$ are conjugate in N, and we argue that in fact no two distinct members of $\mathcal{P}$ can be conjugate in G. This follows because if P and P^g are in $\mathcal{P}$, then by the result of the previous paragraph, $g \in N$.

If $\psi \in \mathrm{I}_\pi(G|\theta)$, then since $KQ \lhd G$, Lemma 6.30 guarantees that ψ has a vertex P containing Q as a normal subgroup. Then $P \subseteq N$, and thus P is conjugate in G to some unique member of $\mathcal{P}$. It follows that

$$|\,\mathrm{I}_\pi(G|\theta)| = \sum_{P\in\mathcal{P}} n(G,P,\theta)\,.$$

Similarly, each member of $\mathrm{I}_\pi(N|\varphi)$ has a unique vertex in $\mathcal{P}$, and thus

$$|\,\mathrm{I}_\pi(N|\varphi)| = \sum_{P\in\mathcal{P}} n(N,P,\varphi)\,.$$

Now $|\,\mathrm{I}_\pi(G|\theta)| = |\,\mathrm{I}_\pi(N|\varphi)|$ by Corollary 6.22, so to prove that $n(G,Q,\theta) = n(N,Q,\varphi)$ as required, it suffices to show that $n(G,P,\theta) = n(N,P,\varphi)$ for all π'-subgroups P such that $Q < P \subseteq N$.

If Q is a full Hall π'-subgroup of G, there is no π'-subgroup P properly containing Q, so there is nothing to prove. We may assume, therefore, that Q is not a Hall π'-subgroup of G, and thus by hypothesis, the Hall π'-subgroups of G are nilpotent. We proceed by induction on $|G:Q|$.

Fix a π'-subgroup P with $Q < P \subseteq N$, and let $H = \mathbf{N}_G(P)$. Then $KP \lhd KH$, and since P is solvable, (KH,K,P,θ) is an OW-configuration. Similarly, (LH,L,P,φ) is an OW-configuration, where we recall that $L = K \cap N$ because (N,L,φ) is the character triple associated with (G,K,Q,θ).

Write $M = K \cap H$, and note that $H \subseteq N$ by the result of the first paragraph. Then $M = K \cap H = K \cap N \cap H = L \cap H$, and thus $\mathbf{C}_K(P) = M = \mathbf{C}_L(P)$. Now let $\eta \in \mathrm{Irr}(M)$ be the image of φ under the Glauberman map $\mathrm{Irr}_P(L) \to \mathrm{Irr}(M)$, and observe that (H,M,η) is the character triple associated with (LH,L,P,φ).

We argue that (H,M,η) is also the character triple associated with (KH,K,P,θ). Since $Q \lhd P$ and Q acts trivially on L, we see that η is the image of φ under the Glauberman map corresponding to the action of P/Q on L. Also, φ is the image of θ under the Glauberman map corresponding to the action of Q on K, and it follows by Theorem 6.2 that η is the image of θ under the Glauberman map corresponding to the action of P on K. We conclude that (H, M, η) is the character triple associated with (KH, K, P, θ), as claimed.

We argue next that

$$n(KH, P, \theta) = n(H, P, \eta) = n(LH, P, \varphi)\,.$$

Since $|KH : P| < |KH : Q| \leq |G : Q|$, the first of these equalities will follow by applying the inductive hypothesis to the OW-configuration (HK, K, P, θ). To justify this, we observe that a Hall π'-subgroup of KH is nilpotent since it is contained in a Hall π'-subgroup of G. We must also check that Theorem 6.29 is valid whenever the relevant index r satisfies $r \leq |KH : K|$. This is clear, however, since $|KH : K| \leq |G : K|$, and by hypothesis Theorem 6.29 holds if $r \leq |G : K|$.

The second equality follows similarly by observing that $|LH : P| < |G : Q|$ and applying the inductive hypothesis to the OW-configuration (LH, L, P, φ). Certainly, a Hall π'-subgroup of LH is nilpotent, so it suffices to check that Theorem 6.29 is valid whenever the relevant index r satisfies $r \leq |LH : L|$. This follows since $|LH : L| \leq |N : L| = |G : K|$.

To complete the proof now, we show that

$$n(G, P, \theta) = n(KH, P, \theta) \qquad \text{and} \qquad n(N, P, \varphi) = n(LH, P, \varphi)\,.$$

Since $KH = \mathbf{N}_G(KP)$ by the Frattini argument, the first of these equalities is a consequence of Theorem 6.29 applied with P, K, and θ in place of Q, B, and β. (Although we have not yet established Theorem 6.29 in general, it applies here by hypothesis, since the relevant index in this case is $|G : K|$.) Similarly, since $LH = \mathbf{N}_N(LP)$, the second of the above equalities follows by Theorem 6.29 applied with N, P, L, and φ in place of G, Q, B, and β. (The relevant index here is $|N : L| = |G : K|$, so this application of Theorem 6.29 is also valid.) ■

Finally, we are ready to prove Theorem 6.29.

Proof of Theorem 6.29. The result is clear if $B = G$, so we can assume that $B < G$, and we proceed by induction on the relevant index $|G : B|$.

By Lemma 3.11, there exists a character triple (G^*, B^*, β^*) isomorphic to (G, B, β), where $B^* \subseteq \mathbf{Z}(G^*)$ and B^* is a π-group. Write $U = BQ$, so $N = \mathbf{N}_G(U)$, and observe that $N^* = \mathbf{N}_{G^*}(U^*)$, where U^* and N^* are the subgroups of G^* corresponding to U and N, respectively.

Now $U^*/B^* \cong U/B$, so U^*/B^* is a π'-group. We can thus write $U^* = N^*Q^*$, where Q^* is a Hall π'-subgroup of U^*, and we observe that $Q^* \cong Q$. Since B and B^* are π-groups and $G/B \cong G^*/B^*$, it follows that the Hall π'-subgroups of G^* are isomorphic to the Hall π'-subgroups of G, and thus if a Hall π'-subgroup of G is nilpotent, the same is true for G^*. If the Hall π'-subgroups of G are not nilpotent, then by hypothesis Q must be a full Hall π'-subgroup of G, and Q is solvable, and it follows that Q^* is a full Hall π'-subgroup of G^* and Q^* is solvable.

By Lemma 6.32, we have $n(G, Q, \beta) = n(G^*, Q^*, \beta^*)$ and $n(N, Q, \beta) = n(N^*, Q^*, \beta^*)$, so it suffices to prove that $n(G^*, Q^*, \beta^*) = n(N^*, Q^*, \beta^*)$. We have seen that the hypotheses of the theorem hold in G^*, with B^*, Q^*, N^*, and β^* in place of B, Q, N, and β, and furthermore the relevant indices $|G : B|$ and $|G^* : B^*|$ are equal. It follows that we can replace G by G^*, and so we can assume that $B \subseteq \mathbf{Z}(G)$. In particular, Q is characteristic in BQ, and so $N = \mathbf{N}_G(Q)$.

Let $M = \mathbf{O}_{\pi'}(G)$. If $M \not\subseteq Q$, then by Lemma 5.31 there cannot exist any member of $\mathrm{I}_\pi(G)$ having Q as a vertex, and thus $n(G, Q, \beta) = 0$. In this case, $Q < QM$, and since QM is a π'-group, we see that Q is not a full Hall π'-subgroup of G. By hypothesis, therefore, QM is nilpotent, and thus $Q < \mathbf{N}_{QM}(Q) = QM \cap N = Q(M \cap N)$, and we see that $M \cap N \not\subseteq Q$. Since $M \cap N \lhd N$, it follows that Q is not a vertex for any member of $\mathrm{I}_\pi(N)$, and thus $n(N, Q, \beta) = 0$. There is nothing further to prove in this case, so we can assume that $M \subseteq Q$.

Write $\overline{G} = G/M$. Lemma 6.31 yields $n(G, Q, \beta) = n(\overline{G}, \overline{Q}, \hat{\beta})$ and $n(N, Q, \beta) = n(\overline{N}, \overline{Q}, \hat{\beta})$, where $\hat{\beta}$ is the extension of β to BM with M contained in the kernel, and $\hat{\beta}$ is viewed as a character of $\overline{B}$. It suffices, therefore, to show that $n(\overline{G}, \overline{Q}, \hat{\beta}) = n(\overline{N}, \overline{Q}, \hat{\beta})$. Now $\overline{N} = \mathbf{N}_{\overline{G}}(\overline{B}\,\overline{Q})$, and it is easy to check that the hypotheses of the theorem hold if G, Q, B, and β are replaced by $\overline{G}$, $\overline{Q}$, $\overline{B}$, and $\hat{\beta}$. Also, if $M > 1$, then $|\overline{G} : \overline{B}| = |G : BM| < |G : M|$, and thus the required equality follows by the inductive hypothesis. We may assume, therefore, that $\mathbf{O}_{\pi'}(G) = 1$.

Now let K/B be a minimal normal subgroup of G/B. If K/B is a π'-group, then since B is central, it follows that $\mathbf{O}_{\pi'}(K) > 1$, and this is a contradiction since $\mathbf{O}_{\pi'}(G) = 1$. It follows that K/B is a π-group, and thus K is a π-group.

Let $\mathcal{X}$ be a set of representatives for the N-orbits of Q-invariant characters in $\mathrm{Irr}(K|\beta)$. For $\sigma \in \mathcal{X}$, write G_σ to denote the stabilizer of σ in G, and let $\mathcal{V}_\sigma$ be the set of members $\psi \in \mathrm{I}_\pi(G)$ that lie over σ and such that Q is a vertex for the Clifford correspondent α of ψ with respect to σ. Then every member of $\mathcal{V}_\sigma$ lies over β and has vertex Q, and by Lemma 6.33, the sets $\mathcal{V}_\sigma$ are disjoint and exhaust the full set of members of $\mathrm{I}_\pi(G|\beta)$ that have vertex Q. We thus have

$$n(G, Q, \beta) = \sum_{\sigma \in \mathcal{X}} |\mathcal{V}_\sigma| = \sum_{\sigma \in \mathcal{X}} n(G_\sigma, Q, \beta),$$

where the second equality holds because by Theorem 5.11 the map $\alpha \mapsto \alpha^G$ is a bijection from the set of members of $\mathrm{I}_\pi(G_\sigma|\sigma)$ having vertex Q onto $\mathcal{V}_\sigma$.

Now write $L = N \cap K$, so $L \lhd N$ and $L = \mathbf{C}_K(Q) \supseteq B$. The Glauberman correspondence maps the set $\mathrm{Irr}_Q(K)$ onto $\mathrm{Irr}(L)$, and since this map is canonical, it carries the N-orbits in $\mathrm{Irr}_Q(K)$ to the N-orbits in $\mathrm{Irr}(L)$.

Suppose that $\sigma \in \mathrm{Irr}_Q(K)$ and $\tau \in \mathrm{Irr}(L)$, and that σ and τ correspond under the Glauberman map. Then σ lies over τ by Corollary 6.3, and since β is G-invariant, it follows that σ lies over β if and only if τ lies over β. We conclude that the image $\mathcal{Y}$ of $\mathcal{X}$ under the Glauberman map is a set of representatives for the N-orbits of members of $\mathrm{Irr}(L)$ that lie over β.

For each member $\tau \in \mathcal{Y}$, let $\mathcal{W}_\tau$ be the set of members of $\mathrm{I}_\pi(N)$ lying over τ for which the Clifford correspondent in $\mathrm{I}_\pi(N_\tau|\tau)$ has vertex Q. (Here, of course, N_τ is the stabilizer of τ in N.) Reasoning as before and applying Lemma 6.33 in the group N, we see that the sets $\mathcal{W}_\tau$ are disjoint and exhaust the set of members of $\mathrm{I}_\pi(N|\beta)$ having vertex Q, and thus

$$n(N, Q, \beta) = \sum_{\tau \in \mathcal{Y}} |\mathcal{W}_\tau| = \sum_{\tau \in \mathcal{Y}} n(N_\tau, Q, \beta)\,.$$

To prove that $n(G, Q, \beta) = n(N, Q, \beta)$ as required, therefore, it suffices to show that $n(G_\sigma, Q, \sigma) = n(N_\tau, Q, \tau)$ whenever $\sigma \in \mathrm{Irr}_Q(K)$ and $\tau \in \mathrm{Irr}(L)$ correspond under the Glauberman map $\mathrm{Irr}_Q(K) \to \mathrm{Irr}(L)$.

Now fix $\sigma \in \mathrm{Irr}_Q(K)$, and $\tau \in \mathrm{Irr}(L)$ as above, and for notational simplicity, write $S = G_\sigma$ and $T = N_\tau$. Note that S and T contain Q and that $T = S \cap N$ since σ and τ uniquely determine each other, and thus $T = \mathbf{N}_S(Q)$.

To complete the proof now, we argue that

$$n(S, Q, \sigma) = n(KT, Q, \sigma) = n(T, Q, \tau)\,.$$

First, since $T = \mathbf{N}_S(Q)$, it follows that $KT = \mathbf{N}_S(KQ)$. Also, $|S : K| \leq |G : K| < |G : B|$, and thus the first equality above follows by the inductive hypothesis applied with S, K, and σ in place of G, B, and β. (The hypotheses on the Hall π'-subgroups of S are clearly satisfied.)

To prove the second equality, we observe that (KT, K, Q, σ) is an OW-configuration. Now

$$\mathbf{N}_{KT}(Q) = KT \cap N = (K \cap N)T = LT = T\,,$$

and $L = \mathbf{C}_K(Q)$. It follows that (T, L, τ) is the character triple associated with (KT, K, Q, σ). The desired equality follows by Theorem 6.34 with KT in place of G, but to apply that result, we must verify that Theorem 6.29 holds whenever the relevant index r satisfies $r \leq |KT : K|$. We see that $|KT : K| \leq |G : K| < |G : B|$, so if $r \leq |KT : K|$, then $r < |G : B|$, and the inductive hypothesis guarantees that Theorem 6.29 holds in this case. ∎

Problems

6.1. Let $P \in \mathrm{Syl}_p(G)$ and suppose that G has a normal Hall p'-subgroup N. Assume that there exists a faithful character $\chi \in \mathrm{Irr}(G)$ such that $p + 1 < \chi(1) < 2p - 2$.

(a) Let $C = \mathbf{C}_N(P)$. If χ_C is irreducible, show that $C = N$.

(b) Now assume that χ_C is not irreducible. Show that χ_C has the form $p\lambda + \theta$, where λ is linear.

(c) Let $g \in G$, and let $H = \langle C, C^g \rangle$. Show that χ_H has a linear constituent, and deduce that $H < N$.

Notes and References

The Glauberman correspondence was constructed in [**27**], and Theorem 4 of that paper shows that the map is canonical. That theorem also includes Theorem 6.1 of this chapter, which is the description of the correspondence in the case where the acting group is a p-group.

The McKay conjecture (but only for simple groups and for $p = 2$) was stated in [**57**] and then with additional information in [**58**]. It seems that the first suggestion that the conjecture might be true for nonsimple groups and for all primes was in [**34**], where it was proved (with a canonical bijection) for those solvable groups in which the normalizer of a Sylow p-subgroup has odd index. (In particular, this established the McKay conjecture for groups of odd order.) What is essentially the proof in [**34**] appears Chapter 8 of this book.

Using an argument related to the proof in [**34**] for odd-order groups, T. R. Wolf proved the McKay conjecture for solvable groups without assuming oddness in [**74**]. In that paper, Wolf appealed to ideas of E. C. Dade to overcome the difficulties caused by the prime 2.

J. Alperin in [**1**] extended the McKay conjecture (for all primes and all finite groups) to a block-theoretic version that is now called the Alperin–McKay conjecture.

In [**50**], the McKay conjecture was reduced to a question about simple groups, and the methods of that paper have yielded proofs of the conjecture in many cases. In particular, the McKay conjecture for $p = 2$ has been established by G. Malle and B. Späth in [**56**].

T. Okuyama and M. Wajima proved the Alperin–McKay conjecture for p-solvable groups using what is essentially the technique of this chapter. A somewhat sketchy version of their proof appears in [**65**], and a more detailed presentation is in [**66**]. Other (less elementary) proofs or sketches of proofs for p-solvable groups appeared in [**13**] and [**29**].

Corollary 6.18, which is the analogue of the Gallagher character count, is included in Theorem 6.2 of [**44**].

The Alperin weight conjecture was formulated in [**2**]. The version of the AWC stated in this chapter is a special case of Alperin's version, from which we have removed the block theory. Our Theorem 6.23, which is a strong form of the π-analogue of the AWC for π-separable groups, appeared in [**49**].

The conclusion of Corollary 6.26 is true even if the Hall subgroup Q is not assumed to be solvable. This more general result is Corollary 1.16 of [**76**].

Chapter 7

Fully Ramified Abelian Sections

7A

If L and K are normal subgroups of a group G with $L \subseteq K$, we say that K/L is a **normal section** of G, and we recall that a normal section K/L is said to be a **chief factor** of G if $L < K$ and there exists no subgroup $N \triangleleft G$ such that $L < N < K$. Since a group is solvable if and only if all of its chief factors are abelian, it is reasonable when studying the character theory of solvable groups to investigate the interactions between characters and abelian normal sections. In this chapter, we initiate such an investigation.

We begin by reviewing some notation. Let χ be a character of some group G, and suppose that X and Y are subgroups of G such that χ vanishes (i.e., has the value zero) on all elements of $G - X$ and on all elements of $G - Y$. It is clear that χ vanishes on $G - (X \cap Y)$, and it follows that there is a unique smallest subgroup $\mathbf{V}(\chi)$ with the property χ vanishes on $G - \mathbf{V}(\chi)$. We refer to $\mathbf{V}(\chi)$ as the **vanishing-off** subgroup of χ, and we observe that $\mathbf{V}(\chi) \triangleleft G$ because χ is a class function on G. Note that an alternative description of the vanishing-off subgroup is that $\mathbf{V}(\chi) = \langle x \in G \mid \chi(x) \neq 0 \rangle$.

Lemma 7.1. *Let $\varphi \in \mathrm{Irr}(L)$, where $L \triangleleft K$ and K/L is abelian, and let $\theta \in \mathrm{Irr}(K|\varphi)$. Then*

(a) $\mathrm{Irr}(K|\varphi) = \{\lambda\theta \mid \lambda \in \mathrm{Irr}(K/L)\}$.

(b) *All members of* $\mathrm{Irr}(K|\varphi)$ *have the same vanishing-off subgroup* V.

(c) $VL = K$ *if and only if* $\theta_L = \varphi$.

Proof. The set on the right-hand side of the equation in (a) is contained in the set on the left because if λ is a linear character of K/L, then $\lambda\theta$ is irreducible, and it clearly lies over φ. To prove the opposite containment, observe that every linear character of K/L is a constituent of $(1_L)^K$, and since the degree of $(1_L)^K$ is $|K/L| = |\operatorname{Irr}(K/L)|$, it follows that $(1_L)^K$ equals the sum of the members of $\operatorname{Irr}(K/L)$. Now φ is a constituent of θ_L, and hence every member of $\operatorname{Irr}(K|\varphi)$ is a constituent of $(\theta_L)^K$. We have

$$(\theta_L)^K = (1_L\theta_L)^K = (1_L)^K\theta = \sum_{\lambda\in\operatorname{Irr}(K/L)} \lambda\theta\,,$$

and it follows that every member of $\operatorname{Irr}(K|\varphi)$ has the form $\lambda\theta$ for some linear character λ of K/L. This establishes (a), and (b) follows since the value of a linear character is never zero.

Now $VL = K$ if and only if there is no subgroup W such that θ vanishes on $K - W$ and $L \subseteq W < K$. To prove (c), therefore, we must show that such a subgroup W exists if and only if θ_L is reducible. If $L \subseteq W < K$ and θ vanishes on $K - W$, we have

$$|W|[\theta_W, \theta_W] = \sum_{x\in W} |\theta(x)|^2 = \sum_{x\in K} |\theta(x)|^2 = |K|[\theta,\theta] = |K|\,,$$

so $[\theta_W, \theta_W] = |K : W| > 1$. In this case, θ_W is reducible, so θ_L is reducible, as required.

Conversely, assume that θ_L is reducible. Since θ is irreducible, there exists some (not necessarily unique) smallest subgroup X, with $L \subseteq X \subseteq K$ and such that θ_X is irreducible. By hypothesis, θ_L is reducible, so $L < X$, and since K/L is abelian, there exists a subgroup Y such that $L \subseteq Y < X$, where $|X : Y|$ is a prime number. Now θ_Y is reducible, and also $Y \triangleleft K$ because K/L is abelian.

Let α be an irreducible constituent of θ_Y. Then $\alpha(1) < \theta(1)$ because θ_Y is reducible, and it follows that θ_X is not an extension of α. Since θ_X is irreducible, we can apply Lemma 1.8 to deduce that α is not the unique irreducible constituent of θ_Y. The stabilizer T of α in K is thus a proper subgroup.

By the Clifford correspondence, θ is induced from some character of T. Also, $T \triangleleft K$ since K/L is abelian, and it follows that θ vanishes on $K - T$. ∎

Lemma 7.2. *Let $L \triangleleft K$, and suppose that $\theta \in \operatorname{Irr}(K)$ lies over $\varphi \in \operatorname{Irr}(L)$. Write $\theta_L = e\sum\varphi_i$, as in Clifford's theorem, where the characters φ_i for $1 \le i \le t$ are the distinct K-conjugates of φ in $\operatorname{Irr}(L)$. Then $e^2 \le |K : L|$, and the following are equivalent.*

(1) $e^2 = |K : L|$.

(2) *φ is invariant in K, and θ vanishes on $K - L$.*

(3) *φ is invariant in K, and φ^K is a multiple of θ.*

Proof. We have

$$|K| = |K|[\theta,\theta] = \sum_{x\in K} |\theta(x)^2| \geq \sum_{x\in L} |\theta(x)|^2 = |L|[\theta_L,\theta_L] = |L|e^2t\,,$$

and thus $e^2t \leq |K:L|$, with equality if and only if θ vanishes on $K-L$. In particular, $e^2 \leq |K:L|$, as required. If $e^2 = |K:L|$, it follows that $t=1$, and so φ is invariant in K. In this case, equality holds above, so θ vanishes on $K-L$. This shows that (1) implies (2).

Now assume (2), so φ is invariant, and thus $t=1$. Since θ vanishes on $K-L$, it follows that $|K:L| = e^2t = e^2$, and this proves (1). Also, since $t=1$, we have $\theta(1) = e\varphi(1)$, so

$$\varphi^K(1) = |K:L|\varphi(1) = e^2\varphi(1) = e\theta(1)\,.$$

By Frobenius reciprocity, θ occurs with multiplicity e as a constituent of φ^K, and it follows that φ^K can have no irreducible constituent other than θ. This proves (3).

Finally, assume (3), and observe that since θ occurs with multiplicity e in φ^K, we must have $\varphi^K = e\theta$. Since $t=1$ by assumption, $\theta(1) = e\varphi(1)$, and we have

$$e^2\varphi(1) = e\theta(1) = \varphi^K(1) = |K:L|\varphi(1)\,,$$

and thus (1) holds. ∎

The integer e in Lemma 7.2 is sometimes referred to as the "ramification", and since it is maximized when $e^2 = |K:L|$, we say that θ and φ are **fully ramified** with respect to K/L if the three equivalent conditions of the lemma hold.

Theorem 7.3. *Let $L \lhd K$, where K/L is abelian, and let $\varphi \in \mathrm{Irr}(L)$, where φ is invariant in K. Then there exists a subgroup R of K with $L \subseteq R$ and such that the following hold.*

(a) *$R = L\mathbf{V}(\alpha)$ for all $\alpha \in \mathrm{Irr}(K|\varphi)$.*

(b) *If $\beta \in \mathrm{Irr}(R|\varphi)$, then $\beta_L = \varphi$.*

(c) *If $\alpha \in \mathrm{Irr}(K|\varphi)$, then α is fully ramified with respect to K/R.*

Proof. Since K/L is abelian, Lemma 7.1(b) guarantees that $\mathbf{V}(\alpha)$ is independent of the choice of $\alpha \in \mathrm{Irr}(K|\varphi)$, and thus the subgroup $L\mathbf{V}(\alpha)$ is also independent of the choice of α. Let R be this subgroup, so $L \subseteq R$ and (a) is clear.

Let $\beta \in \mathrm{Irr}(R|\varphi)$. We show that β_L is irreducible, and thus $\beta_L = \varphi$, as required for (b). Otherwise, by Lemma 7.1(c), there exists a subgroup W, where $L \subseteq W < R$ and such that β vanishes on $R-W$. Since K/L is abelian, R and W are normal in K, and it follows that every conjugate of β in K also vanishes on $R - W$.

Now choose $\alpha \in \mathrm{Irr}(K|\beta)$, so α lies over φ, and thus $R = L\mathbf{V}(\alpha)$ by (a). Since α_R is a sum of K-conjugates of β and each of these conjugates vanishes on $R - W$, it follows that α vanishes on $R - W$. Also, α vanishes on $K - R$ because R contains $\mathbf{V}(\alpha)$, and it follows that α vanishes on $K - W$. Then $R = L\mathbf{V}(\alpha) \subseteq W < R$, and this contradiction shows that β_L is irreducible, as required.

To prove (c), let $\alpha \in \mathrm{Irr}(K|\varphi)$. Since α_L is a multiple of φ, each irreducible constituent of α_R lies over φ, and we let β be one of these constituents. Then β is an extension of φ by (b), and hence by the Gallagher correspondence, each irreducible constituent of α_R has the form $\lambda\beta$ for some uniquely determined irreducible character λ of R/L. Also, we note that λ must be linear since R/L is abelian.

Let
$$\mathcal{X} = \{\lambda \in \mathrm{Irr}(R/L) \mid \lambda\beta \text{ is a constituent of } \alpha_R\},$$
and let Λ be the sum of the members of $\mathcal{X}$. By Clifford's theorem, the irreducible constituents of α_R have equal multiplicities, and therefore $\alpha_R = e\Lambda\beta$ for some integer e.

We argue now that $\mathcal{X}$ is a subgroup of the group of linear characters of R/L. To see this, suppose that $\mu, \nu \in \mathcal{X}$, so $\mu\beta$ and $\nu\beta$ are irreducible constituents of α_R. These characters are thus conjugate to β in K, and hence there exist $x, y \in K$ such that $\beta^x = \beta\nu$ and $\beta^y = \beta\mu$. Then
$$\beta^{xy} = (\beta\nu)^y = \beta\mu\nu^y = \beta\mu\nu,$$
where the final equality holds because K/L is abelian, and so $\nu^y = \nu$. Now β^{xy} is an irreducible constituent of α_R, and we conclude that $\mu\nu \in \mathcal{X}$. Thus $\mathcal{X}$ is a multiplicatively closed subset of the group of linear characters of R/L, and it follows that $\mathcal{X}$ is a subgroup, as claimed. If $\lambda \in \mathcal{X}$ is arbitrary, therefore, we have $\lambda\Lambda = \Lambda$.

Now for $\lambda \in \mathcal{X}$, we have
$$\alpha_R = e\Lambda\beta = e\lambda\Lambda\beta = \lambda\alpha_R,$$
and it follows that α_R vanishes on $R - \ker(\lambda)$. Since α vanishes on $K - R$, we see that α vanishes on $K - \ker(\lambda)$, and thus $R = L\mathbf{V}(\alpha) \subseteq \ker(\lambda)$. Then λ is the principal character of R/L, and therefore $\mathcal{X}$ is the trivial group. We conclude that β is the unique irreducible constituent of α_R, and thus β is invariant in K. Since α vanishes on $K - R$, it follows by Lemma 7.2 that α is fully ramified with respect to K/R, as required. ∎

Note that by Theorem 7.3(a), the subgroup R is uniquely determined by K and φ. It is also true that R is the unique subgroup containing L

and contained in K such that both conclusion (b) and conclusion (c) of Theorem 7.3 hold.

Corollary 7.4. *Let K/L be an abelian chief factor of G, and suppose that $\theta \in \mathrm{Irr}(K)$ lies over $\varphi \in \mathrm{Irr}(L)$. Let T be the stabilizer of φ in G, and assume either that θ is invariant in G or that $G = KT$. Then in fact, $G = KT$, and one of the following holds.*

(1) $\varphi^K = \theta$ *and* $K \cap T = L$.

(2) $\theta_L = \varphi$.

(3) θ *and* φ *are fully ramified with respect to* K/L.

Proof. If θ is invariant in G, then G permutes the set of irreducible constituents of θ_L, and since the action of K on this set is transitive, we have $G = KT$ in all cases.

Now let S be the stabilizer of φ in K, and let $\alpha \in \mathrm{Irr}(S|\varphi)$ be the Clifford correspondent of θ with respect to φ, so $\alpha^K = \theta$. Also, let $R = L\mathbf{V}(\alpha)$, and note that $L \subseteq R \subseteq S \subseteq K$. Now S is uniquely determined by the choice of φ, and by Theorem 7.3 applied in the group T, it is also true that R is uniquely determined by φ. Then T normalizes both S and R, and since K/L is abelian, K also normalizes S and R, and thus S and R are normal in $KT = G$. By assumption, however, K/L is a chief factor of G, so each of S and R is either L or K. We see now that there are exactly three possibilities: $L = R = S$ or $L = R$ and $K = S$ or $K = R = S$.

Suppose first that $L = R = S$. Then $\alpha = \varphi$, so $\varphi^K = \alpha^K = \theta$. Also $T \cap K = S = L$, and (1) holds.

Next, suppose that $K = R = S$, so $\alpha = \theta$. In this case, it follows by Theorem 7.3(b) that $\theta_L = \varphi$ and (2) holds.

Finally, suppose that $L = R$ and $K = S$. Then $\alpha = \theta$, and by Theorem 7.3(c), this character is fully ramified with respect to $S/R = K/L$. Since θ lies over φ, we see that φ is also fully ramified with respect to K/L, and (3) holds. ■

It is not hard to prove Corollary 7.4 directly, without first establishing Theorem 7.3. We shall need Theorem 7.3 in Chapter 8, however, and that is why we have presented it here.

Suppose now that χ is a nonlinear irreducible character of some group G. Since the restriction of χ to the trivial subgroup of G is reducible, there exists a subgroup $L \lhd G$ maximal with the property that χ_L is reducible, and we let φ be an irreducible constituent of χ_L. Clearly, $L < G$, so we can choose a chief factor K/L of G, and we observe that χ_K is irreducible, and we write $\chi_K = \theta$

One might think that there is little of interest in this very general situation. If G is solvable, however, or more generally, if K/L is abelian, there is much that can be said. Since θ is invariant in G and $\theta_L \neq \varphi$, it follows by Corollary 7.4 that one of two alternatives must occur: either $\theta = \varphi^K$ and L is the stabilizer of φ in K, or else θ and φ are fully ramified with respect to K/L. We shall see that there are interesting consequences in both cases.

Suppose that $\varphi^K = \theta$, so L is the stabilizer of φ in K. If T is the full stabilizer of φ in G, we have $K \cap T = L$, and since θ is G-invariant, we have seen that $KT = G$. Then T/L is a complement for K/L in G/L, and in this situation, we refer to T as a **complement for K relative to L in G**. Furthermore, by the Clifford correspondence, induction defines a natural bijection from $\mathrm{Irr}(T|\varphi)$ onto $\mathrm{Irr}(G|\varphi)$, and in particular, $|\mathrm{Irr}(T|\varphi)| = |\mathrm{Irr}(G|\varphi)|$.

The remaining possibility is that θ and φ are fully ramified with respect to K/L. Although this case is much more subtle, we shall see in this chapter that a similar conclusion holds if we are willing to assume that $|K/L|$ is odd. In that case too, there exists a complement U for K relative to L in G such that $|\mathrm{Irr}(U|\varphi)| = |\mathrm{Irr}(G|\varphi)|$. We shall also see that in this situation, no member of $\mathrm{Irr}(U|\varphi)$ can induce irreducibly to G and no member of $\mathrm{Irr}(G|\varphi)$ can restrict irreducibly to U. In Chapter 8, we will construct an actual correspondence between $\mathrm{Irr}(U|\varphi)$ and $\mathrm{Irr}(G|\varphi)$.

Now let K/L be a normal section of a group G, and suppose that there exist characters $\varphi \in \mathrm{Irr}(L)$ and $\theta \in \mathrm{Irr}(K|\varphi)$ such that φ and θ are fully ramified with respect to K/L. Then $\theta_L = e\varphi$ and $\varphi^K = e\theta$, where $e^2 = |K : L|$, and in particular, each of θ and φ uniquely determines the other. It follows that if either of these characters is G-invariant, then both of them are, and in this situation, we sometimes say that the section K/L is "fully ramified". (Of course, this is an imprecise use of language since it ignores the characters θ and φ.)

As we have seen, fully ramified abelian normal sections of a group G arise naturally, especially if G is solvable. The following theorem allows us to exploit this situation.

Theorem 7.5. *Let K/L be a nontrivial abelian normal section of a group G, and suppose either that $|K : L|$ is odd or that $|K : L|$ and $|G : K|$ are relatively prime. Let $\varphi \in \mathrm{Irr}(L)$, and assume that φ is invariant in G and that it is fully ramified with respect to K/L. Then there exists a complement U for K relative to L in G such that the following hold.*

(a) $|\mathrm{Irr}(G|\varphi)| = |\mathrm{Irr}(U|\varphi)|$.

(b) *If $L < K$ and $\chi \in \mathrm{Irr}(G|\varphi)$, then χ_U is not irreducible.*

(c) *If $L < K$ and $\xi \in \mathrm{Irr}(U|\varphi)$, then ξ^G is not irreducible.*

By Theorem 7.5, there exists a complement U for K relative to L in G, where U has certain special properties, and it should be clear that each conjugate of U in G is also a relative complement that enjoys the same special properties. If all complements for K relative to L in G are conjugate, therefore, we can apply Theorem 7.5 to an arbitrary relative complement, and so it is convenient to have a condition that is sufficient to guarantee that all such relative complements actually are conjugate.

Lemma 7.6. *Let K/L be an abelian normal section of G, and suppose that there exists a normal subgroup M of G containing K and such that*

(1) *$|M : K|$ and $|K : L|$ are relatively prime, and*

(2) *$\mathbf{C}_{K/L}(M)$ is trivial.*

Then K/L has a complement X/L in M/L, and writing $H = \mathbf{N}_G(X)$, we have the following.

(a) *H is a complement for K relative to L in G.*

(b) *H is the unique complement for K relative to L in G that contains X.*

(c) *All complements for K relative to L in G are conjugate to H.*

Proof. Since K/L is a normal Hall subgroup of M/L, the existence of the complement X/L is guaranteed by the Schur–Zassenhaus theorem applied in the group M/L. Also, if Y/L is an arbitrary complement for K/L in M/L, it follows by the conjugacy part of the Schur–Zassenhaus theorem that Y/L is conjugate to X/L in M/L, and thus Y is conjugate to X in G.

Note that $H/L = \mathbf{N}_{G/L}(X/L)$. Since the complements for K/L in M/L are conjugate, we can apply the Frattini argument for $M/L \triangleleft G/L$ to deduce that $G/L = (M/L)(H/L)$, and thus $G = MH = KXH = KH$.

We argue next that $K \cap H = L$. Let $D = K \cap H$, so $D \triangleleft H$, and thus $[D, X] \subseteq [D, H] \subseteq D$. Also, $D \subseteq H = \mathbf{N}_G(X)$, so $[D, X] \subseteq X$. Then $[D, X] \subseteq D \cap X \subseteq K \cap X = L$, and hence X centralizes D/L. Since we are assuming that K/L is abelian, K also centralizes D/L, and thus $M = KX$ centralizes D/L. Then $D/L \subseteq \mathbf{C}_{K/L}(M)$, so it follows by (2) that $D = L$, and this proves (a).

Suppose now that U is an arbitrary complement for K relative to L in G. Write $Y = U \cap M$, so $Y \triangleleft U$, and thus $U \subseteq \mathbf{N}_G(Y)$. We see that Y/L is a complement for K/L in M/L, and hence Y is conjugate to X.

If $U \supseteq X$, then $Y = U \cap M \supseteq X$, and it follows that $Y = X$, and thus $U \subseteq \mathbf{N}_G(Y) = \mathbf{N}_G(X) = H$. Since U and H are complements for K relative to L in G, we see that $|U| = |H|$, and we deduce that $U = H$, proving (b).

Returning now to the case where U is an arbitrary complement for K relative to L in G, we have seen that $Y = U \cap M$ is conjugate to X. Some conjugate U^g of U thus contains X, and since U^g is a complement for K relative to L in G, it follows by (b) that $U^g = H$, and this establishes (c). ∎

We will prove Theorem 7.5 in the next section, but first, as motivation, we present an application.

Recall that a character χ is said to be "homogeneous" if it has a unique irreducible constituent, or equivalently, $\chi = m\theta$ for some integer $m \geq 1$ and irreducible character θ. Recall also that by definition, a character $\chi \in \mathrm{Irr}(G)$ is quasi-primitive if χ_N is homogeneous for every normal subgroup N of G, and χ is primitive if it is not induced from any proper subgroup. It is easy to see using the Clifford correspondence that primitive characters must be quasi-primitive, but in general, quasi-primitive characters need not be primitive. (Indeed, it is a triviality that every irreducible character of a simple group is quasi-primitive, but it is certainly not true that every such character is primitive. For example the irreducible character of degree 5 of the alternating group A_5 is imprimitive.)

Theorem 7.7. *Let G be π-separable, where $2 \notin \pi$, and let $\chi \in \mathrm{Irr}(G)$ be quasi-primitive, where $\chi(1)$ is a π-number. Then χ is primitive.*

In particular, if G is solvable and $\chi \in \mathrm{Irr}(G)$ is quasi-primitive and has odd degree, we can apply Lemma 7.6 by taking π to be the set of prime divisors of $\chi(1)$, and it follows that χ is primitive. (This is a special case of a theorem of T. Berger, which asserts that if G is solvable, then *every* quasi-primitive member of $\mathrm{Irr}(G)$ is primitive. Berger's result appears as Theorem 11.33 of [**CTFG**].)

To prove Theorem 7.7 and other applications of Theorem 7.5, it is often useful to consider a maximal subgroup H of a π-separable group G, so we present a brief review of some relevant facts. Let $L = \mathrm{core}_G(H)$, so $L \subseteq H < G$. We can thus consider a chief factor K/L of G, and we see that $KH = G$ because $K \nsubseteq H$ and H is maximal in G. Since we are assuming that G is π-separable, K/L is either a π-group or a π'-group. Also, $|G : H| = |KH : H| = |K : K \cap H|$, and since this number divides $|K : L|$, we deduce that $|G : H|$ is either a π-number or a π'-number.

Recall now that a π-separable group is said to be π-solvable if every π-chief factor of G is solvable. If G is π-solvable, therefore, every π-chief factor is an abelian p-group for some prime $p \in \pi$. (Note that by the odd-order theorem, this will automatically be the case if $|G|$ is odd or if $2 \notin \pi$.) If G is π-solvable, it follows that G is p-solvable for all primes $p \in \pi$.

The following technical lemma collects some facts about maximal subgroups of p-solvable groups. When we apply this result here and in Chapter 8, we will need only the special case where the normal subgroup N is the whole group G. The full generality of the lemma will be used in Chapter 9, however, and since only a slight extra effort is needed to obtain the additional generality, we have decided to present the expanded version here.

Lemma 7.8. *Let H be a maximal subgroup of G, where G is p-solvable and p divides $|G : H|$. Also, let $N \triangleleft G$, and assume that $N \cap H$ is not normal in G. Let $L = \mathrm{core}_G(N \cap H)$, and let K/L be a chief factor of G with $K \subseteq N$. The following then hold.*

(a) *K/L is an abelian p-group and $K < N$. Also, H is a complement for K relative to L in G.*
(b) *$\mathbf{C}_N(K/L) = K$.*
(c) *Let M/K be a chief factor of G with $M \subseteq N$. Then M/K is a p'-group and $\mathbf{C}_{K/L}(M)$ is trivial.*
(d) *All complements for K relative to L in G are conjugate to H in G.*

Proof. By hypothesis, $N \cap H$ is not normal in G, and thus $L < N \cap H \subseteq N$. It follows that there exists a chief factor K/L of G with $K \subseteq N$ as in the statement of the lemma. Also, if $K \subseteq H$, then $N \cap K$ is a normal subgroup of G contained in $N \cap H$, and thus $K \subseteq \mathrm{core}_G(N \cap H) = L < K$, and since this is a contradiction, it follows that $K \not\subseteq H$.

Since H is maximal in G, we have $KH = G$, and thus

$$|G : H| = |KH : H| = |K : (K \cap H)|,$$

and this divides $|K : L|$ because $L \subseteq K \cap H$. By hypothesis, p divides $|G : H|$, and thus p divides $|K : L|$. Since K/L is a chief factor of the p-solvable group G, we see that K/L is an abelian p-group, and thus $K \cap H \triangleleft K$. Also $K \cap H \triangleleft H$ because $K \triangleleft G$, and we conclude that $K \cap H \triangleleft KH = G$.

Now $L \subseteq K \cap H < K$ and $K \cap H \triangleleft G$, and since K/L is a chief factor of G, we have $K \cap H = L$. It follows that H is a complement for K relative to L in G, as required. Now $K \subseteq N$, and to complete the proof of (a), we show that $K < N$. Otherwise, $K = N$, and thus $N \cap H = K \cap H = L$, and this is a contradiction since by hypothesis, $N \cap H$ is not normal in G.

Now let $C = \mathbf{C}_N(K/L)$, so $C \triangleleft G$. Also, $K \subseteq C$ because $K \subseteq N$ and K/L is abelian. We have $[(C \cap H), K] \subseteq [C, K] \subseteq L \subseteq C \cap H$, and thus K normalizes $C \cap H$. Also, H normalizes $C \cap H$ because $C \triangleleft G$, and thus $C \cap H \triangleleft G$. Then $C \cap H \subseteq \mathrm{core}_G(N \cap H) = L$, and we deduce that $C \cap H = L$. Finally, since $KH = G$ and $K \subseteq C$, Dedekind's lemma yields $C = K(C \cap H) = KL = K$, as required for (b).

Now $K < N$ by (a), so we can choose a chief factor M/K of G with $M \subseteq N$. Write $F/L = \mathbf{C}_{K/L}(M)$, so $L \subseteq F \subseteq K$ and $F \lhd G$. If $F = K$, then $M \subseteq \mathbf{C}_N(K/L) = K$, and since this is false, we deduce that $F = L$, and thus $\mathbf{C}_{K/L}(M)$ is trivial.

Since K/L is abelian, the action of M on K/L induces an action of M/K, and we have $\mathbf{C}_{K/L}(M/K) = \mathbf{C}_{K/L}(M)$, which is trivial. Now K/L is a nontrivial p-group, and it follows that M/K is not a p-group. Since M/K is a chief factor of G and G is p-solvable, we conclude that M/K is a p'-group, proving (c).

For (d), let M be as in (c). Condition (1) of Lemma 7.6 is satisfied because K/L is a p-group by (a) and M/K is a p'-group by (c). Condition (2) holds since $\mathbf{C}_{K/L}(M)$ is trivial by (c), and thus we can apply Lemma 7.6 to conclude that all complements for K relative to L in G are conjugate. By (a), the subgroup H is one of these relative complements, and (d) follows. ∎

Proof of Theorem 7.7. Suppose that the quasi-primitive character $\chi \in \mathrm{Irr}(G)$ is not primitive. We can thus write $\chi = \xi^G$, where ξ is a character of some maximal subgroup H of G. Now $|G : H| = \chi(1)/\xi(1)$, so $|G : H|$ is divisible by some prime divisor p of $\chi(1)$, and thus $p \in \pi$ since by hypothesis, $\chi(1)$ is a π-number.

We are assuming that G is π-separable and $2 \notin \pi$, and thus G is p-solvable. We want to apply Lemma 7.8 with $N = G$, so we must show that H is not normal in G. Otherwise, it follows from the quasi-primitivity of χ that χ_H is a multiple of ξ, and since $[\chi_H, \xi] = [\chi, \xi^G] =]\chi, \chi] = 1$, we deduce that $\chi_H = \xi$. This is a contradiction, however, since $\chi(1)/\xi(1) = |G : H| > 1$.

Write $L = \mathrm{core}_G(H)$, and let K/L be a chief factor of G. Then K/L is an abelian p-group by Lemma 7.8(a), and in particular, $|K : L|$ is odd. Also, by Lemma 7.8(d), all complements for K relative to L in G are conjugate to H.

Now χ_K and χ_L are homogeneous, and we let θ and φ, respectively, be their unique irreducible constituents. Then θ and φ are invariant in G, so we can apply Corollary 7.4. The first of the three alternatives of that result does not hold since the stabilizer T of φ does not satisfy the condition that $K \cap T = L$. The second alternative also fails since if $\theta_L = \varphi$, then Lemma 2.11(a) guarantees that χ_H is irreducible, and this is false because χ is induced from $H < G$.

By Corollary 7.4, therefore, φ is fully ramified with respect to K/L. Also, K/L is abelian of odd order, and thus by Theorem 7.5, there is a complement U for K relative to L in G such that no member of $\mathrm{Irr}(U|\varphi)$ induces irreducibly to G. Since $\xi^G = \chi$, we see that U cannot be conjugate to H, and this is a contradiction. ∎

7B

We begin work toward a proof of Theorem 7.5 with an elementary commutator calculation.

Lemma 7.9. *Suppose that elements x and y of an arbitrary group commute with their commutator $[x,y] = x^{-1}y^{-1}xy$. Then*

$$[x^m, y] = [x,y]^m = [x, y^m]$$

for all integers m.

Proof. A general and easily checked commutator identity asserts that $[ab,c] = [a,c]^b[b,c]$, and it follows that $[x^2,y] = [x,y]^x[x,y] = [x,y]^2$, where the second equality follows because x centralizes $[x,y]$. Similarly, since x^2 also centralizes $[x,y]$, we see that

$$[x^3,y] = [xx^2,y] = [x,y]^{x^2}[x^2,y] = [x,y]^3 ,$$

and proceeding by induction, we obtain $[x^m,y] = [x,y]^m$ for all positive integers m.

Now $[x,y] = [y,x]^{-1}$, so for $m > 0$, we have

$$[x,y^m] = [y^m,x]^{-1} = [y,x]^{-m} = [x,y]^m ,$$

where the second equality follows by interchanging the roles of x and y in the calculation of the previous paragraph.

To handle negative exponents, observe that

$$1 = [xx^{-1},y] = [x,y]^{x^{-1}}[x^{-1}y] = [x,y][x^{-1},y] ,$$

so $[x^{-1},y] = [x,y]^{-1}$, and in particular, x^{-1} and y both centralize $[x^{-1},y]$. By the previous calculation, therefore,

$$[x^{-m},y] = [x^{-1},y]^m = [x,y]^{-m} ,$$

for positive m. A similar argument shows that $[x,y^{-m}] = [x,y]^{-m}$, and the proof is complete. ∎

Given a group K and a subgroup $L \lhd K$, we say that K is a **GCR-group** with respect to L if every coset of L in K contains an element u such that $L \cap \langle u \rangle = 1$. (We have written GCR as an abbreviation for good coset representatives.)

As might be expected, a key step in our proof of Theorem 7.5 is to replace the character triple (G, L, φ) with an isomorphic triple (G^*, L^*, φ^*), where $L^* \subseteq \mathbf{Z}(G^*)$. In fact, we require somewhat more than this: we want the subgroup $K^* \lhd G^*$ corresponding to K to be a GCR-group with respect to L^*.

The following shows how the GCR assumption will be used.

Lemma 7.10. *Let $L \subseteq \mathbf{Z}(K)$, where K/L is abelian, and suppose that K is a GCR-group with respect to L. Then there exists an automorphism τ of K such that*

(a) $\tau^2 = 1$,

(b) τ *acts trivially on* L, *and*

(c) τ *inverts the elements of* K/L.

Proof. Given a cyclic subgroup C of K/L, choose a generator Lu for C. Then

$$C = \langle Lu \rangle = \langle u \rangle L/L \cong \langle u \rangle/(L \cap \langle u \rangle)\,,$$

so there is a natural homomorphism from $\langle u \rangle$ onto C with kernel $L \cap \langle u \rangle$. By hypothesis, we can choose u so that $L \cap \langle u \rangle = 1$, and thus the map $\langle u \rangle \to C$ is an isomorphism. It follows that the map $x \mapsto Lx$ from $\langle u \rangle$ to C is a bijection.

By hypothesis, K/L is abelian, so the fundamental theorem of abelian groups allows us to write

$$K/L = C_1 \times C_2 \times \cdots \times C_r\,,$$

where the C_i are cyclic subgroups of K/L. By the result of the previous paragraph, we can choose cyclic subgroups $U_i \subseteq K$ such that each element of C_i is uniquely of the form Lx for some element $x \in U_i$. Since $L \subseteq \mathbf{Z}(K)$ and every coset Ls is uniquely a product of elements of the cyclic direct factors C_i, we can write each element $s \in K$ uniquely as

$$s = a \prod_{i=1}^{r} x_i\,,$$

where $a \in L$ and $x_i \in U_i$. We can now define τ as a permutation of the elements of K by setting

$$s^{\tau} = a \prod_{i=1}^{r} x_i^{-1}\,,$$

where s is as above. In particular, we see that $\tau^2 = 1$ and that τ fixes the elements of L.

We will show that τ is a homomorphism, and that will imply that $\tau \in \mathrm{Aut}(K)$. Also, by definition, τ inverts the elements of a generating set for the abelian group K/L, so once we show that τ is an automorphism, it will follow that τ inverts all elements of K/L, and that will complete the proof.

Given $s, t \in K$, write

$$s = a \prod_{i=1}^{r} x_i \qquad \text{and} \qquad t = b \prod_{i=1}^{r} y_i\,,$$

where $a, b \in L$ and $x_i, y_i \in U_i$. Since K/L is abelian, we can write

$$\Big(\prod_{i=1}^{r} x_i\Big)\Big(\prod_{i=1}^{r} y_i\Big) = c\prod_{i=1}^{r}(x_i y_i)\,,$$

for some element $c \in L$, and our next task is to compute the correction factor c. To obtain the right-hand side of the above equation from the left-hand side, we must move each y_i toward the left, crossing over all x_j with $j > i$. Since

$$x_j y_i = y_i x_j [x_j, y_i]\,,$$

we see that each such crossing introduces the factor $[x_j, y_i]$. Since each of these commutators lies in L, and hence is central, we have

$$c = \prod_{j>i}[x_j, y_i]\,.$$

Similarly,

$$\Big(\prod_{i=1}^{r} x_i^{-1}\Big)\Big(\prod_{i=1}^{r} y_i^{-1}\Big) = d\prod_{i=1}^{r}(x_i^{-1} y_i^{-1})\,,$$

where $d \in L$ and

$$d = \prod_{j>i}[x_j^{-1}, y_i^{-1}]\,.$$

Since $K' \subseteq L \subseteq \mathbf{Z}(K)$, Lemma 7.9 yields $[x^{-1}, y^{-1}] = [x, y]$, and thus the correction factors c and d above, are equal.

Using the assumption that $L \subseteq \mathbf{Z}(K)$ once again, we have

$$st = abc\prod_{i=1}^{r}(x_i y_i)\,.$$

Since U_i is abelian, $(x_i y_i)^{-1} = x_i^{-1} y_i^{-1}$, and thus by definition,

$$(st)^{\tau} = abc\prod_{i=1}^{r}(x_i^{-1} y_i^{-1})\,.$$

Also,

$$s^{\tau} t^{\tau} = ab\Big(\prod_{i=1}^{r} x_i^{-1}\Big)\Big(\prod_{i=1}^{r} y_i^{-1}\Big) = abd\prod_{i=1}^{r}(x_i^{-1} y_i^{-1})\,,$$

and since $c = d$, we have $(st)^{\tau} = s^{\tau} t^{\tau}$, and the proof is complete. ∎

In order to show how GCR-groups can be constructed, we digress briefly to discuss central products. Given distinct subgroups $H_i \subseteq G$ for $1 \le i \le r$, suppose that H_i and H_j centralize each other whenever $i \ne j$. Then $\prod H_i$ is a subgroup, and if this product is the whole group G, we say that G is the **central product** of its subgroups H_i. Note that in this case, $H_i \lhd G$ for all i and $H_i \cap H_j \subseteq \mathbf{Z}(G)$ if $i \ne j$.

Now suppose that G is the central product of two subgroups H and K, and let $D = H \cap K$. Then $D \subseteq \mathbf{Z}(G)$, and thus $D \subseteq \mathbf{Z}(H)$ and $D \subseteq \mathbf{Z}(K)$. Viewing H and K as abstract groups, therefore, we see that there exist subgroups $U \subseteq \mathbf{Z}(H)$ and $V \subseteq \mathbf{Z}(K)$ such that $U \cong V$, where, in fact, $U = D = V$ in the group G. Conversely, we have the following.

Lemma 7.11. *Let H and K be groups having central subgroups U and V, respectively, where $U \cong V$. Then there exists a group G and injective homomorphisms*

$$\alpha : H \to G \qquad \text{and} \qquad \beta : K \to G$$

such that G is the central product of the subgroups $\alpha(H)$ and $\beta(K)$ and

$$\alpha(U) = \alpha(H) \cap \beta(K) = \beta(V)\,.$$

In fact, the group G and the maps α and β can be constructed so that the composite map

$$U \cong \alpha(U) = \beta(V) \cong V$$

is the given isomorphism from U to V.

In the setting of Lemma 7.11, it is customary to identify H and K with their isomorphic copies $\alpha(H)$ and $\beta(K)$, and thus H and K are viewed as subgroups of G, and G is the central product of these subgroups. Then $H \cap K$ is identified with both the subgroup U of H and the subgroup V of K.

The group G constructed as in Lemma 7.11 is usually referred to as the "central product of H and K with U and V identified". This can be misleading, however, because the isomorphism type of G may depend on the particular isomorphism between U and V that is given. The group G, however, is uniquely determined if this isomorphism is specified.

Proof of Lemma 7.11. Let $\Gamma = H \times K$ be the external direct product, and let $\Delta = U \times V$, so Δ is a central subgroup of Γ. Given an isomorphism $\delta : U \to V$, define

$$N = \{(u, v) \in \Delta \mid \delta(u) = v^{-1}\}\,.$$

If (u_1, v_1) and (u_2, v_2) lie in N, then since V is abelian, we have

$$\delta(u_1u_2) = \delta(u_1)\delta(u_2) = v_1^{-1}v_2^{-1} = (v_1v_2)^{-1}\,,$$

and thus (u_1u_2, v_1v_2) lies in N, and it follows that N is a subgroup. In fact, $N \triangleleft \Gamma$ since $\Delta \subseteq \mathbf{Z}(\Gamma)$.

Let $G = \Gamma/N$. Using overbars to denote the natural homomorphism from Γ onto G, we define α and β by setting

$$\alpha(h) = \overline{(h, 1)} \qquad \text{and} \qquad \beta(k) = \overline{(1, k)}$$

for $h \in H$ and $k \in K$. It is clear that α and β are homomorphisms, and we observe that if $\alpha(h) = 1$, then $(h, 1) \in N$, and thus $h \in U$ and $\delta(h) = 1$, and since δ is an isomorphism, it follows that $h = 1$. Also, if $\beta(k) = 1$, then $(1, k) \in N$, and thus $\delta(1) = k^{-1}$, and we have $k = 1$. We conclude that α and β are injective, as required. Also, it is clear that the subgroups $\alpha(H)$ and $\beta(K)$ centralize each other and that $G = \alpha(H)\beta(K)$, and thus G is the central product of the subgroups $\alpha(H)$ and $\beta(K)$.

Now if $u \in U$, let $\delta(u) = v$. Then $(u, 1) = (u, v^{-1})(1, v)$, and since $(u, v^{-1}) \in N$, it follows that

$$\alpha(u) = \overline{(u, 1)} = \overline{(1, v)} = \beta(v) = \beta(\delta(u)),$$

and thus $\alpha(U) = \beta(V)$, and this subgroup is contained in $\alpha(H) \cap \beta(K)$. This calculation also shows that the composite map $U \cong \alpha(U) = \beta(V) \cong V$ is the isomorphism δ, as required.

To complete the proof, it suffices to show that $\alpha(H) \cap \beta(K) \subseteq \alpha(U)$, so suppose that $h \in H$ and $k \in K$ and $\alpha(h) = \beta(k)$. Then

$$\overline{(h, 1)} = \overline{(1, k)},$$

and thus $(h, k^{-1}) \in N$. It follows that $h \in U$, as required. ∎

We will also need some information about the characters of central products.

Lemma 7.12. *Let $G = HK$ be a central product. Then there is a map $f : \chi \mapsto (\alpha, \beta)$ from* $\mathrm{Irr}(G)$ *into the set* $\mathrm{Irr}(H) \times \mathrm{Irr}(K)$ *such that the following hold.*

(a) *If $f(\chi) = (\alpha, \beta)$, then χ_H is a multiple of α and χ_K is a multiple of β.*

(b) *The pair (α, β) lies in the image of f if and only if α and β lie over a common linear character of $H \cap K$.*

(c) *If $f(\chi) = (\alpha, \beta)$, then $\chi(hk) = \alpha(h)\beta(k)$ for all $h \in H$ and $k \in K$.*

(d) *The map f is injective.*

Proof. Since H and K centralize each other, H and K are normal in G, and all members of $\mathrm{Irr}(H)$ and $\mathrm{Irr}(K)$ are G-invariant. If $\chi \in \mathrm{Irr}(G)$, it follows that χ_H and χ_K are multiples of characters $\alpha \in \mathrm{Irr}(H)$ and $\beta \in \mathrm{Irr}(K)$, respectively, and thus there is a unique map f satisfying (a).

Now $H \cap K \subseteq \mathbf{Z}(G)$, so if $\chi \in \mathrm{Irr}(G)$, then $\chi_{H \cap K}$ is a multiple of some linear character λ. If $f(\chi) = (\alpha, \beta)$, then χ lies over α and β, and it follows that α and β each lie over λ.

Conversely, suppose that $\alpha \in \mathrm{Irr}(H)$ and $\beta \in \mathrm{Irr}(K)$ lie over a common linear character λ of $H \cap K$. Since $G = HK$, we have

$$(\alpha^G)_K = (\alpha_{H\cap K})^K = (\alpha(1)\lambda)^K\,,$$

and β is an irreducible constituent of this character because we are assuming that β lies over λ. It follows that some irreducible constituent χ of α^G lies over β, and thus $f(\chi) = (\alpha, \beta)$, and this completes the proof of (b).

Now consider the external direct product $\Gamma = H \times K$. Define the map $\pi : \Gamma \to G$ by setting $\pi(h, k) = hk$, and observe that π is a homomorphism from Γ onto G.

Given an arbitrary character $\chi \in \mathrm{Irr}(G)$, we see that the function π followed by χ defines a character $\psi \in \mathrm{Irr}(\Gamma)$. Since ψ is an irreducible character of the direct product $H \times K$, there exist characters $\alpha \in \mathrm{Irr}(H)$ and $\beta \in \mathrm{Irr}(K)$ such that $\psi = \alpha \times \beta$. If $h \in H$ and $k \in K$, therefore, we have

$$\chi(hk) = \chi(\pi(h,k)) = \psi(h,k) = (\alpha \times \beta)(h,k) = \alpha(h)\beta(k)\,.$$

To prove (c) now, it suffices to check that $f(\chi) = (\alpha, \beta)$. For $h \in H$, we have $\chi(h) = \alpha(h)\beta(1)$, so $\chi_H = \beta(1)\alpha$. Similarly, $\chi_K = \alpha(1)\beta$, and thus $f(\chi) = (\alpha, \beta)$, as required. In particular, it follows from (c) that the pair (α, β) uniquely determines χ, and thus f is injective. ∎

Theorem 7.13. *Let K/L be an abelian normal section of G, and suppose $\varphi \in \mathrm{Irr}(L)$ is G-invariant. Then there exists a character triple (G^*, L^*, φ^*) isomorphic to (G, L, φ), where $L^* \subseteq \mathbf{Z}(G^*)$ and φ^* is faithful, and such that the subgroup $K^* \subseteq G^*$ corresponding to K is a GCR-group with respect to L^*.*

Proof. We can replace (G, L, φ) by an isomorphic triple, so by Theorem 1.13, we can assume that $L \subseteq \mathbf{Z}(G)$ and that φ is faithful, and in particular, L is cyclic. Let L^* be cyclic with order equal to $|K|$, and let G^* be the central product of G with L^*, where L is identified with the subgroup of the same order in L^*.

Working in G^*, we have $L \subseteq L^*$, and since L^* is cyclic and φ is faithful, φ has a faithful extension $\varphi^* \in \mathrm{Irr}(L^*)$. Also, $L^* \subseteq \mathbf{Z}(G^*)$, so φ^* is invariant in G^*, and thus (G^*, L^*, φ^*) is a character triple.

To see that (G^*, L^*, φ^*) is isomorphic to (G, L, φ), we observe first that $G^* = L^*G$, so $G^*/L^* = L^*G/L^* \cong G/(G \cap L^*) = G/L$. For each intermediate subgroup X with $L \subseteq X \subseteq G$, we see that L^*X is the corresponding subgroup X^* with $L^* \subseteq X^* \subseteq G^*$. We must construct a bijection $\xi \mapsto \xi^*$ from $\mathrm{Irr}(X|\varphi)$ onto $\mathrm{Irr}(X^*|\varphi^*)$ such that when these $*$-maps are extended linearly, the "restriction property" holds. (See the discussion preceding Lemma 1.12.)

Now X^* is a central product of L^* and X, and φ^* lies over φ. Given $\xi \in \mathrm{Irr}(X|\varphi)$, it follows by Lemma 7.12 that there is a character $\xi^* \in \mathrm{Irr}(X^*)$ such that $\xi^*(zx) = \varphi^*(z)\xi(x)$ for $z \in L^*$ and $x \in X$. In particular, taking $x = 1$, we have $\xi^*(z) = \varphi^*(z)\xi(1)$, and thus ξ^* lies over φ^*, as wanted. It is easy to see by Lemma 7.12 that the map $\xi \mapsto \xi^*$ is a bijection, and it is routine to check that the required restriction property holds. The character triple (G^*, L^*, φ^*) is thus isomorphic to (G, L, φ).

It remains to show that $K^* = L^*K$ is a GCR-group with respect to L^*. Each coset of L^* in K^* has the form L^*k for some element $k \in K$, so given $k \in K$, it suffices to find an element $t \in L^*$ such that $L^* \cap \langle tk \rangle = 1$.

Let m be the order of the coset L^*k as an element of K^*/L^*. Then m is the smallest positive integer such that $k^m \in L^*$, and thus $m = |\langle k \rangle : L^* \cap \langle k \rangle|$. We conclude that m divides $|K : L^* \cap \langle k \rangle|$, and this is equal to $|L^* : L^* \cap \langle k \rangle|$ because $|L^*| = |K|$. Since L^* is cyclic, it follows that every element of $L^* \cap \langle k \rangle$ is an mth power in L^*. Now $k^m \in L^* \cap \langle k \rangle$, and it follows that there exists an element $t \in L^*$ such that $t^m = k^{-m}$. Since L^* centralizes K, we conclude that $(tk)^m = 1$.

To prove that $L^* \cap \langle tk \rangle = 1$, we show that the identity is the only power of tk that lies in L^*. Assuming that $(tk)^n \in L^*$ for some integer n and working in L^*K/L^*, we have

$$(L^*k)^n = (L^*tk)^n = L^*(tk)^n = L^*,$$

and since m is the order of L^*k, we conclude that m divides n. Then $(tk)^n = 1$, as required. ∎

By Theorem 7.13, we see that to prove Theorem 7.5, it suffices to assume that $L \subseteq \mathbf{Z}(G)$ and φ is faithful, and that K is a GCR-group with respect to L. We recall now that Theorem 7.5 has a pair of alternative hypotheses: either $|K : L|$ is odd or $|K : L|$ and $|G : K|$ are relatively prime. In the latter case, the GCR condition is irrelevant, and the following easy result yields a complement U for K in G relative to L, where U satisfies a useful commutator condition.

Theorem 7.14. *Let $L \subseteq K \triangleleft G$, where $L \subseteq \mathbf{Z}(G)$, and assume that $|K : L|$ and $|G : K|$ are relatively prime. Then there exists a complement for K relative to L in G. Also, for every such relative complement U and every choice of elements $u \in U$ and $k \in K$ such that $[k, u] \in L$, we have $[k, u] = 1$.*

Proof. By the Schur–Zassenhaus theorem, K/L has a complement U/L in G/L, so U is a complement for K relative to L. Now let $u \in U$ and $k \in K$, and suppose that $[k, u] \in L$. Since $L \subseteq \mathbf{Z}(G)$, we see by Lemma 7.9 that $[k^m, u] = [k, u]^m = [k, u^m]$ for all integers m.

In particular, taking $m = |K : L|$, we have $k^m \in L$, and thus $[k, u]^m = [k^m, u] = 1$. Also, if $n = |G : K| = |U : L|$, then $u^n \in L$, so $[k, u]^n = [k, u^n] = 1$. Since m and n are relatively prime, we conclude that $[k, u] = 1$, as required. ■

To prove Theorem 7.5, we also need to consider the case where $|G : K|$ and $|K : L|$ are not relatively prime, and where we assume instead that $|K : L|$ is odd. The following theorem shows that in that case too, we can find a complement U for K relative to L in G, where U satisfies the commutator condition of the previous theorem. To do this, however, we must assume that K is a GCR-group with respect to L and that L is cyclic, where $L = \mathbf{Z}(K)$. In that case, we also obtain some additional information that will be needed later.

Theorem 7.15. *Let $L = \mathbf{Z}(K)$, where $K \lhd G$ and K/L is abelian of odd order. Suppose that $L \subseteq \mathbf{Z}(G)$ and that L is cyclic, and assume that K is a GCR-group with respect to L. Let $\tau \in \mathrm{Aut}(K)$ where $\tau^2 = 1$ and such that τ acts trivially on L and inverts the elements of K/L, and let*

$$U = \{u \in G \mid (x^u)^\tau = (x^\tau)^u \text{ for all } x \in K\}.$$

The following then hold.

(a) *U is a complement for K relative to L in G.*

(b) *If $[k, u] \in L$, where $u \in U$ and $k \in K$, then $[k, u] = 1$.*

(c) *τ has a unique extension $\sigma \in \mathrm{Aut}(G)$ such that σ centralizes U.*

(d) *If σ is as in (c), then $\sigma^2 = 1$ and σ acts trivially on G/L.*

Note that by Lemma 7.10, there necessarily exists an automorphism τ of K with the properties assumed in the statement of Theorem 7.15, so the theorem is not vacuous. In particular, the complement U and the automorphism σ having the stated properties actually do exist. We say that the subgroup U constructed as in Theorem 7.15 is a **distinguished** complement for K relative to L in G, and we mention that the distinguished relative complements are all conjugate in G.

Although distinguished complements satisfy the commutator condition of Theorem 7.15(b), it is not true in general that every relative complement that satisfies this condition must be distinguished. This weaker commutator condition is all that is required in this chapter, but in Chapter 8 we will use more fully the fact that distinguished complements exist.

We need part (c) of the following lemma for the proof of Theorem 7.15. Parts (a) and (b) are used to prove (d), which will be needed later.

Lemma 7.16. *Let $L = \mathbf{Z}(K)$, where L is cyclic and K/L is abelian. For each subgroup X with $L \subseteq X \subseteq K$, write $\mathcal{A}(X)$ to denote the group consisting of those automorphisms of X that act trivially on both X/L and L. The following then hold.*

(a) *For every choice of X, each automorphism in $\mathcal{A}(X)$ is induced by conjugation by some element of K.*

(b) $|\mathcal{A}(X)| = |X : L|$.

(c) $\mathcal{A}(K) = \mathrm{Inn}(K)$.

(d) $|K : \mathbf{C}_K(X)| = |X : L|$.

Proof. Since L is cyclic, we can choose a faithful linear character ν of L. For each automorphism $\sigma \in \mathcal{A}(X)$, define the function $\lambda_\sigma : X \to \mathbb{C}$ by setting $\lambda_\sigma(x) = \nu(x^{-1}x^\sigma)$. (This makes sense because $x^{-1}x^\sigma$ lies in L since σ acts trivially on X/L.)

If $x, y \in X$, we compute that

$$(xy)^{-1}(xy)^\sigma = y^{-1}(x^{-1}x^\sigma)y^\sigma = (x^{-1}x^\sigma)(y^{-1}y^\sigma)\,,$$

where the second equality holds because $x^{-1}x^\sigma \in L \subseteq \mathbf{Z}(X)$. It follows that $\lambda_\sigma(xy) = \lambda_\sigma(x)\lambda_\sigma(y)$, and thus λ_σ is a linear character of X. If $z \in L$, then $z^{-1}z^\sigma = 1$ because σ fixes the elements of L, and thus $L \subseteq \ker(\lambda_\sigma)$, and we can view λ_σ as a linear character of X/L. Let $f_X : \mathcal{A}(X) \to \mathrm{Irr}(X/L)$ be the map defined by $f(\sigma) = \lambda_\sigma$.

We show now that the map f_X is injective, and thus

$$|\mathcal{A}(X)| \leq |\,\mathrm{Irr}(X/L)| = |X : L|\,.$$

To see this, suppose $\lambda_\sigma = \lambda_\tau$, where $\sigma, \tau \in \mathcal{A}(X)$. Since ν is faithful, it follows that $x^{-1}x^\sigma = x^{-1}x^\tau$ for all $x \in X$, and hence $x^\sigma = x^\tau$. The automorphisms σ and τ are therefore equal, as required.

Now $\mathrm{Inn}(K) \subseteq \mathcal{A}(K)$, and since $L = \mathbf{Z}(K)$, we have

$$|K : L| = |\,\mathrm{Inn}(K)| \leq |\mathcal{A}(K)| \leq |K : L|\,,$$

where the last inequality holds by the result of the previous paragraph applied in the case $X = K$. We conclude that $\mathcal{A}(K) = \mathrm{Inn}(K)$, and this proves (c). It also follows that the map $f_K : \mathcal{A}(K) \to \mathrm{Irr}(K/L)$ is surjective.

Let $\mathcal{B}(X)$ be the subset of $\mathcal{A}(X)$ consisting of the restrictions of the inner automorphisms of K to X, and observe that

$$|\mathcal{B}(X)| \leq |\mathcal{A}(X)| \leq |\,\mathrm{Irr}(X/L)|\,.$$

We will show that f_X carries $\mathcal{B}(X)$ onto $\mathrm{Irr}(X/L)$, and it will follow that $\mathcal{B}(X) = \mathcal{A}(X)$, proving (a). Another consequence is that

$$|\mathcal{A}(X)| = |\,\mathrm{Irr}(X/L)| = |X : L|\,,$$

and this proves (b).

Given $\alpha \in \mathrm{Irr}(X/L)$, we seek an element $\sigma \in \mathcal{B}(X)$ such that $\lambda_\sigma = \alpha$. Since K/L is abelian, we can extend α to a character $\beta \in \mathrm{Irr}(K/L)$, and since we have seen that f_K carries $\mathcal{A}(K)$ onto $\mathrm{Irr}(K/L)$, there exists a member $\tau \in \mathcal{A}(K) = \mathrm{Inn}(K)$ such that $\lambda_\tau = \beta$.

Now τ is an inner automorphism and $X \lhd K$, and it follows that τ stabilizes X. Let σ be the restriction of τ to X, and observe that σ lies in $\mathcal{B}(X)$. Now λ_σ is the restriction of λ_τ to X, and since $\lambda_\tau = \beta$, we have $\lambda_\sigma = \beta_X = \alpha$. This shows that f_X carries $\mathcal{B}(X)$ onto $\mathrm{Irr}(X/L)$, as wanted, and this completes the proofs of (a) and (b).

Finally, for (d), observe that the map that carries an element $k \in K$ to the automorphism of X induced by conjugation by k is a homomorphism. By (a), the image of this map is $\mathcal{A}(X)$, and since its kernel is $\mathbf{C}_K(X)$, we see that $|K : \mathbf{C}_K(X)| = |\mathcal{A}(X)| = |X : L|$, where the second equality follows by (b). ∎

Proof of Theorem 7.15. Given $g \in G$, let $\hat{g}$ be the automorphism of K induced by conjugation by g, and note that the map $g \mapsto \hat{g}$ is a homomorphism from G into $\mathrm{Aut}(K)$. For subgroups $X \subseteq G$, write $\hat{X}$ to denote the image of X under this map, so in particular, $\hat{K} = \mathrm{Inn}(K) \cong K/L$, and thus $|\hat{K}|$ is odd.

Observe that conjugation by τ in $\mathrm{Aut}(K)$ inverts the elements of $\hat{K}$. To see this, let $k \in K$, so if $x \in K$, we have

$$x^{(\hat{k})^\tau} = x^{\tau^{-1}\hat{k}\tau} = ((x^{\tau^{-1}})^k)^\tau = ((x^{\tau^{-1}})^\tau)^{k^\tau} = x^{k^\tau} = x^{k^{-1}} = x^{(\hat{k})^{-1}},$$

where the penultimate equality holds because $k^\tau = k^{-1}z$, for some element $z \in L = \mathbf{Z}(K)$. It follows that $(\hat{k})^\tau = (\hat{k})^{-1}$, as claimed.

Observe that U is the full preimage in G of the centralizer of τ in $\mathrm{Aut}(K)$, and in particular U is a subgroup of G. Also, if $k \in K \cap U$, then the inner automorphism $\hat{k}$ of K commutes with τ, and thus

$$\hat{k} = (\hat{k})^\tau = (\hat{k})^{-1},$$

and it follows that k^2 centralizes K, so $k^2 \in \mathbf{Z}(K) = L$. Since $|K : L|$ is odd by hypothesis, we conclude that $k \in L$, and thus $K \cap U = L$, as required.

Now let $g \in G$ be arbitrary. We argue that the automorphism $[\hat{g}, \tau]$ of K is inner. By Lemma 7.16(c), it suffices to show that $[\hat{g}, \tau]$ acts trivially on both L and K/L. Since both $\hat{g}$ and τ centralize L, it is clear that their commutator acts trivially on L, so we consider the action of $[\hat{g}, \tau]$ on an element $x \in K/L$. We have

$$x^{\tau\hat{g}} = (x^{-1})^g = (x^g)^{-1} = x^{\hat{g}\tau},$$

and thus $[\hat{g}, \tau]$ fixes x, as required, and we conclude that $[\hat{g}, \tau]$ lies in $\hat{K}$. Now $|\hat{K}| = |K : L|$ is odd, so every element of $\hat{K}$ is a square. We can thus write

$\hat{g}^{-1}(\hat{g})^{\tau} = [\hat{g}, \tau] = (\hat{k})^2$ for some element $k \in K$, and hence $(\hat{g})^{\tau} = \hat{g}(\hat{k})^2$. It follows that

$$(\hat{g}\hat{k})^{\tau} = (\hat{g})^{\tau}(\hat{k})^{\tau} = \hat{g}(\hat{k})^2(\hat{k})^{-1} = \hat{g}\hat{k}\,,$$

and since $\hat{g}\hat{k}$ commutes with τ in $\mathrm{Aut}(K)$, it follows that gk lies in U, and thus $g \in KU$, This completes the proof of (a).

For (b), assume that $[k, u] \in L$, where $k \in K$ and $u \in U$. Then $k^u = kz$ for some element $z \in L$, and thus $(k^{-1})^u = z^{-1}k^{-1}$. Also $k^{\tau} = k^{-1}y$, where $y \in L$. Now $L \subseteq \mathbf{Z}(G)$ and τ acts trivially on L, and so we have

$$\begin{aligned} k^{\tau u} &= (k^{-1}y)^u = z^{-1}k^{-1}y \quad \text{and} \\ k^{u\tau} &= (kz)^{\tau} = k^{-1}yz\,. \end{aligned}$$

Since $u \in U$, we have $k^{\tau u} = k^{u\tau}$, and we deduce that $z = z^{-1}$, so $z^2 = 1$. Now $(k^2)^u = k^2z^2 = k^2$, so u centralizes k^2. Also $k \in \langle L, k^2 \rangle$ because $|K : L|$ is odd, and it follows that u centralizes k, and thus $[k, u] = 1$, as wanted.

For (c), we must show that τ can be extended to an automorphism σ of G, where σ acts trivially on U. Since $KU = G$, every element of G has the form ku, where $k \in K$ and $u \in U$, and so we must have $(ku)^{\sigma} = k^{\tau}u$. Using this as the definition of σ, we must show that σ is a well-defined automorphism of G.

If $au = bv$ with $a, b \in K$ and $u, v \in U$, then to see that σ is well defined, we must show that $a^{\tau}u = b^{\tau}v$. Now $a^{-1}b = uv^{-1}$, so this element lies in $K \cap U = L$, and so we can write $b = ay$, where $y \in L$. Then $u = yv$, and we have

$$b^{\tau}v = (ay)^{\tau}v = a^{\tau}yv = a^{\tau}u\,,$$

so σ is well defined.

It is clear that σ is a bijection from G onto itself, so to show that σ is an automorphism, it suffices to check that it is a homomorphism. Given $u, v \in U$ and $a, b \in K$, then since $\hat{u}$ and τ commute, we have

$$\begin{aligned} ((au)(bv))^{\sigma} = (ab^{u^{-1}}uv)^{\sigma} = (ab^{u^{-1}})^{\tau}uv &= a^{\tau}(b^{\tau})^{u^{-1}}uv \\ &= (a^{\tau}u)(b^{\tau}v) \\ &= (au)^{\sigma}(bv)^{\sigma}\,, \end{aligned}$$

and it follows that $\sigma \in \mathrm{Aut}(G)$, as required. Finally, it is clear that $\sigma^2 = 1$, and since $G = KU$ and σ acts trivially on U and stabilizes K, we see that σ acts trivially on G/K. ∎

We need another purely group-theoretic result before we return to character theory.

Lemma 7.17. *Let K/L be an abelian normal section of G, where $L = \mathbf{Z}(K)$ and L is cyclic, and let U be a complement for K relative to L in G. Let $\mathcal{X}$ be the set of elements $x \in G$ such that $[k, x] = 1$ whenever $k \in K$ and $[k, x] \in L$, and suppose that $U \subseteq \mathcal{X}$. The following then hold.*

(a) *If $u \in U$, then $\mathcal{X} \cap Ku$ is the set of elements of Ku that are K-conjugate to elements of Lu. Also, $|\mathcal{X} \cap Ku| = |L||K : C|$, where $C/L = \mathbf{C}_{K/L}(Lu)$.*

(b) *$\mathcal{X}$ is exactly the set of elements of G that are conjugate to to some element of U.*

Proof. Observe that K acts by conjugation on the coset Ku because if $k \in K$, then $(Ku)^k = k^{-1}Kuk = Kuk = uKk = uK = Ku$. It follows that K permutes the cosets of L contained in Ku, and we see that C is the stabilizer of Lu in this action. The number of K-conjugates of Lu is thus $|K : C|$, and hence the total number of elements contained in the K-conjugates of Lu is $|L||K : C|$.

Now $L \subseteq U \subseteq \mathcal{X}$, and thus $Lu \subseteq \mathcal{X}$. Since $\mathcal{X}$ is invariant under conjugation by elements of K, all of the K-conjugates of the elements of Lu lie in $\mathcal{X}$. We deduce that $\mathcal{X} \cap Ku$ contains all K-conjugates of elements of Lu, and thus $|\mathcal{X} \cap Ku| \geq |L||K : C|$. Also, if equality holds, then $\mathcal{X} \cap Ku$ exactly the set of K-conjugates of elements of Lu.

To obtain the reverse inequality, let $x \in \mathcal{X} \cap Ku$, and write $x = au$, where $a \in K$. Now C/L centralizes Lu in G/L, so if $c \in C$, we have $[c, u] \in L$. Since $c \in K$ and $u \in \mathcal{X}$, it follows from the definition of the set $\mathcal{X}$ that $[c, u] = 1$, so u centralizes c.

Now K/L is abelian and $a, c \in K$, so c centralizes a modulo L. Since c centralizes u, it follows that c centralizes au modulo L. Then $[c, au] \in L$, and since $c \in K$ and $au \in \mathcal{X}$, we have $[c, au] = 1$, and we deduce that c centralizes au.

We now have $u, au \in \mathbf{C}_G(C)$, and thus $a \in \mathbf{C}_K(C)$. Then $x = au$ lies in the coset $\mathbf{C}_K(C)u$, and we conclude that $|\mathcal{X} \cap Ku| \leq |\mathbf{C}_K(C)u| = |\mathbf{C}_K(C)|$.

Now $L \subseteq C \subseteq K$, so it follows by Lemma 7.16(d) that $|K : \mathbf{C}_K(C)| = |C : L|$, and thus

$$|\mathcal{X} \cap Ku| \leq |\mathbf{C}_K(C)| = \frac{|K||L|}{|C|} = |L||K : C|\,.$$

Then $|\mathcal{X} \cap Ku| = |L||K : C|$, and (a) follows.

By (a), every element of $\mathcal{X} \cap Ku$ is conjugate to an element of Lu, and since $Lu \subseteq U$, every such element is conjugate to an element of U. Every member of $\mathcal{X}$ lies in some coset of K, and every such coset has the form Ku for some element $u \in U$. It follows that every member of $\mathcal{X}$ is conjugate to an element of U, as required for (b). Conversely, since $U \subseteq \mathcal{X}$ and $\mathcal{X}$ is invariant under conjugation by elements of G, we see that every conjugate of an element of U lies in $\mathcal{X}$. ∎

Theorem 7.18. *Let $L = \mathbf{Z}(K)$, where $K \lhd G$ and K/L is abelian, and assume that $L \subseteq \mathbf{Z}(G)$. Also, let $\varphi \in \mathrm{Irr}(L)$, where φ is faithful and is fully ramified with respect to K/L. Suppose that U is a complement for K relative to L in G and that $[k, u] = 1$ whenever $k \in K$ and $u \in U$ are elements such that $[k, u] \in L$. Finally, let π be the permutation character of the conjugation action of G on the elements of K/L. The following then hold.*

(a) *If $\chi \in \mathrm{Irr}(G|\varphi)$, then $(\chi_U)^G = \pi\chi$, and $\chi(g) = 0$ if g lies in no conjugate of U.*

(b) *If ξ is a class function on U, then $(\xi^G)_U = \pi_U\xi$.*

(c) $|\mathrm{Irr}(G|\varphi)| = |\mathrm{Irr}(U|\varphi)|$.

(d) *If $L < K$ and $\chi \in \mathrm{Irr}(G|\varphi)$, then χ_U is not irreducible.*

(e) *If $L < K$ and $\xi \in \mathrm{Irr}(U)$, then ξ^G is not irreducible.*

Proof. First, we show that elements of U that are conjugate in G are conjugate in U. Suppose that $u, u^g \in U$, and write $g = kv$, where $k \in K$ and $v \in U$. Since $(u^k)^v = u^g \in U$, it follows that $u^k \in U$, and thus $[k, u^{-1}] = u^k u^{-1} \in U$. Also, $[k, u^{-1}] \in K$ because $K \lhd G$, and thus $[k, u^{-1}] \in K \cap U = L$. By hypothesis, therefore, $[k, u^{-1}] = 1$, and hence u and k commute. Then $u^g = (u^k)^v = u^v$, so u and u^g are conjugate in U, as claimed.

Let ξ be a class function on U. If $u \in U$ and $g \in G$ with $gug^{-1} \in U$, then gug^{-1} and u are conjugate in U, and thus $\xi(gug^{-1}) = \xi(u)$. It follows that

$$\xi^G(u) = \frac{1}{|U|} \sum_{\substack{g \in G \\ gug^{-1} \in U}} \xi(gug^{-1}) = \frac{m}{|U|}\xi(u),$$

where m is the number of elements $g \in G$ such that $gug^{-1} \in U$. Now $gug^{-1} \in U$ if and only if $Ugu = Ug$, so m is the total number of elements in those right cosets of U that are fixed by right multiplication by u. By Lemma 2.34, the right multiplication action of U on the right cosets of U in G is permutation isomorphic to the conjugation action of U on K/L, and hence the number of right cosets of U in G fixed by u is $\pi(u)$. Then $m = |U|\pi(u)$, and thus $(\xi^G)_U = \pi_U\xi$, proving (b).

If $L < K$, then the permutation action of U on the elements of K/L has at least two orbits (counting the orbit of the identity of K/L), and thus the principal character of U occurs with multiplicity at least 2 in π_U. If

$\xi \in \mathrm{Irr}(U)$, therefore, we have

$$[\xi^G, \xi^G] = [(\xi^G)_U, \xi] = [\pi_U \xi, \xi] \geq 2[\xi, \xi] = 2\,,$$

and hence ξ^G is not irreducible, proving (e).

Now let $\chi \in \mathrm{Irr}(G|\varphi)$. To prove (a), we show first that χ vanishes on all elements of G that are not conjugate to elements of U. By hypothesis, U is contained in the set $\mathcal{X}$ of Lemma 7.17, and also, L is cyclic since φ is faithful. We can thus apply Lemma 7.17 to deduce that if $g \in G$ and g lies in no conjugate of U, then $g \notin \mathcal{X}$, and hence there exists an element $k \in K$ such that $[k, g] \in L$ but $[k, g] \neq 1$. Then $g^{-1}g^k = [g, k]$ is some nonidentity element $z \in L$, and thus $g^k = gz$, and we have

$$\chi(g) = \chi(g^k) = \chi(gz) = \chi(g)\varphi(z)\,,$$

where the last equality holds because χ_L is a multiple of the linear character φ. (To see this, consider a representation affording χ.) Since φ is faithful and $z \neq 1$, we have $\varphi(z) \neq 1$, and thus $\chi(g) = 0$, as wanted.

If g is not conjugate to an element of U, therefore, we have

$$(\chi_U)^G(g) = 0 = \pi(g)\chi(g)\,,$$

so to prove (a), it suffices to observe that $(\chi_U)^G$ and $\pi\chi$ agree on elements of U, and this follows since by (b), we have

$$((\chi_U)^G)_U = \pi_U \chi_U\,.$$

If $L < K$, then

$$[\chi_U, \chi_U] = [\chi, (\chi_U)^G] = [\chi, \pi\chi] \geq 2[\chi, \chi] = 2\,,$$

so χ_U is not irreducible, and this proves (d).

Finally, to prove (c), let V and W be the complex vector spaces spanned by members of $\mathrm{Irr}(G|\varphi)$ and $\mathrm{Irr}(U|\varphi)$, respectively. Since these sets of characters are linearly independent, we have

$$\dim V = |\,\mathrm{Irr}(G|\varphi)| \qquad \text{and} \qquad \dim W = |\,\mathrm{Irr}(U|\varphi)|\,,$$

and hence it suffices to prove that $\dim V = \dim W$. If $\xi \in \mathrm{Irr}(U|\varphi)$, then every irreducible constituent of ξ^G lies over φ, so induction defines a linear transformation from W into V, and similarly, restriction defines a linear transformation from V into W. We complete the proof by showing that each of these maps has trivial kernel. First, if $\xi \in W$ and $\xi^G = 0$, then by (b), we have $\pi_U \xi = 0$, and since π is never zero, we conclude that $\xi = 0$, as required. Also, if $\chi \in V$ and $\chi_U = 0$, then $\pi\chi = 0$ by (a), and thus $\chi = 0$. ∎

We have now established everything needed for the proof of Theorem 7.5.

Proof of Theorem 7.5. We can replace (G, L, φ) by an isomorphic character triple, so by Theorem 7.13, it suffices to assume that $L \subseteq \mathbf{Z}(G)$, that φ is faithful, and that K is a GCR-group with respect to L.

In order to apply Theorem 7.15, we must establish that $L = \mathbf{Z}(K)$, so let $Z = \mathbf{Z}(K)$, and note that $L \subseteq Z$. Since φ is fully ramified with respect to K/L, there is a unique character $\theta \in \mathrm{Irr}(K|\varphi)$, and $\theta_L = e\varphi$, where $e^2 = |K : L|$. Also, $\theta(1) = e$ because φ is linear. Now θ_Z is a multiple of some linear character λ, so $\theta_Z = \theta(1)\lambda = e\lambda$, and hence $e^2 \leq |K : Z|$ by Lemma 7.2. Then $|K : L| = e^2 \leq |K : Z|$, and it follows that $Z = L$, as wanted.

By Theorem 7.14 and Theorem 7.15, there exists a complement U for K relative to L in G such that $[k, u] = 1$ whenever $k \in K$ and $u \in U$ are elements such that $[k, u] \in L$. We can now apply Theorem 7.18 to complete the proof. ∎

7C

In this section, we present some additional applications of Theorem 7.5. We begin with the following preliminary result.

Lemma 7.19. *Let $L \lhd K$ where K/L abelian, and let M act on K and stabilize L. Assume that $\mathbf{C}_{K/L}(M)$ is trivial, and let $\varphi \in \mathrm{Irr}(L)$ be M-invariant. Let $C \lhd M$, and assume that $|M : C|$ and $|K : L|$ are relatively prime. Assume also that C acts trivially on K/L and on $\mathrm{Irr}(K|\varphi)$. Then there is a unique M-invariant character θ lying in $\mathrm{Irr}(K|\varphi)$.*

Proof. By Lemma 7.1, the group Γ of linear characters of K/L acts transitively by multiplication on $\mathrm{Irr}(K|\varphi)$. Also, M acts on $\mathrm{Irr}(K|\varphi)$ because φ is M-invariant, and M acts on the group Γ because M acts on K/L,

By hypothesis, C acts trivially on K/L, so C acts trivially on Γ, and thus M/C acts on Γ. Also, M/C acts $\mathrm{Irr}(K|\varphi)$ because by assumption, the action of C on this set is trivial. Now Γ acts transitively on $\mathrm{Irr}(K|\varphi)$, and we have $|\Gamma| = |K/L|$. We are assuming that $|K/L|$ and $|M/C|$ are relatively prime, so we are in the situation of Glauberman's lemma (Lemma 1.4).

It is routine to check that the hypotheses of Glauberman's lemma are satisfied, and we conclude that the set of (M/C)-invariant members of $\mathrm{Irr}(K|\varphi)$ is nonempty and that the subgroup $\mathbf{C}_\Gamma(M/C)$ acts transitively on this set. Equivalently, the set of M-invariant members of $\mathrm{Irr}(K|\varphi)$ is nonempty and the subgroup $\mathbf{C}_\Gamma(M)$ acts transitively on it. We will show that $\mathbf{C}_\Gamma(M)$ is trivial, and it follows that there is a unique M-invariant member $\theta \in \mathrm{Irr}(K|\varphi)$, as wanted.

Given an M-invariant linear character λ of K/L, we show that λ is principal. For $m \in M$ and $x \in K/L$, we compute

$$\lambda(x^{-1}x^m) = \lambda(x^{-1})\lambda(x^m) = \lambda(x^{-1})\lambda(x) = 1\,,$$

where the second equality folds because λ is M-invariant. It follows that $[(K/L), M] \subseteq \ker(\lambda)$, and thus to prove that λ is the trivial character of K/L, it suffices to show that $[(K/L), M] = K/L$. Equivalently, we show that $[(K/L), (M/C)] = K/L$.

Now $|M/C|$ and $|K/L|$ are relatively prime and K/L is abelian, so Fitting's lemma (Theorem 4.34 of [**FGT**]) yields

$$K/L = [(K/L), (M/C)] \times \mathbf{C}_{K/L}(M/C)\,.$$

Since $\mathbf{C}_{K/L}(M)$ is trivial by hypothesis, we have $K/L = [(K/L), M/C]$, as required. ∎

We can now add an additional conclusion to Lemma 7.8.

Corollary 7.20. *Assume the hypotheses and notation of Lemma 7.8, and assume that $\varphi \in \mathrm{Irr}(L)$ is M-invariant, where M is as in Lemma 7.8(c). Then $\mathrm{Irr}(K|\varphi)$ contains a unique M-invariant character.*

Proof. By Lemma 7.8(a), we know that K/L is an abelian p-group. Also, M/K is a p'-group and $\mathbf{C}_{K/L}(M) = 1$ by Lemma 7.8(c). Now $|M : K|$ and $|K : L|$ are relatively prime, and K acts trivially on K/L because K/L is abelian. Also, of course, K acts trivially on $\mathrm{Irr}(K|\varphi)$, so we can apply Lemma 7.19 with K in the role of C, and we conclude that $\mathrm{Irr}(K|\varphi)$ contains a unique M-invariant character, as required. ∎

Next, we use Theorem 7.5 to establish the following technical result.

Theorem 7.21. *Let H be a maximal subgroup of a p-solvable group G, where p divides $|G : H|$. Let $L = \mathrm{core}_G(H)$, and assume that either $p \neq 2$ or that $|H : L|$ is odd. Let $\varphi \in \mathrm{Irr}(L)$, and suppose that some member of $\mathrm{Irr}(H|\varphi)$ induces irreducibly to G. Then φ is not invariant in G.*

Proof. Suppose that φ is invariant in G, and let $\xi \in \mathrm{Irr}(H|\varphi)$, where ξ^G is irreducible. If $H \triangleleft G$, then $L = H$ and $\xi = \varphi$, so ξ is invariant in G. Then $(\xi^G)_H$ is a multiple of ξ, and we can write $(\xi^G)_H = e\xi$ for some integer e. We have

$$|G : H| = \xi^G(1)/\xi(1) = e = [(\xi^G)_H, \xi] = [\xi^G, \xi^G] = 1\,,$$

and this is a contradiction. It follows that H is not normal in G, and so the hypotheses of Lemma 7.8 hold with $N = G$.

Let K/L be a chief factor of G. By Lemma 7.8(a), we see that K/L is a p-group and $K < G$. Also, H is a complement for K relative to L in G, and by Lemma 7.8(d), all such relative complements are conjugate to H.

Let M/K be a chief factor of G, as in Lemma 7.8(c). By Corollary 7.20, there exists a unique M-invariant character $\theta \in \mathrm{Irr}(K|\varphi)$, and since each of φ, M, and K is G-invariant, it follows that θ is also G-invariant.

Now ξ_L is a multiple of φ, so the set of irreducible constituents of $(\xi_L)^K$ is exactly $\mathrm{Irr}(K|\varphi)$. Also, $(\xi_L)^K = (\xi^G)_K$, and since ξ^G is irreducible, it follows that the irreducible constituents of $(\xi^G)_K$ are conjugate in G, and thus G acts transitively on the set $\mathrm{Irr}(K|\varphi)$. Since θ is G-invariant, however, we see that θ is the unique member of $\mathrm{Irr}(K|\varphi)$, and thus φ is fully ramified with respect to K/L by Lemma 7.2.

Now K/L is a p-group, so if $p \neq 2$, then $|K : L|$ is odd. Otherwise, we are assuming that $|H : L|$ is odd, and since $|G : K| = |H : L|$, we see that $|K : L|$ and $|G : K|$ are relatively prime. We can thus apply Theorem 7.5 to deduce that there exists a complement U for K in G relative to L with the property that no member of $\mathrm{Irr}(U|\varphi)$ induces irreducibly to G. We know that U is conjugate to H, however, and thus no member of $\mathrm{Irr}(H|\varphi)$ induces irreducibly to G, and this is a contradiction. ∎

We can now present the first of our applications in this section. Recall that the Fitting subgroup $\mathbf{F}(G)$ of a group G is the unique largest normal nilpotent subgroup of G.

Theorem 7.22 (Navarro). *Let H be maximal among nilpotent subgroups of G, and suppose that some character of H induces irreducibly to G. Then $\mathbf{F}(G) \subseteq H$.*

Proof. We can certainly assume that $H < G$, and we proceed by induction on $|G|$. Let $\xi \in \mathrm{Irr}(H)$, where ξ^G is irreducible. If $H \subseteq M < G$, then H is a maximally nilpotent subgroup of M and ξ^M is irreducible because $(\xi^M)^G = \xi^G$. By the inductive hypothesis, therefore, $\mathbf{F}(M) \subseteq H$. If $H\mathbf{F}(G) < G$, we can apply this reasoning with $M = H\mathbf{F}(G)$, and we deduce that $\mathbf{F}(H\mathbf{F}(G)) \subseteq H$. Since $\mathbf{F}(G) \subseteq \mathbf{F}(H\mathbf{F}(G))$, we are done in this case, and hence we can assume that $H\mathbf{F}(G) = G$, and in particular, G is solvable.

Again, assume that $H \subseteq M < G$. Since $H\mathbf{F}(G) = G$, Dedekind's lemma yields $M = H(\mathbf{F}(G) \cap M)$. Now $\mathbf{F}(G) \cap M \subseteq \mathbf{F}(M)$, and as we have seen, $\mathbf{F}(M) \subseteq H$. It follows that $\mathbf{F}(G) \cap M \subseteq H$, and we conclude that $M = H$. This proves that H is a maximal subgroup of G.

Now let $L = \mathrm{core}_G(H)$, so L is a normal nilpotent subgroup of G, and hence $L \subseteq \mathbf{F}(G)$. If $L = \mathbf{F}(G)$, we are done, so we can assume that $L < \mathbf{F}(G)$

and we choose a chief factor K/L of G with $K \subseteq \mathbf{F}(G)$. We work now to derive a contradiction.

Now H is not normal in G since otherwise, $H \subseteq \mathbf{F}(G)$, and since by hypothesis H is maximally nilpotent, we have $H = \mathbf{F}(G) \supseteq K$, which is not the case. Since G is solvable, we can now apply Lemma 7.8 (with $N = G$), where we take p to be some prime divisor of $|G : H|$.

By Lemma 7.8(a), we see that $K < G$ and K/L is an abelian p-group. Also H is a complement for K relative to L in G, and thus $G/K \cong H/L$, so G/K is nilpotent.

Let M/K be a chief factor of G, so M/K is a q-group for some prime q. Also, $q \neq p$ by Lemma 7.8(c), and thus $\mathbf{O}_p(G/K)$ is trivial. Since G/K is nilpotent, it follows that p does not divide $|G : K|$, and so we can apply Theorem 7.21 to deduce that no character of H that induces irreducibly to G can lie over a G-invariant irreducible character of L.

Recall now that $\xi \in \mathrm{Irr}(H)$ and ξ^G is irreducible. Since H is nilpotent, we can write $\xi = \alpha\beta$, where α is p-special and β is p'-special. Let μ and ν be irreducible constituents of α_L and β_L, respectively, so μ is p-special and ν is p'-special.

Since $|K : L|$ is a power of p, we have $K = LP$, where P is the Sylow p-subgroup of K. Now ν is determined by its restriction to the Hall p'-subgroup of L, and since P centralizes this subgroup, it follows that P stabilizes ν, and thus ν is invariant in K.

Now ν has a canonical extension $\delta \in \mathrm{Irr}(K)$, and δ is invariant under the stabilizer of ν in H. It follows by Lemma 2.11(a) that restriction defines a bijection from $\mathrm{Irr}(G|\delta)$ onto $\mathrm{Irr}(H|\nu)$, and thus there is some character $\gamma \in \mathrm{Irr}(G|\delta)$ such that $\beta = \gamma_H$.

We have

$$\xi^G = (\alpha\beta)^G = (\alpha\gamma_H)^G = \alpha^G\gamma\,,$$

and since ξ^G is irreducible, we deduce that α^G is irreducible. Since α lies over μ, it follows by our earlier reasoning that μ cannot be invariant in G.

We derive a contradiction now by showing that μ actually is invariant in G. To see this, let Q be the Sylow p-subgroup of L and write $C = \mathbf{C}_G(Q)$, so C contains a full Hall p'-subgroup of H. Also, $C \triangleleft G$ because $Q \triangleleft G$.

We saw previously that p does not divide $|G : K|$. Since $|G : K| = |H : L|$, it follows that H/L is a p'-group, and since L contains the Hall p'-subgroup of H, we have $H \subseteq LC \triangleleft G$. Now H is not normal in G, so H must be properly contained in LC, and since H is maximal in G, we deduce that $LC = G$. Now μ is uniquely determined by its restriction to Q, and thus μ is stabilized by C, and it follows that μ is invariant in $LC = G$, as claimed. ∎

Corollary 7.23. *Let $H \subseteq G$, where H is nilpotent and G is solvable, and write $F = \mathbf{F}(G)$. Suppose that $\chi \in \mathrm{Irr}(G)$ is induced from a character of H, and assume that χ_F is homogeneous. Then G is nilpotent.*

We need two standard facts about Fitting subgroups. First, by Theorem 2.2 of [**FGT**], the subgroups of $\mathbf{F}(G)$ are exactly the subnormal nilpotent subgroups of G. Also, if G is solvable, then $\mathbf{C}_G(\mathbf{F}(G)) \subseteq \mathbf{F}(G)$. (See Problem 3B.14 and the note following it in [**FGT**].)

Proof of Corollary 7.23. Since χ is induced from every subgroup containing H, we can replace H by a maximally nilpotent subgroup. We can thus assume that H is a maximally nilpotent subgroup, and hence $F \subseteq H$ by Theorem 7.22.

Let $E/F = \mathbf{F}(G/F)$. Now $(E \cap H)/F \lhd\lhd E/F$ because E/F is nilpotent, and thus $E \cap H \lhd\lhd E \lhd G$. Then $E \cap H$ is a nilpotent subnormal subgroup of G, and it follows that $E \cap H \subseteq \mathbf{F}(G) = F$, and we have $E \cap H = F$.

Now let θ be the unique irreducible constituent of χ_F. Then (G, F, θ) is a character triple, and hence there exists an isomorphic triple (G^*, F^*, θ^*), where $F^* \subseteq \mathbf{Z}(G^*)$. Let E^* and H^* be the subgroups of G^* corresponding to E and H, respectively. Then $E^*/F^* \cong E/F$ and $H^*/F^* \cong H/F$, and thus E^*/F^* and H^*/F^* are nilpotent, and since F^* is central, it follows that E^* and H^* are nilpotent. Also, since $E^* \lhd G^*$, we have $E^* \subseteq \mathbf{F}(G^*)$.

The character $\chi^* \in \mathrm{Irr}(G^*)$ is induced from the nilpotent subgroup H^*, so it is induced from some maximally nilpotent subgroup M^* of G^*, where $M^* \supseteq H^*$. Then $E^* \subseteq \mathbf{F}(G^*) \subseteq M^*$ by Theorem 7.22, and hence $H^*E^* \subseteq M^*$. It follows that H^*E^* is nilpotent, and thus $H^* \lhd\lhd H^*E^*$, and we conclude that $H \lhd\lhd HE$.

Since H is nilpotent, we have $H \subseteq \mathbf{F}(HE)$, and thus $H = \mathbf{F}(HE)$ because H is maximally nilpotent in G. Then $H \lhd HE$, and since also $E \lhd HE$ and $H \cap E = F$, we conclude that H/F centralizes E/F. Now $E/F = \mathbf{F}(G/F)$ and G is solvable by hypothesis, and therefore $H/F \subseteq E/F$. We conclude that $H \subseteq E$ and thus $F = H \cap E = H$.

Now $\chi \in \mathrm{Irr}(G)$ is induced from some character ξ of H lying over a G-invariant character $\theta \in \mathrm{Irr}(F)$. Since $H = F$, we have $\xi = \theta$, and thus χ is induced from a G-invariant character of F. It follows that $F = G$, and thus G is nilpotent. ∎

We conclude this chapter with some applications concerning π-special characters and B_π-characters. The following result is a converse for Corollary 4.37(b).

Theorem 7.24. *Let $H \subseteq G$, where $|G|$ is odd, and let π be a set of primes. Suppose that $\psi \in \mathrm{Irr}(H)$ and that $\psi^G = \chi \in \mathrm{B}_\pi(G)$. Then $\psi \in \mathrm{B}_\pi(H)$.*

Proof. We proceed by induction on $|G|$, and we observe that there is no loss if we assume that H is a maximal subgroup of G. Let $L = \mathrm{core}_G(H)$, and let φ be an irreducible constituent of ψ_L. Since $|G|$ is odd, G is solvable, and so we can apply Theorem 7.21 to deduce that φ is not invariant in G.

Let T be the stabilizer of φ in G, so $T < G$. Write $S = T \cap H$, and note that S is the stabilizer of φ in H. Let $\xi \in \mathrm{Irr}(S|\varphi)$ be the Clifford correspondent of ψ with respect to φ, so $\xi^H = \psi$, and hence $\xi^G = \chi$.

Let $\xi^T = \eta$, so $\eta^G = (\xi^T)^G = (\xi^H)^G = \psi^G = \chi$, and thus η is irreducible. Also, $\eta \in \mathrm{Irr}(T|\varphi)$, and hence η is the Clifford correspondent of χ with respect to φ.

Since $\chi \in \mathrm{B}_\pi(G)$, it follows by Corollary 4.37(c) that $\eta \in \mathrm{B}_\pi(T)$. Now $\xi^T = \eta \in \mathrm{B}_\pi(T)$ and $T < G$, so $\xi \in \mathrm{B}_\pi(S)$ by the inductive hypothesis. Since $\xi^H = \psi$, it follows by another application of Corollary 4.37(c) that $\psi \in \mathrm{B}_\pi(H)$, as wanted. ■

As in Theorem 7.24, let $\psi \in \mathrm{Irr}(H)$, where $H \subseteq G$ and $|G|$ is odd, but suppose we replace the condition that ψ^G lies in $\mathrm{B}_\pi(G)$ with the stronger assumption that ψ^G is π-special. By Theorem 7.24, ψ must lie in $\mathrm{B}_\pi(H)$, but in this case, we know $\psi^G(1)$ is a π-number, so $\psi(1)$ is also π-number, and it follows by Theorem 4.12 that ψ is π-special.

We consider now what happens if we drop the assumption that $|G|$ is odd, and we assume instead that G is π-separable. As above, suppose that $H \subseteq G$, and let $\psi \in \mathrm{Irr}(H)$, where ψ^G is π-special. If $2 \in \pi$, then by Theorem 2.29 the conclusion that ψ is π-special remains valid, but if $2 \notin \pi$, then although ψ may not be π-special, Theorem 2.29 guarantees that there is a certain linear character λ of H with $\lambda^2 = 1_H$ and such that $\lambda\psi$ is π-special. (In fact, λ is the π-standard sign character $\delta_{(G,H)}$.) Note that if $|G|$ is odd, then λ is the principal character, so ψ is π-special, and we see that this special case of Theorem 2.29 is exactly the consequence of Theorem 7.24 that was discussed in the previous paragraph.

In the case where $2 \notin \pi$, we have the following converse.

Theorem 7.25. *Let $H \subseteq G$, where G is π-separable and $|G : H|$ is a π-number, and suppose that $2 \notin \pi$. Let $\psi \in \mathrm{Irr}(H)$, and assume that $\psi^G = \chi$ is irreducible and that $\delta_{(G,H)}\psi$ is π-special. Then χ is π-special.*

There is a case of Theorem 7.25 that we have already seen: the result holds if the irreducible induction of ψ to G is actually a Clifford induction. More precisely, if H is the stabilizer in G of some character θ of a normal subgroup and $\psi \in \mathrm{Irr}(H|\theta)$, then, of course, ψ^G is irreducible by the Clifford correspondence. Assuming that G is π-separable and $|G : H|$ is a π-number, where $2 \notin \pi$, it follows by Theorem 2.37 that if $\delta_{(G,H)}\psi$ is π-special, then

ψ^G is π-special. This fact will be used in the following proof, which parallels the proof of Theorem 7.24.

Proof of Theorem 7.25. Since $\delta_{(G,G)}$ is the trivial character, the result is obvious if $H = G$. We can thus assume that $H < G$, and we proceed by induction on $|G|$.

We argue first that it suffices to prove the theorem in the case where H is a maximal subgroup of G. To see this, choose a maximal subgroup M of G with $H \subseteq M$, and let $\zeta = \psi^M$, so $\zeta^G = \chi$ and ζ is irreducible. Since we are temporarily assuming the result for maximal subgroups, we know that if $\delta_{(G,M)}\zeta$ is π-special, then $\chi = \zeta^G$ is π-special, as required.

Now

$$\delta_{(G,M)}\zeta = \delta_{(G,M)}\psi^M = ((\delta_{(G,M)})_H\psi)^M ,$$

so to show that $\delta_{(G,M)}\zeta$ is π-special, it suffices by the inductive hypothesis to show that $\delta_{(M,H)}(\delta_{(G,M)})_H\psi$ is π-special. By Lemma 2.33(a), we have $\delta_{(M,H)}(\delta_{(G,M)})_H = \delta_{(G,H)}$, so $\delta_{(M,H)}((\delta_{(G,M)})_H\psi) = \delta_{(G,H)}\psi$, and this is π-special by hypothesis.

We can now assume that H is maximal in G, and we let p be a prime divisor of $|G : H|$, so $p \in \pi$. Now G is π-separable, and $2 \notin \pi$, so G is π-solvable, and thus G is p-solvable.

Write $L = \mathrm{core}_G(H)$, let φ be an irreducible constituent of ψ_L. Since $p \neq 2$ and ψ^G is irreducible, it follows by Theorem 7.21 that φ is not invariant in G, and we let T be the stabilizer of φ in G, so $T < G$.

Write $S = T \cap H$, so S is the stabilizer of φ in H. Let $\xi \in \mathrm{Irr}(S|\varphi)$ be the Clifford correspondent of ψ with respect to φ. Then $\xi^H = \psi$, and we write $\xi^T = \eta$. Then $\eta^G = (\xi^T)^G = (\xi^H)^G = \psi^G = \chi$, and thus η is irreducible. Also, $\eta \in \mathrm{Irr}(T|\varphi)$, so η is the Clifford correspondent of χ with respect to φ.

Now $\psi(1)$ is a π-number since $\psi(1) = (\delta_{(G,H)}\psi)(1)$ and by hypothesis, $\delta_{(G,H)}\psi$ is π-special. Since $\psi^G = \chi$, it follows that $\chi(1) = |G : H|\psi(1)$ is a π-number, and since $\chi = \xi^G$ we see that $|G : S|$ divides $\chi(1)$. Then $|G : S|$ is a π-number, and hence $|G : T|$ and $|T : S|$ are π-numbers.

Since $|G : T|$ is a π-number and η is the Clifford correspondent of χ with respect to φ, we see by Theorem 2.37 that if $\delta_{(G,T)}\eta$ is π-special, then χ is π-special, and we are done. To complete the proof, therefore, it suffices to show that $\delta_{(G,T)}\eta$ is π-special.

Now $\delta_{(G,T)}\eta = ((\delta_{(G,T)})_S\xi)^T$. Also $T < G$ and since $|T : S|$ is a π-number, it follows by the inductive hypothesis that to show that $\delta_{(G,T)}\eta$ is π-special, it suffices to check that $\delta_{(T,S)}(\delta_{(G,T)})_S\xi$ is π-special.

We have

$$\delta_{(T,S)}(\delta_{(G,T)})_S = \delta_{(G,S)} = \delta_{(H,S)}(\delta_{(G,H)})_S$$

by Lemma 2.33(a), so to complete the proof, it suffices to show that $\delta_{(H,S)}(\delta_{(G,H)})_S\xi$ is π-special. Since $\xi^H = \psi$, we have $((\delta_{(G,H)})_S\xi)^H = \delta_{(G,H)}\psi$, and by hypothesis, this character is π-special. It follows by Theorem 2.37 that $\delta_{(H,S)}(\delta_{(G,H)})_S\xi$ is π-special, as wanted. ■

Next, we consider irreducible restrictions of π-special characters. Suppose that $H \subseteq G$, where G is π-separable, and let $\chi \in \mathrm{Irr}(G)$ be π-special. If $\chi_H = \psi \in \mathrm{Irr}(H)$, it is clear that $\psi(1)$ and $o(\psi)$ are π-numbers, so it is reasonable to guess that ψ must be π-special. If $|G|$ is odd, this is true by Corollary 2.27, but it is false in general. (For a counterexample, consider $G = GL(2,3)$. Each of the two degree 2 faithful irreducible characters of G is $\{2\}$-special, and each restricts irreducibly to a subgroup H of index 4, but these restrictions are not $\{2\}$-special.)

Using results of this chapter, however, we can show that if $2 \notin \pi$, then irreducible restrictions of π-special characters must be π-special.

Theorem 7.26. *Let $H \subseteq G$, where G is π-separable and $2 \notin \pi$. Suppose that $\chi \in \mathrm{Irr}(G)$ is π-special and that χ_H is irreducible. Then χ_H is π-special.*

Since the statement of Theorem 7.24 does not involve induced characters, it may be surprising that our proof depends heavily on properties of π-standard sign characters. In particular, we need the following lemma concerning these objects.

Lemma 7.27. *Let K/L be a normal section of a π-separable group G, and let H be a complement for K relative to L in G. Assume that K/L is a π-group and $2 \notin \pi$. Let $K \subseteq X \subseteq G$, and write $Y = X \cap H$. Then $(\delta_{(G,X)})_Y = \delta_{(H,Y)}$.*

Proof. First, observe that Lemma 2.33(a) yields

$$(\delta_{(G,X)})_Y\delta_{(X,Y)} = \delta_{(G,Y)} = (\delta_{(G,H)})_Y\delta_{(H,Y)}\,,$$

so it suffices to prove that $\delta_{(X,Y)} = (\delta_{(G,H)})_Y$. By Lemma 2.35, we have

$$\delta_{(G,H)} = \det(((1_H)^G)_H) \qquad \text{and} \qquad \delta_{(X,Y)} = \det(((1_Y)^X)_Y)\,.$$

Also,

$$(((1_H)^G)_H)_Y = ((1_H)^G)_Y = (((1_H)^G)_X)_Y = (((1_H)_Y)^X)_Y = ((1_Y)^X)_Y\,,$$

where the third equality holds because $XH = G$. Taking determinants, we obtain

$$(\delta_{(G,H)})_Y = \det((((1_H)^G)_H))_Y = \det(((1_Y)^X)_Y) = \delta_{(X,Y)}\,,$$

as wanted. ■

Proof of Theorem 7.26. We can assume that H is a maximal subgroup of G, and we proceed by induction on G.

If $H \lhd G$, then χ_H is π-special, and we are done. Also, if $|G : H|$ is a π'-number, then Gajendragadkar's restriction theorem (Theorem 2.10) guarantees that χ_H is π-special, and so we are done in this case too. We can assume, therefore, that H is not normal and that $|G : H|$ is divisible by some prime $p \in \pi$. Also, since $2 \notin \pi$ and G is π-separable, we see that G is p-solvable.

Let $L = \mathrm{core}_G(H)$, and let K/L be a chief factor of G. By Lemma 7.8 (with $N = G$), we see that $K < G$, that K/L is an abelian p-group, and that H is a complement for K relative to L in G. Also, every such relative complement is conjugate to H.

Now let θ be an irreducible constituent of χ_K, let T be the stabilizer of θ in G, and let $\eta \in \mathrm{Irr}(T|\theta)$ be the Clifford correspondent of χ with respect to θ. Since $\eta^G = \chi$ and χ is π-special, it follows by Theorem 2.37 that $\delta_{(G,T)}\eta$ is π-special.

Write $\psi = \chi_H$, so ψ is irreducible by hypothesis, and let $S = T \cap H$ and $\xi = \eta_S$. We have $TH = G$, and thus

$$\xi^H = (\eta_S)^H = (\eta^G)_H = \chi_H = \psi\,,$$

and hence ξ is irreducible. Also, $|H : S| = \chi(1)/\xi(1)$, and this is a π-number because χ is π-special

Now $\delta_{(G,T)}\eta$ is π-special, and by Lemma 7.27, we have

$$(\delta_{(G,T)}\eta)_S = (\delta_{(G,T)})_S\eta_S = \delta_{(H,S)}\xi\,,$$

and this is irreducible because $\delta_{(H,S)}$ is linear and ξ is irreducible. If $T < G$, therefore, it follows by the inductive hypothesis that $\delta_{(H,S)}\xi$ is π-special. Since $\xi^H = \psi$, Theorem 2.37 guarantees that ψ is π-special, as required.

We can now assume that $T = G$, so θ is invariant in G, and we let φ be an irreducible constituent of θ_L. Then K acts transitively on the set of G-conjugates of φ, so if V is the stabilizer of φ in G, we have $G = KV$. We consider the three alternatives of Theorem 7.3.

First, if $K \cap V = L$, then V is a complement for K relative to L in G, and thus V is conjugate to H. Replacing φ by a conjugate, therefore, we can assume that H is the stabilizer of φ in G. This is a contradiction because χ lies over φ but χ is not induced from H because by hypothesis, χ_H is irreducible.

Next, we argue that φ cannot be fully ramified with respect to K/L. Otherwise, we can apply Theorem 7.5, and we see that the subgroup U whose existence is guaranteed by that theorem must be conjugate to H. Since χ_H is irreducible, this is a contradiction by Theorem 7.5(c).

We conclude by Theorem 7.3 that $\theta_L = \varphi$. Now let M be a maximal normal subgroup of G containing K, and let $N = M \cap H$. Then $M = NK$ and $N \cap K = L$, and since θ is invariant in M, we can apply Lemma 2.11(a) to deduce that restriction defines a bijection from $\mathrm{Irr}(M|\theta)$ to $\mathrm{Irr}(N|\varphi)$.

Let α be an irreducible constituent of χ_M, and write $\beta = \alpha_N$. Since α is π-special and β is irreducible, it follows by the inductive hypothesis that β is π-special, and since β lies under χ, it also lies under ψ.

If G/M is a π-group, then H/N is a π-group, and since β is π-special, it follows by Theorem 2.4(a) that ψ is π-special, as wanted. In the remaining case, G/M is a π'-group, so H/N is a π'-group, and since $\psi(1)$ is a π-number, it follows that ψ_N is irreducible, and thus $\psi_N = \beta$. To show that ψ is π-special, therefore, it suffices to show that ψ is the canonical extension of β to H. This true, however, since $\psi = \chi_H$ and thus $o(\psi)$ divides $o(\chi)$, which is a π-number. ∎

As a consequence, we obtain the following.

Corollary 7.28. *Let $H \subseteq G$, where G is π-separable and $2 \notin \pi$. Let $\psi \in \mathrm{Irr}(H)$ be π-special, and suppose that ψ has an extension to G. Then ψ has a π-special extension to G.*

Proof. Let U be a Hall π-subgroup of H, and let V be a Hall π-subgroup of G containing U. If $\zeta \in \mathrm{Irr}(G)$ is an extension of ψ, then $\zeta_U = \psi_U$, and since ψ is π-special, ψ_U is irreducible by Gajendragadkar's restriction theorem (Theorem 2.10). Now $U \subseteq V$ and ζ_U is irreducible, and it follows that ζ_V is irreducible. By Corollary 3.15, there exists a π-special character $\chi \in \mathrm{Irr}(G)$ such that $\chi_V = \zeta_V$, and thus $\chi_U = \zeta_U$ is irreducible, and hence χ_H is irreducible. It follows by Theorem 7.26 that χ_H is π-special. Since χ_H and ψ are π-special characters of H that have equal restrictions to the Hall π-subgroup U of H, it follows by the injectivity of the map in Theorem 2.10 that $\chi_H = \psi$, as wanted. ∎

Recall that Theorem 3.6 asserts that if H is a Hall π-subgroup of a π-separable group G and ψ is a character of H such that $\psi(x) = \psi(y)$ whenever $x, y \in H$ are conjugate in G, then ψ is the restriction to H of some character χ of G. If the character ψ is irreducible, then, of course, χ must also be irreducible. In this situation, Corollary 3.15 applies, and we can conclude that ψ is the restriction to H of a π-special character of G.

In the case where $2 \notin \pi$, the following application of Theorem 7.26 allows us to obtain a similar conclusion even if we relax the requirement that H is a Hall π-subgroup of G. It suffices to assume that $|G : H|$ is a π'-number.

Corollary 7.29. *Let G be π-separable, where $2 \notin \pi$, and suppose $H \subseteq G$, where $|G : H|$ is a π'-number. Let $\psi \in \mathrm{Irr}(H)$ be π-special, and assume that $\psi(x) = \psi(y)$ whenever $x, y \in H$ are π-elements that are conjugate in G. Then ψ is the restriction to H of a π-special character of G.*

Proof. Let U be a Hall π-subgroup of H, and observe that U is also a Hall π-subgroup of G. Let $\alpha = \psi_U$, so α is irreducible by Theorem 2.10. Also, if $x, y \in U$ are conjugate in G, then by hypothesis $\alpha(x) = \psi(x) = \psi(y) = \alpha(y)$. So α extends to G by Theorem 3.6, and it follows by Corollary 3.15 that α has a π-special extension $\chi \in \mathrm{Irr}(G)$. Since $\chi_U = \alpha$ is irreducible, we see that χ_H is irreducible, and hence χ_H is π-special by Theorem 7.26. Now χ_H and ψ are π-special characters of H that agree on the Hall π-subgroup U of H, and thus $\chi_H = \psi$ by the injectivity in Theorem 2.10. ∎

We have not, as yet, made use of Theorem 7.5(a), which asserts (in the notation of the theorem) that $|\mathrm{Irr}(G|\varphi)| = |\mathrm{Irr}(U|\varphi)|$. The reason for this is that this character-count equality is too weak, and much more is true. In Chapter 8, we will strengthen this result, and then we will present applications.

Problems

7.1. Assuming the hypotheses of Theorem 7.3, let $L \subseteq S \subseteq K$, and assume that $\beta_L = \varphi$ for all characters $\beta \in \mathrm{Irr}(S|\varphi)$. Assume also that α is fully ramified with respect to K/S for all characters $\alpha \in \mathrm{Irr}(K|\varphi)$. Show that S is the subgroup R of the theorem.

7.2. Assuming the hypotheses of Theorem 7.15, let A be the subset of $\mathrm{Aut}(K)$ consisting of automorphisms τ where $\tau^2 = 1$ and such that τ acts trivially on L and it inverts the elements of K/L.

(a) Show that A is contained in a single coset of $\mathrm{Inn}(K)$ in $\mathrm{Aut}(K)$, and thus $|A| \leq |K : L|$.

(b) Show that A is a single orbit in the conjugation action of $\mathrm{Inn}(K)$ on $\mathrm{Aut}(K)$, and deduce that $|A| = |K : L|$.

(c) Show that all distinguished complements for K relative to L in G are conjugate in G.

Notes and References

Theorem 7.5 of this chapter is essentially Theorem 1.6 of [**39**], and in some sense, it is included in Theorem 9.1 of [**34**]. The proof given here of Theorem 7.5 is a hybrid, consisting of arguments from both of these papers.

The theorem of T. R. Berger that quasi-primitive characters of solvable groups are primitive appeared in [**4**].

Theorem 7.13 is Theorem 8.2 of [**34**].

Theorem 7.22 and Corollary 7.23 of G. Navarro appear in [**60**].

Theorem 7.24 appears as Theorem 8.6 of [**47**].

Theorem 7.25 and Theorem 7.26 are Theorem B and Theorem A of [**42**], respectively.

Chapter 8

Fully Ramified Sections and Character Correspondences

8A

We now continue our study of an abelian normal section K/L of a group G, where $\varphi \in \mathrm{Irr}(L)$ is invariant in G and is fully ramified with respect to K/L. Recall that Theorem 7.5(a) asserts that if either $|K : L|$ is odd or $|K : L|$ and $|G : K|$ are relatively prime, then there exists a complement U for K relative to L in G such that the sets $\mathrm{Irr}(G|\varphi)$ and $\mathrm{Irr}(U|\varphi)$ have equal cardinalities. In Chapter 7, we made no attempt to construct a specific correspondence between these two sets of characters, but here we remedy that deficiency and, at least in the case where $|G|$ is odd, we construct a canonical correspondence.

Our correspondence is well-behaved with respect to subgroups W that are intermediate between K and G, but unfortunately the introduction of the subgroup W complicates the statement of our result. To more easily understand Theorem 8.1, therefore, the reader might wish to pretend at first that $W = G$ and $V = U$. (In fact, this will be the case for most applications.)

Theorem 8.1. *Let $L \subseteq K \triangleleft G$, where $L \triangleleft G$ and K/L is abelian, and assume either that $|K : L|$ is odd or that $|K : L|$ and $|G : K|$ are relatively prime. Also, let $\varphi \in \mathrm{Irr}(L)$, where φ is invariant in G and is fully ramified with* *respect to K/L. Then there exists a complement U for K relative to L in G and a character ψ of U with $L \subseteq \ker(\psi)$ such that the following hold.*

(a) *If $K \subseteq W \subseteq G$ and $V = W \cap U$, then for each character $\chi \in \mathrm{Irr}(W|\varphi)$, there exists a unique character $\xi \in \mathrm{Irr}(V|\varphi)$ such that $\chi_V = \psi_V \xi$. Also, the map $\chi \mapsto \xi$ is a bijection from $\mathrm{Irr}(W|\varphi)$ onto $\mathrm{Irr}(V|\varphi)$.*

(b) *$|\psi(u)|^2 = |\mathbf{C}_{K/L}(u)|$ for all elements $u \in U$. In particular, $\psi(u)$ is never zero, and $\psi(1) = e$, where $e^2 = |K : L|$.*

(c) *If $|W : L|$ is odd, then the principal character 1_V is the unique irreducible character of V having odd multiplicity as a constituent of ψ_V.*

Taking $W = G$ and $V = U$, we see that Theorem 8.1(a) yields a bijection from $\mathrm{Irr}(G|\varphi)$ onto $\mathrm{Irr}(U|\varphi)$ such that if χ maps to ξ, then $\chi(1) = e\xi(1)$, where $e = \psi(1) = |K : L|^{1/2}$. Also, by (c), it follows that if $|G : L|$ is odd, we can write $\psi = 1_U + 2\delta$, where δ is some character of U (or $\delta = 0$ in the degenerate case where $K = L$). If ξ is the image of χ in this situation, we have $\chi_U = \psi\xi = \xi + 2\delta\xi$, and thus ξ is the unique irreducible constituent of χ_U that has odd multiplicity. In other words, if $|G : L|$ is odd, the map $\chi \mapsto \xi$ can be described very simply and without reference to the somewhat mysterious character ψ. If $\chi \in \mathrm{Irr}(G|\varphi)$, then the image of χ is the unique member of $\mathrm{Irr}(U)$ that occurs with odd multiplicity as a constituent of χ_U. In particular, we see that in this case, our map is canonical.

In fact, by imposing additional conditions, one can guarantee that the character ψ is uniquely determined. This makes it possible to define canonical correspondences in Theorem 8.1 without assuming that $|G : L|$ is odd, but we will not pursue that here. Some caution is appropriate, however, because if $|G : L|$ is not odd, the principal character 1_U may not be a constituent of ψ, and thus the irreducible character ξ corresponding to χ can fail to be a constituent of χ_U. An example of this phenomenon can be constructed by taking G to be the semidirect product of an extraspecial group K of order 5^3 and exponent 5, acted on by a group S isomorphic to $SL(2,3)$, where $\mathbf{C}_K(S) = L = \mathbf{Z}(K)$. There is a unique conjugacy class of complements for K relative to L in G, so we can take U to be any one of these relative complements. A faithful linear character φ of L is fully ramified with respect to K/L, and there exist characters $\chi \in \mathrm{Irr}(G|\varphi)$ having degree 5. If χ is one of these characters, one can compute that χ_U has exactly two irreducible constituents: one of degree 2 and one of degree 3. In this case $e = 5$, so the character ξ corresponding to χ must have degree $\chi(1)/e = 1$, and hence ξ is not a constituent of χ_U.

We will prove Theorem 8.1 following the strategy of Chapter 7, using character-triple isomorphisms. In particular, we will use Theorem 7.13 to reduce to the case where L is central in G and φ is faithful, and where K is a GCR-group with respect to L. (Recall that this means that every coset of L in K contains an element u such that $L \cap \langle u \rangle = 1$.)

Unfortunately, there is a complication in the present situation that we have not encountered previously. According to the definition in Chapter 1, a character-triple isomorphism is a collection of bijections having certain stated properties. In Theorem 1.13, we asserted (without proof) that given a character triple (G, N, φ), there exists an isomorphic triple (G^*, N^*, φ^*) such that N^* is central in G^* and φ^* is faithful. The construction of the character-triple isomorphism whose existence is asserted in Theorem 1.13 appears in the proof of Theorem 11.28 of [**CTFG**], and according to that proof, this isomorphism enjoys a useful additional property, which we call the "multiplication property".

Although this multiplication property is part of the definition of a character-triple isomorphism as given in [**CTFG**], it was not included in the definition given here in Chapter 1, and up to this point, this property has not been needed. The multiplication property is essential for our proof of Theorem 8.1, however, and so we discuss it now.

Let (G, N, φ) and (G^*, N^*, φ^*) be isomorphic character triples. Suppose that $N \subseteq X \subseteq H$, and write X^* to denote the subgroup of G^* corresponding to X, so $N^* \subseteq X^* \subseteq G^*$ and X/N is carried to X^*/N^* be the given isomorphism from G/N onto G^*/N^*. Let ξ be an admissible character of X, and let ξ^* be the admissible character of X^* corresponding to ξ. (Thus ξ_N is a multiple of φ, and $\chi^*_{N^*}$ is a multiple of φ^*.) The multiplication property of character-triple isomorphisms asserts that for every choice of X and ξ, and every character ψ of X/N, we have $(\xi\psi)^* = \xi^*\psi^*$, where ψ^* is the character of X^*/N^* corresponding to ψ under the isomorphism from X/N to X^*/N^*. (Note that $\xi\psi$ is an admissible character of X, so $(\xi\psi)^*$ is defined.)

As we have said, the character-triple isomorphisms whose existence is asserted in Theorem 1.13 satisfy the multiplication property. In the proof of Theorem 7.13, however, there was another (much more elementary) construction of a character-triple isomorphism, and so to justify using Theorem 7.13 in the present context, we should check that the character-triple isomorphism constructed there also satisfies the multiplication property. This is fairly clear, however, and we omit the proof. It follows that to prove Theorem 8.1, we can indeed apply Theorem 7.13, so we can assume that $L \subseteq \mathbf{Z}(G)$, that φ is faithful, and that K is a GCR-group with respect to L.

If $|K : L|$ is odd in this situation, Theorem 7.15 guarantees the existence of an automorphism σ of G such that $\sigma^2 = 1$, where σ acts trivially on L and on G/K, and σ inverts the elements of K/L. The subgroup $U = \mathbf{C}_G(\sigma)$ is a complement for K relative to L in G, and recall that we refer to U as a "distinguished" complement. If $|K : L|$ is even, then by hypothesis, $|K : L|$ and $|G : K|$ are relatively prime, and although there is no distinguished complement defined in this case, the Schur–Zassenhaus theorem guarantees that there does exist a complement U for K relative to L in G, and in fact, U is unique up to conjugation in G.

To establish Theorem 8.1, therefore, we can assume that we are given a 6-tuple $(G, K, U, L, \theta, \varphi)$ satisfying (1) through (5) below.

(1) $L \subseteq \mathbf{Z}(G)$.

(2) $L \subseteq K \lhd G$, where K/L is abelian and K is a GCR-group with respect to L.

(3) Either $|K : L|$ is odd, or else $|K : L|$ and $|G : K|$ are relatively prime.

(4) U complements K relative to L in G, and if $|K : L|$ is odd, then U is distinguished.

(5) $\varphi \in \operatorname{Irr}(L)$ is faithful and fully ramified with respect to K/L, and $\theta \in \operatorname{Irr}(K|\varphi)$.

To avoid a tedious repetition of hypothesis, we shall say that the tuple $(G, K, U, L, \theta, \varphi)$ is in the **basic configuration** if all of these conditions hold.

It should now be reasonably clear that Theorem 8.1 is a consequence of the following.

Theorem 8.2. *Assume that* $(G, K, U, L, \theta, \varphi)$ *is in the basic configuration. Then there exists a character* ψ *of* U *with* $L \subseteq \ker(\psi)$ *such that conclusions* (a), (b), *and* (c) *of Theorem* 8.1 *hold.*

In the basic configuration, θ is the unique irreducible character of K that lies over φ, and we know that $\theta(1) = e\varphi(1) = e$, where $e^2 = |K : L|$. Also, if $\chi \in \operatorname{Irr}(G|\varphi)$, then $\chi \in \operatorname{Irr}(G|\theta)$. Assuming the conclusion of Theorem 8.2 for the moment, we see that $\chi_U = \psi\xi$ for some character $\xi \in \operatorname{Irr}(U|\varphi)$. Also, since $\psi(1) = e$ by assertion (b), we have

$$\frac{\chi(1)}{\theta(1)} = \frac{e\xi(1)}{e\varphi(1)} = \frac{\xi(1)}{\varphi(1)},$$

and it follows that χ is an extension of θ if and only if ξ is an extension of φ. In particular, Theorem 8.2 guarantees that if φ is extendible to U, then θ must be extendible to G. As the first step in the proof of Theorem 8.2, we show that this conclusion actually does hold.

Theorem 8.3. *Let $(G, K, U, L, \theta, \varphi)$ be in the basic configuration, and suppose that φ extends to U. Then θ extends to G.*

Proof. Since θ is uniquely determined by φ, and φ is invariant in G, it follows that θ is also invariant in G. By Theorem 1.10, therefore, it suffices to show for each choice of a prime divisor p of $|G:K|$ that θ extends to P, where P/K is some Sylow p-subgroup of G/K.

Fix P, and let $Q = P \cap U$, so $P = KQ$ and $K \cap Q = L$. By hypothesis, φ extends to U, so φ has an extension $\alpha \in \mathrm{Irr}(Q)$, and $\alpha(1) = \varphi(1) = 1$.

Suppose first that $|K:L|$ and $|G:K|$ are relatively prime. Then p does not divide $|K:L|$, so $\alpha^P(1) = |P:Q| = |K:L|$ is not divisible by p. It follows that there is some irreducible constituent δ of α^P such that $\delta(1)$ is not divisible by p. Then $\delta(1)$ is relatively prime to $|P:K|$, and hence δ_K is irreducible by Corollary 1.16. Also, δ lies over φ, so δ lies over θ, and we conclude that $\delta_K = \theta$. In this case, therefore, δ is the required extension of θ to P.

We can now assume that $|K:L|$ is odd, so by hypothesis, U is a distinguished complement, and thus we can write $U = \mathbf{C}_G(\sigma)$, where $\sigma \in \mathrm{Aut}(G)$ with $\sigma^2 = 1$, where σ acts trivially on L and on G/K, and σ inverts the elements of K/L.

Since L is abelian and $L \lhd P$, we can choose a subgroup A with $L \subseteq A \subseteq K$, where A has the largest possible order such that A is abelian and $A \lhd P$. Observe that A is σ-invariant since σ centralizes L and inverts the elements of K/L.

We complete the proof now in two main steps: we show first that the linear character α of Q extends to a character β of AQ, and then we show that β^P has an irreducible constituent δ that extends θ.

Let $N = \{a^{-1}a^{\sigma} \mid a \in A\}$. To prove that α extends to AQ, it suffices to show that N is a normal subgroup of AQ such that $N \cap Q = 1$ and $AQ = NQ$. Now $N \subseteq A$ because A is σ-invariant, and since A is abelian, N is a subgroup, and we have $N = [A, \sigma]$. Now, Q centralizes σ and Q normalizes A, and it follows that Q normalizes N. We conclude that $N \lhd AQ$, as wanted.

Now let B be the Hall $2'$-subgroup of A, so B is characteristic in A, and thus B is σ-invariant. We are assuming that $|K:L|$ is odd, so $|A:L|$ is odd, and thus $A = LB$. Since σ centralizes L, we have

$$N = [A, \sigma] = [LB, \sigma] = [B, \sigma] \subseteq B\,,$$

and thus $|N|$ is odd.

It follows from the definition of N and the fact that $\sigma^2 = 1$ that σ inverts the elements of N, and since $|N|$ is odd, we conclude that $\mathbf{C}_N(\sigma) = 1$. Recall now that $N \subseteq K$ and $K \cap Q = L$, and thus

$$N \cap Q = N \cap K \cap Q = N \cap L \subseteq \mathbf{C}_N(\sigma) = 1\,.$$

To complete the proof that α extends to AQ, we show that $NQ = AQ$. Now $A \subseteq K$, so σ inverts the elements of A/L, and thus σ inverts the elements of A/NL. Also, σ centralizes the elements of A/NL, because $N = [A, \sigma]$. Since A/NL has odd order, we conclude that A/NL is the trivial group, and thus $NL = A$. Then $NQ = NLQ = AQ$, as required.

We now know that α has an extension to a linear character β of AQ, and we see that $\beta^P(1) = |P : AQ| = |K : A|$. Each irreducible constituent of β^P lies over φ and hence over θ, and it follows that $(\beta^P)_K = m\theta$ for some integer m. Since $\theta(1) = e$, we have

$$m = \frac{\beta^P(1)}{\theta(1)} = \frac{|K : A|}{e}\,.$$

We will show that m is not divisible by p, and thus there exists an irreducible constituent δ of β^P such that $\delta(1)/\theta(1)$ is not divisible by p. Since $\delta(1)/\theta(1)$ divides $|P : K|$ and $|P : K|$ is a power of p, it will follow that $\delta(1)/\theta(1) = 1$, and hence δ is the required extension of θ to P.

To compute $m = |K : A|/e$, let $C = \mathbf{C}_K(A)$, and note that $A \subseteq C$ since A is abelian. Also, $C \lhd P$ because P normalizes both K and A, and we conclude that $C/A \lhd P/A$, and furthermore, C/A is abelian because K/L is abelian.

We argue that p does not divide $|C : A|$. Otherwise, there exists a minimal normal subgroup M/A of P/A, where M/A is a p-group and $M \subseteq C$. The conjugation action of K on M/A is trivial because K/L is abelian and

$$L \subseteq A \subseteq M \subseteq C \subseteq K\,,$$

and it follows that the p-group P/K acts on the nontrivial p-group M/A. We conclude that $\mathbf{C}_{M/A}(P) > 1$.

Since M/A is minimal normal in P/A, we see that P acts trivially on M/A, and thus $|M/A| = p$, and in particular, M/A is cyclic. Now $M \subseteq C = \mathbf{C}_K(A)$, so $A \subseteq \mathbf{Z}(M)$. Then M is abelian, and this contradicts the choice of A. We conclude that p does not divide $|C : A|$, as claimed.

Recall that $|K : L| = e^2$, and write $f = |A : L|$. Then $|K : C| = f$ by Lemma 7.16(d), and thus

$$\frac{e^2}{f^2} = \frac{|K : L|}{f^2} = \frac{|K : C||C : A||A : L|}{f^2} = |C : A|\,,$$

and since this is not divisible by p, it follows that e/f is an integer not divisible by p. Also,

$$\frac{|K:A|}{e} = \frac{|K:L|}{e|A:L|} = \frac{e^2}{ef} = \frac{e}{f},$$

so $|K:A|/e$ is an integer not divisible by p, as required ■

Next, we recall some facts that were established in Chapter 7.

Corollary 8.4. *Let $(G, K, U, L, \theta, \varphi)$ be in the basic configuration, and let $\chi \in \operatorname{Irr}(G|\varphi)$. Then*

(a) $(\chi_U)^G = \pi\chi$, *where π is the permutation character of G acting by conjugation on the elements of K/L.*

(b) *χ has the value 0 on all elements of G that are not conjugate to elements of U.*

(c) *If $L < K$, then χ_U cannot be irreducible.*

(d) *If $k \in K$ and $u \in U$ with $[k, u] \in L$, then $[k, u] = 1$.*

Proof. Suppose that $k \in K$ and $u \in U$ are elements such that $[k, u] \in L$. In the case where $|K : L|$ and $|G : K|$ are relatively prime, it follows by Theorem 7.14 that $[k, u] = 1$, and by Theorem 7.15 we obtain the same conclusion in the case where $|K : L|$ is odd. This establishes (d), and assertions (a), (b), and (c) follow by Theorem 7.18(a) and (d). ■

We begin work now toward a proof of Theorem 8.2 by constructing the character ψ in the case where φ is extendible to U. (Recall that in this case, θ extends to G by Theorem 8.3.) Let $(G, K, U, L, \theta, \varphi)$ be in the basic configuration, and suppose that χ and ξ are extensions of θ and φ to G and U, respectively. Since $\chi_L = \theta_L = e\varphi$, it follows that every irreducible constituent of χ_U lies over φ. By the Gallagher correspondence (Theorem 1.7), every such constituent has the form $\beta\xi$ for some character $\beta \in \operatorname{Irr}(U/L)$ (where β is viewed as a character of U). Summing over the irreducible constituents of χ_U, we see that $\chi_U = \psi\xi$ for some character ψ of U, where $L \subseteq \ker(\psi)$. (Note that ψ is not necessarily irreducible. In fact, ψ *cannot* be irreducible except in the degenerate case where $L = K$. To see this, observe that ξ is linear, so if ψ is irreducible, then $\chi_U = \psi\xi$ is irreducible, and this contradicts Corollary 8.4(c).)

If $(G, K, U, L, \theta, \varphi)$ is in the basic configuration and φ extends to U, we will say that a character ψ of U constructed as above is **associated** with $(G, K, U, L, \theta, \varphi)$. Also, we observe that $\psi(1) = \chi(1)/\xi(1) = \theta(1)/\varphi(1) = e$, where $e^2 = |K : L|$.

Although by definition, an associated character ψ is a character of U with kernel containing L, we often view it as a character of U/L, and since there is a natural isomorphism $U/L \cong G/K$, we can also view ψ as a character of G/K or of G.

Given the extensions χ and ξ of θ and φ as before, we have seen that $\chi_U = \psi\xi$ for some associated character ψ, and since ξ is linear, it follows that ψ is uniquely determined by the extensions χ and ξ. (Alternatively, the uniqueness of ψ is a consequence of the uniqueness in the Gallagher correspondence.) Of course, the construction of the associated character ψ depends on the choice of the extensions χ and ξ of θ and φ, so in general, ψ is not uniquely determined by the basic configuration $(G, K, U, L, \theta, \varphi)$. By the Gallagher correspondence, however, all possible extensions of θ have the form $\lambda\chi$ for linear characters λ of G/K, and similarly for extensions of φ to U. It follows that if we replace χ and ξ by other extensions of θ and φ, then the associated character ψ will be replaced by a character of the form $\mu\psi$, where μ is some linear character of U/L. The following is thus immediate.

Lemma 8.5. *Let $(G, K, U, L, \theta, \varphi)$ be in the basic configuration, and suppose that φ extends to U. Then the set of characters of U associated with $(G, K, U, L, \theta, \varphi)$ forms an orbit under multiplication by the group of linear characters of U/L. In particular, if $u \in U$, then $|\psi(u)|^2$ is independent of the choice of the particular associated character ψ.*

To compute $|\psi(u)|^2$, where $u \in U$ and ψ is an associated character as in Lemma 8.5, we need the following general character-theoretic fact.

Lemma 8.6. *Let $N \lhd G$, and suppose that χ_N is irreducible for some character $\chi \in \mathrm{Irr}(G)$. Then for each element $g \in G$, we have*

$$\sum_{x \in Ng} |\chi(x)|^2 = |N|\,.$$

Proof. Define the function α on G/N by

$$\alpha(Ng) = \frac{1}{|N|} \sum_{x \in Ng} |\chi(x)|^2\,.$$

It is clear that α is a class function on the group G/N, and our goal is to show that α is the principal character. Since the irreducible characters of G/N span the space of class functions, we can write $\alpha = \sum[\alpha, \tau]\tau$, where τ runs over $\mathrm{Irr}(G/N)$. It suffices, therefore, to show that $[\alpha, \tau] = 1$ if τ is principal, and that $[\alpha, \tau] = 0$, otherwise.

We compute

$$\begin{aligned}[\alpha, \tau] &= \frac{1}{|G/N|}\frac{1}{|N|}\sum_{\substack{Ng\in G/N\\ x\in Ng}} \chi(x)\overline{\chi(x)}\,\overline{\tau(x)} \\ &= \frac{1}{|G|}\sum_{x\in G}\chi(x)\overline{\chi(x)}\,\overline{\tau(x)} \\ &= [\chi, \chi\tau],\end{aligned}$$

so if τ is principal, we have $[\alpha, \tau] = [\chi, \chi] = 1$, as wanted. If τ is nonprincipal, however, then by the Gallagher correspondence, $\chi\tau$ is irreducible and different from χ, and thus $[\alpha, \tau] = [\chi, \chi\tau] = 0$. ∎

Theorem 8.7. *Let $(G, K, U, L, \theta, \varphi)$ be in the basic configuration. Assume that φ extends to U, and let ψ be an associated character. Then for each element $u \in U$, we have $|\psi(u)|^2 = |\mathbf{C}_{K/L}(Lu)|$. In particular, $\psi(u)$ is nonzero for all elements $u \in U$, and $\overline{\psi}\psi = \pi_U$, where π is the permutation character of the action of G, or equivalently, of G/L, on the elements of K/L.*

Proof. By definition, there exist extensions χ of θ and ξ of φ such that $\chi_U = \psi\xi$, and we observe that $|\psi(u)| = |\chi(u)|$ since ξ is linear.

Let $t \in Ku$. By Corollary 8.4(b), we see that $\chi(t) = 0$ if t is not conjugate in G to an element of U, and we proceed to compute $|\chi(t)|$ in the case where t is conjugate to an element of U. By Corollary 8.4(d), we see that U is contained in the set $\mathcal{X}$ of elements $x \in G$ such that $[k, x] = 1$ whenever $k \in K$ and $[k, x] \in L$. We can thus apply Lemma 7.17(b) to deduce that $\mathcal{X}$ is exactly the set of elements of G that are conjugate to elements of U. In particular, if t is conjugate to an element of U, then $t \in \mathcal{X} \cap Ku$, and it follows by Lemma 7.17(a) that t is conjugate to some element of Lu, say au, where $a \in L$. Then $\chi(x) = \chi(au) = \varphi(a)\chi(u)$, where the second equality holds because χ_L is a multiple of the linear character φ. We conclude that $|\chi(x)| = |\chi(u)|$.

Also by Lemma 7.17, we see that the number of elements $t \in Ku$ that lie in $\mathcal{X}$, or equivalently, that are conjugate to an element of U, is exactly $|L||K : C|$, where $C/L = \mathbf{C}_{K/L}(Lu)$. It follows by Lemma 8.6 that

$$|K| = \sum_{t\in Ku} |\chi(t)|^2 = |K : C||L||\chi(u)|^2 .$$

Then

$$|\psi(u)|^2 = |\chi(u)|^2 = |C/L| = |\mathbf{C}_{K/L}(Lu)|,$$

as wanted. ∎

We have seen that if $L < K$ in the basic configuration $(G, K, U, L, \theta, \varphi)$, then an associated character ψ cannot be irreducible. Theorem 8.7 provides an alternative argument for this. We have

$$[\psi, \psi] = [1_U, \overline{\psi}\psi] = [1_U, \pi_U],$$

and this is equal to the number of orbits in the conjugation action of U on the elements of K/L. Assuming that $L < K$, there are at least two such orbits: the orbit of the identity and at least one more. It follows that $[\psi, \psi] \geq 2$, so ψ is not irreducible.

Theorem 8.8. *Let $(G, K, U, L, \theta, \varphi)$ be in the basic configuration, and assume that φ extends to U and that $|K : L|$ is odd. Let ψ be an associated character, viewed as a character of U. Then there exist characters ψ_0 and ψ_1 of U and a linear character λ of U with the following properties.*

(a) $\psi = \psi_0 + \psi_1$.

(b) *If $L \subseteq V \subseteq U$, where $|V : L|$ is odd, then $(\psi_1)_V = \lambda_V + (\psi_0)_V$.*

(c) *If V is as in* (b), *then $\psi_V = \lambda_V + 2(\psi_0)_V$, and so λ_V is the unique odd-multiplicity irreducible constituent of ψ_V.*

(d) *L is contained in the kernels of ψ_0, ψ_1, and λ.*

Proof. Since $|K : L|$ is odd, it follows by assumption that U is a distinguished complement for K relative to L in G, and thus there exists an automorphism σ of G, where $\sigma^2 = 1$ and such that σ centralizes U and inverts the elements of K/L.

Let G^* be the semidirect product of G by a group $\langle\sigma\rangle$ of order 2. Write $U^* = U\langle\sigma\rangle$, and observe that in fact, $U^* = U \times \langle\sigma\rangle$. It follows that $(G^*, K, U^*, L, \theta, \varphi)$ is in the basic configuration, and that φ extends to U^*. Also, if ψ^* is an associated character for $(G^*, K, U^*, L, \theta, \varphi)$, then $(\psi^*)_U$ is an associated character for $(G, K, U, L, \theta, \varphi)$, and thus $(\psi^*)_U = \mu\psi$ for some linear character μ of U/L. Since μ extends to a linear character μ^* of U^*/L, we can replace ψ^* by $\psi^*\overline{\mu^*}$, and thus we can assume that ψ^* is an extension of ψ.

Since $\sigma \in \mathbf{Z}(U^*)$ and σ has order 2, it follows that for every irreducible character α of U^*, either $\alpha(\sigma) = \alpha(1)$ or $\alpha(\sigma) = -\alpha(1)$. We can thus decompose $\psi^* = (\psi^*)_+ + (\psi^*)_-$, where $(\psi^*)_+$ is the sum (with appropriate multiplicities) of the irreducible constituents α of ψ^* such that $\alpha(\sigma) = \alpha(1)$, and $(\psi^*)_-$ is the sum of the irreducible constituents α of ψ^* such that $\alpha(\sigma) = -\alpha(1)$. Since $(\psi^*)_U = \psi$, restriction to U yields a decomposition $\psi = \psi_+ + \psi_-$, and it is easy to see that for all $u \in U$, we have

$$\psi^*(u\sigma) = (\psi^*)_+(u\sigma) + (\psi^*)_-(u\sigma) = {\psi^*}_+(u) - {\psi^*}_-(u) = \psi_+(u) - \psi_-(u).$$

If $u \in U$ has odd order modulo L, we argue that $\mathbf{C}_{K/L}(u\sigma) = 1$. To see this, suppose $x \in K/L$ and $x^{u\sigma} = x$. Then $x^u = x^\sigma = x^{-1}$, and so $x^{u^2} = x$. Since u has odd order modulo L, we have $u \in \langle L, u^2 \rangle$, and it follows that u centralizes x. Then $x = x^u = x^{-1}$, and since by assumption, K/L has odd order, this implies that $x = 1$, as required.

Continuing to assume that $u \in U$ has odd order modulo L, we have

$$1 = |\mathbf{C}_{K/L}(u\sigma)| = |\psi^*(u\sigma)|^2 = |\psi_+(u) - \psi_-(u)|^2 ,$$

where the second equality follows by Theorem 8.7.

Now let $L \subseteq V \subseteq U$, where $|V : L|$ is odd. It follows that

$$[(\psi_+ - \psi_-)_V, (\psi_+ - \psi_-)_V] = 1 ,$$

and thus $(\psi_+ - \psi_-)_V$ is plus-or-minus an irreducible character τ of V. In fact, τ must be linear since $\tau(1) = |\psi_-(1) - \psi_+(1)| = 1$, and we conclude that the degrees of ψ_+ and ψ_- differ by 1. We now rename the characters ψ_+ and ψ_- so that the one of these having the smaller degree is ψ_0 and the other is ψ_1. Since $\psi = \psi_0 + \psi_1$, we have established (a).

We have $(\psi_1)_V - (\psi_0)_V = \tau$, and since $\psi = \psi_1 + \psi_0$, we conclude that

$$\psi_V = (\psi_1)_V + (\psi_0)_V = \tau + 2(\psi_0)_V ,$$

and thus τ is the unique irreducible constituent of ψ_V having odd multiplicity.

To complete the proof of (b), we must find a linear character λ of U such that $\lambda_V = \tau$, and we require that the construction of λ must be independent of the choice of the subgroup V.

Write $\beta = \det(\psi)$ and $\gamma = \det(\psi_0)$, and let $\lambda = \beta\gamma^{-2}$. Given the subgroup V as above, we have $\psi_V = \tau + 2(\psi_0)_V$, where τ is a linear character, and it follows that $\beta_V = \tau(\gamma^2)_V$. Then $\tau = (\beta\gamma^{-2})_V = \lambda_V$, as required. Assertion (c) is now clear since $\psi_V = \lambda_V + 2(\psi_0)_V$.

Finally, since $\psi = \psi_0 + \psi_1$ and $L \subseteq \ker(\psi)$, it is clear that L is contained in the kernels of ψ_0 and ψ_1. To see that L is contained in $\ker(\lambda)$, it suffices to observe that we can take $V = L$ in (c). ■

Corollary 8.9. *Let $(G, K, U, L, \theta, \varphi)$ be in the basic configuration, where $|K : L|$ is odd, and assume that φ extends to U. Then there is a unique associated character ψ such that the following hold.*

(a) *Whenever $L \subseteq V \subseteq U$, where $|V : L|$ is odd, the principal character 1_V is the unique odd-multiplicity irreducible constituent of ψ_V.*

(b) *The determinantal order $o(\psi)$ is odd.*

Proof. Let ψ be an arbitrary associated character for $(G, K, U, L, \theta, \varphi)$, and let λ be the linear character of U whose existence is guaranteed by Theorem 8.8. Since $L \subseteq \ker(\lambda)$, it follows by Lemma 8.5 that $\lambda^{-1}\psi$ is also associated with $(G, K, U, L, \theta, \varphi)$, and we see that this character has the property required by (a).

We change notation now and assume that ψ satisfies (a), and we recall that every associated character has the form $\nu\psi$, where ν is a linear character of U/L. Now $\nu\psi$ satisfies (a) if and only if $\nu_V = 1_V$ for every subgroup V such that $L \subseteq V \subseteq U$ and $|V : L|$ is odd. It follows that $\nu\psi$ satisfies (a) precisely when ν lies in the group X of linear characters of U/L that have 2-power order.

Write $\det(\psi) = \alpha\beta$, where α has 2-power order and β has odd order, and observe that α lies in X. To complete the proof, we must show that there is a unique member $\nu \in X$ such that $\det(\nu\psi)$ has odd order.

We have $\det(\nu\psi) = \nu^e\alpha\beta$, where $e = \psi(1)$, and we see that this has odd order if and only if $\nu^e\alpha$ is principal, or equivalently, $\nu^e = \alpha^{-1}$. Recall now that $e^2 = |K : L|$ so e is odd. Now X is a 2-group, so as required, there is a unique element $\nu \in X$ such that $\nu^e = \alpha^{-1}$. ∎

In the case where $|K : L|$ is odd, we refer to the unique associated character ψ for $(G, K, U, L, \theta, \varphi)$ described in Corollary 8.9 as the **canonical associated character**, and we mention that ψ has a number of additional interesting properties. We present some of these in the problems at the end of the chapter.

8B

To prove Theorem 8.2, it is essential to show that if $(G, K, U, L, \theta, \varphi)$ is in the basic configuration, then even without assuming that φ extends to U, there exist characters ψ of U/L that behave like associated characters. To do this, we construct a larger group that has the given group G as a homomorphic image.

Let $(G, K, U, L, \theta, \varphi)$ be in the basic configuration, and let $\hat{U}$ be an isomorphic copy of U, where the "hat" denotes a specific fixed isomorphism from U onto $\hat{U}$. For subgroups $X \subseteq U$, we write $\hat{X}$ to denote the image of X under this isomorphism.

Let $\hat{U}$ act on K via $k^{\hat{u}} = k^u$ for elements $k \in K$ and $u \in U$, and let G^* be the semidirect product of K by $\hat{U}$ with respect to this action. View K and $\hat{U}$ as subgroups of G^*, so $K \triangleleft G^*$ and $\hat{U}$ is a complement for K in G^*. Every element of G^* is thus uniquely of the form $\hat{u}k$, where $u \in U$ and

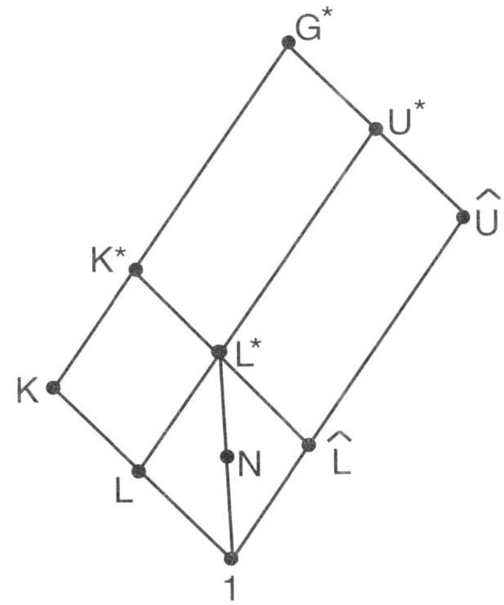

Figure 8.1

$k \in K$, and it is easy to check that the multiplication rule in G^* is

$$(\hat{u}x)(\hat{v}y) = \widehat{uv}x^v y\,,$$

for $u, v \in U$ and $x, y \in K$.

Next, let K^*, L^*, and U^* be the subgroups of G^* defined by $K^* = K\hat{L}$, $L^* = L\hat{L}$, and $U^* = L\hat{U}$, and observe that U^* is a complement for K^* relative to L^* in G^* and that L^* is central in G^* (see Figure 8.1).

We construct a homomorphism f from G^* onto G by setting $f(\hat{u}k) = uk$ for $u \in U$ and $k \in K$. It is clear that f is unambiguously defined and that it maps G^* onto G. The fact that f is a homomorphism follows easily from the multiplication rule stated previously.

Write $N = \ker(f)$, and note that $\hat{u}k \in N$ if and only if $uk = 1$ for $u \in U$ and $k \in K$. This condition forces $u, k \in U \cap K = L$, and so $N = \{\hat{l}l^{-1} \mid l \in L\}$, and in particular, $N \subseteq L^*$. Also, $N \cap K = 1$, and thus f carries K isomorphically to K. Also, it follows that $U^* = N\hat{U}$, so $f(U^*) = f(\hat{U}) = U$.

Now K^* and L^* are direct products of K and L, respectively, with $\hat{L}$, so K^* is a GCR-group with respect to L^*. Also, θ and φ have natural extensions θ^* and φ^* to K^* and L^*, respectively, where $\hat{L}$ is contained in the kernels of these extensions. We see, therefore, that θ^* lies over φ^* and that φ^* is fully ramified with respect to K^*/L^*.

We argue that $(G^*, K^*, U^*, L^*, \theta^*, \varphi^*)$ is in the basic configuration. Since $|G^* : K^*| = |\hat{U} : \hat{L}| = |U : L| = |G : K|$ and $|K^* : L^*| = |K : L|$, we see that either these indices are relatively prime or else $|K^* : L^*|$ is odd, as required.

It remains to check that if $|K^* : L^*|$ is odd, then U^* is a distinguished complement for K^* relative to L^* in G^*.

In this case, $|K : L|$ is odd, so G has an automorphism σ such that $\sigma^2 = 1$, where σ acts trivially on G/K and on L, and σ inverts the elements of K/L. Also, since U is distinguished, we can assume that σ acts trivially on U.

We can define a bijection τ from G^* to itself by setting $(\hat{u}k)^\tau = \hat{u}k^\sigma$. Since σ acts trivially on U, it follows easily from the multiplication rule that τ is an automorphism of G^*, and a further easy calculation shows that τ inverts the elements of K^*/L^* and that it centralizes G^*/K^* and L^*. Also, τ centralizes U^*, and hence U^* is a distinguished complement, as required. Then $(G^*, K^*, U^*, L^*, \theta^*, \varphi^*)$ is in the basic configuration, as claimed.

Next, we observe that $U^* = L \times \hat{U}$, so φ^* extends to U^*, and thus we can choose an associated character ψ for $(G^*, K^*, U^*, L^*, \theta^*, \varphi^*)$. In particular, ψ is a character of U^* and $L^* \subseteq \ker(\psi)$. Now $\ker(f) = N \subseteq L^* \subseteq \ker(\psi)$, and since $f(U^*) = U$, we can view ψ as a character of U with kernel containing $f(L^*) = L$.

We can now prove Theorem 8.2, and we recall that this will also establish Theorem 8.1.

Proof of Theorem 8.2. Let $(G^*, K^*, U^*, L^*, \theta^*, \varphi^*)$ be as above, so it is in the basic configuration and φ^* extends to U^*. We can thus choose an associated character ψ of $(G^*, K^*, U^*, L^*, \theta^*, \varphi^*)$, and if $|K : L|$ is odd, then $|K^* : L^*|$ is odd, and we can assume that ψ is the canonical associated character.

By the definition of an associated character, there exist extensions ζ and η of θ^* and φ^* to G^* and U^*, respectively, where $\zeta_{U^*} = \psi\eta$. As we have seen, ψ can be viewed as a character of U with kernel containing L, and we must show that if $K \subseteq W \subseteq G$ and $V = K \cap U$, then there is a bijection from $\mathrm{Irr}(W|\theta)$ to $\mathrm{Irr}(V|\varphi)$, where $\chi \mapsto \xi$ if $\chi_V = \psi_V\xi$. It should be clear that there is no loss if we assume that $W = G$ and $V = U$.

We argue first that if $\chi \in \mathrm{Irr}(G^*|\theta)$, then there exists a unique character $\xi \in \mathrm{Irr}(U^*|\varphi)$ such that $\chi_{U^*} = \psi\xi$, and furthermore, that the map $\chi \mapsto \xi$ that this defines is a bijection from $\mathrm{Irr}(G^*|\theta)$ to $\mathrm{Irr}(U^*|\varphi)$.

Since ζ is an extension of θ^* to G^* and θ^* is an extension of θ, we see that $\zeta_K = \theta$. By the Gallagher correspondence, therefore, we can write $\chi = \zeta\beta$ for some character $\beta \in \mathrm{Irr}(G^*/K)$. We thus have

$$\chi_{U^*} = \zeta_{U^*}\beta_{U^*} = \psi\eta\beta_{U^*}\,.$$

Now U^* complements K relative to L in G^*, and it follows that $\beta_{U^*} \in \mathrm{Irr}(U^*/L)$. Writing $\xi = \eta\beta_{U^*}$, we see that ξ is irreducible by the Gallagher correspondence (or alternatively, because η is linear.) Then $\xi \in \mathrm{Irr}(U^*|\varphi)$, and we have $\chi_{U^*} = \psi\xi$, as wanted. Also, since ψ is nonvanishing, ξ is uniquely determined by χ, and this defines a map $\chi \mapsto \xi$ from $\mathrm{Irr}(G^*|\theta)$ to $\mathrm{Irr}(U^*|\varphi)$.

Corollary 8.4(b) tells us that ζ vanishes on elements of G^* that are not conjugate to elements of U^*, and hence the same is true for $\chi = \zeta\beta$. It follows that χ is determined by its restriction to U^*, and thus our map $\chi \mapsto \xi$ is injective. To see that this map is also surjective, let $\xi \in \mathrm{Irr}(U^*|\varphi)$. Since $\eta_L = \varphi$, we can write $\xi = \eta\gamma$ for some character $\gamma \in \mathrm{Irr}(U^*/L)$. Now $\gamma = \beta_{U^*}$ for some character $\beta \in \mathrm{Irr}(G^*/K)$ and setting $\chi = \zeta\beta$, we see that $\chi \in \mathrm{Irr}(G^*|\theta)$, and that $\chi_{U^*} = \psi\xi$, so $\chi \mapsto \xi$ and our map is a bijection.

Recall now that we have constructed a homomorphism f from G^* onto G, and we saw that $f(U^*) = U$. We can thus identify $\mathrm{Irr}(G)$ with $\mathrm{Irr}(G^*/N)$ and $\mathrm{Irr}(U)$ with $\mathrm{Irr}(U^*/N)$, where $N = \ker(f)$. Also, since $f(K) = K$ and $f(L) = L$, we see that according to this identification, the members of $\mathrm{Irr}(G|\theta)$ correspond to those members of $\mathrm{Irr}(G^*|\theta)$ with kernel containing N and, similarly, the members of $\mathrm{Irr}(U|\varphi)$ correspond to those members of $\mathrm{Irr}(U^*|\varphi)$ with kernel containing N.

As we have seen, there is a bijection from $\mathrm{Irr}(G^*|\theta)$ onto $\mathrm{Irr}(U^*|\varphi)$ such that $\chi \mapsto \xi$ precisely when $\chi_{U^*} = \psi\xi$. Also, $N \subseteq L^* \subseteq \ker(\psi)$, and we see that if $\chi \mapsto \xi$, then $N \subseteq \ker(\chi)$ if and only if $N \subseteq \ker(\xi)$. It follows that we have a bijection from $\mathrm{Irr}(G|\theta)$ to $\mathrm{Irr}(U|\varphi)$. Furthermore, since $N \subseteq \ker(\psi)$, we can view ψ as a character of $f(U^*) = U$, and thus for $\chi \in \mathrm{Irr}(G|\theta)$ and $\xi \in \mathrm{Irr}(U|\varphi)$, we have $\chi \mapsto \xi$ if and only if $\chi_U = \psi\xi$, as required.

We have seen that the additional properties of the character ψ asserted by Theorem 8.2 hold, and this completes the proof. ∎

Next, we describe the inverse of the bijection $\chi \mapsto \xi$ of Theorem 8.1.

Corollary 8.10. *In the situation of Theorem* 8.1, *suppose that* $\chi \in \mathrm{Irr}(G|\varphi)$ *and* $\xi \in \mathrm{Irr}(U|\varphi)$, *where* $\chi_U = \psi\xi$. *Then* χ *is the unique member of* $\mathrm{Irr}(G|\varphi)$ *such that* $\xi^G = \overline{\psi}\chi$, *where we view* $\overline{\psi}$ *as a character of* G. *In particular, if* $|G:L|$ *is odd, then* χ *is the unique irreducible constituent of* ξ^G *having odd multiplicity.*

Proof. It suffices to assume that $(G, K, U, L, \theta, \varphi)$ is in the basic configuration. To prove that $\xi^G = \overline{\psi}\chi$, we show that these characters agree on U. (This is sufficient because both ξ^G and $\overline{\psi}\chi$ vanish on elements of G that are not conjugate to elements of U.) We must prove, therefore, that $(\xi^G)_U = \overline{\psi}\chi_U$.

By Corollary 8.4(a), we have

$$\pi\chi = (\chi_U)^G = (\psi\xi)^G = \psi\xi^G \,,$$

where π is the permutation character corresponding to the conjugation action of G on the elements of K/L, and where on the right, we are viewing ψ as a character of G. Since $\pi_U = \overline{\psi}\psi$ by Theorem 8.7, we can restrict to U to obtain

$$\overline{\psi}\psi\chi_U = \pi_U\chi_U = \psi(\xi^G)_U\,,$$

and since ψ is nonvanishing, we can cancel ψ to obtain $\overline{\psi}\chi_U = (\xi^G)_U$, as required.

We have now shown that $\xi^G = \overline{\psi}\chi$, and we observe that this equation uniquely determines χ because $\overline{\psi}$ is nonvanishing. Also, if $|G : L|$ is odd, it follows by Theorem 8.1 that the principal character is the unique odd-multiplicity irreducible constituent of ψ. The same is therefore true for $\overline{\psi}$, and thus χ is the unique odd-multiplicity irreducible constituent of ξ^G.

8C

In this section, we present two closely related applications of Theorem 8.1. The first of these is the Isaacs correspondence, which can be viewed as "the other half" of the Glauberman correspondence. To explain this, we suppose that A acts on G via automorphisms, where A and G are groups having relatively prime orders. Let $C = \mathbf{C}_G(A)$, and recall that we write $\mathrm{Irr}_A(G)$ to denote the set of A-invariant characters in $\mathrm{Irr}(G)$.

If A is solvable, the Glauberman correspondence is a canonical bijection from $\mathrm{Irr}_A(G)$ to $\mathrm{Irr}(C)$. (This was discussed in some detail in Chapter 6, and the relevant definitions and proofs can be found in Chapter 13 of [**CTFG**].) If A is not solvable, then $|A|$ is even by the odd-order theorem, and thus $|G|$ is odd (and G is solvable). If $|G|$ is odd (whether or not A is solvable), there is a completely different construction that also yields a canonical bijection from $\mathrm{Irr}_A(G)$ onto $\mathrm{Irr}(C)$. ∎

Theorem 8.11 (Isaacs correspondence)**.** *Let A act on G via automorphisms, and write $C = \mathbf{C}_G(A)$. Assume that $|A|$ and $|G|$ are relatively prime and that $|G|$ is odd. Then there exists a canonical correspondence from $\mathrm{Irr}_A(G)$ onto $\mathrm{Irr}(C)$.*

If A is solvable *and* $|G|$ is odd, both the Glauberman correspondence and the Isaacs correspondence are defined, and each of them yields a canonical bijection from $\mathrm{Irr}_A(G)$ onto $\mathrm{Irr}(C)$. The constructions of these maps are completely different, but nevertheless, a remarkable theorem of T. R. Wolf shows that the two maps are identical when they are both defined. Unfortunately, we are unable to present a proof of Wolf's theorem in this book.

Also in this section, we revisit the McKay conjecture, which, we recall, asserts that $|\mathrm{Irr}_{p'}(G)| = |\mathrm{Irr}_{p'}(N)|$, where p is a prime and N is the normalizer in G of a Sylow p-subgroup of G. (We are using the notation $\mathrm{Irr}_\pi(X)$

to denote the set of those irreducible characters of a group X that have degrees that are π-numbers.) In Theorem 6.11, we proved the π-analogue of the McKay conjecture for a π-separable group G in which a Hall π-subgroup is solvable. In other words, $|\operatorname{Irr}_{\pi'}(G)| = |\operatorname{Irr}_{\pi'}(N)|$ in this case.

Our proof of Theorem 6.11 relies on the Okuyama–Wajima technique, and so it does not provide a canonical correspondence. (In fact, even if G is solvable and π consists of a single prime p, there does not always exist a canonical correspondence between $\operatorname{Irr}_{p'}(G)$ and $\operatorname{Irr}_{p'}(N)$.) As an application of Theorem 8.1, however, we show that if $|G|$ is odd, then there is a canonical bijection from $\operatorname{Irr}_{\pi'}(G)$ to $\operatorname{Irr}_{\pi'}(N)$. Also, we establish some additional information about the degrees of corresponding characters.

Theorem 8.12. *Suppose that $|G|$ is odd, and let $N = \mathbf{N}_G(A)$, where A is a Hall π-subgroup of G. Then there is a canonical bijection from $\operatorname{Irr}_{\pi'}(G)$ onto $\operatorname{Irr}_{\pi'}(N)$. Also, if ξ is the image of χ under this map, then ξ is a constituent of χ_N and $\xi(1)$ divides $\chi(1)$. Furthermore, if π consists of just one prime p, then $\chi(1) \equiv \pm\xi(1) \bmod p$.*

We mention that the Isaacs–Navarro extension of the McKay conjecture proposes that for *all* finite groups G and all primes p, there exists a (not necessarily canonical) bijection from $\operatorname{Irr}_{p'}(G)$ onto $\operatorname{Irr}_{p'}(N)$ such that if $\chi \mapsto \xi$, then $\chi(1) \equiv \pm\xi(1) \bmod p$. (Note that this conjecture is really just a statement about the numbers of characters in the sets $\operatorname{Irr}_{p'}(G)$ and $\operatorname{Irr}_{p'}(N)$ that have certain degrees.) By Theorem 8.12, the Isaacs–Navarro conjecture is true for odd-order groups.

We begin with some preliminary results that we will use to prove Theorems 8.11 and 8.12.

Lemma 8.13. *Let A act via automorphisms on a group K, and let $L \lhd K$, where L is A-invariant and $|K : L|$ and $|A|$ are relatively prime. Let $\theta \in \operatorname{Irr}_A(K)$. The following then hold.*

(a) *Some irreducible constituent of θ_L is A-invariant.*

(b) *If $\mathbf{C}_{K/L}(A) = K/L$ or, equivalently, the action of A on K/L is trivial, then every irreducible constituent of θ_L is A-invariant.*

(c) *If $\mathbf{C}_{K/L}(A) = 1$, then exactly one irreducible constituent φ of θ_L is A-invariant. Furthermore, if K/L is abelian in this case, then the map $\theta \mapsto \varphi$ is a bijection from $\operatorname{Irr}_A(K)$ onto $\operatorname{Irr}_A(L)$.*

It follows by Lemma 4.15(b) that the "dual" of (b) is also true: if A acts trivially on K/L and $\varphi \in \operatorname{Irr}_A(L)$, then every irreducible constituent of φ^G is A-invariant. Also, we mention that in (c), it is not really necessary to assume that K/L is abelian in order to conclude that the map $\theta \mapsto \varphi$ is a bijection. The abelian case, however, is all that we shall need. Note that Lemma 8.13(c) and Lemma 7.19 have a nontrivial intersection.

Proof of Lemma 8.13. By Clifford's theorem, the group K/L acts transitively on the set of irreducible constituents of θ_L. Also, A acts on this set because θ is A-invariant. Furthermore, A acts on K/L, and it is routine to check that these three actions are compatible in the sense of Glauberman's lemma (Lemma 1.4). It follows by Glauberman's lemma that the set of A-invariant irreducible constituents of θ_L is nonempty, and this proves (a). Glauberman's lemma also tells us that the set of A-invariant irreducible constituents of θ_L is transitively permuted by $\mathbf{C}_{K/L}(A)$, and since for (b) we are assuming that $\mathbf{C}_{K/L}(A) = K/L$, it follows that every irreducible constituent of θ_L is A-invariant.

For (c), assume that $\mathbf{C}_{K/L}(A)$ is trivial. Since $\mathbf{C}_{K/L}(A)$ acts transitively on the set of A-invariant irreducible constituents of θ_L, it follows that θ_L has a unique A-invariant irreducible constituent φ, and this proves the first part of (c). In particular, $\theta \mapsto \varphi$ is a well-defined map from $\mathrm{Irr}_A(K)$ to $\mathrm{Irr}_A(L)$, so to complete the proof, we assume that K/L is abelian, and we show that this map is a bijection.

Let $\varphi \in \mathrm{Irr}_A(L)$. By Lemma 7.1(a), the group Λ of linear characters of K/L acts transitively on the set of irreducible constituents of φ^K, and since φ is A-invariant, we see that A also acts on this set. Furthermore, A acts on the group Λ, and in this case too, it is routine to check that the Glauberman-lemma compatibility condition is satisfied. We conclude that some irreducible constituent of φ^K is A-invariant, and this proves that our map is surjective.

Finally, for injectivity, we must prove that φ^K has a unique A-invariant irreducible constituent, so it suffices to show that $\mathbf{C}_\Lambda(A)$ is trivial. To see this, suppose that λ is a nonprincipal A-invariant linear character of K/L. The linearity of λ guarantees that $[K, A] \subseteq \ker(\lambda) < K$, and since also $L \subseteq \ker(\lambda)$, we see that $[K/L, A]$ is a proper subgroup of K/L. It follows by Fitting's theorem (Theorem 4.34 of [**FGT**]) that $\mathbf{C}_{K/L}(A)$ is nontrivial, and this contradiction completes the proof. ■

For our next several results, we consider 4-tuples (G, K, L, A), where A is a Hall subgroup of G and K/L is an abelian normal section of G such that $|K : L|$ and $|A|$ are relatively prime. In addition, we assume that $KA \triangleleft G$ and that $\mathbf{C}_{K/L}(A)$ is trivial. In this situation, we will say that (G, K, L, A) is a **standard 4-tuple**.

Next, we present some elementary properties of standard 4-tuples.

Lemma 8.14. *Let (G, K, L, A) be a standard 4-tuple, and let $H = \mathbf{N}_G(LA)$. Then the following hold.*

(a) *H is a complement for K relative to L in G.*

(b) *H is the only complement for K relative to L in G that contains A.*

(c) *All complements for K relative to L in G are conjugate in G to H.*

Proof. It suffices to observe that the hypotheses of Lemma 7.6 are satisfied with KA in place of M. Certainly, $K \subseteq KA \triangleleft G$, so we must check that $|K : L|$ and $|KA : K|$ are relatively prime and that $\mathbf{C}_{K/L}(KA)$ is trivial. The first of these conditions is satisfied because $|KA : K|$ divides $|A|$, and by assumption, $|A|$ is relatively prime to $|K : L|$. Also, since K/L is abelian, K acts trivially on K/L, and thus $\mathbf{C}_{K/L}(KA) = \mathbf{C}_{K/L}(A)$, and this is trivial by assumption. ∎

Lemma 8.15. *Let (G, K, L, A) be a standard 4-tuple, and let $H = \mathbf{N}_G(LA)$, so H is a complement for K relative to L in G.*

(a) *There is a bijection from $\mathrm{Irr}_A(K)$ onto $\mathrm{Irr}_A(L)$ such that $\theta \in \mathrm{Irr}_A(K)$ maps to $\varphi \in \mathrm{Irr}_A(L)$ precisely when θ lies over φ.*

(b) *Conjugation by elements of H permutes the sets $\mathrm{Irr}_A(K)$ and $\mathrm{Irr}_A(L)$, and if $\theta \in \mathrm{Irr}_A(K)$ lies over $\varphi \in \mathrm{Irr}_A(L)$, then θ and φ have the same stabilizer in H.*

Proof. By hypothesis, $|A|$ and $|K : L|$ are relatively prime and $\mathbf{C}_{K/L}(A)$ is trivial, so it follows by Lemma 8.13(c) that each character $\theta \in \mathrm{Irr}_A(K)$ lies over a unique character $\varphi \in \mathrm{Irr}_A(L)$. This defines a map $\theta \mapsto \varphi$, and since K/L is abelian, Lemma 8.13(c), guarantees that this map is a bijection from $\mathrm{Irr}_A(K)$ onto $\mathrm{Irr}_A(L)$. This proves (a).

Since $LA \triangleleft H$, conjugation by elements of H permutes the sets of LA-invariant characters of K and of L. The action of L on $\mathrm{Irr}(K)$ and $\mathrm{Irr}(L)$ is trivial, however, so the LA-invariant characters of K and L are exactly the A-invariant characters of these subgroups. It follows that conjugation by elements of H permutes each of the sets $\mathrm{Irr}_A(K)$ and $\mathrm{Irr}_A(L)$.

Now if $\theta \in \mathrm{Irr}_A(K)$ lies over $\varphi \in \mathrm{Irr}_A(L)$, then θ and φ uniquely determine each other by (a). Also, the action of H is compatible with the "lying over" relation, and it follows that θ and φ have the same stabilizers in H. ∎

Corollary 8.16. *Let A be a Hall π-subgroup of G. Let $K \triangleleft G$, and assume that $KA \triangleleft G$. Also, suppose that $K \subseteq N \triangleleft G$. Finally, let $\chi \in \mathrm{Irr}_A(N)$, where $\chi(1)$ is a π'-number. Then every irreducible constituent of χ_K is A-invariant.*

Proof. Write $B = N \cap KA$. Since KA/K is a normal Hall π-subgroup of G/K, we see that B/K is a normal Hall π-subgroup of N/K. Also, KA/B centralizes N/B, so A acts trivially on the π'-group N/B. By hypothesis, $\chi \in \mathrm{Irr}(N)$ is A-invariant, and hence by Lemma 8.13(b), every irreducible constituent of χ_B is A-invariant.

Since the degree of χ is a π'-number and $B \lhd N$, the degrees of the irreducible constituents of χ_B are also π'-numbers, and since B/K is a π-group, these constituents restrict irreducibly to K. We have seen that the irreducible constituents of χ_B are A-invariant, and since they restrict irreducibly to K, it follows that every irreducible constituent of χ_K is A-invariant. ∎

We need one more preliminary result.

Lemma 8.17. *Let (G, K, L, A) be a standard 4-tuple, and suppose that the Hall subgroup A of G is actually a Sylow p-subgroup. Suppose that $L \subseteq V \subseteq K$, where V is A-invariant. Then $|K : V| \equiv 1 \equiv |V : L| \bmod p$.*

Proof. Let $H = \mathbf{N}_G(LA)$. Then H is a complement for K relative to L in G by Lemma 8.14(a), and thus $KA \cap H = LA$. It follows that $LA = \mathbf{N}_{KA}(LA)$, so LA/L is a self-normalizing Sylow p-subgroup of KA/L, so by Sylow's theorem, the index of LA/L in every subgroup of KA/L that contains it is congruent to 1 modulo p. We deduce that $|V : L| = |VA : LA| \equiv 1 \bmod p$, and similarly, $|K : L| \equiv 1 \bmod p$. Since $|K : L| = |K : V||V : L|$, it follows that $|K : V| \equiv 1 \bmod p$, as required. ∎

Next, we introduce some temporary notation. If A acts on a group X and π is a set of primes, we write $\mathrm{Irr}_{\pi' A}(X)$ to denote the set of those A-invariant irreducible characters χ of X such that $\chi(1)$ is a π'-number. In other words, $\mathrm{Irr}_{\pi' A}(X) = \mathrm{Irr}_A(X) \cap \mathrm{Irr}_{\pi'}(X)$.

Now suppose that $Y \subseteq X \subseteq G$, and assume that the π-subgroup A of G normalizes both X and Y. It may happen that for some character $\alpha \in \mathrm{Irr}_{\pi' A}(X)$, there is a unique character $\beta \in \mathrm{Irr}_{\pi' A}(Y)$ such that the multiplicity of β as a constituent of α_Y is odd. In this situation, we write $(\alpha)f_{XY} = \beta$, and we say that "f_{XY} is defined for α". If f_{XY} is defined for all members of $\mathrm{Irr}_{\pi' A}(X)$, we simply say that "f_{XY} is defined". Using this notation, we can now state a result that contains most of what is needed to prove both Theorems 8.11 and 8.12.

Theorem 8.18. *Let (G, K, L, A) be a standard 4-tuple, where A is a Hall π-subgroup of G. Also, suppose $K \subseteq N \lhd G$, where $|N : L|$ is odd. Write $H = \mathbf{N}_G(LA)$ and $M = N \cap H$. Then*

(a) *f_{NM} is defined and is a bijection from $\mathrm{Irr}_{\pi' A}(N)$ onto $\mathrm{Irr}_{\pi' A}(M)$.*

(b) *Let* $\chi \in \mathrm{Irr}_{\pi' A}(N)$, *and suppose that* $(\chi)f_{NM} = \xi$. *Then* $\xi(1)$ *divides* $\chi(1)$. *Also, if* π *consists of a single prime* p, *then* $\chi(1) \equiv \pm\xi(1) \bmod p$.

Proof. We proceed by induction on $|G|$. By hypothesis, $N \triangleleft G$, so $M \triangleleft H$, and thus M is A-invariant. Since H is a complement for K relative to L in G by Lemma 8.14(a), it follows that M is a complement for K relative to L in N, and thus $|N : M| = |K : L|$. Also, by Lemma 8.15(b), we see that H permutes the sets $\mathrm{Irr}_A(K)$ and $\mathrm{Irr}_A(L)$, and it follows that M permutes each of these sets.

If $\xi \in \mathrm{Irr}_{\pi' A}(M)$, then by Corollary 8.16 (applied with L in place of K) the irreducible constituents of ξ_L lie in $\mathrm{Irr}_A(L)$, and by Clifford's theorem these constituents form an M-orbit. We show next that if $\chi \in \mathrm{Irr}_{\pi' A}(N)$, then the set of members of $\mathrm{Irr}_A(L)$ that lie under χ is nonempty, and this set is also an M-orbit.

Given $\chi \in \mathrm{Irr}_{\pi' A}(N)$, let $\mathcal{X}$ be the set of irreducible constituents of χ_K, and observe that $\mathcal{X} \subseteq \mathrm{Irr}_A(K)$ by Corollary 8.16. By Clifford's theorem, $\mathcal{X}$ is an N-orbit, and since $N = KM$, we see that $\mathcal{X}$ is also an M-orbit. By Lemma 8.15(a), "lying over" defines a bijection from $\mathrm{Irr}_A(K)$ onto $\mathrm{Irr}_A(L)$, and we let $\mathcal{Y}$ be the image of $\mathcal{X}$ under this map. Since the actions of M on $\mathrm{Irr}_A(K)$ and $\mathrm{Irr}_A(L)$ are compatible with the lying-over bijection, it follows that $\mathcal{Y}$ is also an M-orbit.

Every member of $\mathcal{Y}$ lies under χ, and in fact, if φ is an arbitrary member of $\mathrm{Irr}_A(L)$ that lies under χ, then φ must lie under some member of $\mathcal{X}$, and thus $\varphi \in \mathcal{Y}$. It follows that $\mathcal{Y}$ is the full set of members of $\mathrm{Irr}_A(L)$ that lie under χ, and we conclude that the members of $\mathrm{Irr}_A(L)$ that lie under χ form an M-orbit, as claimed.

If $\chi \in \mathrm{Irr}_{\pi' A}(N)$ and χ lies over $\xi \in \mathrm{Irr}_{\pi' A}(M)$, we deduce that χ and ξ lie over exactly the same set of members of $\mathrm{Irr}_A(L)$, and we see that this set of characters forms an M-orbit. (But note that we have not yet shown that each character $\chi \in \mathrm{Irr}_{\pi' A}(N)$ actually does lie over some character $\xi \in \mathrm{Irr}_{\pi' A}(M)$.)

Let $\mathcal{R}$ be a set of representatives for the M-orbits on $\mathrm{Irr}_A(L)$, and observe that each member of $\mathrm{Irr}_{\pi' A}(N)$ and of $\mathrm{Irr}_{\pi' A}(M)$ lies over a unique member of $\mathcal{R}$. Each of the sets $\mathrm{Irr}_{\pi' A}(N)$ and $\mathrm{Irr}_{\pi' A}(M)$, therefore, is partitioned by the subsets consisting of the characters in these sets that lie over the various members of $\mathcal{R}$.

Given an arbitrary member $\varphi \in \mathcal{R}$, we will show for each character $\chi \in \mathrm{Irr}_{\pi' A}(M)$, where χ lies over φ, that $(\chi)f_{NM} = \xi$ is defined, and that ξ also lies over φ. Also, we shall see that this defines a bijection from

$\mathrm{Irr}_{\pi' A}(N|\varphi)$ onto $\mathrm{Irr}_{\pi' A}(M|\varphi)$. It will follow that f_{NM} is defined and is a bijection from $\mathrm{Irr}_{\pi' A}(N)$ onto $\mathrm{Irr}_{\pi' A}(M)$, as required for (a).

Fix $\varphi \in \mathcal{R}$, and let T be the stabilizer of φ in G. Observe that $LA \subseteq T$, and write $K_0 = K \cap T$, so K_0 is A-invariant and $L \subseteq K_0 \subseteq K$. We argue next that (T, K_0, L, A) is a standard 4-tuple.

Since $KA \lhd G$, we see that

$$K_0A = (K \cap T)A = KA \cap T \lhd T\,.$$

Now K_0/L is an abelian normal section of T, and $|K_0 : L|$ is a π'-number. Also, A is a Hall π-subgroup of T and $\mathbf{C}_{K_0/L}(A)$ is trivial. This shows that (T, K_0, L, A) is a standard 4-tuple, as claimed, and furthermore, writing $H_0 = H \cap T$, we see that $H_0 = \mathbf{N}_T(LA)$.

Let $N_0 = N \cap T$ and $M_0 = M \cap T$, and observe that $K_0 \subseteq N_0 \lhd T$ and

$$M_0 = M \cap T = N \cap H \cap T = N_0 \cap H_0\,.$$

Also, $|N_0 : L|$ is odd since by hypothesis $|N : L|$ is odd. The 4-tuple (T, K_0, L, A) thus satisfies the hypotheses of the theorem, with N_0 and M_0 in place of N and M.

The Clifford correspondence guarantees that induction defines bijections from $\mathrm{Irr}(N_0|\varphi)$ to $\mathrm{Irr}(N|\varphi)$ and from $\mathrm{Irr}(M_0|\varphi)$ to $\mathrm{Irr}(M|\varphi)$. Under these maps, A-invariant characters correspond to A-invariant characters, and we argue that in fact these Clifford bijections carry $\mathrm{Irr}_{\pi' A}(N_0)$ and $\mathrm{Irr}_{\pi' A}(M_0)$ onto $\mathrm{Irr}_{\pi' A}(N)$ and $\mathrm{Irr}_{\pi' A}(M)$, respectively. For this purpose, it suffices to show that $|N : N_0|$ and $|M : M_0|$ are π'-numbers.

First, $|N : N_0| = |N : N \cap T| = |NT : T|$, and this divides $|G : T|$, which is a π'-number because $A \subseteq T$. Also,

$$|N : M||M : M_0| = |N : M_0| = |N : N_0||N_0 : M_0| = |N : N_0||K_0 : L|\,,$$

and since the quantity on the right is a π'-number, it follows that $|M : M_0|$ is also a π'-number, as wanted. In addition, we obtain the following equality, which will be needed later:

$$\frac{|N : N_0|}{|M : M_0|} = \frac{|N : M|}{|K_0 : L|} = \frac{|K : L|}{|K_0 : L|} = |K : K_0|\,.$$

Assume now that $T < G$, so by the inductive hypothesis, $f_{N_0M_0}$ is a bijection from $\mathrm{Irr}_{\pi'A}(N_0)$ onto $\mathrm{Irr}_{\pi'A}(M_0)$. If $\tau \in \mathrm{Irr}_{\pi'A}(N_0)$ and $\sigma = (\tau)f_{N_0M_0}$, then $[\tau_{M_0}, \sigma]$ is odd, and in particular, this integer is nonzero, and thus τ lies over σ. By our earlier reasoning, it follows that τ and σ lie over exactly the same sets of A-invariant characters of L, and in particular τ lies over φ if and only if σ lies over φ. It follows that $f_{N_0M_0}$ defines a bijection from $\mathrm{Irr}_{\pi'A}(N_0|\varphi)$ onto $\mathrm{Irr}_{\pi'A}(M_0|\varphi)$.

Now let $\chi \in \mathrm{Irr}_{\pi'A}(N|\varphi)$, and let $\tau \in \mathrm{Irr}_{\pi'A}(N_0|\varphi)$ be the Clifford correspondent of χ. Also, let $\sigma \in \mathrm{Irr}_{\pi'A}(M_0|\varphi)$, and write $\xi = \sigma^M$, so $\xi \in \mathrm{Irr}_{\pi'A}(M|\varphi)$ and σ is the Clifford correspondent of ξ. We will show that if $(\tau)f_{N_0M_0} = \sigma$, then $(\chi)f_{MN} = \xi$, and it will follow that f_{MN} defines a function from $\mathrm{Irr}_{\pi'A}(N|\varphi)$ into $\mathrm{Irr}_{\pi'A}(M|\varphi)$.

We argue first that $[\chi_M, \xi] = [\tau_{M_0}, \sigma]$. To see this, we begin by showing that τ is the unique member of $\mathrm{Irr}(N_0)$ that simultaneously lies under χ and over σ. If $\alpha \in \mathrm{Irr}(N_0)$ has this property, then α lies over φ because σ lies over φ, and hence by the Clifford correspondence, α^N is irreducible. We are assuming that α lies under χ, however, so $\alpha^N = \chi$, and thus α is the Clifford correspondent of χ with respect to φ. We conclude that $\alpha = \tau$, as wanted.

Now

$$[\chi_M, \xi] = [\chi_M, \sigma^M] = [\chi_{M_0}, \sigma] = [\chi_{N_0}, \sigma^{N_0}] = [\chi_{N_0}, \tau][\tau, \sigma^{N_0}],$$

where the final equality holds because τ is the only irreducible character of N_0 that can simultaneously be a constituent of χ_{N_0} and σ^{N_0}. Since $\chi = \tau^N$, we have $[\chi_{N_0}, \tau] = 1$, and it follows that

$$[\chi_M, \xi] = [\tau, \sigma^{N_0}] = [\tau_{M_0}, \sigma],$$

as claimed.

Assuming now that $(\tau)f_{N_0M_0} = \sigma$, recall that we want to show that $(\chi)f_{NM} = \xi$. We have $[\chi_M, \xi] = [\tau_{M_0}, \sigma]$, and this is an odd number by the definition of $f_{N_0M_0}$. We must also show that if $\eta \in \mathrm{Irr}_{\pi'A}(M)$ and η is different from ξ, then η occurs with even multiplicity as a constituent of χ_M. We can certainly assume that $[\chi_M, \eta] \neq 0$, so χ lies over η, and we have seen that in this situation, χ and η lie over exactly the same members of $\mathrm{Irr}_A(L)$. Since χ lies over φ, it follows that η also lies over φ. Now if $\gamma \in \mathrm{Irr}_{\pi'A}(M_0|\varphi)$ is the Clifford correspondent of η, then $\gamma \neq \sigma$ since $\eta \neq \xi$. Also $[\chi_M, \eta] = [\tau_{M_0}, \gamma]$, and since $\gamma \neq \sigma$, we see that this quantity is even. Then $[\chi_M, \eta]$ is even, and we conclude $(\chi)f_{NM} = \xi$, as claimed.

Continuing to assume that $T < G$, we now know that f_{NM} defines a function mapping $\mathrm{Irr}_{\pi'A}(N|\varphi)$ into $\mathrm{Irr}_{\pi'A}(M|\varphi)$. In fact, this function is a bijection, as required, because it is the composition of three bijections: first, the Clifford correspondence from $\mathrm{Irr}_{\pi'A}(N|\varphi)$ onto $\mathrm{Irr}_{\pi'A}(N_0|\varphi)$; then the bijection $f_{N_0M_0}$ from $\mathrm{Irr}_{\pi' A}(N_0|\varphi)$ onto $\mathrm{Irr}_{\pi' A}(M_0|\varphi)$; and finally, the Clifford correspondence from $\mathrm{Irr}_{\pi' A}(M_0|\varphi)$ onto $\mathrm{Irr}_{\pi' A}(M|\varphi)$.

Next, we prove (b) in this case. If χ, ξ, τ, and σ are as before, we have

$$\chi(1) = |N : N_0|\tau(1) \qquad \text{and} \qquad \xi(1) = |M : M_0|\sigma(1)\,,$$

and so

$$\frac{\chi(1)}{\xi(1)} = \frac{\tau(1)}{\sigma(1)}\frac{|N : N_0|}{|M : M_0|}\,.$$

By the inductive hypothesis, $\sigma(1)$ divides $\tau(1)$, and if $\pi = \{p\}$, then $\tau(1) \equiv \pm\sigma(1) \bmod p$. It suffices, therefore, to show that $|N : N_0|/|M : M_0|$ is an integer, and that this integer is congruent to 1 mod p if $\pi = \{p\}$.

We saw previously that

$$\frac{|N : N_0|}{|M : M_0|} = |K : K_0|\,,$$

and thus $|N : N_0|/|M : M_0|$ is an integer. Also, if $\pi = \{p\}$, then since K_0 is A-invariant and $L \subseteq K_0 \subseteq K$, Lemma 8.17 guarantees that $|K : K_0| \equiv 1$ mod p, as required.

We can now assume that $T = G$, and thus φ is invariant in G. In particular, φ is invariant in K, so by Theorem 7.3, there is a subgroup V with $L \subseteq V \subseteq K$, and such that every member of $\mathrm{Irr}(K|\varphi)$ is fully ramified with respect to K/V, and every member of $\mathrm{Irr}(V|\varphi)$ is an extension of φ. Theorem 7.3 also guarantees that V is uniquely determined by φ, and since φ is G-invariant, we deduce that $V \triangleleft G$.

Let θ be the unique member of $\mathrm{Irr}_A(K)$ that lies over φ. (We are appealing to Lemma 8.13(c) here.) Then θ is the unique KA-invariant member of $\mathrm{Irr}(K)$ lying over φ, and since K, KA, and φ are all G-invariant, it follows that θ is invariant in G. Also, θ is fully ramified with respect to K/V, and so θ_V has a unique irreducible constituent μ, and we see that μ is invariant in G and $\mu_L = \varphi$.

Let $W = MV$ and $J = HV$, and observe that W is A-invariant. It is easy to check that $J \cap N = W$ and $J \cap K = V$ (see Figure 8.2). In particular, J is a complement for K relative to V in G, and in fact, up to G-conjugacy, J is the unique complement for K relative to V in G. To see this, observe that (G, K, V, A) is a standard 4-tuple and that J contains A. The uniqueness (up to conjugacy) then follows by Lemma 8.14(c).

By hypothesis, $|N : L|$ is odd, so $|K : V|$ is odd. Since θ and μ are fully ramified with respect to K/V and they are invariant in G, we can appeal to Theorem 8.1. According to that result, there is a complement U for K relative to V in G having certain special properties, and since U is conjugate to J in G, we can assume without loss that $U = J$.

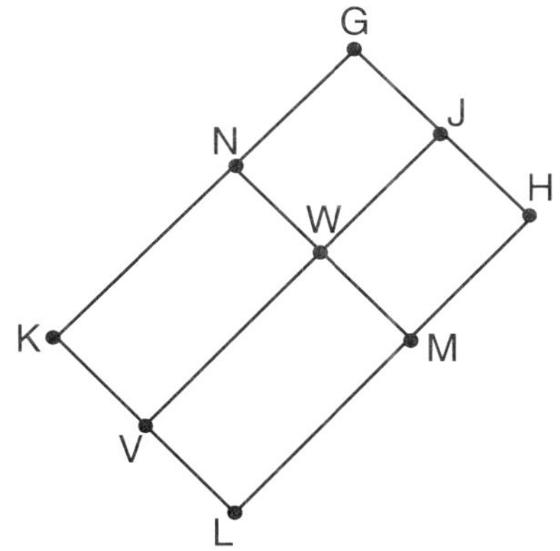

Figure 8.2

Since $|N : V|$ is odd, it follows by Theorem 8.1 that there is a bijection from $\text{Irr}(N|\mu)$ onto $\text{Irr}(W|\mu)$ such that if $\chi \mapsto \zeta$, then ζ is the unique odd-multiplicity irreducible constituent of χ_W, and thus we can write $\chi_W = \zeta + 2\Delta$, where either Δ is zero or it is a character of W.

Since χ and ζ uniquely determine each other, we see that χ is A-invariant if and only if ζ is A-invariant. Theorem 8.1 also guarantees that in this situation, $\chi(1) = e\zeta(1)$, where $e^2 = |K : V|$, and in particular, e is a π'-number. It follows that $\chi \in \text{Irr}_{\pi' A}(N)$ if and only if $\zeta \in \text{Irr}_{\pi' A}(W)$.

Now, M is a complement for V relative to L in W, and since $\mu_L = \varphi$ and μ and φ are invariant in W and M, respectively, it follows by Lemma 2.11(a) that restriction defines a bijection from $\text{Irr}(W|\mu)$ to $\text{Irr}(M|\varphi)$. This yields a bijection from $\text{Irr}_{\pi' A}(W|\mu)$ onto $\text{Irr}_{\pi' A}(M|\varphi)$ such that if $\zeta \mapsto \xi$, then $\zeta_M = \xi$.

The composition of the bijections

$$\text{Irr}_{\pi' A}(N|\mu) \to \text{Irr}_{\pi' A}(W|\mu) \to \text{Irr}_{\pi' A}(M|\varphi)$$

is a bijection from $\text{Irr}_{\pi' A}(N|\mu)$ onto $\text{Irr}_{\pi' A}(M|\varphi)$ such that if $\chi \mapsto \zeta \mapsto \xi$, then

$$\chi_M = \zeta_M + 2\Delta_M = \xi + 2\Delta_M\,,$$

and thus ξ is the unique odd-multiplicity irreducible constituent of χ_M. In particular, we have $(\chi)f_{MN} = \xi$.

To complete the proof of (a), it suffices to show that

$$\text{Irr}_{\pi' A}(N|\mu) = \text{Irr}_{\pi' A}(N|\varphi)\,.$$

Since μ lies over φ, it is clear that every irreducible character of N that lies over μ also lies over φ. We must show, therefore, that if $\chi \in \text{Irr}_{\pi' A}(N)$ and χ lies over φ, then χ lies over μ. To see this, observe that by Corollary 8.16, the irreducible constituents of χ_K are A-invariant. Now χ lies over φ, and since θ is the unique A-invariant irreducible character of K that lies over φ, it follows that χ lies over θ. Since θ lies over μ, we conclude that χ lies over μ, as required.

To complete the proof of (b) now, we recall that if $\chi \mapsto \zeta \mapsto \xi$, then $\chi(1) = e\zeta(1) = e\xi(1)$, where $e^2 = |K : V|$, and thus $\xi(1)$ divides $\chi(1)$. Also, if $\pi = \{p\}$, then $e^2 = |K : V| \equiv 1 \bmod p$ by Lemma 8.17, and hence $e \equiv \pm 1 \bmod p$. Then $\chi(1) \equiv \pm\xi(1)$, as required. ∎

We can now prove the following somewhat enhanced version of Theorem 8.11.

Theorem 8.19. *Let A act on G via automorphisms, and write $C = \mathbf{C}_G(A)$. Assume that $|A|$ and $|G|$ are relatively prime and that $|G|$ is odd. Then there exists a canonical correspondence from* $\mathrm{Irr}_A(G)$ *onto* $\mathrm{Irr}(C)$. *Also, if ξ is the image of χ under this map, then ξ is a constituent of χ_C and $\xi(1)$ divides $\chi(1)$. Furthermore, if A is a p-group for some prime p, then $\chi(1) \equiv \pm\xi(1) \bmod p$.*

Proof. We can assume that $C < G$. Let $K = [G, A]$, and observe that A acts trivially on G/K and that K is the unique smallest A-invariant normal subgroup of G with this property. Also, $KC = G$ by Lemma 4.28 of [**FGT**]. Let $L = K'$, and write $M = LC$.

Clearly, $C \subseteq M$ and M is A-invariant, and we will show that $M < G$. In addition, we will see that if $\chi \in \mathrm{Irr}_A(G)$, then there is a unique character $\xi \in \mathrm{Irr}_A(M)$ such that ξ has odd multiplicity as a constituent of χ_M. Then $\chi \mapsto \xi$ is a well-defined map, and we will see that it is a bijection from $\mathrm{Irr}_A(G)$ onto $\mathrm{Irr}_A(M)$. We will also see that $\xi(1)$ divides $\chi(1)$, and that if A is a p-group, then $\chi(1) \equiv \pm\xi(1) \bmod p$.

If $M > C$, we can repeat this construction with M in place of G. Continuing like this, we obtain a canonically defined chain of subgroups

$$G = G_0 > G_1 > \cdots > G_r = C\,,$$

where $G_{i+1} = [G_i, A]'C$ for $0 \leq i < r$. Also, we have canonical bijections from $\mathrm{Irr}_A(G_i)$ onto $\mathrm{Irr}_A(G_{i+1})$. The composition of these maps is a canonical bijection from $\mathrm{Irr}_A(G)$ onto $\mathrm{Irr}(C)$ having the required properties.

First, to see that $M < G$, note that $K > 1$ because A does not act trivially on G. Also, by assumption, $|G|$ is odd, so K is solvable, and thus $L = K' < K$. Then A does not act trivially on G/L, and it follows that $M = LC < G$.

Let Γ be the semidirect product of G by A, and view G and A as subgroups of Γ. Then $G \lhd \Gamma$ and A is a Hall π-subgroup of Γ, where π is the set of prime divisors of $|A|$.

We argue that (Γ, K, L, A) is a standard 4-tuple. First, K/L is an abelian normal section of Γ because K and L are A-invariant. Since A and C centralize each other, we see that C normalizes KA, and thus KA is a normal subgroup of $CKA = GA = \Gamma$, as required. Clearly, $|A|$ and $|K : L|$ are relatively prime, so it remains to show that $\mathbf{C}_{K/L}(A)$ is trivial.

Since $K = [G, A]$ and $|A|$ and $|K|$ are relatively prime, it follows by Lemma 4.29 of [**FGT**] that $[K, A] = K$, and thus $[(K/L), A] = K/L$. Now K/L is abelian and has order relatively prime to $|A|$, so we can apply Fitting's lemma to deduce that $K/L = [(K/L), A] \times \mathbf{C}_{K/L}(A)$, and it follows that $\mathbf{C}_{K/L}(A)$ is trivial. This completes the proof that (Γ, K, L, A) is standard.

We argue next that MA is a complement for K relative to L in Γ. Since $KC = G$ and $C \subseteq M$, we have $KM = G$, and thus $K(MA) = GA = \Gamma$. To show that $K \cap MA = L$, observe that $G \cap MA = M(G \cap A) = M$, and thus

$$K \cap MA = K \cap G \cap MA = K \cap M = K \cap LC = L(K \cap C) = L\,,$$

where the final equality holds because $\mathbf{C}_{K/L}(A)$ is trivial, so $K \cap C \subseteq L$.

Since MA is a complement for K in Γ relative to L and MA contains A, it follows by Lemma 8.14(b) that $MA = \mathbf{N}_\Gamma(LA)$, and so for consistency with our earlier notation, we write $H = MA$.

Now $G \cap H = G \cap MA = M$ and $|G : L|$ is odd, so we can apply Theorem 8.19 to Γ, with G in the role of N. Since all irreducible characters of G have degrees that are π'-numbers, we see that $\mathrm{Irr}_{\pi' A}(G) = \mathrm{Irr}_A(G)$, and similarly, $\mathrm{Irr}_{\pi' A}(M) = \mathrm{Irr}_A(M)$. Theorem 8.18 thus guarantees that f_{GM} is defined, and it is the desired bijection from $\mathrm{Irr}_A(G)$ onto $\mathrm{Irr}_A(M)$. ■

Proof of Theorem 8.12. Recall that A is a Hall π-subgroup of G and $N = \mathbf{N}_G(A)$. We can assume that $N < G$, and we let K be the unique smallest normal subgroup of G such that G/K has a normal Hall π-subgroup. (In other words, $K = \mathbf{O}^{\pi'\pi}(G)$.) Then $KA/K \triangleleft G/K$, and thus $KA \triangleleft G$. Since $|G|$ is odd, G is solvable, and thus by the Frattini argument, $G = (KA)N = KN$. Let $L = K'$, and write $M = LN$.

Clearly, $N \subseteq M$, and we will show that $M < G$. In addition, we will see that if $\chi \in \mathrm{Irr}_{\pi'}(G)$, then there is a unique character $\xi \in \mathrm{Irr}_{\pi'}(M)$ such that ξ has odd multiplicity as a constituent of χ_M. Then $\chi \mapsto \xi$ is a well-defined map, and we will see that it is a bijection from $\mathrm{Irr}_{\pi'}(G)$ onto $\mathrm{Irr}_{\pi'}(M)$. We will also see that $\xi(1)$ divides $\chi(1)$ and that if A is a p-group, then $\chi(1) \equiv \pm\xi(1) \bmod p$.

If $M > N$, we can repeat this construction with M in place of G. Continuing like this, we obtain a canonically defined chain of subgroups

$$G = G_0 > G_1 > \cdots > G_r = N\,,$$

where $G_{i+1} = \mathbf{O}^{\pi'\pi}(G_i)'N$ for $0 \leq i < r$. Also, we have canonical bijections from $\mathrm{Irr}_{\pi'}(G_i)$ onto $\mathrm{Irr}_{\pi'}(G_{i+1})$. The composition of these maps is a canonical bijection from $\mathrm{Irr}_{\pi'}(G)$ onto $\mathrm{Irr}_{\pi'}(N)$ having the required properties.

First, to see that $M < G$, note that $K > 1$ because G does not have a normal Hall π-subgroup. Also, $L = K' < K$ since G is solvable. Then G/L does not have a normal Hall π-subgroup, and it follows that $M = LN < G$.

We argue next that (G, K, L, A) is a standard 4-tuple. Certainly, K/L is an abelian normal section of G, and we have already observed that $KA \triangleleft G$, as required. Also, $\mathbf{O}^{\pi}(K) = K$, and since K/L is abelian, we see that K/L must be a π'-group.

Both K and N normalize $L[K, A]$, and thus $L[K, A] \triangleleft KN = G$. The image of A in $G/L[K, A]$ is centralized by the image of K in this factor group, and since the image of A is certainly normalized by N, it follows that the image of A in $G/L[K, A]$ is a normal Hall π-subgroup. By the definition of K, we have $L[K, A] = K$, and hence $[(K/L), A] = K/L$. Since K/L is abelian and has order relatively prime to $|A|$, it follows by Fitting's lemma that $\mathbf{C}_{K/L}(A) = 1$, and this completes the proof that (G, K, L, A) is standard.

Next, we argue that M is a complement for K relative to L in G. We have $G = KN$ and $N \subseteq M$, so $G = KM$, and we must show that $K \cap M = L$. Now $M = LN$ normalizes LA, so writing $H = \mathbf{N}_G(LA)$, we have $M \subseteq H$. It follows by Lemma 8.14(a) that H is a complement for K relative to L in G, and thus in particular $K \cap H = L$. It follows that $K \cap M = L$, and thus M is a complement for K relative to L in G, as claimed. Also, it follows by Lemma 8.14(b) that $M = H$.

Since $|G|$ is odd, we can apply Theorem 8.18 with $N = G$ and $M = H$. All members of $\mathrm{Irr}(G)$ and $\mathrm{Irr}(M)$ are A-invariant because $A \subseteq M \subseteq G$, and thus $\mathrm{Irr}_{\pi'A}(G) = \mathrm{Irr}_{\pi'}(G)$ and $\mathrm{Irr}_{\pi'A}(M) = \mathrm{Irr}_{\pi'}(M)$. Theorem 8.18 thus guarantees that f_{GM} is the desired bijection from $\mathrm{Irr}_{\pi'}(G)$ onto $\mathrm{Irr}_{\pi'}(M)$, and this completes the proof. ■

8D

We now discuss some further applications of Theorem 8.1 to groups of odd order. To motivate our first result, suppose that $\chi \in \mathrm{Irr}(G)$ and $H \subseteq G$. If $H \triangleleft G$, then by Clifford's theorem, $\xi(1)$ divides $\chi(1)$ for every irreducible constituent ξ of χ_H. If we drop the assumption that H is normal in G, however, then, except for the two obvious inequalities

$$\xi(1) \leq \chi(1) \qquad \text{and} \qquad \chi(1) \leq |G : H|\xi(1)\,,$$

there seems to be nothing that can be said in general relating $\chi(1)$ to the degrees of the irreducible constituents of χ_H.

Again assume that $H \triangleleft G$, and let $\xi \in \mathrm{Irr}(H)$. If χ is an irreducible constituent of ξ^G, then $\chi(1)/\xi(1)$ divides $|G:H|$, and thus $\chi(1)$ divides $|G:H|\xi(1) = \xi^G(1)$. In the general case, however, if H is not assumed to be normal in G, there is nothing that can be said relating the degrees of χ and ξ except for the trivialities

$$\chi(1) \le \xi^G(1) \qquad \text{and} \qquad \chi(1) \ge \xi(1)\,.$$

For groups of odd order, however, we have the following.

Theorem 8.20. *Let $H \subseteq G$, where $|G|$ is odd.*

(a) *If $\chi \in \mathrm{Irr}(G)$, then there exists an irreducible constituent ξ of χ_H such that $\xi(1)$ divides $\chi(1)$.*

(b) *If $\xi \in \mathrm{Irr}(H)$, then there exists an irreducible constituent χ of ξ^G such that $\chi(1)$ divides $\xi^G(1)$.*

Of course, since G has odd order in Theorem 8.20, it follows that G is solvable. Solvability by itself, however, without assuming that $|G|$ is odd, is not sufficient to obtain the divisibility assertions of the theorem.

Proof of Theorem 8.20. It should be clear that we can assume that H is a maximal subgroup of G and that H is not normal in G. We proceed by induction on $|G|$.

Let $L = \mathrm{core}_G(H)$, and let K/L be a chief factor of G. By assumption, $|G|$ is odd, so G is solvable, and thus we can apply Lemma 7.8 with $N = G$. By Lemma 7.8(a), we see that $K < G$ and H is a complement for K relative to L in G. Also, by Lemma 7.8(d), all such relative complements are conjugate to H in G.

To prove (a), let $\chi \in \mathrm{Irr}(G)$. Let θ be an irreducible constituent of χ_K, and suppose first θ is not invariant in G. Let T be the stabilizer of θ in G, so $K \subseteq T < G$. Write $S = T \cap H$, and observe that $TH = G$ and $|G:T| = |H:S|$.

Let $\eta \in \mathrm{Irr}(T|\theta)$ be the Clifford correspondent of χ. By the inductive hypothesis with T in place of G, we can apply (a) to deduce that η_S has an irreducible constituent ψ such that $\psi(1)$ divides $\eta(1)$. Since $H < G$, the inductive hypothesis allows us to apply (b) with H in place of G, and thus we can choose an irreducible constituent ξ of ψ^H such that $\xi(1)$ divides $\psi^H(1) = |H:S|\psi(1)$.

Now

$$\chi_H = (\eta^G)_H = (\eta_S)^H\,,$$

and since ψ is a constituent of η_S and ξ is a constituent of ψ^H, it follows that ξ is a constituent of χ_H. Also,

$$\frac{\chi(1)}{\xi(1)} = \frac{|G:T|\eta(1)}{\xi(1)} = \frac{|G:T|\eta(1)}{\psi(1)}\frac{\psi(1)}{\xi(1)} = \frac{\eta(1)}{\psi(1)}\frac{|H:S|\psi(1)}{\xi(1)}$$
$$= \frac{\eta(1)}{\psi(1)}\frac{\psi^H(1)}{\xi(1)},$$

and since this is a product of two integers, we deduce that $\xi(1)$ divides $\chi(1)$ as required for (a).

Next, we prove (a) in the case where θ is invariant in G. Since K/L is an abelian chief factor of G and $\theta \in \mathrm{Irr}(K)$ is G-invariant, it follows by Corollary 7.4 that there are exactly three alternatives:

(1) $\theta = \varphi^K$ for some character $\varphi \in \mathrm{Irr}(L)$;

(2) θ_L is irreducible; or

(3) θ is fully ramified with respect to K/L.

Also, if (1) holds, then Corollary 7.4 guarantees that the stabilizer of φ in G is a complement for K relative to L in G.

First, assume (1). Since all complements for K in G relative to L are conjugate to H, we can replace φ by an appropriate conjugate in G, and thus we can assume that the stabilizer of φ in G is H.

Let $\xi \in \mathrm{Irr}(H)$ be the Clifford correspondent of χ with respect to φ. Then $\xi^G = \chi$, so ξ is a constituent of χ_H and $\chi(1) = |G:H|\xi(1)$, and this is divisible by $\xi(1)$, as wanted.

Next, assume (2). Then θ_L is irreducible, and we write $\theta_L = \varphi$. Since θ is G-invariant and χ lies over θ, it follows by Lemma 2.11(a) that χ_H is irreducible. In this case, we are forced to take $\xi = \chi_H$, so $\xi(1) = \chi(1)$, and hence $\xi(1)$ divides $\chi(1)$, as needed.

Finally, assume (3). Then θ is fully ramified with respect to K/L, and we let φ be the unique irreducible constituent of θ_L. Since all complements for K relative to L in G are conjugate, we can assume in this case that H is the subgroup U of Theorem 8.1, and thus χ_H has a unique odd-multiplicity irreducible constituent ξ. Also, $\chi(1) = e\xi(1)$, where $e^2 = |K:L|$, and thus $\xi(1)$ divides $\chi(1)$. This completes the proof of (a).

For (b), we are given $\xi \in \mathrm{Irr}(H)$, and we let φ be an irreducible constituent of ξ_L. Suppose first that the stabilizer T of φ in G is a proper subgroup of G, and let $S = T \cap H$, so S is the stabilizer of φ in H.

Let $\psi \in \mathrm{Irr}(S|\varphi)$ be the Clifford correspondent of ξ. By the inductive hypothesis with T in place of G, we can choose an irreducible constituent η of ψ^T such that $\eta(1)$ divides $\psi^T(1) = |T:S|\psi(1)$. Since T is the stabilizer of φ in G and η lies over ψ, which lies over φ, it follows by the Clifford correspondence that η^G is irreducible, and we write $\chi = \eta^G$. Then χ lies over ψ, and since $\psi^H = \xi$, it follows that χ lies over ξ.

We have

$$\frac{\xi^G(1)}{\chi(1)} = \frac{|G:H|\xi(1)}{\chi(1)} = \frac{|G:H||H:S|\psi(1)}{|G:T|\eta(1)} = \frac{|T:S|\psi(1)}{\eta(1)} = \frac{\psi^T(1)}{\eta(1)},$$

and since this is an integer, it follows that $\chi(1)$ divides $\xi^G(1)$, as required for (b).

We can now assume that φ is invariant in G, so we can apply Corollary 7.4, and in this case, we see that there are just two possibilities: either φ extends to K or φ is fully ramified with respect to K/L.

Suppose first that φ extends to K. Since K/L is abelian, it follows by Lemma 7.1(a) that all members of $\mathrm{Irr}(K|\varphi)$ are extensions of φ. We will show that in this case, ξ has an extension $\chi \in \mathrm{Irr}(G)$, and thus $\chi(1) = \xi(1)$, and this divides $\xi^G(1)$, as wanted.

To prove that ξ extends to G, recall that $K < G$, and let M/K be a chief factor of G. It follows by Lemma 7.8(c) (with $N = G$), that $|M : K|$ and $|K : L|$ are relatively prime and that $\mathbf{C}_{K/L}(M)$ is trivial. Since φ is M-invariant, Lemma 8.13(c) (or Corollary 7.20) guarantees that there is a unique M-invariant character θ in the set $\mathrm{Irr}(K|\varphi)$, and because of the uniqueness, θ is G-invariant.

Recall that θ is an extension of φ, so we deduce by Lemma 2.11(a) that restriction defines a bijection from $\mathrm{Irr}(G|\theta)$ onto $\mathrm{Irr}(H\ \varphi)$, and thus ξ has an extension $\chi \in \mathrm{Irr}(G|\theta)$, as wanted.

Finally, we can assume that φ is fully ramified with respect to K/L, so as before, we can apply Theorem 8.1 with $H = U$. It follows that there is a character $\chi \in \mathrm{Irr}(G|\varphi)$ such that ξ is the unique odd-multiplicity irreducible constituent of χ_H. Then χ is a constituent of ξ^G, and we have $\chi(1) = e\xi(1)$, where $e^2 = |K : L| = |G : H|$. Then

$$\frac{\xi^G(1)}{\chi(1)} = \frac{|G:H|\xi(1)}{\chi(1)} = \frac{e^2\xi(1)}{e\xi(1)} = e,$$

and this is an integer, as required for (b). ■

A careful analysis of the foregoing proof of Theorem 8.20 shows that if the given character χ of G lies in $\mathrm{B}_\pi(G)$ for some set π of primes, then the irreducible constituent ξ of χ_H with degree dividing the degree of χ will lie (or can be chosen to lie) in $\mathrm{B}_\pi(H)$. This is true because in most of the cases considered in the proof, χ uniquely determines ξ, so if χ is fixed by the magic field automorphism σ, then ξ will also be fixed by σ. Similarly, in the proof of (b), if the given character ξ of H lies in $\mathrm{B}_\pi(H)$, then the irreducible constituent χ of ξ^G with degree dividing $\xi^G(1)$ will lie (or can be chosen to lie) in $\mathrm{B}_\pi(G)$.

These observations, which we will not prove, yield the following.

Corollary 8.21. *Let $H \subseteq G$, where $|G|$ is odd, and let π be a set of primes.*

(a) *If $\varphi \in \mathrm{I}_\pi(G)$, then φ_H has a constituent $\mu \in \mathrm{I}_\pi(H)$ such that $\mu(1)$ divides $\varphi(1)$.*

(b) *If $\mu \in \mathrm{I}_\pi(H)$, then μ^G has a constituent $\varphi \in \mathrm{I}_\pi(G)$ such that $\varphi(1)$ divides $\mu^G(1)$.*

Proof. For (a), we can assume that H is maximal in G. Let $\chi \in \mathrm{B}_\pi(G)$ with $\chi^0 = \varphi$. (This is possible by Theorem 5.1.) By Theorem 8.20(a) and the above discussion, χ_H has a constituent $\xi \in \mathrm{B}_\pi(H)$ such that $\xi(1)$ divides $\chi(1)$. Writing $\mu = \xi^0$, we have $\mu \in \mathrm{I}_\pi(H)$ and μ is a constituent of $(\chi_H)^0 = (\chi^0)_H = \varphi_H$. Also $\mu(1) = \xi(1)$, and this divides $\chi(1) = \varphi(1)$. The proof of (b) is similar, except that we appeal to Theorem 8.20(b) in place of Theorem 8.20(a). ∎

Recall that if G is p-solvable, and $\pi = p'$, then $\mathrm{I}_\pi(G)$ is exactly the set $\mathrm{IBr}(G)$ of irreducible Brauer characters of G for the prime p. (See the discussion in Chapter 3.) If $|G|$ is odd, therefore, it follows by Corollary 8.21 that the divisibility conclusions that were stated for ordinary irreducible characters in Theorem 8.20 also hold for irreducible Brauer characters.

8E

Let H be a Hall π'-subgroup of an odd-order group G. We saw in Corollary 4.39 that if $\chi \in \mathrm{B}_\pi(G)$, then $[\chi_H, 1_H]$ is odd. We can now prove that the converse of this statement is also true.

Theorem 8.22. *Suppose that $|G|$ is odd. Let H be a Hall π'-subgroup of G for some set π of primes, and let $\chi \in \mathrm{Irr}(G)$.*

(a) *$[\chi_H, 1_H]$ is odd if and only if $\chi \in \mathrm{B}_\pi(G)$.*

(b) *If χ is π-special, then 1_H is the unique odd-multiplicity irreducible constituent of χ_H.*

Proof. If $\chi \in \mathrm{B}_\pi(G)$, then $[\chi, 1_H]$ is odd by Corollary 4.39, so to prove (a), we can assume that $[\chi_H, 1_H]$ is odd, and we show that $\chi \in \mathrm{B}_\pi(G)$. Also, since $\chi(1)$ is odd, we see that to establish (b), it suffices to show that if χ is π-special, then $[\chi_H, \alpha]$ is even for all nonprincipal characters $\alpha \in \mathrm{Irr}(H)$.

If $H = G$, then (a) is clear because by assumption, $[\chi_G, 1_G]$ is odd, and hence it is nonzero, and we have $\chi = 1_G \in \mathrm{B}_\pi(G)$, as wanted. Also,

(b) is clear in this case, because G is a π'-group, and hence the given π-special character χ must be principal, and thus $[\chi, \alpha] = 0$ for nonprincipal characters $\alpha \in \mathrm{Irr}(H)$.

We can now assume that $H < G$, and we proceed by induction on $|G|$. Since G is not a π'-group, $\mathbf{O}^{\pi'}(G) > 1$, and writing $K = \mathbf{O}^{\pi'}(G)$, we have $KH = G$. (Note that possibly $K = G$.) Let θ be an irreducible constituent of χ_K, and let T be the stabilizer of θ in G.

First, we prove (a) in the case where $T < G$. Since $KH = G$ and $T \supseteq K$, we have $TH = G$, and we write $J = T \cap H$. Then $|T : J| = |G : H|$, and since this is a π-number, it follows that J is a Hall π'-subgroup of T. Let $\psi \in \mathrm{Irr}(T|\theta)$ be the Clifford correspondent of χ, and observe that

$$[\psi_J, 1_J,] = [(\psi_J)^H, 1_H] = [(\psi^G)_H, 1_H] = [\chi_H, 1_H] \equiv 1 \bmod 2\,,$$

where the congruence holds by assumption. It follows by the inductive hypothesis that $\psi \in \mathrm{B}_\pi(T)$, and since $\psi^G = \chi$, Corollary 4.37(b) yields $\chi \in \mathrm{B}_\pi(G)$, as required.

To prove (a), therefore, we can assume that $T = G$, and for (b), we argue that we *must* have $T = G$. This is because $|G : K|$ is a π'-number, and since we are assuming that χ is π-special, $\chi(1)$ is a π-number, and hence χ_K is irreducible by Corollary 1.16. Then $\chi_K = \theta$, and thus θ is invariant in G.

Since K is nontrivial, we can choose a subgroup $L \lhd G$ such that K/L is a chief factor of G. Now K/L is not a π'-group because $K = \mathbf{O}^{\pi'}(G)$, and because G is solvable, we conclude that K/L is an abelian π-group.

Let $U = HL$, so U/L is a Hall π'-subgroup of G/L, and thus U is a complement for K relative to L in G. Since the Hall π'-subgroups of G/L are conjugate, it follows that every complement for K relative to L in G is conjugate to U in G.

Since θ is invariant in G, it follows by Corollary 7.4 that there are exactly three possibilities: either θ_L is irreducible, θ is fully ramified with respect to K/L, or $\theta = \varphi^G$ for every irreducible constituent φ of θ_L. We consider these alternatives in turn.

First, suppose that θ_L is irreducible, and write $\varphi = \theta_L$. Since θ is invariant in G, it follows by Lemma 2.11(a) that restriction defines a bijection from $\mathrm{Irr}(G|\theta)$ onto $\mathrm{Irr}(U|\varphi)$. In particular, χ_U is irreducible, and we write $\xi = \chi_U$. Also, since the restriction map is a bijection, it follows that χ is the unique member of $\mathrm{Irr}(G|\theta)$ that lies over ξ.

Since $H \subseteq U$, we see that

$$[\xi_H, 1_H] = [\chi_H, 1_H] \equiv 1 \bmod 2\,,$$

so by the inductive hypothesis, $\xi \in \mathrm{B}_\pi(U)$. Also, since φ lies under χ, we have $\varphi \in \mathrm{B}_\pi(L)$, and thus $\theta \in \mathrm{B}_\pi(K)$ because K/L is a π-group. It follows that both θ and ξ are fixed by the magic field automorphism σ, and hence χ is also fixed by σ because ξ and θ uniquely determine χ. We conclude that $\chi \in \mathrm{B}_\pi(G)$, as required for (a).

To prove (b) in this case, suppose that χ is π-special. Since $\chi_U = \xi$ and $|G|$ is odd, it follows by Corollary 2.27 that ξ is π-special, and hence if $\alpha \in \mathrm{Irr}(H)$ is nonprincipal, the inductive hypothesis guarantees that

$$[\chi_H, \alpha] = [\xi_H, \alpha] \equiv 0 \bmod 2\,,$$

as required.

Next, suppose that θ is fully ramified with respect to K/L. By Theorem 8.1, there is a complement for K relative to L in G having certain special properties, and since all complements for K relative to L in G are conjugate to U, we see that U enjoys these properties. In particular, χ_U has a unique irreducible constituent ξ such that $[\chi_U, \xi]$ is odd, and thus for all characters $\alpha \in \mathrm{Irr}(H)$, we have

$$[\chi_H, \alpha] \equiv [\xi_H, \alpha] \bmod 2\,.$$

Also, χ and ξ uniquely determine each other, so χ is fixed by the magic field automorphism σ if and only if ξ is fixed by σ, and it follows that $\chi \in \mathrm{B}_\pi(G)$ if and only if $\xi \in \mathrm{B}_\pi(U)$.

For (a), we are assuming that $[\chi_H, 1_H]$ is odd, and thus $[\xi_H, 1_H]$ is also odd. By the inductive hypothesis, therefore, $\xi \in \mathrm{B}_\pi(U)$, and it follows that $\chi \in \mathrm{B}_\pi(G)$, as wanted.

For (b), we are assuming that χ is π-special. Then $\chi \in \mathrm{B}_\pi(G)$, so $\xi \in \mathrm{B}_\pi(U)$. Also $\chi(1)$ is a π-number, and since $\xi(1)$ divides $\chi(1)$ by Theorem 8.1, we see that $\xi(1)$ is also a π-number. It follows by Theorem 4.12 that ξ is π-special, and hence by the inductive hypothesis, we see that if $\alpha \in \mathrm{Irr}(H)$ is nonprincipal, then

$$[\chi_H, \alpha] \equiv [\xi_H, \alpha] \equiv 0 \bmod 2\,,$$

as required.

Finally, assume that θ is induced from each irreducible constituent of θ_L. Since $|H|$ and $|K : L|$ are relatively prime and θ is H-invariant, it follows by Lemma 8.13(a) that θ_L has an H-invariant irreducible constituent φ. Then φ is invariant in U, and since φ^K is irreducible, it follows that U is the full stabilizer of φ in G, and we let $\xi \in \mathrm{Irr}(U|\varphi)$ be the Clifford correspondent of χ.

The irreducible constituents of χ_L form a G-orbit $\mathcal{O}$, and K acts transitively on $\mathcal{O}$ because each member of $\mathcal{O}$ lies under $\theta \in \mathrm{Irr}(K)$. Also, U acts on $\mathcal{O}$, so $\mathcal{O}$ is partitioned into U-orbits $\mathcal{O}_i$ for $0 \le i \le r$, and we number these so that $\mathcal{O}_0 = \{\varphi\}$.

Since every member of $\mathcal{O}$ has the form φ^x for some element $x \in K$, we can write $\mathcal{O}_i = \{\varphi^x \mid x \in X_i\}$, where X_i is some subset of K. Given a subscript i, we argue that the set

$$\{\varphi^{x^{-1}} \mid x \in K \text{ and } \varphi^x \in \mathcal{O}_i\}$$

is one of the U-orbits in $\mathcal{O}$.

To see this, we must show that if φ^x and φ^y lie in $\mathcal{O}_i$, where $x, y \in K$, then $\varphi^{x^{-1}}$ and $\varphi^{y^{-1}}$ lie in the same U-orbit. We have $\varphi^y = \varphi^{xu}$ for some element $u \in U$, and thus xuy^{-1} stabilizes φ. Then $xuy^{-1} \in U$, and we write $v = xuy^{-1}$, so $v \in U$. Then $y^{-1} = u^{-1}x^{-1}v$, and since u^{-1} stabilizes φ, we have

$$\varphi^{y^{-1}} = \varphi^{u^{-1}x^{-1}v} = \varphi^{x^{-1}v} .$$

Also, $v \in U$, so this character lies in the U-orbit containing $\varphi^{x^{-1}}$, as required.

We have now constructed an involutory permutation $i \mapsto i^*$ on the set of subscripts $\{0, 1, \ldots, r\}$, where for $x \in K$, we have $\varphi^x \in \mathcal{O}_i$ if and only if $\varphi^{x^{-1}} \in \mathcal{O}_{i^*}$. Clearly, $0^* = 0$, and we argue that if $i \neq 0$, then $i^* \neq i$. To see this, suppose $i^* = i$, so for some element $x \in K$, both φ^x and $\varphi^{x^{-1}}$ lie in the U-orbit $\mathcal{O}_i$.

We have $\varphi^{x^{-1}} = \varphi^{xu}$ for some element $u \in U$, so xux stabilizes φ, and thus $xux \in U$. Then $(xu)^2 \in U$, and since $|G|$ is odd, it follows that $xu \in U$, and thus $x \in U$ and $\varphi^x = \varphi$. Then $\varphi = \varphi^x \in \mathcal{O}_i$, and thus $i = 0$, as required.

Each irreducible constituent of χ_U lies over the members of exactly one of the orbits $\mathcal{O}_i$, so we can write $\chi_U = \sum \Delta_i$, where Δ_i is the sum (counting multiplicities) of the irreducible constituents of χ_U that lie over the members of $\mathcal{O}_i$. (Note that $\Delta_0 = \xi$ since ξ is the unique irreducible constituent of χ_U that lies over φ, and ξ has multiplicity 1 as a constituent of χ_U.)

Fix an element $x \in K$, and write $V = U \cap U^x$. Since U^x is the stabilizer of φ^x in G, we see that V is the stabilizer of φ^x in U. We claim that $x^{-1} \in \mathbf{N}_G(V)$, and hence also $x \in \mathbf{N}_G(V)$. To see this, let $v \in V$, so $v \in U^x$ and thus $v^{x^{-1}} \in U$. Also, $v \in U$, so $[v, x^{-1}] = v^{-1}v^{x^{-1}} \in U$. Now $x \in K$, so $[v, x^{-1}] \in K$, and we have $[v, x^{-1}] \in U \cap K = L \subseteq V$. Then $[V, x^{-1}] \subseteq V$, and thus x^{-1} normalizes V, as claimed. Then

$$V = V^{x^{-1}} = (U \cap U^x)^{x^{-1}} = U^{x^{-1}} \cap U ,$$

and it follows that V is also the stabilizer of $\varphi^{x^{-1}}$ in U.

Now suppose that $\mathcal{O}_i$ is the U-orbit containing φ^x. Each irreducible constituent of Δ_i lies over φ^x, so we can apply the Clifford correspondence to each of these constituents, and we deduce that there exists a (not necessarily irreducible) character Λ_i of V such that $(\Lambda_i)^U = \Delta_i$. Also, Λ_i is the sum (with appropriate multiplicities) of all of those irreducible constituents of $(\Delta_i)_V$ that lie over φ^x. Note that these are exactly the irreducible constituents of χ_V that lie over φ^x.

Similarly, since V is also the stabilizer of $\varphi^{x^{-1}}$ in U, we can write $\Delta_{i^*} = (\Lambda_{i^*})^U$, where Λ_{i^*} is the sum of those irreducible constituents of χ_V that lie over $\varphi^{x^{-1}}$. Now x normalizes V and x stabilizes the character χ_V, and since $(\varphi^{x^{-1}})^{x^2} = \varphi^x$, it follows that $(\Lambda_{i^*})^{x^2} = \Lambda_i$.

Let $D = H \cap V$. Since $LH = U$, we have $VH = U$, so $|V : D| = |U : H|$, and thus D is a Hall π'-subgroup of V. Also, D^{x^2} is a Hall π'-subgroup of $V^{x^2} = V$, and hence $D^{x^2 v} = D$ for some element $v \in V$, and we write $g = x^2 v$.

Now g normalizes both D and V, and we see that $(\Lambda_{i^*})^g = \Lambda_i$, and it follows that $[(\Lambda_{i^*})_D, 1_D] = [(\Lambda_i)_D, 1_D]$. Also,

$$[(\Delta_i)_H, 1_H] = [((\Lambda_i)^U)_H, 1_H] = [((\Lambda_i)_D)^H, 1_H] = [(\Lambda_i)_D, 1_D]\,,$$

and similarly,

$$[(\Delta_{i^*})_H, 1_H] = [(\Lambda_{i^*})_D, 1_D]\,.$$

We deduce that

$$[(\Delta_i)_H, 1_H] = [(\Delta_{i^*})_H, 1_H]\,.$$

To prove (a) now, recall that we are assuming that $[\chi_H, 1_H]$ is odd. Observe that

$$[\chi_H, 1_H] = [\xi_H, 1_H] + \sum_{i=1}^{r} [(\Delta_i)_H, 1_H]\,.$$

For $i > 0$, we have shown that the contributions of Δ_i and Δ_{i^*} to the above sum are equal, and thus

$$[\xi_H, 1_H] \equiv [\chi_H, 1_H] \equiv 1 \bmod 2\,.$$

Then $\xi \in \mathrm{B}_\pi(U)$ by the inductive hypothesis, and since $\chi = \xi^G$, Corollary 4.37(b) guarantees that $\chi \in \mathrm{B}_\pi(G)$, as required.

To prove (b), we assume that χ is π-special, and we work to show that $[\chi_H, \alpha] \equiv 0 \bmod 2$ for every nonprincipal irreducible character α of H.

First, we argue that ξ is π-special. Now $L \triangleleft G$ and $\varphi \in \mathrm{Irr}(L)$ lies under χ, and it follows that φ is π-special. Then both χ and φ are fixed by the magic field automorphism σ, and since χ and ψ uniquely determine ξ, we conclude that ξ is fixed by σ, and thus $\xi \in \mathrm{B}_\pi(U)$. Also, $\xi(1)$ divides the π-number $\chi(1)$, and hence ξ is π-special by Theorem 4.12.

If $\alpha \in \mathrm{Irr}(H)$ is nonprincipal, therefore, the inductive hypothesis yields $[\xi_H, \alpha] \equiv 0 \bmod 2$, so it suffices to show that

$$\sum_{i=1}^{r} [(\Delta_i)_H, \alpha] \equiv 0 \bmod 2$$

for all $\alpha \in \mathrm{Irr}(H)$. We will establish this by proving that

$$[(\Delta_i)_H, \alpha] \equiv [(\Delta_{i^*})_H, \alpha] \bmod 2\,,$$

for $1 \leq i \leq r$.

Recall now that $\Delta_i = (\Lambda_i)^U$, and so

$$[(\Delta_i)_H, \alpha] = [((\Lambda_i)^U)_H, \alpha] = [((\Lambda_i)_D)^H, \alpha] = [(\Lambda_i)_D, \alpha_D]\,,$$

and similarly with i^* in place of i. It therefore suffices to show that

$$[(\Lambda_i)_D, \beta] \equiv [(\Lambda_{i^*})_D, \beta] \bmod 2$$

for all $\beta \in \mathrm{Irr}(D)$. We saw previously that

$$[(\Lambda_i)_D, 1_D] = [(\Lambda_{i^*})_D, 1_D]\,,$$

so it is enough to prove that the above congruence holds for all nonprincipal characters $\beta \in \mathrm{Irr}(D)$.

We will show that Λ_i and Λ_{i^*} are irreducible, and that in fact, they are π-special. The inductive hypothesis will then yield

$$[(\Lambda_i)_D, \beta] \equiv 0 \equiv [(\Lambda_{i^*})_D, \beta] \bmod 2$$

for nonprincipal characters $\beta \in \mathrm{Irr}(D)$, and this will complete the proof.

Recall that U^x is the stabilizer of φ^x in G, and let $\zeta \in \mathrm{Irr}(U^x|\varphi^x)$ be the Clifford correspondent of χ. We saw that ξ is π-special, and similar reasoning shows that ζ is also π-special. Since $|U^x : L| = |U : L|$ is a π'-number, it follows that ζ_L is irreducible, and thus ζ_V is irreducible. In fact, ζ_V is π-special by Corollary 2.27.

Next, we argue that $\Lambda_i = \zeta_V$ (assuming that $\varphi^x \in \mathcal{O}_i$). Recall that Λ_i is the sum (counting multiplicities) of all irreducible constituents of χ_V that lie over φ^x. Also, ζ is the unique irreducible constituent of the restriction $\chi_{(U^x)}$ that lies over φ^x. Since ζ_V is irreducible, it follows that ζ_V is the unique irreducible constituent of χ_V that lies over φ^x, and thus $\zeta_V = \Lambda_i$, as claimed. We conclude that Λ_i is π-special, and similarly, Λ_{i^*} is π-special, and this completes the proof. ∎

It is natural to ask whether or not the necessary condition for χ to be π-special in Theorem 8.22(b) is also sufficient, but unfortunately, the answer is not at all clear. As we shall see, this is equivalent to asking the following question: Given a Hall π'-subgroup H of an odd-order group G, and given a character $\chi \in \mathrm{Irr}(G)$ with the property that $\chi_H = 1_H + 2\Delta$ for some character Δ of H, is it true that $\chi(1)$ is a π-number?

If χ has the stated property, it follows that $[\chi_H, 1_H]$ is odd, so Theorem 8.22(a) guarantees that $\chi \in \mathrm{B}_\pi(G)$. If it can be proved that $\chi(1)$ must be a π-number, it would follow that χ is π-special, and this would establish the converse of Theorem 8.22(b). It is interesting to note that in this formulation, the question mentions neither π-special characters nor B_π-characters. Also, we see that it would suffice to provide an answer in the case where H is a Sylow subgroup.

We conclude this chapter with some applications of Theorem 8.22.

Corollary 8.23. *Let $\chi \in \mathrm{Irr}(G)$ be primitive, where $|G|$ is odd, and let H be a Hall subgroup of G. Then χ_H has a unique odd-multiplicity irreducible constituent.*

Proof. Let π be the set of prime divisors of $|G : H|$, so H is a Hall π'-subgroup of G. By Theorem 2.9, we can write $\chi = \alpha\beta$, where α is π-special and β is π'-special. By Theorem 8.22(b), we have $\alpha_H = 1_H + 2\Delta$, where Δ is a character of H or is zero, and thus $\chi_H = \alpha_H\beta_H = \beta_H + 2\Delta\beta_H$. Also, β_H is irreducible by Lemma 2.20, and thus β_H is the unique odd-multiplicity irreducible constituent of χ_H. ■

Corollary 8.24. *Let H be a Hall π'-subgroup of an odd-order group G. Then the number of odd-multiplicity irreducible constituents of $(1_H)^G$ is equal to the number of conjugacy classes of π-elements of G.*

Proof. By Frobenius reciprocity, a character $\chi \in \mathrm{Irr}(G)$ has odd multiplicity as a constituent of $(1_H)^G$ if and only if $[\chi_H, 1_H]$ is odd, and by Corollary 8.21(a), this happens precisely when $\chi \in \mathrm{B}_\pi(G)$. The number of odd-multiplicity irreducible constituents of $(1_H)^G$, therefore, is equal to $|\mathrm{B}_\pi(G)|$. By Theorem 5.1, we have $|\mathrm{B}_\pi(G)| = |\,\mathrm{I}_\pi(G)|$, and by Theorem 3.3 this is the number of π-classes in G. ■

Corollary 8.25. *Let $P \in \mathrm{Syl}_p(G)$, where G has odd order. Then the odd-multiplicity irreducible constituents of $(1_P)^G$ form a set of lifts for the irreducible p-Brauer characters of G.*

Proof. By Theorem 5.1, the set $\mathrm{B}_{p'}(G)$ is a set of lifts for the irreducible p'-partial characters of G, and these are exactly the irreducible p-Brauer characters of G. (This is essentially Theorem 3.2.) ■

Problems

8.1. Let $(G, K, U, L, \theta, \varphi)$ be in the basic configuration, where φ extends to G and $|K : L|$ is odd. Show that the canonical associated character ψ has values in the cyclotomic field $\mathbb{Q}_{|K|}$, and deduce that if $u \in U$ has order modulo L that is coprime to $|K|$, then $\psi(u) = \pm\sqrt{|\mathbf{C}_{K/L}(u)|}$.

8.2. In Problem 8.1, assume that U/L is cyclic of order m and that its action on K/L is Frobenius. (Recall that this means that $\mathbf{C}_{K/L}(x) = 1$ for every nonidentity element x of U/L, and note that it follows that m divides $e^2 - 1$, where $|K/L| = e^2$.) Show that if λ and μ are any two linear characters of U/L, then the multiplicities $]\psi, \lambda]$ and $[\psi, \mu]$ differ by at most 1.

8.3. In Problem 8.2, show that $\mathrm{Irr}(U/L)$ can be partitioned into two sets X and Y such that

$$\psi = t \sum_{\lambda \in X} \lambda + (t+1) \sum_{\lambda \in Y} \lambda ,$$

for some nonnegative integer t. Write $a = |X|$ and $b = |Y|$, and deduce that

$$\begin{aligned} a + b &= m, \\ at + b(t+1) &= e, \text{ and} \\ at^2 + b(t+1)^2 &= [\psi, \psi] = \frac{e^2 + m - 1}{m} . \end{aligned}$$

Solve these equations to show that either $t = (e-1)/m$ and $b = 1$ or $t = (e+1)/m - 1$ and $a = 1$. Deduce that m divides either $e - 1$ or $e + 1$ and that $\psi = k\rho \pm \lambda$, where $k = (e \mp 1)/m$ and ρ is the regular character of U/L, and where λ is a linear character of U/L such that λ^2 is principal.

8.4. Suppose that m is odd in Problem 8.3, and let x be a nonidentity element of U/L. If m divides $e - 1$, show that $\psi(x) = 1$, and if m divides $e + 1$, show that $\psi(x) = -1$.

8.5. Let A act on G, where $|A|$ is relatively prime to $|G|$ and $|G|$ is odd, and let $\chi \in \mathrm{Irr}_A(G)$. If $H \subseteq G$ is A-invariant, show that some irreducible constituent of χ_H is A-invariant.

Hints: Note that Lemma 8.13(a) handles the case where H is normal in G. Some of the ideas in the proof of Theorem 8.20 are relevant. In particular, one can work by induction on $|G|$, and there is no loss in assuming that H is maximal among A-invariant proper subgroups of G. Let $L = \mathrm{core}_G(H)$ and let K/L be a chief factor of the semidirect product GA. Let $\varphi \in \mathrm{Irr}_A(L)$, and let T be the stabilizer of φ in G. If $T < G$, apply the inductive hypothesis.

Note: The assertion of this problem remains true without the assumption that $|G|$ is odd.

Notes and References

Theorem 8.1 is essentially Theorem 9.1 of [**34**]. The proof in this chapter, however, has components that appear in various other papers.

Theorem 8.3 is closely related to Theorem 6.1 of [**38**]. The basic idea of the proof is a generalization of an argument of E. C. Dade.

The technique in Section 8B for dealing with the case where the character φ does not extend to U appears in Theorem 5.3 of [**38**].

Theorem 8.11, which is the Isaacs correspondence, first appeared as Theorem 10.8 of [**34**].

The theorem of T. R. Wolf, asserting that the Glauberman and Isaacs correspondences agree when both are defined appears in [**75**].

Theorem 8.12 is a strong form of Theorem 10.9 of [**34**].

The Isaacs–Navarro extension of the McKay conjecture was stated in [**51**]. This paper also includes a corresponding extension of the Alperin–McKay block-theoretic conjecture.

Theorem 8.20 occurs in [**48**].

Theorem 8.22(a) is Theorem C of [**46**] and Theorem 8.22(b) is Theorem D. Also Corollary 8.23 is Corollary E of that paper.

Part 3

M-Groups

Chapter 9

M-Groups and Monomial Characters

9A

This chapter is primarily concerned with some of the deeper properties of M-groups and monomial characters, but we begin with a fairly comprehensive review of the basics. Much of this material (and more) can also be found in [**CTFG**].

First, we recall that an irreducible character χ of G is said to be "primitive" if there do not exist a proper subgroup H of G and a character α of H such that $\alpha^G = \chi$. (In particular, all linear characters are primitive.) If $\chi \in \mathrm{Irr}(G)$ is primitive, we have seen that it is automatically quasi-primitive, which means that for each normal subgroup N of G, the restriction χ_N is homogeneous, or equvalently, there exists a character $\theta \in \mathrm{Irr}(N)$ such that $\chi_N = e\theta$ for some positive integer e. To see that a primitive character $\chi \in \mathrm{Irr}(G)$ must be quasi-primitive, recall that the Clifford correspondence guarantees that χ is induced from the stabilizer of an irreducible constituent θ of χ_N. If χ is primitive, therefore, the stabilizer of θ cannot be a proper subgroup, and thus θ is invariant in G. It follows by Clifford's theorem that $\chi_N = e\theta$ for some integer e, as wanted.

It is a triviality that an arbitrary irreducible character χ of G is induced from some character of G (namely χ, itself) so there must be subgroups H of G minimal with the property that there exists a character α of H such that $\alpha^G = \chi$. The minimality of H implies that α is primitive, and so we see that every irreducible character χ of G is induced from a primitive character of some subgroup. In general, however, neither the subgroup nor the primitive character is uniquely determined by χ.

A character χ of G is **monomial** if it is induced from a linear character of some (not necessarily proper) subgroup. (Although the definition does not require that a monomial character should be irreducible, it is only irreducible monomial characters that concern us here.) If every irreducible character of some group G is monomial, we say that G is an **M-group**. Since linear characters are clearly monomial, all abelian groups are M-groups, and more generally, we shall see all nilpotent groups are M-groups.

In some sense, linear characters are easy to understand; they are simply homomorphisms into the multiplicative group of the complex numbers. One might suppose, therefore, that monomial characters, which are constructed by inducing linear characters of subgroups, should take second place after linear characters on the ease-of-understanding scale. Paradoxically, however, monomial characters play a central role in some of the deepest theorems and most intractable conjectures in ordinary character theory.

We begin our study with Taketa's theorem, which asserts that M-groups are solvable. (This justifies the inclusion of a chapter on M-groups in a book on the character theory of solvable groups.) To state the full strength of Taketa's result, we use the notation $\mathrm{cd}(G)$ for the set $\{\chi(1) \mid \chi \in \mathrm{Irr}(G)\}$ of irreducible character degrees of G. Also, we write $G^{(i)}$ to denote the ith term of the derived series of the group G. Thus, for example, $G^{(1)} = G'$ and $G^{(2)} = G''$, and we recall that G is solvable if and only if $G^{(n)} = 1$ for some integer n. The **derived length** $\mathrm{dl}(G)$ of a nontrivial solvable group G, is the smallest integer n such that $G^{(n)} = 1$.

Theorem 9.1 (Taketa). *Let G be an M-group. Then G is solvable and* $\mathrm{dl}(G) \le |\,\mathrm{cd}(G)|$. *More precisely, suppose that the members of* $\mathrm{cd}(G)$ *are* $1 = f_1 < f_2 < \cdots < f_r$, *where* $r = |\,\mathrm{cd}(G)|$. *Then if* $\chi \in \mathrm{Irr}(G)$ *and* $\chi(1) \le f_i$, *we have* $G^{(i)} \subseteq \ker(\chi)$.

We need an easy preliminary result.

Lemma 9.2. *Let $H \subseteq G$, and let α be a character of H. Then* $\ker(\alpha^G) \subseteq \ker(\alpha)$.

Proof. For elements $x \in G$, recall that by definition,

$$\alpha^G(x) = \frac{1}{|H|} \sum_{g \in G} \alpha^0(gxg^{-1}),$$

where α^0 is the function on G equal to α for elements of H and equal to zero otherwise.

Given $k \in \ker(\alpha^G)$, we have

$$\frac{1}{|H|}\sum_{g\in G}\alpha^0(gkg^{-1}) = \alpha^G(k) = \alpha^G(1) = \frac{1}{|H|}\sum_{g\in G}\alpha^0(g1g^{-1}) = \frac{1}{|H|}\sum_{g\in G}\alpha(1),$$

and since $|\alpha^0(gkg^{-1})| \leq \alpha(1)$, a term-by-term comparison of the equal sums on the left- and right-hand side of the above equation shows that $\alpha^0(gkg^{-1}) = \alpha(1)$ for all $g \in G$. Taking $g = 1$, we conclude that $\alpha^0(k) = \alpha(1)$, and since $\alpha(1) \neq 0$, we see that $k \in H$. Then $\alpha(k) = \alpha^0(k) = \alpha(1)$, and so $k \in \ker(\alpha)$, as required. ■

Proof of Theorem 9.1. First, we show by induction on i that if $\chi \in \mathrm{Irr}(G)$ with $\chi(1) = f_i$, then $G^{(i)} \subseteq \ker(\chi)$. If $i = 1$, then $\chi(1) = f_1 = 1$, so χ is linear, and thus $G^{(1)} = G' \subseteq \ker(\chi)$, as required.

Now suppose $i > 1$. Since G is an M-group, we can write $\chi = \lambda^G$, where λ is a linear character of some subgroup $H \subseteq G$, and we observe that

$$|G : H| = |G : H|\lambda(1) = \lambda^G(1) = \chi(1) = f_i\,.$$

Let $\Xi = (1_H)^G$, so $\Xi(1) = |G : H| = f_i$. Also,

$$[\Xi, 1_G] = [(1_H)^G, 1_G] = [1_H, 1_H] = 1\,,$$

so Ξ has a principal constituent. If ψ is any nonprincipal irreducible constituent of Ξ, therefore, we have $\psi(1) \leq \Xi(1) - 1 = f_i - 1$. Then $\psi(1) = f_j$, for some subscript $j \leq i - 1$, and thus

$$G^{(i-1)} \subseteq G^{(j)} \subseteq \ker(\psi)\,,$$

where the second containment holds by the inductive hypothesis. It follows that $G^{(i-1)}$ is contained in the kernel of every irreducible constituent of Ξ, and hence

$$G^{(i-1)} \subseteq \ker(\Xi) = \ker((1_H)^G) \subseteq \ker(1_H) = H\,,$$

where the second containment follows by Lemma 9.2.

We have

$$G^{(i)} = (G^{(i-1)})' \subseteq H' \subseteq \ker(\lambda)\,,$$

so λ lies over the principal character of $G^{(i)}$, and since χ lies over λ, we conclude that χ lies over the principal character of $G^{(i)}$. Finally, since $G^{(i)}$ is a normal subgroup, we have $G^{(i)} \subseteq \ker(\chi)$, as required.

Now let $\chi \in \mathrm{Irr}(G)$ with $\chi(1) \leq f_i$. Then $\chi(1) = f_j$ for some subscript $j \leq i$, and thus $G^{(i)} \subseteq G^{(j)} \subseteq \ker(\chi)$, as wanted. In particular, for an arbitrary character $\chi \in \mathrm{Irr}(G)$, we have $\chi(1) \leq f_r$, and it follows that $G^{(r)} \subseteq \ker(\chi)$. The intersection of the kernels of all members of $\mathrm{Irr}(G)$ is trivial, however, so we have $G^{(r)} = 1$, and thus G is solvable with derived length at most r. ■

If G is an M-group, then by Taketa's theorem $\mathrm{dl}(G) \le |\mathrm{cd}(G)|$, so it is natural to ask whether or not this "Taketa inequality" holds more generally for solvable groups G other than M-groups. This remains open, but a result of T. R. Berger shows that the Taketa inequality holds for (solvable) groups of odd order, and D. Gluck proved that the weaker inequality $\mathrm{dl}(G) \le 2|\mathrm{cd}(G)|$ always holds for solvable groups. There is also some evidence that for large solvable groups, derived-length bounds that are stronger than the Taketa inequality might hold.

By Taketa's theorem, M-groups must be solvable, but it is certainly not true that every solvable group is an M-group. (The smallest solvable group that is not an M-group is $SL(2,3)$, which has order 24.) We consider the question now of what conditions can be imposed on a solvable group that are sufficient to guarantee that it is an M-group. For example, as we have already mentioned, nilpotent groups are M-groups. More generally, we shall see that supersolvable groups are M-groups. We digress briefly to review the definition.

Recall that a group G is solvable if there exists a series of normal subgroups

$$1 = G_0 \subseteq G_1 \subseteq G_2 \subseteq \cdots \subseteq G_r = G\,,$$

such that each of the factors G_i/G_{i-1} is abelian. Also, G is nilpotent if there exists such a normal series where each factor G_i/G_{i-1} is central in G/G_{i-1}. In the same spirit, we say that G is **supersolvable** if there exists a normal series with the property that each factor G_i/G_{i-1} is cyclic. Clearly, supersolvable groups are solvable, and it is also true, but perhaps not quite so obvious, that finite nilpotent groups are supersolvable. To prove this, we proceed by induction on the order of the group. If G is nilpotent and nontrivial, then $\mathbf{Z}(G)$ is nontrivial, so there exists a nontrivial central cyclic group C. Now G/C is nilpotent, so it is supersolvable by the inductive hypothesis, and since C is cyclic, it follows easily that G is supersolvable.

It is routine to check that all subgroups and factor groups of a supersolvable group are supersolvable, and another useful fact is that chief factors of supersolvable groups have prime order or, equivalently, minimal normal subgroups of supersolvable groups have prime order. An easy argument proving this is as follows.

Let $\{G_i \mid 0 \le i \le r\}$ be a normal series for G, where the factors G_i/G_{i-1} are cyclic, and let N be minimal normal in G. Let j be maximal such that $N \not\subseteq G_j$, so $N \cap G_j = 1$ because N is minimal normal. Then $N \cong NG_j/G_j$, and since $N \subseteq G_{j+1}$, we see that NG_j/G_j is a subgroup of the cyclic group G_{j+1}/G_j. Then N is cyclic, and we conclude that N has prime order, as claimed.

We shall see that supersolvable groups are M-groups (and thus nilpotent groups are M-groups) but there is an even larger class of groups that we can prove are M-groups.

Theorem 9.3 (Huppert). *Assume that G has a solvable normal subgroup N such that all Sylow subgroups of N are abelian and G/N is supersolvable. Then G is an M-group.*

In particular, by taking $N = 1$ in Theorem 9.3, we deduce that supersolvable groups are M-groups. Also, it follows by taking $N = G$ that solvable groups in which all Sylow subgroups are abelian are M-groups.

Theorem 9.3 is an easy consequence of the following,

Lemma 9.4. *Assume that G has a solvable normal subgroup N such that all Sylow subgroups of N are abelian and G/N is supersolvable. Then*

(a) *All subgroups and factor groups of G inherit the properties assumed for G.*

(b) *If A is maximal among abelian normal subgroups of G, then $A = \mathbf{C}_A(G)$.*

Proof. For (a), let $H \subseteq G$, and write $M = N \cap H$. Then $M \triangleleft H$, and since $M \subseteq N$, we see that M is solvable and all Sylow subgroups of M are abelian. Also,

$$H/M = H/(N \cap H) \cong NH/N \subseteq G/N\,,$$

so H/M is supersolvable, and thus H satisfies the stated assumptions on G, as required.

Now let $K \triangleleft G$, and observe that $NK/K \triangleleft G/K$. Also, NK/K is a homomorphic image of N, so NK/K is solvable, and all of its Sylow subgroups are abelian. Also,

$$\frac{G/K}{NK/K} \cong \frac{G}{NK}\,,$$

and since G/NK is a homomorphic image of the supersolvable group G/N, we see that $(G/K)/(NK/K)$ is supersolvable. This completes the proof of (a).

For (b), let A be maximal among abelian normal subgroups of G, and write $C = \mathbf{C}_G(A)$. Then $A \subseteq C \triangleleft G$, and we complete the proof by showing that $A = C$. Otherwise, we can choose a chief factor B/A of G with $B \subseteq C$, and we work to derive a contradiction.

Since $B \subseteq C = \mathbf{C}_G(A)$, we have $A \subseteq \mathbf{Z}(B)$. Also, by the maximality of A, we see that B is not abelian, and it follows that there is no abelian subgroup U of B such that $AU = B$, and in particular, B/A cannot be cyclic.

We argue next that $B \subseteq AN$. Otherwise $A \subseteq B \cap AN < B$, and since B/A is a chief factor of G and $B \cap AN \lhd G$, we have $B \cap AN = A$. Then

$$\frac{B}{A} = \frac{B}{B \cap AN} \cong \frac{BAN}{AN} = \frac{BN}{AN},$$

and since this is a G-isomorphism and B/A is a chief factor of G, it follows that BN/AN is also a chief factor of G. Now $N \subseteq AN \subseteq BN$, so BN/AN can be viewed as a chief factor of the supersolvable group G/N. Then BN/AN has prime order, and so B/A also has prime order, and this is a contradiction since B/A is not cyclic.

We have shown that $B \subseteq AN$, and thus $B = A(B \cap N)$ by Dedekind's lemma. Since B/A is a p-group for some prime p, it follows that $B = AU$, where U is a Sylow p-subgroup of $B \cap N$. By hypothesis, however, the Sylow p-subgroups of N are abelian, and it follows that U is abelian, and this is our final contradiction. ∎

Proof of Theorem 9.3. Let $\chi \in \mathrm{Irr}(G)$, and choose a subgroup H of G minimal such that there exists a character $\alpha \in \mathrm{Irr}(H)$ such that $\alpha^G = \chi$. Then α is primitive, and we let $K = \ker(\alpha)$, so $K \lhd H$. Then H/K satisfies the stated assumptions on G by Lemma 9.4(a), and by Lemma 9.4(b), there exists a self-centralizing normal abelian subgroup A of H/K.

Now α remains primitive when viewed as a character of H/K, and thus α_A is a multiple of some linear character. Also, α is faithful as a character of H/K, and thus A is central in H/K. Since A is self-centralizing, however, we have $A = H/K$, and thus H/K is abelian. We conclude that α is linear, and thus $\chi = \alpha^G$ is monomial. ∎

In some sense, the M-groups constructed using Theorem 9.3 are not at all typical of M-groups in general. If G satisfies the hypotheses of Theorem 9.3 and $H \subseteq G$, then H also satisfies the hypotheses of Theorem 9.3, and thus H is also an M-group. In general, however, subgroups of M-groups need not be M-groups. In fact, other than the fact that they are solvable, subgroups of M-groups enjoy no special properties whatsoever; *every* finite solvable group is a subgroup of some M-group. To prove this result of E. C. Dade, we digress to consider certain wreath products.

Let A be a nontrivial finite group. Let r be a positive integer, and construct the direct product B of r copies of A. We think of B as the set of r-tuples $(a_1, a_2, \ldots, a_r)$, where the entries a_i lie in A, and where multiplication is componentwise, and we let t be the "cycle-right" automorphism of B, so

$$(a_1, a_2, \ldots, a_r)^t = (a_r, a_1, a_2, \ldots, a_{r-1}).$$

Write $T = \langle t \rangle$, so $T \subseteq \mathrm{Aut}(B)$ and T is cyclic of order r. The **wreath product** W of A by the cyclic group C of order r is the semidirect product of B by T, and we write $W = A \wr C$. We view B and T as subgroups of W, so $W = TB$, where $B \triangleleft W$ and $B \cap T = 1$. The subgroup B is called the **base group** of the wreath product, and conjugation by t in W induces the cycle-right automorphism of B.

Lemma 9.5. *Let A be an M-group, and let C be cyclic of prime order p. Then the wreath product $W = A \wr C$ is an M-group.*

Proof. Given $\chi \in \mathrm{Irr}(W)$, we must show that χ is monomial. Let B be the base group of the wreath product W, so B has prime index p in W, and thus if θ is an irreducible constituent of χ_B, then by Lemma 1.8, either $\chi = \theta^W$ or $\chi_B = \theta$. Now B is a direct product of p copies of the M-group A, and it follows easily that B is an M-group, and thus θ is monomial. If $\chi = \theta^G$, it is clear that χ is monomial, as required.

We can now assume that $\chi_B = \theta$, and thus θ is invariant in W. Then $\theta = \varphi \times \varphi \times \cdots \times \varphi$, where $\varphi \in \mathrm{Irr}(A)$. Since, A is an M-group, $\varphi = \lambda^A$ for some linear character λ of a subgroup H of A.

Let $K \subseteq B$ be the subgroup consisting of the p-tuples $(h_1, h_2, \ldots, h_p)$, where $h_i \in H$. Then K is the direct product of p copies of H, and hence K has a linear character $\mu = \lambda \times \lambda \times \cdots \times \lambda$, and it is easy to check that $\mu^B = \theta$.

Let $T \subseteq W$ be as in the discussion preceding the statement of the lemma, and observe that T normalizes K and stabilizes μ. Let $L = KT$, so $|L : K| = p$, and μ is invariant in L. By Lemma 1.8, we see that μ has an extension to a linear character ν of L.

Now $W = BL$ and $B \cap L = K$, and thus

$$(\nu^W)_B = (\nu_K)^B = \mu^B = \theta\,,$$

so ν^W is an extension of θ to W. Recall that χ is also an extension of θ to W, so it follows by Gallagher's theorem that there exists a linear character β of W/B such that $\chi = \beta\nu^W = (\beta_L\nu)^W$. Since $\beta_L\nu$ is linear, we see that χ is monomial, as required. ∎

Lemma 9.6. *Let $N \triangleleft G$, where G/N is cyclic of order r. If N is isomorphic to a subgroup of some group A, then G is isomorphic to a subgroup of the wreath product $W = A \wr C$, where C is cyclic of order r.*

Proof. We can assume without loss that $N \subseteq A$, and we choose $x \in G$ such that $G = N\langle x\rangle$. Let B be the base group of W, and define $\sigma : N \to B$ by

$$\sigma(n) = (n^{x^{r-1}}, n^{x^{r-2}}, \ldots, n^{x^2}, n^x, n)\,.$$

Then σ is an isomorphism from N into B because each of the maps $n \mapsto n^{x^i}$ is an isomorphism from N into A.

Recall that $W = TB$, where $T = \langle t \rangle$, and conjugation by t on elements of B induces the cycle-right map on B. Now $x^r \in N \subseteq A$, and for notational convenience, we write $x^r = y$. We define $b = (y, 1, 1, \ldots, 1)$, so $b \in B$.

We claim that

$$\sigma(n^x) = \sigma(n)^{tb}$$

for all $n \in N$. To see this, we compute

$$\begin{aligned}\sigma(n)^{tb} = ((n^{x^{r-1}}, n^{x^{r-2}}, \ldots, n^{x^2}, n^x, n)^t)^b &= (n, n^{x^{r-1}}, n^{x^{r-2}}, \ldots, n^{x^2}, n^x)^b \\ &= (n^{x^r}, n^{x^{r-1}}, n^{x^{r-2}}, \ldots, n^{x^2}, n^x) \\ &= \sigma(n^x)\,,\end{aligned}$$

as claimed.

The element tb of W thus normalizes the subgroup $\sigma(N)$, and the conjugation action of tb on $\sigma(N)$ corresponds via the isomorphism σ to the conjugation action of x on N in G. We complete the proof by showing that G is isomorphic to $N\langle tb \rangle$ via an isomorphism that agrees with σ on N and that carries x to tb. To do this, it suffices to show that $\sigma(x^r) = (tb)^r$. Writing $x^r = y$, as before, and recalling that $t^r = 1$, we have

$$\begin{aligned}(tb)^r &= t^r b^{t^{r-1}} b^{t^{r-2}} \cdots b^t b \\ &= (1, \ldots, 1, y)(1, 1, \ldots, y, 1) \cdots (1, y, 1, \ldots, 1)(y, 1, \ldots, 1) \\ &= (y, y, \ldots, y) \\ &= \sigma(y)\,,\end{aligned}$$

as required. ■

Theorem 9.7 (Dade). *Let G be solvable. Then G is isomorphic to a subgroup of some M-group.*

Proof. We can assume that G is nontrivial, and we proceed by induction on $|G|$. Let N be a maximal normal subgroup of G, so G/N is a solvable simple group, and thus $|G : N|$ is some prime number p. By the inductive hypothesis, N is isomorphic to a subgroup of some M-group A, and thus Lemma 9.6 guarantees that G is isomorphic to a subgroup of $W = A \wr C$, where C is cyclic of order p. The result now follows because W is an M-group by Lemma 9.5. ■

Although by Theorem 9.7, every solvable group can be isomorphically embedded in an M-group, it may be that there are restrictions on the embedding. For example, not every solvable group can be embedded as a Hall subgroup of an M-group. This is because by Corollary 5.16, every primitive character of a Hall subgroup of an M-group must be linear, and thus if G is a solvable group that has a nonlinear primitive character, then G cannot be embedded as a Hall subgroup of an M-group.

The fact that every primitive character of a Hall subgroup of an M-group must be linear suggests that perhaps Hall subgroups of M-groups must actually be M-groups. This was an open question for many years, but eventually, an example of an M-group having a Hall subgroup that is not an M-group was constructed by H. Fukushima. It is not hard to prove, however, that *normal* Hall subgroups of M-groups are always M-groups.

Theorem 9.8 (Dornhoff). *Let G be an M-group, and suppose that N is a normal Hall subgroup of G. Then N is an M-group.*

Proof. Given $\theta \in \mathrm{Irr}(N)$, we work to show that θ is monomial. Let $\chi \in \mathrm{Irr}(G|\theta)$, so χ is monomial because G is an M-group. There exist, therefore, a subgroup $H \subseteq G$ and a linear character λ of H such that $\lambda^G = \chi$. Write $\psi = \lambda^{NH}$, so $\psi^G = \chi$, and thus ψ is irreducible and lies under χ.

We have $\psi(1) = |NH : H| = |N : N \cap H|$, and thus $\psi(1)$ divides $|N|$. Since N is a Hall subgroup of G, it follows that $\psi(1)$ is relatively prime to $|G : N|$, and hence $\psi(1)$ is relatively prime to $|NH : N|$. By Corollary 1.16, therefore, ψ_N is irreducible, and we write $\varphi = \psi_N$. Then

$$\varphi = (\lambda^{NH})_N = (\lambda_{N\cap H})^N ,$$

and thus φ is monomial.

Now $\varphi \in \mathrm{Irr}(N)$ and φ lies under ψ, which lies under χ. Also $\theta \in \mathrm{Irr}(N)$ and θ lies under χ, so by Clifford's theorem, φ and θ are conjugate in G. Since φ is monomial, it follows that θ is also monomial, as required. ∎

9B

By Theorem 9.8, a normal Hall subgroup of an M-group G must be an M-group, so it is natural to ask if the assumption that the normal subgroup is a Hall subgroup is really essential. Is it true, in other words, that an arbitrary normal subgroup of an M-group must be an M-group? A counterexample was discovered independently by E. C. Dade and R. van der Waall, and so the answer is "no": normal subgroups of M-groups need not be M-groups.

The smallest known M-group having a normal subgroup that is not an M-group has order $2^9{\cdot}7$. The construction of this counterexample by Dade and by van der Waall involves the prime 2 in a fundamental way, and this suggests that perhaps normal subgroups of odd-order M-groups must be M-groups. No proof of this is known, however, and no counterexample has been constructed.

Some caution is appropriate when considering this problem because what may seem to be an obvious approach is doomed to failure. Given $N \lhd G$ and $\theta \in \mathrm{Irr}(N)$, then even if every irreducible character of G lying over θ is monomial, it does not follow that θ is monomial. (And assuming that $|G|$ is odd does not help.)

Lemma 9.9. *There exists an odd-order group G having a normal subgroup N such that N has a nonmonomial irreducible character θ, and yet every member of* $\mathrm{Irr}(G|\theta)$ *is monomial.*

Sketch of proof. Let E be nonabelian of order 5^3 and exponent 5. Then $\mathrm{Aut}(E)$ has a subgroup C of order 3, where C acts trivially on $Z = \mathbf{Z}(E)$, and no subgroup with index 5 in E is normalized by C. Let F be nonabelian of order 3^3, and let $K \lhd F$, where $|K| = 3^2$. Let F act on E, where K acts trivially and F/K acts like the group C, and finally, construct the semidirect product $G = EF$, so $|G| = 3^3{\cdot}5^3$.

Let $\varphi \in \mathrm{Irr}(E)$ have degree 5, and observe that φ is invariant in G because F acts trivially on Z and φ vanishes on the elements of $E - Z$. Let α be an extension of φ to G, and observe that G/E has an irreducible character β of degree 3. Let $\chi = \alpha\beta$, so $\chi(1) = 15$ and $\chi \in \mathrm{Irr}(G)$ by Gallagher's theorem.

We argue next that χ is monomial. Since K acts trivially on E, there is an abelian subgroup J of index 5 in EK. Then $|G : J| = 15 = \chi(1)$, and since J is abelian, χ_J has a linear constituent λ. It follows that $\lambda^G = \chi$, so χ is monomial, as claimed.

Now let $L \lhd F$, where $|L| = 3^2$ and $L \neq K$, and let $N = EL$. Then χ_N is a sum of three irreducible characters of degree 5, and we let θ be one of these. Then $\theta^G = \chi$, and so the monomial character χ is the unique irreducible character of G lying over θ. It suffices to show, therefore, that θ is not monomial.

If θ is monomial, there exists a subgroup $H \subseteq N$ such that $|N : H| = \theta(1) = 5$. Then $|E : E \cap H| = 5$, and $E \cap H$ is normalized by both E and H. It follows that $E \cap H \lhd N$, and thus L normalizes $E \cap H$. The action of L on E is the action of $L/(L \cap K)$, and this is the action of C. This is a contradiction, however, because C normalizes no subgroup having index 5 in E. ∎

We shall see in Corollary 9.16, however, that if $\theta \in \mathrm{Irr}(N)$, where $N \lhd G$ and G is p-solvable for some odd prime p, then if even one of the members of $\mathrm{Irr}(G|\theta)$ is monomial *and* has p-power degree, then θ is guaranteed to be monomial. In the example described in the proof of Lemma 9.9, therefore, it is essential that the character χ does not have prime power degree. (Recall that we had $\chi(1) = 15$.)

Another possible approach to finding a proof that normal subgroups of odd-order M-groups must be M-groups would be to show that in an odd-order M-group every irreducible character is supermonomial. (Recall from Chapter 2 that a character $\chi \in \mathrm{Irr}(G)$ is **supermonomial** if every primitive character that induces χ is linear or, equivalently, every character that induces χ is monomial.) If it is true that every irreducible character of an odd-order M-group must actually be supermonomial, it follows easily that normal subgroups of odd-order M-groups are M-groups.

To see this, observe that it suffices to assume that $|G : N|$ is prime. If $\theta \in \mathrm{Irr}(N)$, there are two possibilities: either θ^G is irreducible, and we write $\theta^G = \chi$; or else θ extends to an irreducible character χ of G. In the first case, θ is monomial because χ is supermonomial. In the second case, $\chi = \lambda^G$ for some linear character λ of a subgroup H of G, where $H \not\subseteq N$. Then $NH = G$, and thus

$$(\lambda_{N\cap H})^N = (\lambda^G)_N = \chi_N = \theta\,,$$

and hence θ is monomial.

Unfortunately, there is very little evidence that irreducible characters of odd-order M-groups must be supermonomial. In Theorem 9.14, however, we will prove that a monomial character of odd prime-power degree of a solvable group must be supermonomial, and that, at least, is suggestive.

Although it is not known if normal subgroups of odd-order M-groups must be M-groups, it is possible that something even stronger might be true. If $N \lhd G$, where G is an M-group, perhaps it is enough to assume that *at least one* of $|N|$ or $|G/N|$ is odd in order to conclude that N is an M-group. There is some slight evidence for this since if $N \lhd G$, where G is an M-group, and if one of $|N|$ or $|G/N|$ is odd, then every primitive character of N is linear. (Of course, this is what we would expect if N actually is an M-group.) We prove the following, which is slightly stronger.

Theorem 9.10. *Let $N \lhd G$, where G is an M-group, and suppose $\theta \in \mathrm{Irr}(N)$ is primitive. Then $\theta(1)$ is a power of 2, and if either $|G : N|$ or $|N|$ is odd, then θ is linear.*

Our proof of Theorem 9.10 relies on the following result.

Theorem 9.11. *Let G be p-solvable, where $p \neq 2$, and let $M \lhd G$, where $|G : M|$ is a power of p. Suppose that $\chi \in \mathrm{Irr}(G)$ is primitive and has p-power degree. Then χ_M is irreducible and primitive.*

The fact that χ_M is irreducible in Theorem 9.11 is essentially trivial, and this part of the conclusion holds even if $p = 2$. The primitivity of χ_M, however, lies much deeper, and we know no proof of this that uses only character theoretic techniques. Also, we mention that the conclusion in Theorem 9.11 that χ_M is primitive can fail if $p = 2$; there is a counterexample with $|G| = 2^8{\cdot}7$ and $|G : M| = 2$.

Our proof of Theorem 9.11 relies on the theory of symplectic modules over finite fields, and we have decided to defer the argument to Chapter 10, where the relevant machinery is introduced. (This result appears as Theorem 10.2.)

We shall also need the following for our proof of Theorem 9.10.

Theorem 9.12. *Let $N \lhd G$, where G is p-solvable and $p \neq 2$, and suppose that $\gamma \in \mathrm{Irr}(N)$ is primitive and has p-power degree. Let $\chi \in \mathrm{Irr}(G|\gamma)$, and assume that $\chi = \psi^G$ for some character ψ of a subgroup $H \subseteq G$, where $|N : N \cap H|$ is a power of p. Then $\mathbf{O}^p(N) \subseteq H$.*

Proof. Since $|N : N \cap H|$ is a power of p, we see that if $N \cap H \lhd G$, then $\mathbf{O}^p(N) \subseteq N \cap H \subseteq H$, as wanted. We can thus assume that $N \cap H$ is not normal in G, and in particular, $N \cap H \neq N$. Then $N \not\subseteq H$, so $H < G$, and we proceed by induction on $|G|$.

We argue first that it suffices to prove the theorem in the case where H is maximal in G. To see this, let $H \subseteq M$, where M is a maximal subgroup of G. Write $\zeta = \psi^M$, so $\zeta^G = \chi$, and in particular ζ is irreducible. By hypothesis, $|N : N \cap H|$ is a power of p, and thus $|N : N \cap M|$ is also a power of p, and since we are temporarily assuming the result for maximal subgroups, we have $\mathbf{O}^p(N) \subseteq M$.

Now $\psi^M = \zeta$ and $|M| < |G|$, and we propose to apply the inductive hypothesis with M in place of G and ζ in place of χ, and with $\mathbf{O}^p(N)$ in place of N. Since $\mathbf{O}^p(N) \lhd G$ and $\mathbf{O}^p(N) \subseteq M$, we have $\mathbf{O}^p(N) \lhd M$ as needed. Also, $|\mathbf{O}^p(N) : \mathbf{O}^p(N) \cap H|$ divides $|N : N \cap H|$, which is a power of p.

The only remaining condition that we must check in order to apply the inductive hypothesis in this situation is that ζ should lie over some p-power-degree primitive character of $\mathbf{O}^p(N)$. By Theorem 9.11, we see that $\gamma_{\mathbf{O}^p(N)}$ is primitive. Also, χ lies over γ and $\mathbf{O}^p(N) \lhd G$, and thus the irreducible constituents of $\chi_{\mathbf{O}^p(N)}$ are the G-conjugates of $\gamma_{\mathbf{O}^p(N)}$. Since χ lies over ζ, it follows that ζ lies over some G-conjugate of $\gamma_{\mathbf{O}^p(N)}$. This conjugate is primitive and has p-power degree, as wanted. The inductive hypothesis thus yields $\mathbf{O}^p(N) = \mathbf{O}^p(\mathbf{O}^p(N)) \subseteq H$, as required.

We can now assume that H is maximal in G, and we recall that $N \not\subseteq H$, so $NH = G$. Then $|G : H| = |NH : H| = |N : N \cap H|$, and since this is a power of p, we can apply Lemma 7.8. As in that lemma, we let $L = \mathrm{core}_G(N \cap H)$, and we choose a chief factor K/L of G with $K \subseteq N$.

Since γ is primitive, γ_L is a multiple of some character $\varphi \in \mathrm{Irr}(L)$, and since χ lies over γ, it follows that χ lies over φ. Now χ lies over ψ, and hence ψ lies over some G-conjugate φ^g of φ. Replacing γ by γ^g and φ by φ^g, we can assume that ψ lies over φ.

Now let T be the stabilizer of φ in G, and observe that since φ is invariant in N, we have $N \subseteq T$. Write $S = T \cap H$, so S is the stabilizer of φ in H. Both χ and ψ lie over φ, so we can consider the Clifford correspondents $\eta \in \mathrm{Irr}(T|\varphi)$ for χ and $\xi \in \mathrm{Irr}(S|\varphi)$ for ψ.

Now $\xi^H = \psi$, so $(\xi^T)^G = (\xi^H)^G = \psi^G = \chi$, and thus ξ^T is an irreducible character of T that lies over φ and induces χ, and it follows that $\xi^T = \eta$. Assuming that $T < G$, we propose to use the inductive hypothesis with T in place of G and S in place of H, and with η in place of χ and ξ in place of ψ. Now $N \cap S = N \cap (T \cap H) = N \cap H$, and thus $|N : N \cap S| = |N : N \cap H|$, and this is a power of p, as required.

In order to apply the inductive hypothesis, we must check one further condition: that η lies over some p-power degree primitive character of N. Now χ lies over η, so η lies over some G-conjugate of γ, and this conjugate is primitive and has p-power degree, as needed. By the inductive hypothesis, therefore, we can conclude that $\mathbf{O}^p(N) \subseteq S \subseteq H$, as wanted.

We can now assume that $T = G$, so φ is G-invariant, and we work to obtain a contradiction. Since $K < N$ by Lemma 7.8(a), we can choose a chief factor M/K of G with $M \subseteq N$, as in Lemma 7.8(c). Now Corollary 7.20 guarantees that there is a unique M-invariant member θ of $\mathrm{Irr}(K|\varphi)$, and since φ is G-invariant, it follows that θ is also G-invariant.

We can write $\psi_L = a\varphi$ for some positive integer a. Also, H is a complement for K relative to L in G by Lemma 7.8(a), and thus

$$\chi_K = (\psi^G)_K = (\psi_L)^K = a\varphi^K ,$$

and we see that χ lies over every member of $\mathrm{Irr}(K|\varphi)$. In particular, χ lies over θ, and since θ is G-invariant, it follows that χ_K is a multiple of θ. We conclude that θ is the unique member of $\mathrm{Irr}(K|\varphi)$, and so by Lemma 7.2, we see that φ is fully ramified with respect to K/L.

By Lemma 7.8(d), every complement for K relative to L in G is conjugate to H. Also, K/L is a nontrivial abelian p-group, and in particular, $|K : L|$ is odd. It follows by Theorem 7.5(c) that no member of $\mathrm{Irr}(H|\varphi)$ can induce irreducibly to G, and this is a contradiction since $\psi^G = \chi$. ∎

We can now prove the main theorem of this section.

Proof of Theorem 9.10. We are given a normal subgroup N of an M-group G, and a primitive character $\theta \in \mathrm{Irr}(N)$. Given a prime p dividing $\theta(1)$, we must show that $p = 2$.

Since G is an M-group, it is solvable by Taketa's theorem, and hence N is solvable. Since θ is primitive, it follows by Theorem 2.9 that we can write $\theta = \gamma\delta$, where γ is p-special and δ is p'-special. We are assuming that p divides $\theta(1)$, so it follows that $\gamma(1)$ is a p-power exceeding 1. Also, γ is primitive since if $\gamma = \mu^N$, where μ is a character of some subgroup $J < N$, then

$$\theta = \gamma\delta = \mu^N\delta = (\mu\delta_J)^N ,$$

and this contradicts the primitivity of θ.

Let $\chi \in \mathrm{Irr}(G)$ lie over γ. Since G is an M-group, we can write $\chi = \psi^G$, where ψ is a linear character of some subgroup H of G. Let $\zeta = \psi^{NH}$ so $\zeta^G = \chi$, and thus ζ is irreducible. Since χ lies over γ, each irreducible constituent of χ_N is G-conjugate to γ, and hence each such constituent is p-special, primitive, and nonlinear. Since χ lies over ζ, it follows that the irreducible constituents of ζ_N are all p-special, primitive, and nonlinear. Now

$$\zeta_N = (\psi^{NH})_N = (\psi_{N\cap H})^N ,$$

so the irreducible constituents of $(\psi_{N\cap H})^N$ are all p-special, and it follows by Theorem 2.29 that $|N : N \cap H|$ is a power of p.

Now assume that $p \neq 2$. It follows by Theorem 9.12 that $\mathbf{O}^p(N) \subseteq N \cap H$, and thus since γ lies over the linear character $\psi_{N\cap H}$, it follows that $\gamma_{\mathbf{O}^p(N)}$ has a linear constituent. Since γ is primitive and $|N : \mathbf{O}^p(N)|$ is a power of p, we can apply the easy part of Theorem 9.11 to deduce that $\gamma_{\mathbf{O}^p(N)}$ is irreducible, and it follows that γ is linear. This is a contradiction, however, and we conclude that $p = 2$.

To complete the proof, we must derive a contradiction if either $|N|$ is odd or $|G : N|$ is odd. First, $\gamma \in \mathrm{Irr}(N)$, and thus $\gamma(1)$ divides $|N|$. Since $\gamma(1)$ is a nontrivial power of 2, this cannot happen in the case where $|N|$ is odd.

Assume now that $|G : N|$ is odd. Since $\zeta = \psi^{NH}$, we see that $\zeta(1) = |NH : H|\psi(1) = |NH : H| = |N : N \cap H|$, and we have seen that this is a power of $p = 2$. Since $|G : N|$ is odd, it follows that ζ_N is irreducible. We saw previously that the irreducible constituents of ζ_N are primitive and nonlinear, and it follows that ζ_N is a primitive nonlinear character of N. Also, $\zeta_N = (\psi_{N\cap H})^N$, and since ζ_N is primitive, it follows that $N \cap H$ cannot be a proper subgroup of N. Then $N = N \cap H$ and thus $\zeta_N = \psi_N$. It follows that ζ_N is linear, and this is the desired contradiction. ∎

As a consequence of Theorem 9.10, we obtain the following strong form of a result of I. A. Chubarov.

Corollary 9.13. *Let $N \triangleleft G$ where G is an M-group, and suppose that every proper subgroup of N is an M-group. If either $|N|$ is odd or $|G : N|$ is odd, then N is an M-group.*

Proof. Let $\theta \in \mathrm{Irr}(N)$, and assume first that θ is not primitive. Then $\theta = \varphi^N$, where φ is an irreducible character of some proper subgroup of N.

By hypothesis, φ is monomial, and thus θ is monomial, as wanted. We can thus assume that θ is primitive, so θ is linear by Theorem 9.10, and thus θ is monomial. ∎

In his original version of Corollary 9.13, Chubarov assumed that both $|N|$ and $|G:N|$ are odd, but in fact, as we have seen, it suffices to assume only that one of these two numbers is odd. We mention that in another paper, Chubarov argued that every normal subgroup of an odd-order M-group must be an M-group. Unfortunately, however, his argument was not correct, and this remains an open problem.

9C

In this section we establish some conditions sufficient to guarantee that a monomial irreducible character is supermonomial.

Theorem 9.14. *Let G be p-solvable, where $p \neq 2$, and suppose that $\chi \in \mathrm{Irr}(G)$ is monomial and has p-power degree. Then χ is supermonomial.*

The following decidedly nontrivial result (which is essentially due to Dade) lies at the heart of Theorem 9.14.

Theorem 9.15. *Let G be p-solvable, where $p \neq 2$, and suppose that $\chi \in \mathrm{Irr}(G)$ is monomial and has p-power degree. Let $N \lhd G$, and let $\theta \in \mathrm{Irr}(N)$ lie under χ. Then the Clifford correspondent of χ with respect to θ is monomial.*

We can view the conclusion of Theorem 9.15 as a kind of "partial supermonomiality" for the character χ. True supermonomiality, of course, would require that *every* character that induces χ should be monomial. The assertion of Theorem 9.15 is weaker, however, because it applies only to Clifford induction.

As was the case for Theorem 9.11, Theorem 9.15 relies on the theory of symplectic modules, and we defer the proof to Chapter 10, where the result appears as Theorem 10.1.

Proof of Theorem 9.14. We are given that χ is monomial and has p-power degree, and we must show that if $\alpha^G = \chi$, where α is a character of some subgroup $H \subseteq G$, then α is monomial. If $H = G$, then $\alpha = \chi$, so α is monomial and there is nothing further to prove. We can assume, therefore, that $H < G$, and we proceed by induction on $|G|$.

We argue next that it suffices to prove the result in the case where H is a maximal subgroup of G. To deduce the general result from this, let $H \subseteq M$, where M is maximal in G. Write $\beta = \alpha^M$, so $\beta^G = \chi$. Since we are temporarily assuming that the theorem holds in the case where H is a maximal subgroup, it follows that β is monomial. Also, $\beta(1)$ divides $\chi(1)$, so $\beta(1)$ is a power of p, and by the inductive hypothesis applied with M in place of G and β in place of χ, we conclude that α is monomial, as required.

Assuming now that H is a maximal subgroup of G, let $L = \mathrm{core}_G(H)$, and choose $\varphi \in \mathrm{Irr}(L)$, where φ lies under α. Now $\chi(1)$ is a power of p and $\alpha^G = \chi$, so $|G : H| = \chi(1)/\alpha(1)$ is a power of p, and in particular, p divides $|G : H|$. Since $p \neq 2$ and α^G is irreducible, we can apply Theorem 7.21 to deduce that φ is not invariant in G.

Let T be the stabilizer of φ in G, so $T < G$. Write $S = T \cap H$, and observe that S is the stabilizer of φ in H. Let $\xi \in \mathrm{Irr}(S|\varphi)$ and $\eta \in \mathrm{Irr}(T|\varphi)$ be the Clifford correspondents of α and χ, respectively. We have $\eta^G = \chi$, and thus $\eta(1)$ divides $\chi(1)$, and it follows that η has p-power degree.

Now
$$(\xi^T)^G = \xi^G = (\xi^H)^G = \alpha^G = \chi\,,$$
so ξ^T is an irreducible character of T that lies over φ and induces χ, and it follows that $\xi^T = \eta$. Theorem 9.15 guarantees that η is monomial, and thus η is supermonomial by the inductive hypothesis applied with T in place of G. Since $\xi^T = \eta$, it follows that ξ is monomial, and thus $\alpha = \xi^H$ is monomial, as required. ∎

Corollary 9.16 (Dade). *Let $N \lhd\lhd G$, where G is p-solvable with $p \neq 2$, and let $\chi \in \mathrm{Irr}(G)$ be monomial and have p-power degree. Then every irreducible constituent of χ_N is monomial.*

Proof. Working by induction on $|G|$, it suffices to prove that if M is a maximal normal subgroup of G containing N, then the irreducible constituents of χ_M are monomial. Since the irreducible constituents of χ_M have p-power degree, the theorem will follow by the inductive hypothesis applied with M in place of G.

To show that the irreducible constituents of χ_M are monomial, assume first that χ_M is irreducible. Write $\chi = \lambda^G$, where λ is a linear character of some subgroup $H \subseteq G$, and note that $|G : H| = \chi(1)$. Since $\lambda^G = \chi$ is irreducible, it follows that λ^{MH} is irreducible and lies under χ. By assumption, however, χ_M is irreducible, so χ_{MH} is irreducible, and thus $\chi_{MH} = \lambda^{MH}$. We have $|G : H| = \chi(1) = \lambda^{MH}(1) = |MH : H|$, and thus $MH = G$. Then $\chi_M = (\lambda^G)_M = (\lambda_{M\cap H})^M$, and thus χ_M is monomial, as required.

We can now assume that χ_M is not irreducible, so $|G : M|$ is not relatively prime to $\chi(1)$. By hypothesis, $\chi(1)$ is a power of p, so p divides $|G : M|$, and thus $|G : M| = p$ because G is p-solvable and M is a maximal normal subgroup.

Let θ be an irreducible constituent of χ_M, and observe that $\chi_M \neq \theta$ because we are assuming that χ_M is not irreducible. By Lemma 1.8, the number of distinct conjugates of θ in G is $|G : M|$, and it follows that M is the full stabilizer of θ in G. Then θ^G is irreducible by the Clifford correspondence, and we conclude that $\theta^G = \chi$. Since χ is monomial and has p-power degree, χ must be supermonomial by Theorem 9.14, and thus θ is monomial, as required. ∎

By Theorem 9.14, we know that if $\chi \in \mathrm{Irr}(G)$ is monomial and $\chi(1)$ is a power of an odd prime p, then χ is guaranteed to be supermonomial. If we assume that $|G|$ is odd, there is a different condition that is also sufficient to guarantee that a monomial character χ is supermonomial, and we discuss that next.

We are assuming that $|G|$ is odd, so G is π-separable for every set π of primes, and thus we can apply the theory of π-partial characters introduced in Chapter 3. Recall that a π-partial character of G is, by definition, the restriction χ^0 of some character χ of G to the set G^0 of π-elements of G. Also, a π-partial character φ of G is irreducible if it is not possible to write φ as a sum of two (nonzero) π-partial characters, and the set of irreducible π-partial characters of G is denoted $\mathrm{I}_\pi(G)$.

Recall that $\mathrm{B}_\pi(G)$ is a canonically defined subset of $\mathrm{Irr}(G)$, and the map $\chi \mapsto \chi^0$ is a bijection from $\mathrm{B}_\pi(G)$ onto $\mathrm{I}_\pi(G)$. (This is Theorem 5.1.) Also, if $H \subseteq G$ and $\beta \in \mathrm{B}_\pi(H)$, where β^G is irreducible, then in the case where $|G|$ is odd, Corollary 4.37(b) guarantees that β^G lies in $\mathrm{B}_\pi(G)$.

Theorem 9.17. *Suppose that $|G|$ is odd, and let $\chi \in \mathrm{Irr}(G)$ be monomial. Then χ is supermonomial if there exists a prime p such that the restriction χ^0 of χ to the set of p-elements of G lies in $\mathrm{I}_p(G)$. In particular, if χ is monomial and lies in $\mathrm{B}_p(G)$, then χ is supermonomial.*

Note that when we discussed p-Brauer characters from the point of view of π-partial characters, we took π to be the set of all primes different from p. In Theorem 9.17, however, we are doing exactly the opposite: we are taking $\pi = \{p\}$.

We need some additional notation. Continuing to assume that $|G|$ is odd, we fix a prime p, and we observe that G is p-solvable. Write $\mathcal{P}(G)$ to denote the set of pairs (U, μ), where U is a subgroup of G and μ is an irreducible p-partial character of U such that $\mu(1)$ is a power of p. Also, given $\varphi \in \mathrm{I}_p(G)$, it follows by Corollary 5.5 that there exists a pair $(U, \mu) \in \mathcal{P}(G)$ such that $\mu^G = \varphi$, and we write $\mathcal{P}(\varphi)$ to denote the set of such pairs (U, μ).

Next, we define a partial order on $\mathcal{P}(G)$ by setting $(U, \mu) \le (V, \nu)$ if $U \subseteq V$ and $\mu^V = \nu$. Observe that if $(U, \mu) \le (V, \nu)$ and $\varphi \in \mathrm{I}_p(G)$, then $\mu^G = \varphi$ if and only if $\nu^G = \varphi$. In other words, if $(U, \mu) \le (V, \nu)$, then (U, μ) lies in $\mathcal{P}(\varphi)$ if and only if (V, ν) lies in $\mathcal{P}(\varphi)$.

Finally, we shall say that a p-partial character φ of G is monomial if there exists a subgroup $H \subseteq G$ and a linear p-partial character $\lambda \in \mathrm{I}_p(H)$ such that $\lambda^G = \varphi$. Also, if $(U, \mu) \in \mathcal{P}(G)$ and μ is monomial, we say that the pair (U, μ) is monomial.

Lemma 9.18. *Suppose that $|G|$ is odd and that $(U, \mu) \le (V, \nu)$, where (U, μ) and (V, ν) lie in $\mathcal{P}(G)$. Then μ is monomial if and only if ν is monomial.*

Proof. Since $\nu = \mu^V$, it is clear that if μ is monomial, then ν is also monomial. Conversely, assume that ν is monomial, and write $\nu = \lambda^V$, where λ is a linear p-partial character of some subgroup $H \subseteq V$. Let $\sigma \in \mathrm{B}_p(H)$ with $\sigma^0 = \lambda$, and write $\beta = \sigma^V$. Then

$$\beta^0 = (\sigma^V)^0 = (\sigma^0)^V = \lambda^V = \nu \in \mathrm{I}_p(V)\,,$$

and it follows that $\beta \in \mathrm{Irr}(V)$. Also, $|V|$ is odd and $\sigma \in \mathrm{B}_p(H)$, and we conclude that $\beta \in \mathrm{B}_p(V)$.

Observe that β is monomial because $\beta = \sigma^V$ and σ is linear. Also, $\beta(1) = \nu(1)$ is a power of p, and thus β is supermonomial by Theorem 9.14.

Let $\alpha \in \mathrm{B}_p(U)$ with $\alpha^0 = \mu$. Then

$$(\alpha^V)^0 = (\alpha^0)^V = \mu^V = \nu\,,$$

and thus $\alpha^V \in \mathrm{Irr}(V)$. Also, since $|U|$ is odd and $\alpha \in \mathrm{B}_p(U)$, it follows $\alpha^V \in \mathrm{B}_p(V)$.

Each of α^V and β lies in $\mathrm{B}_p(V)$, and we have

$$(\alpha^V)^0 = \nu = \beta^0\,,$$

and it follows that $\alpha^V = \beta$. We have seen, however, that β is supermonomial, and thus α is monomial, and we can write $\alpha = \tau^U$, where τ is a linear character of some subgroup of U. Then

$$\mu = \alpha^0 = (\tau^U)^0 = (\tau^0)^U\,,$$

and thus μ is a monomial p-partial character, as required. ∎

Theorem 9.19. *Suppose that $|G|$ is odd, and let $\varphi \in \mathrm{I}_p(G)$. If φ is monomial, then all members of $\mathcal{P}(\varphi)$ are monomial, and otherwise, no member of $\mathcal{P}(\varphi)$ is monomial.*

Proof. If some member (U, μ) of $\mathcal{P}(\varphi)$ is monomial, then since $\varphi = \mu^G$, it is clear that φ is monomial. Conversely, if φ is monomial, write $\varphi = \lambda^G$, where λ is a linear p-partial character of some subgroup W of G. Since $\lambda(1) = 1$ is a power of p, we see that $(W, \lambda) \in \mathcal{P}(\varphi)$, and of course, (W, λ) is monomial because λ is linear. It remains to show that if some member of $\mathcal{P}(\varphi)$ is monomial, then every member of $\mathcal{P}(\varphi)$ is monomial. We must show, in other words, that it is not possible to have two pairs (U, μ) and

(V, ν) in $\mathcal{P}(\varphi)$ such that one of them is monomial and the other is not. We prove this by induction on $|G|$.

By Lemma 9.18, we can replace each of (U, μ) and (V, ν) by a maximal member of $\mathcal{P}(\varphi)$. If $\varphi(1)$ is a power of p, then (G, φ) is the unique maximum member of $\mathcal{P}(\varphi)$, so in this case $(U, \mu) = (G, \varphi) = (V, \nu)$, and there is nothing further to prove. We can assume, therefore, that $\varphi(1)$ is not a power of p, and it follows by Lemma 5.12 that there exists a subgroup $N \lhd G$ such that the irreducible constituents of φ_N are not G-invariant, and their common degree is a power of p. (Recall that Corollary 5.7 guarantees that the set of irreducible constituents of φ_N forms an orbit under the conjugation action of G.)

Since $\mu^G = \varphi$ and the degrees of the irreducible constituents of φ_N are powers of p, it follows by Lemma 5.21 that $|NU : U|$ is a power of p. Also, μ^{NU} is irreducible because $\mu^G = \varphi$ is irreducible. Furthermore, $\mu^{NU}(1) = |NU : U|\mu(1)$, and since this is a power of p, we see that (NU, μ^{NU}) lies in $\mathcal{P}(\varphi)$. We have $(U, \mu) \leq (NU, \mu^{NU})$, and it follows by the maximality of (U, μ) that $NU = U$. Similarly $NV = V$, and thus N is contained in both U and V.

Now $\mu^G = \varphi$, so φ lies over μ by Lemma 5.8, and thus φ lies over the irreducible constituents of μ_N, and similarly φ lies over the irreducible constituents of ν_N. By Corollary 5.7, all irreducible constituents of φ_N are conjugate in G, so replacing the pair (U, μ) by a conjugate if necessary, we can assume that μ_N and ν_N have a common irreducible constituent θ.

Let T be the stabilizer of θ in G, and note that $T < G$. By Theorem 5.11, there exists a unique member $\zeta \in \mathrm{I}_p(T|\theta)$ such that $\zeta^G = \varphi$. Also, writing $X = T \cap U$, we see by Theorem 5.11 that there is a unique member $\xi \in \mathrm{I}_p(X|\theta)$ such that $\xi^U = \mu$, and in particular, ξ has p-power degree. Now $\xi^G = (\xi^U)^G = \mu^G = \varphi$, and we argue that $\xi^T = \zeta$.

We have

$$(\xi^T)^G = \xi^G = (\xi^U)^G = \mu^G = \varphi,$$

so ξ^T is irreducible. Also, ξ^T lies over ξ by Lemma 5.8, and hence ξ^T lies over θ, and it follows that $\xi^T = \zeta$, as claimed. Since $\xi(1)$ is a power of p, we see that $(X, \xi) \in \mathcal{P}(\zeta)$. and $(X, \xi) \leq (U, \mu)$. Similarly, there exists a pair $(Y, \eta) \in \mathcal{P}(\zeta)$ such that $(Y, \eta) \leq (V, \nu)$.

Now (X, ξ) and (Y, η) lie in $\mathcal{P}(\zeta)$, and by Lemma 9.18, one of these pairs is monomial, and the other is not. Since $T < G$, this contradicts the inductive hypothesis, and the proof is complete. ∎

Corollary 9.20. *Suppose that $|G|$ is odd and that $\varphi \in \mathrm{I}_p(G)$ is monomial. If $H \subseteq G$ and $\sigma \in \mathrm{I}_p(H)$ with $\sigma^G = \varphi$, then σ is monomial.*

Proof. We have seen that $\mathcal{P}(\sigma)$ is nonempty, so we can choose $(U, \mu) \in \mathcal{P}(\sigma)$. Then $\mu^H = \sigma$, so $\mu^G = \sigma^G = \varphi$, and thus $(U, \mu) \in \mathcal{P}(\varphi)$. Since φ is monomial by hypothesis, Theorem 9.19 guarantees that μ is monomial, and thus $\sigma = \mu^G$ is monomial, as wanted. ■

Lemma 9.21. *Let $\xi \in \mathrm{Irr}(G)$, where ξ is primitive and $|G|$ is odd. Fix a prime p, and assume that the restriction ξ^0 of ξ to the p-elements of G is an irreducible monomial p-partial character of G. Then ξ is linear.*

Proof. Suppose that ξ is not linear, and let q be a prime divisor of $\xi(1)$. By Theorem 2.9, we can write $\xi = \alpha\beta$, where α is q-special and β is q'-special, and we observe that α is not linear because we are assuming that q divides $\xi(1)$. Restricting to p-elements, we see that $\xi^0 = \alpha^0\beta^0$, and it follows that α^0 must be irreducible as a p-partial character.

Now let $P \in \mathrm{Syl}_p(G)$. We argue that the principal character 1_P of P cannot be a constituent of $(\alpha^0)_P = \alpha_P$. Otherwise, 1_P is a constituent of the smallest possible degree of $(\alpha^0)_P$, and thus 1_P is a Fong character for the irreducible p-partial character α^0. This would imply that α^0 is the principal p-partial character of G, and this is not the case because α is not linear.

Clearly, α_P has values in some cyclotomic field $\mathbb{Q}_a$, where a is a power of p. Also, since α is q-special, all values of α lie in a cyclotomic field $\mathbb{Q}_b$, where b is a power of q. If $q \neq p$, therefore, α_P is rational valued, and so complex conjugation permutes the irreducible constituents of α_P.

Now $\alpha(1)$ is odd, so some irreducible constituent of α_P is fixed by complex conjugation. Since $|G|$ is odd, however, the principal character 1_P is the unique real irreducible character of P, and thus 1_P is a constituent of α_P. This is a contradiction, however, and we conclude that $q = p$, and thus p is the unique prime divisor of $\xi(1)$, and it follows that the p'-special factor β of ξ is linear.

By hypothesis, ξ^0 is a monomial p-partial character of G, so we can write $\alpha^0\beta^0 = \xi^0 = \lambda^G$, where λ is a linear p-partial character of some subgroup $H \subseteq G$. Since β is linear, we have

$$\alpha^0 = \lambda^G\overline{\beta^0} = (\lambda(\overline{\beta^0})_H)^G\,,$$

and thus α^0 is monomial.

We can now write $\alpha^0 = (\mu^G)^0$, where μ is some linear character in $\mathrm{B}_p(H)$, and since α^0 is irreducible, it follows that μ^G is irreducible. Since $\mu \in \mathrm{B}_p(H)$ and $|G|$ is odd, we deduce that μ^G lies in $\mathrm{B}_p(G)$. Also, $\alpha \in \mathrm{B}_p(G)$ because α is p-special, and since $\alpha^0 = (\mu^G)^0$, we conclude that $\mu^G = \alpha$. Then

$$(\mu\beta_H)^G = \mu^G\beta = \alpha\beta = \xi\,,$$

and since ξ is primitive, we have $H = G$ and $\mu\beta = \xi$. This is a contradiction since we assumed that ξ is not linear. ∎

Proof of Theorem 9.17. We are given a monomial character $\chi \in \mathrm{Irr}(G)$, where $|G|$ is odd and $\chi^0 \in \mathrm{I}_p(G)$. To show that χ is supermonomial, we suppose that $\chi = \xi^G$, where ξ is a primitive character of some subgroup H of G, and we prove that ξ is linear.

Now $(\xi^0)^G = (\xi^G)^0 = \chi^0$, and since χ^0 is irreducible by hypothesis, it follows that ξ^0 is irreducible. Also, χ is monomial, so χ^0 is a monomial irreducible p-partial character of G. It follows by Corollary 9.20 that ξ^0 is monomial. Now Lemma 9.21 guarantees that ξ is linear, as required. ∎

By Corollary 5.16, we know that a primitive character of a Hall subgroup of an M-group must be linear. More generally, it follows by a theorem of Navarro that if ψ is a primitive character of an arbitrary subgroup H of an M-group G, then every prime dividing $\psi(1)$ must divide $|G : H|$. (Note that Corollary 5.16 is an immediate corollary of Navarro's result because if H is a Hall subgroup of G, then every prime divisor of $\psi(1)$ would have to divide both $|G : H|$ and $|H|$, and this implies that $\psi(1) = 1$.)

If ψ is a primitive character of a *maximal* subgroup H of an M-group G, then even more can be said. If we are willing to assume that $|G : H|$ is odd, then by a result of Lewis, $\psi(1)^2$ divides $|G : H|$. In fact, by a theorem of Isaacs and Wilde, it follows in this situation that either $\psi(1) = 1$ or $\psi(1)^2 = |G : H|$. (Note that although in the theorems of Lewis and of Isaacs and Wilde, it is assumed that $|G : H|$ is odd, no oddness assumption is necessary for the result of Navarro, to which we referred in the previous paragraph.)

We will not present proofs of these deep theorems here, but as an application of Theorem 9.17, we obtain a weaker result that is in the same spirit.

Theorem 9.22. *Let H be a maximal subgroup of an odd-order M-group G, and suppose $\psi \in \mathrm{Irr}(H)$ is primitive. Then $\psi(1)$ is a power of the unique prime divisor q of $|G : H|$.*

Proof. Suppose that p is a prime dividing $\psi(1)$ with $p \neq q$. By Theorem 2.9, we can write $\psi = \alpha\beta$, where α is p-special and β is p'-special. Since we are assuming that p divides $\psi(1)$, we see that p must divide $\alpha(1)$, and in particular, α is not linear, Also, since $\psi = \alpha\beta$ is primitive, it follows that α is primitive, and in particular, α is not monomial.

Let $L = \mathrm{core}_G(H)$, so $L < G$. Let K/L be a chief factor of G, and observe that $K \not\subseteq H$. Then $KH = G$ and $|G : H| = |KH : H| = |K : H \cap K|$, and this divides $|K : L|$. It follows that K/L is an abelian q-group, and $K \cap H = L$.

Since α is primitive, α_L is a multiple of some character $\sigma \in \mathrm{Irr}(L)$, and thus H is contained in the stabilizer T of σ in G. Also, because H is maximal in G, we see that either $T = H$ or $T = G$.

Suppose first that $T = H$. By the Clifford correspondence, α^G is irreducible, and we write $\alpha^G = \chi$. Now α is p-special, so $\alpha \in \mathrm{B}_p(G)$, and since by hypothesis, $|G|$ is odd, it follows that χ lies in $\mathrm{B}_p(G)$.

Since G is an M-group, χ is monomial, and thus χ is supermonomial by Theorem 9.17. Since $\alpha^G = \chi$, it follows that α is monomial, and this is a contradiction.

We can now assume that $T = G$, and in particular, σ is invariant in K. Now σ lies under the p-special character α, so σ is p-special, and since K/L is a q-group and $q \neq p$, we see that σ has a canonical extension $\tau \in \mathrm{Irr}(K)$. The stabilizer of σ in G fixes τ, so it follows by Lemma 2.11(a) that there exists $\chi \in \mathrm{Irr}(G)$ such that $\chi_H = \alpha$. Now χ is monomial, and it follows that α is monomial, and again we have a contradiction. ∎

Problems

9.1. Let $G = AB$, where $[A, B] = 1$. If A and B are M-groups, show that G is an M-group.

9.2. Let G be an M-group, and suppose that $N \lhd G$, where N and G/N are nilpotent. Show that every Hall subgroup of G is an M-group.

9.3. Let $A \lhd G$, where A is abelian, and suppose that $\chi \in \mathrm{Irr}(G)$ is monomial. Show that there exists a subgroup H of G and a linear character $\tau \in \mathrm{Irr}(H)$ such that $A \subseteq H$ and $\tau^G = \chi$.

Hint: Let λ be a linear character of a subgroup J of G, where $\lambda^G = \chi$, and let $\psi = \lambda^{AJ}$. Show that the irreducible constituents of ψ_A occur with multiplicity 1.

9.4. Let $N \lhd G$ and suppose that $\chi \in \mathrm{Irr}(G)$ is monomial and that χ_N has a linear constituent λ. Let T be the stabilizer of λ in G, and let $\eta \in \mathrm{Irr}(T|\lambda)$ be the Clifford correspondent of χ. Show that η is monomial.

9.5. Suppose that every irreducible character of G has the form α^G, where α is a character of some subgroup J, where $J/\ker(\alpha)$ is solvable. Show that G is solvable.

Hint: Let $\chi \in \mathrm{Irr}(G)$ have minimum degree such that $G/\ker(\chi)$ is nonsolvable, and let L and α be as in the statement of the problem. Write $J^{(\infty)}$ to denote the final term of the derived series of J, and use ideas from the proof of Taketa's theorem to show that $J^{(\infty)} \lhd G$ and that $G/J^{(\infty)}$ is solvable.

9.6. Let $N \lhd G$, where N is nilpotent, and suppose that $\chi \in \operatorname{Irr}(G)$ is monomial. Let $\theta \in \operatorname{Irr}(N)$ lie under χ, let T be the stabilizer of θ in G, and let $\eta \in \operatorname{Irr}(T|\theta)$ be the Clifford correspondent of χ with respect to θ. Prove that η is monomial by assuming that G is a counterexample of minimum possible order and carrying out the following steps.

(a) Observe that χ is faithful. Let $A \subseteq N$ be maximal such that A is abelian and $A \lhd G$. Let α be a linear character of some subgroup J, where $\alpha^G = \chi$, and $A \subseteq J$, and suppose that J is chosen so that $|NJ|$ is as small as possible. Show that $NJ = G$.

(b) Let $J \subseteq S < G$ and write $\xi = \alpha^S$. Show that ξ_U is homogeneous, where $U = N \cap S$.

(c) Let $\lambda = \alpha_A$. Show that λ is invariant in G, and deduce that $A = \mathbf{Z}(G)$ and that N/A is abelian. (**Hint**: If X is a nilpotent group and $X/\mathbf{Z}(X)$ is not abelian, then X has a noncentral characteristic abelian subgroup.) Conclude that $A = M$.

(d) Let $\varphi = \alpha_M$, and let R be as in Theorem 7.3, applied to $M \lhd N$. Show that $R = N$, and deduce that $|\operatorname{Irr}(N|\varphi)| = |N : M|$. Conclude that $|G : T| = |N : M|$, and observe that this yields a contradiction.

Notes and References

K. Taketa's theorem on the solvability and derived length of M-groups appears in [**71**]. Taketa also wrote a number of other papers on M-groups, and sadly, in some of his later papers, there are "theorems" that are simply false.

T. R. Berger's proof of the Taketa inequality for odd-order groups appears in [**4**], and D. Gluck's inequality is proved in [**28**]. The earliest result in this direction for general solvable groups is the inequality $\operatorname{dl}(G) \leq 3|\operatorname{cd}(G)| - 2$, which appears in [**35**].

Theorem 9.3 appears in [**31**].

Dade's theorem that every finite solvable group is embedded in an M-group is Satz 18.11 of Huppert's book [**33**], where it is attributed to an oral communication from Dade.

Theorem 9.8 is Theorem 4.1 of [**16**].

An example of an M-group having a normal subgroup that is not an M-group can be found in [**12**] and the same example appears in [**72**].

Theorems 9.10, 9.11, and 9.12 appear in [**37**] as Theorem A, Theorem B, and Theorem C, respectively.

Corollary 9.13 is in [**9**]. Chubarov's erroneous proof that odd-order normal subgroups of M-groups are M-groups is in [**8**].

Theorem 9.14 appears as Theorem 10.1 in [**47**].

Theorem 9.15 is essentially due to Dade, and it appears in [**14**]. The main result of an earlier, preprint version of this paper, was Corollary 9.16 of this chapter. Isaacs pointed out to Dade that Theorem 9.15 was a consequence of Dade's argument, and (with appropriate credit) that result appeared in the published version of the paper.

Theorem 9.17 is a strong form of Theorem 10.2 of [**47**].

The result of Navarro that asserts that every prime divisor of the degree of a primitive character of a subgroup of an M-group must divide the index of the subgroup is Theorem B of [**59**].

The theorems concerning primitive characters of maximal subgroups of M-groups to which we referred are in [**55**] and [**54**].

Problem 9.6 is a somewhat weaker form of Theorem 3.1 in [**67**].

Chapter 10

Symplectic Modules and Character Theory

10A

In this chapter we introduce some techniques that are quite different from anything that we have presented previously. After we develop the necessary theory, we use it to establish the two theorems (Theorems 9.11 and 9.15) whose proofs were deferred from Chapter 9, and which, as far as we know, cannot be proved using only more standard character-theory methods.

First, we recall the two theorems. The first of these is essentially due to E. C. Dade, and the second is a result of Isaacs.

Theorem 10.1. *Let G be p-solvable, where $p \neq 2$, and suppose that $\chi \in \mathrm{Irr}(G)$ is monomial and has p-power degree. Let $N \triangleleft G$, and let $\theta \in \mathrm{Irr}(N)$ lie under χ. Then the Clifford correspondent of χ with respect to θ is monomial.*

Theorem 10.2. *Let G be p-solvable, where $p \neq 2$, and let $M \triangleleft G$, where G/M is a p-group. Suppose that $\chi \in \mathrm{Irr}(G)$ is primitive and has p-power degree. Then χ_M is irreducible and primitive.*

The assertion that χ_M is irreducible in Theorem 10.2 is essentially just an exercise, and its proof does not require that p should be odd. The real force of this result is the primitivity of χ_M, and for that conclusion, the assumption that p is odd is necessary. The oddness of p is also necessary for Theorem 10.1.

We start with a review of some elementary module theory. Let F be an arbitrary field, and let V be a finite-dimensional F-vector space on which a finite group G acts via linear operators. In this situation, we say that V

is an FG-module, and we stress that for us, the word "module" will always refer to a finite-dimensional module.

A G-invariant F-subspace U of an FG-module V is referred to as an FG-**submodule**, and it is easy to check that the factor space V/U inherits the structure of an FG-module in a natural way.

If V and W are FG-modules, we say that a linear transformation $\varphi : V \to W$ is an FG-**homomorphism** if $(xg)\varphi = (x\varphi)g$ for all $x \in V$ and $g \in G$. Also, an FG-homomorphism that is both injective and surjective is an FG-**isomorphism**. It is routine to check that the usual homomorphism theorems hold in this setting. For example, if $\varphi : V \to W$ is an FG-homomorphism, then the kernel $\ker(\varphi)$ is a submodule of V and the image $V\varphi$ is a submodule of W. Also, the factor module $V/\ker(\varphi)$ is FG-isomorphic to $V\varphi$.

If V is a nonzero FG-module that has no nonzero proper submodules, we say that V is **simple** as an FG-module, and we note that if V is an arbitrary nonzero FG-module, then by the finite dimensionality of V, it follows that V must contain some simple submodule. (This does not really require V to be finite dimensional; it can also be proved using our standing assumption that G is a finite group.)

If $\mathcal{X}$ is any (possibly infinite) collection of FG-submodules of V, then $\sum \mathcal{X}$, the sum of the members of $\mathcal{X}$, is defined to be the set of all elements of V that can be written as *finite* sums of elements of the various members of $\mathcal{X}$. (If $\mathcal{X}$ is empty, we view $\sum \mathcal{X}$ as the zero submodule of V.)

The following result is fundamental.

Theorem 10.3. *Let V be an FG-module. Then the following are equivalent.*

(1) *V is a direct sum of a (necessarily finite) collection of simple submodules.*
(2) *$V = \sum \mathcal{X}$, where $\mathcal{X}$ is some collection of simple submodules.*
(3) *For each submodule X of V, there exists a submodule Y such that $V = X \dotplus Y$, where the dot indicates that the sum is direct.*

Proof. It is clear that (1) implies (2), so we assume (2), and we work to prove (3). Given a submodule $X \subseteq V$, we must show that X is a direct summand of V, so we choose a (possibly zero) submodule $Y \subseteq V$ such that Y is maximal with the property that $X \cap Y = 0$. (This is possible because V is finite dimensional, but we mention that a Zorn's-lemma argument could be used to establish the existence of Y even without assuming finite dimensionality.)

To prove (3) now, it suffices to show that $X + Y = V$. Otherwise, since $V = \sum \mathcal{X}$, where $\mathcal{X}$ is a family of simple submodules, there exists a member $S \in \mathcal{X}$ such that $S \not\subseteq X + Y$, and since S is simple, we have $(X+Y)\cap S = 0$. Now $Y+S > Y$ because $S \not\subseteq Y$, and hence the maximality of Y guarantees that $X \cap (Y + S)$ must be nonzero. Let x be a nonzero element of $X \cap (Y + S)$, and write $x = y + s$, where $y \in Y$ and $s \in S$. Then $s = x - y \in (X + Y) \cap S = 0$, and hence $s = 0$. We conclude that $x = y$, and thus this element lies in $X \cap Y = 0$. Then $x = 0$, and this contradiction establishes (3)

Finally, assume (3), and choose a collection $\mathcal{X}$ of simple submodules of V such that $\mathcal{X}$ is maximal with the property that $\sum \mathcal{X}$ is direct. (The existence of such a maximal collection is, of course, a consequence of the finite dimensionality of V.) By (3), there exists a submodule Y of V such that $V = (\sum \mathcal{X}) \dotplus Y$. If Y is nonzero, it contains a simple submodule S, and since $(\sum \mathcal{X})\cap S \subseteq (\sum \mathcal{X})\cap Y = 0$, we see that the sum of the collection $\mathcal{X} \cup \{S\}$ is direct. This contradicts the maximality of $\mathcal{X}$, and we deduce that $Y = 0$, so $V = \sum \mathcal{X}$, proving (1). ∎

A module satisfying the three equivalent conditions of Theorem 10.3 is said to be **semisimple**, but condition (3) is sometimes referred to by saying that V is **completely reducible**.

Corollary 10.4. *Submodules of semisimple modules are semisimple.*

Proof. Let $U \subseteq V$, where V is a semisimple FG-module. We prove that U is semisimple by showing that it is completely reducible, so we consider a submodule $X \subseteq U$. By the complete reducibility of V, we can write $V = X \dotplus Y$ for some submodule $Y \subseteq V$. Then $U = X+(U\cap Y)$ by Dedekind's lemma, and this sum is direct because $X \cap (U \cap Y) \subseteq X \cap Y = 0$. ∎

If S is a simple FG-module and φ is a nonzero FG-homomorphism from S into some arbitrary FG-module T, then the simplicity of S guarantees that the kernel $\ker(\varphi) = 0$, and thus S is isomorphic to the submodule $S\varphi$ of T. If also T is simple, then $S\varphi = T$, and thus φ is an isomorphism from S to T. This essentially trivial (but useful) observation is often referred to as "Schur's lemma".

An FG-module is **homogeneous** if it is a direct sum of isomorphic simple submodules. If S is a simple module, we say that a nonzero homogeneous FG-module has **type** S if it is a direct sum of isomorphic copies of S. Although it may not be completely obvious, it is true that the type of a homogeneous module is uniquely determined up to isomorphism. This fact is an immediate consequence of the following.

Lemma 10.5. *Let S be a simple FG-module, and suppose that V is a homogeneous FG-module of type S. Then every simple submodule of V is isomorphic to S.*

Proof. Write $V = \sum^{\cdot} S_i$, where each of the submodules S_i is isomorphic to S, and let π_i be the projection homomorphism from V onto S_i. Given a simple submodule T of V, we see that for some subscript i, the submodule $T\pi_i$ must be nonzero. Since S_i is simple, we have $T\pi_i = S_i$, and thus $T \cong S_i$ by Schur's lemma. Also, $S_i \cong S$, so $T \cong S$, as required. ∎

To state our next result, we introduce a bit of notation. Given an FG-module V and a simple FG-module T, We write $V(T)$ to denote the sum of all submodules of V that are isomorphic to T, and we note that the submodule $V(T)$ of V depends only on the isomorphism type of T.

Theorem 10.6. *Let V be a semisimple FG-module. Then there is a unique direct-sum decomposition $V = \sum^{\cdot} V_i$ such that the submodules V_i are nonzero and homogeneous of pairwise nonisomorphic types. Also, we have the following.*

(a) *Each simple submodule T of V is contained in one of the homogeneous direct summands V_i, and it is the type of V_i.*

(b) *If T is the type of V_i, then $V_i = V(T)$.*

Proof. Since V is semisimple, it is a direct sum of some collection $\mathcal{X}$ of simple submodules. Let $\{S_i \mid 1 \leq i \leq n\}$ be a set of representatives for the isomorphism types occurring among the members of $\mathcal{X}$, so in particular, the S_i are pairwise nonisomorphic. Let V_i be the (necessarily direct) sum of those members of $\mathcal{X}$ that are isomorphic to S_i, so V_i is nonzero, and it is homogeneous of type S_i. Clearly, $V = \sum^{\cdot} V_i$, and this proves the existence of the required submodules V_i.

Suppose now that $V = \sum^{\cdot} V_i$ is an arbitrary decomposition of V such that the summands V_i are nonzero and homogeneous with pairwise nonisomorphic types T_i.

To prove (a), let T be a simple submodule of V, and let π_i be the projection homomorphism from V onto V_i. Since T is nonzero, we see that $T\pi_i$ is nonzero for some subscript i, and since T is simple, it follows that $T\pi_i \cong T$. Now $T\pi_i$ is a simple submodule of V_i, and we conclude by Lemma 10.5 that $T\pi_i \cong T_i$. Then $T \cong T_i$ and we can say that V_i has type T. Also, T is not isomorphic to T_j for $j \neq i$, and it follows that the subscript i is uniquely determined. Then $T\pi_j = 0$ for all subscripts j different from i, and we conclude that $T \subseteq V_i$, as wanted.

For (b), we show that $V_i = V(T_i)$. Since V_i is homogeneous of type T_i, it follows by definition that V_i is a sum of submodules isomorphic to T_i, and thus $V_i \subseteq V(T_i)$. For the reverse containment, it suffices to show that every submodule T of V that is isomorphic to T_i is contained in V_i. We know by

(a) that T is contained in one of the summands V_j, and thus $T \cong T_j$ by Lemma 10.5. Then $T_j \cong T \cong T_i$, so $j = i$, and thus $T \subseteq V_i$, as required.

We argue now that the submodules V_i are exactly the nonzero submodules of V that have the form $V(T)$ for some simple FG-module T. Each V_i is nonzero, and by (b), we have $V_i = V(T_i)$. Conversely, if T is a simple FG-module such that $V(T)$ is nonzero, we must show that $V(T)$ is one of the homogeneous submodules V_i. Since $V(T)$ is nonzero, there exists a submodule S of V isomorphic to T, and by (a), we see that S is isomorphic to one of the type modules T_i. Then $V(T) = V(S) = V(T_i) = V_i$. It follows that the submodules V_i are uniquely determined, as required. ∎

The uniquely determined submodules V_i in Theorem 10.6 are called the **homogeneous components** of the semisimple module V.

Corollary 10.7. *Let V be a semisimple FG-module. Then every FG-submodule of V is the direct sum of its intersections with the homogeneous components of V.*

Proof. Let $U \subseteq V$ be a submodule. Since U is semisimple by Corollary 10.4, it is a sum of simple submodules, and by Theorem 10.6, each of these is contained in one of the homogeneous components of V. Then U is the sum of its intersections with the homogeneous components of V, and this sum is direct since the sum of these homogeneous components is direct. ∎

If $H \subseteq G$ is a subgroup and V is an FG-module, then of course, V can also be viewed as an FH-module, and it should be obvious that even if V is simple as an FG-module, it need not be simple as an FH-module. (In order to avoid visual clutter, we will not introduce any particular notation to distinguish V as an FG-module from V as an FH-module; this will always be clear from the context.)

Now let $N \triangleleft G$, and consider an FG-module V. If U is an FN-submodule of V and $g \in G$, it is routine to check that the subspace Ug of V is N-invariant, so Ug is an FN-submodule. Also, it is clear that if U is simple as an FN-module, then Ug is also a simple FN-module, and thus G acts by right multiplication on the set of simple FN-submodules of V.

If U and T are FN-submodules of V and $\varphi : U \to T$ is an FN-homomorphism, it is easy to check that the map $u \mapsto ((ug^{-1})\varphi)g$ is an FN-homomorphism from Ug to Tg. It follows that the action of G on the set of simple FN-submodules of V carries isomorphic simple modules to isomorphic simple modules. If V is semisimple as an FN-module, therefore, we conclude that right multiplication by elements of G permutes the homogeneous components of V.

Theorem 10.8 (Clifford). *Let $N \lhd G$, and suppose that V is an FG-module. If V is semisimple as an FG-module, then it is also semisimple as an FN-module. In particular, if V is simple as an FG-module, it is semisimple as an FN-module, and in this case, the right-multiplication action of G on the homogeneous FN-components of V is transitive.*

Proof. Assume first that V is simple as an FG-module, and let $S \subseteq V$ be a simple FN-submodule. Then $\sum\{Sg \mid g \in G\}$ is G-invariant, and it follows by the simplicity of V as an FG-module that $V = \sum\{Sg \mid g \in G\}$. Then V is a sum of simple FN-submodules, and hence V is semisimple as an FN-module.

More generally, assume only that V is semisimple as an FG-module. Then V is a sum of simple FG-submodules, and by the first part of the proof, each of these is a sum of simple FN-submodules. It follows that V is a sum of simple FN-submodules, and thus V is semisimple as an FN-module, as required.

We return now to the case where V is simple as an FG-module and $S \subseteq V$ is simple as an FN-module. By Theorem 10.6, each of the simple FN-submodules Sg is contained in exactly one of the homogeneous FN-components of V, and we see that set $\mathcal{X}$ of those homogeneous FN-components of V that contain one of the FN-modules Sg forms an orbit under the right-multiplication action of G. Also, $\sum \mathcal{X} = V$ because $V = \sum\{Sg \mid g \in G\}$. Since V is the direct sum of all of its homogeneous FN-components, we conclude that $\mathcal{X}$ must be the full set of all nonzero homogeneous FN-components of V, and thus the action of G on this set is transitive, as required. ∎

An FG-homomorphism from an FG-module S to itself is called an FG-**endomorphism**, and the set of FG-endomorphisms of S is denoted $\mathrm{End}_{FG}(S)$. Note that $\mathrm{End}_{FG}(S)$ is a ring, where addition is pointwise and multiplication is function composition. If S is simple, then by Schur's lemma, every nonzero FG-endomorphism is an isomorphism, and hence it is invertible, and it follows that $\mathrm{End}_{FG}(S)$ is a division ring. If F is a finite field, then S is actually a finite set, and it follows that the division ring $\mathrm{End}_{FG}(S)$ is finite. By a famous theorem of Wedderburn, multiplication in a finite division ring is commutative, and thus if S is simple, $\mathrm{End}_{FG}(S)$ is a field. (See Theorem 21.20 of [**AGC**].)

Lemma 10.9. *Let S be a simple FG-module, and write $E = \mathrm{End}_{FG}(S)$. If T is an FG-module and $f : S \to T$ is an isomorphism, then the map $\delta \mapsto \delta f$ is a bijection from E onto the set of all FG-homomorphisms from S to T.*

Proof. If $\delta \in E$, then δf is certainly an FG-homomorphism from S to T. If $\delta, \varepsilon \in E$ and $\delta f = \varepsilon f$, then $\delta = \varepsilon$ because f is invertible, and thus the map $\delta \mapsto \delta f$ is injective. To see that this map is surjective, let $g : S \to T$ be an arbitrary FG-homomorphism. Then gf^{-1} is an FG-endomorphism, so it is some element $\delta \in E$, and the result follows. ∎

Our next results apply only in the case where the field F is finite, so in particular, the characteristic of F is some prime number p.

Lemma 10.10. *Let S be a simple FG-module, where F is a finite field, and write $E = \mathrm{End}_{FG}(S)$ and $|E| = q$. Suppose V is an FG-module that is a direct sum of n submodules, each isomorphic to S. Then the total number of simple FG-submodules in V is $(q^n - 1)/(q-1)$.*

Proof. Write $V = \sum_{i=1}^{n} S_i$, where $S_i \cong S$ for $1 \le i \le n$. Given FG-homomorphisms $f_i : S \to S_i$, where not all f_i are zero, write

$$\mathbf{f} = (f_1, f_2, \ldots, f_n)\,,$$

and let

$$W_{\mathbf{f}} = \Big\{ \sum_{i=1}^{n} sf_i \mid s \in S \Big\}\,,$$

so $W_{\mathbf{f}}$ is the image of the FG-homomorphism $s \mapsto \sum sf_i$ from S into V, and hence it is a submodule. Also, $W_{\mathbf{f}}$ is nonzero because at least one of the homomorphisms f_i is nonzero and the sum of the S_i is direct. Since S is simple, it follows that $W_{\mathbf{f}} \cong S$, and in particular $W_{\mathbf{f}}$ is simple.

We argue next that every simple FG-submodule T of V has the form $W_{\mathbf{f}}$ for some n-tuple $\mathbf{f}$ of homomorphisms that are not all zero. Since V is homogeneous of type S, it follows by Lemma 10.5 that $S \cong T$, so we can fix an isomorphism $h : S \to T$. Let $f_i = h\pi_i$, where π_i is the projection map from V onto S_i, and observe that f_i is a (possibly zero) homomorphism from S to S_i. Now $Sf_i = Sh\pi_i = T\pi_i$, and since T is nonzero, at least one of the submodules $T\pi_i$ must be nonzero. The homomorphisms f_i are therefore not all zero, and writing $\mathbf{f} = \{f_1, \ldots, f_n\}$, we show that $T = W_{\mathbf{f}}$.

Given $s \in S$, let $t = sh$, so

$$t = \sum t\pi_i = \sum sh\pi_i = \sum sf_i \in W_{\mathbf{f}}\,.$$

As s runs over S, we see that t runs over T and $\sum sf_i$ runs over $W_{\mathbf{f}}$, and it follows that $T = W_{\mathbf{f}}$, as required.

If $\mathbf{f}$ and $\mathbf{g}$ are n-tuples of homomorphisms f_i and g_i from S to S_i, where not all f_i and not all g_i are zero, we argue that $W_{\mathbf{f}} = W_{\mathbf{g}}$ if and only if there exists $\delta \in E^\times$ such that $g_i = \delta f_i$ for all i.

First, assume that $g_i = \delta f_i$ for all i. Then

$$W_{\mathbf{g}} = \{\sum sg_i \mid s \in S\} = \{\sum s\delta f_i \mid s \in S\} = W_{\mathbf{f}},$$

where the final equality holds because $s\delta$ runs over S as s runs over S.

Conversely, suppose $W_{\mathbf{g}} = W_{\mathbf{f}}$, and call this submodule W. Let α and β be the maps from S to W given by

$$s\alpha = \sum sf_i \qquad \text{and} \qquad s\beta = \sum sg_i .$$

Then α and β are isomorphisms from S onto W, so by Lemma 10.9, there exists an element $\delta \in E^\times$ such that $\beta = \delta\alpha$, and we have

$$\sum sg_i = s\beta = s\delta\alpha = \sum s\delta f_i$$

for each element $s \in S$. Since V is the direct sum of the submodules S_i, it follows from this equation that $sg_i = s\delta f_i$ for each subscript i, and thus $g_i = \delta f_i$ for all i, as wanted.

The number of simple FG-submodules in V is equal to the number of distinct submodules of the form $W_{\mathbf{f}}$, where as usual, not all f_i are zero. To count these, we observe that by Lemma 10.9, the number of homomorphisms f_i from S to S_i is $|E| = q$, and so the total number of n-tuples $(f_1, f_2, \ldots, f_n)$ is q^n. The number of these n-tuples where not all f_i are zero is thus $q^n - 1$.

Since two submodules of the form $W_{\mathbf{f}}$ are identical precisely when the corresponding n-tuples of homomorphisms are proportional via one of the $q-1$ elements of $E^\times$, it follows that the number of distinct submodules $W_{\mathbf{f}}$ is $(q^n - 1)/(q-1)$, as required. ∎

Suppose that T and U are FG-modules, and write $\mathrm{Hom}_F(T, U)$ to denote the F-vector space of F-linear maps from T to U. (Addition and scalar multiplication are defined pointwise.) If $\varphi \in \mathrm{Hom}_F(T, U)$ and $g \in G$, we write φ^g to denote the map $t \mapsto ((tg^{-1})\varphi)g$ for $t \in T$, and it is easy to see that this defines an action of G on $\mathrm{Hom}_F(T, U)$ that respects the vector space structure. We refer to this as the **conjugation action** of G on $\mathrm{Hom}_F(T, U)$, but we note that this is not a genuine conjugation unless $T = U$.

Observe that if $H \subseteq G$, then the H-invariant maps in $\mathrm{Hom}_F(T, U)$ are exactly the FH-homomorphisms from T to U. Also, if $N \triangleleft G$, then the set of N-invariant maps in $\mathrm{Hom}_F(T, U)$ is stabilized by G. In other words, if $\varphi : T \to U$ is an FN-homomorphism, then φ^g is also an FN-homomorphism.

Suppose now that W is a third FG-module, and let $\alpha \in \mathrm{Hom}_F(T, U)$ and $\beta \in \mathrm{Hom}_F(U, W)$. Clearly, $\alpha\beta$ lies in $\mathrm{Hom}_F(T, W)$, and it is easy to check that $(\alpha\beta)^g = \alpha^g\beta^g$. In particular, if $T = U = W$, then the conjugation action of G on $\mathrm{End}_F(T)$ is an action via ring automorphisms. If $N \triangleleft G$, therefore, the conjugation action of G on $\mathrm{End}_{FN}(T)$ is an action by ring automorphisms.

We mention that the proof of the first part of the following result is based on an idea of J. A. Green.

Theorem 10.11. *Let F be a finite field of characteristic p, and let V be a semisimple FG-module. Suppose that $N \triangleleft G$, where G/N is a p-group, and assume for all simple FN-submodules S of V and all elements $g \in G$ that S and Sg are FN-isomorphic. The following then hold.*

(a) *Every simple FG-submodule of V is simple as an FN-module.*

(b) *If T and U are simple FG-submodules of V that are isomorphic as FN-modules, then T and U are isomorphic as FG-modules.*

(c) *The homogeneous FN-components of V are exactly the homogeneous FG-components of V.*

(d) *For each simple FG-submodule T of V, the degree of the field extension $\mathrm{End}_{FG}(T) \subseteq \mathrm{End}_{FN}(T)$ is a power of p.*

(e) *Suppose that $\mathrm{End}_{FG}(T) = \mathrm{End}_{FN}(T)$ for every simple FG-submodule T of V. Then every FN-submodule of V is G-invariant.*

Proof. Suppose that T is a simple FG-submodule of V, so T has no G-invariant nonzero proper subspace. Let S be a simple FN-submodule of T, and observe that the subspace $\sum\{Sg \mid g \in G\}$ is nonzero and G-invariant, so $\sum\{Sg \mid g \in G\} = T$.

Let $\mathcal{X}$ be a subset of $\{Sg \mid g \in G\}$, minimal with the property that $\sum \mathcal{X} = T$, and write $n = |\mathcal{X}|$. It is easy to see that T is the direct sum of the members of $\mathcal{X}$, and since by hypothesis all of the FN-submodules Sg are FN-isomorphic to S, it follows that T is the direct sum of n isomorphic copies of S. By Lemma 10.10, the total number of simple FN-submodules contained in T is $(q^n-1)/(q-1) = 1+q+\cdots+q^{n-1}$, where $q = |\mathrm{End}_{FN}(S)|$. Since q is a power of p, we see that this number is not divisible by p.

Now G acts by right multiplication on the set of simple FN-submodules of T, and N acts trivially on this set. By hypothesis, G/N is a p-group, and it follows that each G-orbit of simple FN-submodules of T has p-power size. Since the total number of such submodules is not divisible by p, there must exist a G-invariant simple FN-submodule in T, and we conclude that this submodule must be all of T. It follows that T is simple as an FN-module, proving (a).

For (b), let T and U be simple FG-submodules of V, and suppose $\varphi : T \to U$ is an FN-isomorphism. By (a), the submodules T and U are simple as FN-modules, and we let E be the field $\mathrm{End}_{FN}(T)$. For each element $g \in G$, the map φ^g is a nonzero FN-homomorphism from T to U, so by Lemma 10.9, we can write $\varphi^g = \delta_g\varphi$ for some nonzero element $\delta_g \in E$, where

δ_g depends on g. If also $h \in G$, we have

$$\delta_{gh}\varphi = \varphi^{gh} = (\varphi^g)^h = (\delta_g\varphi)^h = (\delta_g)^h\varphi^h = (\delta_g)^h\delta_h\varphi\,,$$

and it follows by Lemma 10.9 that $\delta_{gh} = (\delta_g)^h\delta_h$.

We mention that this property of the map $g \mapsto \delta_g$ arises in a number of different contexts, and it has been given a variety of names in the literature. One could say that the map is a "crossed homomorphism" from G into the group $E^\times$ or that it is a "1-cocycle" or that it "satisfies Noether's equations". In any case, it is a standard fact from field theory (see Lemma 23.14 of [**AGC**]) that there exists some element $\alpha \in E^\times$ such that $\delta_g = \alpha^g\alpha^{-1}$ for all $g \in G$. (In other words, the map $g \mapsto \delta_g$ is a "principal crossed homomorphism" or it is a "1-coboundary".)

Let $\alpha \in E^\times$ have the property that $\delta_g = \alpha^g\alpha^{-1}$ for all $g \in G$, and define $\theta = \alpha^{-1}\varphi$. Then θ is an FN-isomorphism from T onto U, and we have

$$\theta^g = (\alpha^{-1}\varphi)^g = (\alpha^{-1})^g\varphi^g = (\alpha^{-1})^g\delta_g\varphi = (\alpha^{-1})^g\alpha^g\alpha^{-1}\varphi = \alpha^{-1}\varphi = \theta\,,$$

for all $g \in G$. It follows that θ is an FG-isomorphism from T to U, as required.

By assumption, V is semisimple as an FG-module, so by Clifford's theorem, V is also semisimple as an FN-module, and thus (c) makes sense. Write $V = \sum V_i$, where the V_i are the homogeneous FG-components of V. Each of the submodules V_i is a direct sum of isomorphic simple FG-modules, and by (a) these summands are simple as FN-modules, and of course, they are also isomorphic as FN-modules. Each of the submodules V_i is thus homogeneous as an FN-module, so to show that the V_i are the homogeneous FN-components of V, it suffices to show that their types as FN-modules are pairwise nonisomorphic.

Let T_i and T_j be simple FG-submodules of V_i and V_j. By Lemma 10.5, we see that T_i and T_j are the types of V_i and V_j as FG-modules. Now T_i and T_j are simple as FN-modules, they are also the types of V_i and V_j as FN-modules. If $i \neq j$, then T_i and T_j are not isomorphic as FG-modules, so by (b) they are also not isomorphic as FN-modules. The V_i are therefore the homogeneous FN-components of V, as required for (c).

For (d), let T be a simple FG-submodule of V, so T is also simple as an FN-module, and thus $\mathrm{End}_{FG}(T) \subseteq \mathrm{End}_{FN}(T)$ is an extension of fields. We have seen that conjugation by G induces an action via automorphisms of G/N on the field $\mathrm{End}_{FN}(T)$, and the fixed field of this action is $\mathrm{End}_{FG}(T)$. It follows by basic Galois theory that the degree of the field extension $\mathrm{End}_{FG}(T) \subseteq \mathrm{End}_{FN}(T)$ is a divisor of $|G/N|$, and by hypothesis, this is a power of p.

Finally, for (e), let W be an FN-submodule of V, and assume that $\mathrm{End}_{FG}(T) = \mathrm{End}_{FN}(T)$ for every simple FG-submodule T of V. Since V is semisimple as an FN-module, it follows by Corollary 10.4 that W is a sum of simple FN-submodules. To prove that W is G-invariant, therefore, we may assume that W is simple as an FN-module.

By Theorem 10.6, we see that W is contained in one of the homogeneous FN-components of V. By (c), this submodule is also one of the homogeneous FG-components of V, and we can assume without loss that this component is all of V. Then V is homogeneous as an FG-module, and thus V is a direct sum of some number, say n, of isomorphic copies of some simple FG-submodule T of V. Since T is also simple as an FN-module, V is a direct sum of n isomorphic copies of T viewed as a simple FN-module.

By assumption, $|\mathrm{End}_{FG}(T)| = |\mathrm{End}_{FN}(T)|$, so by Lemma 10.10, we see that V has equal numbers of simple FG-submodules and simple FN-submodules. Every simple FG-submodule of V is a simple FN-submodule by (a), and it follows that every simple FN-submodule must be a simple FG-submodule. In particular, W is G-invariant, as required. ∎

Let V be an FG-module, and let $N \triangleleft G$. Suppose that S is a simple FN-submodule of V, and assume that $Sg = S$ for every element $g \in G$. Then S is an FG-submodule of V, and in this situation, we have seen that there is a conjugation action of G on the division ring $E = \mathrm{End}_{FN}(S)$. (Recall that if $\delta \in E$ and $g \in G$, then δ^g is the map $s \mapsto ((sg^{-1})\delta)g$ for $s \in S$, and it follows that the conjugation action of N on E is trivial.)

If we assume that F is finite, it is possible to define an action of G on $\mathrm{End}_{FN}(S)$ under a milder assumption. It is not necessary to require that $Sg = S$ for all $g \in G$; we shall see that it suffices to assume that $Sg \cong S$ for all $g \in G$. (The isomorphism here is, of course, an FN-isomorphism.)

Lemma 10.12. *Let V be an FG-module, where F is a finite field. Let $N \triangleleft G$ and suppose that S is a simple FN-submodule of V such that S and Sg are FN-isomorphic for all $g \in G$. Then there is an action of G via automorphisms on the field $E = \mathrm{End}_{FN}(S)$. Also, if g is an element of G such that $Sg = S$, then the action of g agrees with the conjugation action that was defined previously, and in particular the elements of N act trivially.*

Proof. For each element $g \in G$, choose an FN-isomorphism $\varphi_g : Sg \to S$, and let $\alpha_g : S \to S$ be the map $s \mapsto sg\varphi_g$. Clearly, α_g is an F-vector space isomorphism from S to itself, or equivalently, it is a unit of the ring $\mathrm{End}_F(S)$.

Given $n \in N$, let $\hat{n}$ be the map $s \mapsto sn$. Write $\hat{N} = \{\hat{n} \mid n \in N\}$, and observe that $\hat{N}$ is a subgroup of the group of units in the ring $\mathrm{End}_F(S)$. For

$s \in S$, we compute

$$s(\hat{n})^{\alpha_g} = s(\alpha_g)^{-1} n \alpha_g = s(\varphi_g)^{-1} g^{-1} n g \varphi_g = s(\varphi_g)^{-1} n^g \varphi_g = s n^g \,,$$

where the final equality holds because $n^g \in N$ and φ_g is an FN-isomorphism. Then

$$(\hat{n})^{\alpha_g} = \widehat{n^g} \,,$$

and in particular, α_g normalizes $\hat{N}$. Since E is the centralizer of N in the ring $\mathrm{End}_F(S)$, it follows that conjugation by α_g stabilizes E, so it induces a field automorphism of E.

We argue next that the automorphism of E induced by α_g is independent of the choice of the FN-isomorphism φ_g. To see this, suppose we replace φ_g by another FN-isomorphism from Sg to S, and we write β_g to denote the corresponding replacement of α_g. Then for $n \in N$, we have

$$(\hat{n})^{\alpha_g} = \widehat{n^g} = (\hat{n})^{\beta_g} \,,$$

and thus $\alpha_g(\beta_g)^{-1}$ centralizes $\hat{N}$. It follows that $\alpha_g(\beta_g)^{-1}$ lies in E, so $\alpha_g = \mu\beta_g$ for some element $\mu \in E$. Since E is commutative because F is finite, it follows that conjugation by α_g and by β_g induce identical automorphisms of E. Each element $g \in G$, therefore, unambiguously determines an automorphism σ_g of E, where σ_g the map induced by conjugation by α_g.

To see that this is a genuine action of G, we must show that if $g, h \in G$, then $\sigma_{gh} = \sigma_g \sigma_h$ or, equivalently, that $\alpha_{gh}(\alpha_h)^{-1}(\alpha_g)^{-1}$ centralizes E. For $n \in N$, we have

$$((\hat{n})^{\alpha_g})^{\alpha_h} = (\widehat{n^g})^{\alpha_h} = \widehat{n^{gh}} = (\hat{n})^{\alpha_{gh}} \,.$$

Then $\alpha_{gh}(\alpha_h)^{-1}(\alpha_g)^{-1}$ centralizes $\hat{N}$, so this map lies in E, and since E is commutative, it centralizes E, as required.

We show next that if $Sg = S$, then the automorphism σ_g of E is exactly conjugation by g. We have seen that σ_g is independent of the choice of the FN-isomorphism φ_g from Sg to S. Since $Sg = S$, we can take φ_g to be the identity map, and thus α_g is simply multiplication by g, and thus σ_g is conjugation by g. In particular, if $n \in N$, then $Sn = S$, so σ_n is conjugation by n on $E = \mathrm{End}_{FN}(S)$, and this is trivial. ∎

Our next result could be generalized in various directions, but we state and prove it only in the case that we will use.

Lemma 10.13. *Let $N \lhd G$, where $|G : N| = 2$, and let V be an FG-module. Suppose that T is a simple FN-submodule of V, and for some element $g \in G$, assume that T is not FN-isomorphic to Tg. Write $S = T + Tg$. The following then hold.*

(a) *S is a simple FG-submodule of V.*

(b) *Restriction yields an isomorphism from* $\mathrm{End}_{FG}(S)$ *onto* $\mathrm{End}_{FN}(T)$.

(c) *Let* U *be an* FN*-submodule of* V*, and suppose that* $\varphi : T \to U$ *is an* FN*-isomorphism. Then* φ *extends to an* FG*-isomorphism from* S *onto* $U + Ug$.

Note that since T and Tg are not isomorphic in Lemma 10.13, we certainly have $T \neq Tg$, and thus g does not lie in N. Also, by hypothesis, $|G : N| = 2$, so every element of G outside of N has the form ng for some element $n \in N$. Since $T(ng) = Tg$, it follows that every element of G outside N yields the same subspace Tg.

Proof of Lemma 10.13. We have $Tg^2 = T$ since $g^2 \in N$, and thus $Sg = Tg + Tg^2 = Tg + T = S$. The stabilizer of S in G, therefore, includes g, and hence this stabilizer is strictly larger than N. It follows that S is G-invariant, so it is an FG-submodule of V. To see that S is simple as an FG-module, we observe first that since T and Tg are nonisomorphic simple FN-submodules of G, these two FN-submodules are homogeneous and have distinct types, so they are exactly the homogeneous FN-components of S. It follows by Corollary 10.7 that T and Tg are the only proper nonzero FN-submodules of S, and since neither T nor Tg is G-invariant, S has no nonzero proper FG-submodules. Then S is simple as an FG-module, proving (a).

We prove (c) next. Since the sum $T + Tg$ is direct, we can define a map θ from S to $U + Ug$ by setting $(x + yg)\theta = x\varphi + y\varphi g$ for $x, y \in T$. We observe first that θ is an FN-homomorphism because if $n \in N$, we have

$$\begin{aligned}(x + yg)n\theta = (xn + yn^{g^{-1}}g)\theta = xn\varphi + yn^{g^{-1}}\varphi g &= x\varphi n + y\varphi n^{g^{-1}}g \\ &= x\varphi n + y\varphi gn \\ &= (x\varphi + y\varphi g)n \\ &= (x + yg)\theta n\,,\end{aligned}$$

where the third equality holds because $n^{g^{-1}}$ lies in N and φ is an FN-homomorphism.

To see that θ is actually an FG-homomorphism, it suffices to show that $sg\theta = s\theta g$ for all $s \in S$. (This is enough because $G = N \cup Ng$.) Since $g^2 \in N$, we have

$$\begin{aligned}(x + yg)g\theta = (xg + yg^2)\theta = x\varphi g + yg^2\varphi &= x\varphi g + y\varphi g^2 \\ &= (x\varphi + y\varphi g)g \\ &= (x + yg)\theta g\,,\end{aligned}$$

as required, and so θ is an FG-homomorphism. It is clear from its definition that θ maps S onto $U + Ug$, and since S is simple as an FG-module, we see that θ is actually an FG-isomorphism, as required.

For (b), suppose that $\delta \in \mathrm{End}_{FG}(S)$. Now δ is an FN-isomorphism from S to itself, so it must permute the two simple FN-submodules T and Tg of S. Since Tg is not FN-isomorphic to T, however, we cannot have $T\delta = Tg$, and thus $T\delta = T$. It follows that restriction to T defines a ring homomorphism from $\mathrm{End}_{FG}(S)$ to $\mathrm{End}_{FN}(T)$. Since these endomorphism rings are division rings, this restriction homomorphism is injective. Finally, for surjectivity, let $\varphi : T \to T$ be an FN-isomorphism. By (c), there exists an FG-isomorphism $\theta : S \to S$ extending φ, and thus the restriction map carries θ to φ. ∎

We end this section with a brief discussion of "dual" modules. Recall that if V is an F-vector space, then the **dual space** V^* of V is the F-vector space consisting of the F-linear functionals on V. Also, if $\dim(V) = n < \infty$, then also $\dim(V^*) = n$, so V and V^* are isomorphic vector spaces. In general, however, there is no *natural* isomorphism from V to V^*, but there is a natural isomorphism from a finite-dimensional vector space V to its "double dual" $(V^*)^*$. This isomorphism is the map that takes each vector $x \in V$ to the "evaluation" linear functional e_x on V^*, where e_x is defined by the formula $e_x(\lambda) = \lambda(x)$. In all cases, even if V is not finite dimensional, the map $x \mapsto e_x$ is an injective linear transformation from V into its double dual. If V is finite dimensional, however, then the map $x \mapsto e_x$ is also surjective because $\dim(V) = \dim(V^*) = \dim((V^*)^*)$.

If V is an FG-module, then there is a natural action of the group G on the dual space V^*, where for a linear functional λ on V and a group element g, the linear functional λg is defined by the formula $(\lambda g)(xg) = \lambda(x)$, or equivalently, $(\lambda g)(x) = \lambda(xg^{-1})$ for $x \in V$. We say that the dual space V^* together with this action of G forms the **dual module** of V. Also, we say that an FG-module U is **dual** to V if U is FG-isomorphic to the dual module V^*.

The natural vector-space isomorphism from V to $(V^*)^*$ is easily seen to be an FG-module isomorphism, so if U is the dual module for V, then V is isomorphic to the dual module of U. Given FG-modules U and V, therefore, it follows that U is dual to V if and only if V is dual to U. Also, we mention that it is possible for an FG-module V to be "self-dual", which means that it is FG-isomorphic to its own dual module.

Lemma 10.14. *Let U be an FG-module. The following then hold.*

(a) *If U has a submodule with dimension m, then U^* has a submodule with codimension m.*

(b) *If U is simple, then U^* is also simple.*

Actually, much more is true. In fact, there exists a natural bijection from the set of FG-submodules X of U onto the set of FG-submodules Y of U^* such that if $X \mapsto Y$, then the dimension of X equals the codimension of Y.

Proof of Lemma 10.14. Let $X \subseteq U$ be a submodule. Restriction of linear functionals on U to X defines a module homomorphism from U^* onto X^*, and we let Y be the kernel of this map. Then Y is a submodule of U^*, and $U^*/Y \cong X^*$. It follows that $\dim(U^*) - \dim(Y) = \dim(X^*) = \dim(X)$, proving (a).

For (b), let U is a simple FG-module. If U^* is not simple, there exists a submodule X such that $0 < X < U^*$. Now U is dual to U^*, so we can apply (a) with U^* in place of U, and we deduce that U has a submodule Y such that $\dim(U) - \dim(Y) = \dim(X)$. It follows that $0 < \dim(Y) < \dim(U)$, and this contradicts the simplicity of U. ∎

Finally, we mention a connection between dual modules and characters. Suppose that F is the complex number field and S is a simple FG-module. Then S affords a character $\chi \in \mathrm{Irr}(G)$, and one can show that the dual module S^* affords the complex-conjugate character $\overline{\chi}$, and in particular, S is self-dual if and only if χ is real valued. The analogous assertions in prime characteristic are also true. If F is algebraically closed and has prime characteristic p, then a simple FG-module S corresponds to a Brauer character $\varphi \in \mathrm{IBr}(G)$. The dual module S^* corresponds to the complex-conjugate Brauer character $\overline{\varphi}$, and thus S is self-dual if and only if φ is real valued.

10B

We begin this section with a review of some multilinear algebra. Recall that a **bilinear form** on an F-vector space V is a function of two variables on V with values in the field F, where the function is linear in each variable. We will generally use the notation $\langle x, y\rangle$ to denote the value of a bilinear form on a pair of vectors $x, y \in V$.

A bilinear form on V is **alternating** if $\langle x, x\rangle = 0$ for all $x \in V$. Note that if the form $\langle \cdot, \cdot \rangle$ is alternating, then for $x, y \in V$, we have

$$0 = \langle x+y, x+y\rangle = \langle x, x\rangle + \langle x, y\rangle + \langle y, x\rangle + \langle y, y\rangle = \langle x, y\rangle + \langle y, x\rangle\,,$$

and thus $\langle y, x\rangle = -\langle x, y\rangle$. In other words, the form is **skew-symmetric**. Conversely, if a form $\langle \cdot, \cdot \rangle$ is skew-symmetric, we see that $\langle x, x\rangle = -\langle x, x\rangle$, so if the characteristic of F is different from 2, it follows that $\langle x, x\rangle = 0$, and the form is alternating. (In characteristic 2, however, a skew-symmetric form need not be alternating.)

If the form $\langle\cdot,\cdot\rangle$ on V is alternating, then since it is skew-symmetric, we see that $\langle x, y\rangle = 0$ if and only if $\langle y, x\rangle = 0$, and in this case, we say that the vectors x and y are **orthogonal**. More generally, if X and Y are arbitrary subsets of V, we say that X and Y are orthogonal if every vector in X is orthogonal to every vector in Y.

Suppose that the vector space V carries an alternating form, and let $U \subseteq V$ be a subspace. We write $U^\perp$ to denote the subspace of V consisting of all vectors orthogonal to U, so in particular, the subspace $V^\perp$ of V is the set of all vectors $x \in V$ such that $\langle x, v\rangle = 0$ for all $v \in V$. The subspace $V^\perp$ is sometimes called the **radical** of V, and if $V^\perp = 0$, we say that the form on V is **nondegenerate** and that V is a **symplectic** space. Also, we say that a subspace U of V is **nondegenerate** if the restriction of the form on V to U is nondegenerate, or equivalently, $U \cap U^\perp = 0$.

Lemma 10.15. *Let U be a subspace of a finite-dimensional symplectic F-space V, and for each vector $x \in V$, let λ_x be the linear functional on U defined by $\lambda_x(u) = \langle x, u\rangle$. The following then hold.*

(a) *The map $\theta : x \mapsto \lambda_x$ is a linear transformation from V onto U^*, and $\ker(\theta) = U^\perp$.*
(b) $\dim(U) + \dim(U^\perp) = \dim(V)$.
(c) $(U^\perp)^\perp = U$.
(d) *$U + U^\perp = V$ if and only if U is nondegenerate.*

Also, if $V = X + Y$, where X and Y are orthogonal subspaces, then X and Y are nondegenerate.

Proof. Most of (a) is clear, and we need only prove that $\theta : V \to U^*$ is surjective. To see this, observe first that by elementary linear algebra, every linear functional on U is the restriction to U of some linear functional on V. For (a), therefore, it suffices to assume that $U = V$ and to show that θ maps V onto V^*.

If $U = V$, then θ is a linear transformation from V into V^*, and we observe that $\ker(\theta)$ is the radical of V, and this is 0 because by hypothesis, the form on V is nondegenerate. Then θ is injective, and since $\dim(V) = \dim(V^*)$ is finite, it follows that θ is also surjective, proving (a).

We return now to the general case, where U is an arbitrary subspace of V. Since θ is a linear transformation from V onto U^* and $\ker(\theta) = U^\perp$, we have

$$\dim(V) - \dim(U^\perp) = \dim(U^*) = \dim(U),$$

and (b) follows.

It is clear that $U \subseteq (U^\perp)^\perp$. Two applications of (b) yield

$$\dim(U) = \dim(V) - \dim(U^\perp) = \dim((U^\perp)^\perp),$$

and thus $U = (U^\perp)^\perp$, as required for (c).

Now

$$\dim(U+U^\perp) = \dim(U)+\dim(U^\perp)-\dim(U\cap U^\perp) = \dim(V)-\dim(U\cap U^\perp),$$

and thus $U + U^\perp = V$ if and only if $U \cap U^\perp = 0$, and this proves (d).

Finally, suppose that $V = X + Y$, where X and Y are orthogonal subspaces. Then $Y \subseteq X^\perp$, so $V = X + X^\perp$, and it follows by (d) that X is nondegenerate. That Y is nondegenerate follows similarly. ∎

At the opposite extreme from a nondegenerate subspace is a subspace U such that the restriction of the form on V to U is identically zero. In this case, we say that U is **totally isotropic**, and we observe that U is totally isotropic if and only if $U \subseteq U^\perp$.

Lemma 10.16. *Let V be a finite-dimensional symplectic F-space. Then* $\dim(V)$ *is even. Also, if U is a totally isotropic subspace of V, then* $\dim(U) \leq (1/2)\dim(V)$, *and equality holds whenever U is a maximal totally isotropic subspace.*

Proof. Since the zero subspace of V is totally isotropic, V certainly has totally isotropic subspaces. Also, because V is finite dimensional, every totally isotropic subspace is contained in some maximal totally subspace, so there exists a subspace U that is maximal among totally isotropic subspaces of V.

We have $U \subseteq U^\perp$, and we argue that equality holds here. Otherwise, $U < U^\perp$, and we can choose a vector $x \in U^\perp$ such that $x \notin U$. Let $W = U + Fx$, so W is a subspace, and $W \subseteq U^\perp$. Then $U \subseteq W^\perp$ and also, x is contained in $W^\perp$ because x is orthogonal to U and x is orthogonal to Fx since $\langle x, x\rangle = 0$. It follows that $W = U + Fx \subseteq W^\perp$, and hence W is totally isotropic. This contradicts the maximality of U, and we deduce that $U^\perp = U$.

By Lemma 10.15(b), we have

$$\dim(U) = \dim(U^\perp) = \dim(V) - \dim(U),$$

and hence $\dim(U) = (1/2)\dim(V)$, and in particular, $\dim(V)$ is even. Also, since every totally isotropic subspace of V is contained in a maximal totally isotropic subspace, it follows that the dimension of every totally isotropic subspace is at most $(1/2)\dim(V)$, as required. ∎

Now let V be an FG-module that carries a G-invariant alternating form. (To say that the form is "G-invariant" means that $\langle xg, yg\rangle = \langle x, y\rangle$, or equivalently, $\langle xg, y\rangle = \langle x, yg^{-1}\rangle$ for all $x, y \in V$ and $g \in G$.) Note that if $U \subseteq V$ is an arbitrary FG-submodule, it follows that $U^\perp$ is also a submodule.

If the given G-invariant form on the FG-module V is nondegenerate, we say that V is a **symplectic** FG-module. Symplectic modules and their submodules will be the principal objects of study for the remainder of this chapter.

Recall that if V is a symplectic F-vector space (with no group action assumed), then by Lemma 10.16, the largest possible dimension of a totally isotropic subspace is $(1/2)\dim(V)$, and this upper bound is always attained. If V is a symplectic FG-module, however, there may not exist a totally isotropic sub*module* having dimension as large as $(1/2)\dim(V)$, and indeed, V may not have any nonzero totally isotropic submodules at all. A symplectic FG-module V is said to be **hyperbolic** if it has a totally isotropic submodule of the maximum possible dimension, $(1/2)\dim(V)$.

We can now state the theorem of E. C. Dade toward which we are working.

Theorem 10.17 (Dade). *Let V be a symplectic FG-module, where F is a finite field of odd characteristic p and G is p-solvable, and let $H \subseteq G$ be a subgroup having p-power index. If V is hyperbolic as an FH-module, then V is also hyperbolic as an FG-module.*

In other words, if V has an H-invariant totally isotropic subspace of dimension $(1/2)\dim(V)$, then there exists a totally isotropic G-invariant subspace of the same dimension. (The converse of Dade's theorem is, of course, a triviality.) We begin work now toward a proof of Theorem 10.17.

Lemma 10.18. *Let V a symplectic FG-module, Then the maximal totally isotropic submodules of V all have the equal dimensions. Also, V is hyperbolic if and only if this common dimension is $(1/2)\dim(V)$.*

Proof. Let X be a maximal totally isotopic submodule of V. To prove that all such submodules have the same dimension, it suffices to show that $\dim(X) \geq \dim(W)$ for an arbitrary totally isotropic submodule W of V. To see this, write $D = W \cap X^\perp$, and observe that D is a submodule, and it is totally isotropic because it is contained in W. Let $Y = D + X$, so Y is a totally isotropic submodule because each of D and X is totally isotropic and D and X are orthogonal. Since $X \subseteq Y$, the maximality of X guarantees that $X = Y$, and thus $D \subseteq X$.

Now $D \subseteq W \subseteq W^\perp$, so $W \subseteq D^\perp$. Also, since $D \subseteq X$, we have $X^\perp \subseteq D^\perp$, and thus $W + X^\perp \subseteq D^\perp$. Then

$$\begin{aligned}\dim(D^\perp) &\geq \dim(W + X^\perp)\\ &= \dim(W) + \dim(X^\perp) - \dim(W \cap X^\perp)\\ &= \dim(W) + \dim(X^\perp) - \dim(D)\\ &= \dim(W) + (\dim(V) - \dim(X)) - (\dim(V) - \dim(D^\perp))\\ &= \dim(W) - \dim(X) + \dim(D^\perp)\,,\end{aligned}$$

and thus $\dim(X) \geq \dim(W)$, as required.

Now V is hyperbolic precisely when it contains a totally isotropic submodule of dimension $(1/2)\dim(V)$. By Lemma 10.16 no totally isotropic subspace can have larger dimension, and it follows that V is hyperbolic if and only if the common dimension of its maximal totally isotropic submodules is $(1/2)\dim(V)$. ∎

We say that an FG-module that carries a G-invariant alternating form is **anisotropic** if it has no nonzero totally isotropic submodules. Note that if V is anisotropic, then the form on V must be nondegenerate because the radical $V^\perp$ is a totally isotropic FG-submodule, and hence it is zero. It follows that an anisotropic FG-module is automatically a symplectic module.

Now let W be an FG-module that carries a (not necessarily nondegenerate) G-invariant alternating form, and let U be the radical of W. (Recall that this means that $U = W^\perp$.)

For vectors $x, y \in W$, it is easy to see that $\langle x, y\rangle$ depends only on the cosets of U in W that contain x and y, and thus we can define a bilinear form on the FG-module $\overline{W} = W/U$ by setting $\langle \overline{x}, \overline{y}\rangle = \langle x, y\rangle$, where $\overline{x}$ and $\overline{y}$ are the images of x and y in $\overline{W}$. (In the hope that this will not create excessive confusion, we use the notation $\langle\cdot,\cdot\rangle$ for both the original form on W and also for the form induced on $\overline{W}$.) It should be clear that the resulting form on $\overline{W}$ is alternating and nondegenerate, and thus $\overline{W}$ is a symplectic FG-module.

Lemma 10.19. *Let U be a totally isotropic submodule of a symplectic FG-module V, and write $W = U^\perp$, so $U \subseteq W$. Then W/U is a symplectic FG-module, and the following hold.*

(a) *W/U is nonzero if* $\dim(U) < (1/2)\dim(V)$.

(b) *If U is maximal among totally isotropic submodules of V, then W/U is anisotropic.*

(c) *If V is hyperbolic, then W/U is hyperbolic.*

Proof. Let $m = (1/2)\dim(V) - \dim(U)$, and observe that $\dim(W) = \dim(V) - \dim(U)$ by Lemma 10.15(b). Then

$$\dim(W/U) = \dim(W) - \dim(U) = \dim(V) - 2\dim(U) = 2m\,,$$

and thus W/U is nonzero if $m > 0$, proving (a).

We have $W^{\perp} = (U^{\perp})^{\perp} = U$ by Lemma 10.15(c), so U is the radical of W. It follows that the bilinear form $\langle\cdot,\cdot\rangle$ on W induces a nondegenerate alternating bilinear form on W/U, which is therefore a symplectic FG-module, as required.

Now suppose that $U \subseteq X \subseteq W$, where X is a submodule. By the definition of the form on W/U, it follows that X/U is totally isotropic if and only if X is totally isotropic. If U is maximally totally isotropic in V, therefore, then X/U cannot be totally isotropic unless $X/U = 0$, and hence W/U is anisotropic in this case. This establishes (b).

We have

$$\dim(X) = \dim(X/U) + \dim(U) = \dim(X/U) + (1/2)\dim(V) - m\,,$$

so

$$\dim(X) - (1/2)\dim(V) = \dim(X/U) - m = \dim(X/U) - (1/2)\dim(W/U)\,,$$

and it follows that $\dim(X) = (1/2)\dim(V)$ if and only if $\dim(X/U) = (1/2)\dim(W/U)$.

To prove (c) now, suppose that V is hyperbolic. Since U is totally isotropic, we can choose a submodule X containing U such that X is maximally totally isotropic, and it follows by Lemma 10.18 that $\dim(X) = (1/2)\dim(V)$. Also, since $U \subseteq X$, we see that $X \subseteq X^{\perp} \subseteq U^{\perp} = W$, so the foregoing observations apply. In particular, X/U is totally isotropic and $\dim(X/U) = (1/2)\dim(W/U)$, and thus W/U is hyperbolic. ∎

The following shows that to prove Dade's theorem by contradiction, it suffices to assume that the given symplectic FG-module V is anisotropic.

Corollary 10.20. *Let $H \subseteq G$, and suppose that V is a symplectic FG-module that is hyperbolic as an FH-module, and assume that V is not hyperbolic as an FG-module. Then there exists a nonzero anisotropic FG-module V_0 such that V_0 is hyperbolic as an FH-module. Also,* $\dim(V_0) \leq \dim(V)$.

Proof. Let U be a submodule of V maximal with the property that it is totally isotropic, and let $W = U^{\perp}$. We show that the FG-module $V_0 = W/U$ meets the requirements of the corollary. Certainly, $\dim(W/U) \leq \dim(V)$. Also, since V is not hyperbolic, we have $\dim(U) < (1/2)\dim(V)$, and thus W/U is nonzero by Lemma 10.19(a), and W/U is anisotropic by

Lemma 10.19(b). Finally, since V is hyperbolic as an FH-module, it follows by Lemma 10.19(c) that W/U is hyperbolic as an FH-module, as required. ∎

One advantage of reducing Dade's theorem to the case where V is anisotropic is that it allows us to work with semisimple modules. This is because of the following result.

Lemma 10.21. *Let V be an anisotropic FG-module. Then V is a direct sum of pairwise orthogonal simple FG-modules, and in particular, V is semisimple.*

Proof. We can assume that V is not zero, and we proceed by induction on $\dim(V)$. Let S be a simple submodule of V, and note that S is not totally isotropic since by hypothesis, V is anisotropic. Then $S \cap S^\perp < S$, and since S is simple, we have $S \cap S^\perp = 0$. It follows Lemma 10.15(b) that $\dim(S + S^\perp) = \dim(S) + \dim(S^\perp) = \dim(V)$, and thus $S + S^\perp = V$, and furthermore, this sum is direct. By the inductive hypothesis, $S^\perp$ is a direct sum of pairwise orthogonal simple FG-modules, and since each submodule of $S^\perp$ is orthogonal to S, the result follows. ∎

10C

By the results of the previous section, we see that it suffices to study semisimple symplectic FG-modules. We shall need the following.

Lemma 10.22. *Let U be a submodule of a symplectic FG-module V. Then the module $V/U^\perp$ is dual to U.*

Proof. By Lemma 10.15(a), there is a linear transformation θ from V onto U^*, where $\theta(x)$ is the linear functional λ_x on U, defined by the formula $\lambda_x(u) = \langle x, u\rangle$. Clearly, $\ker(\theta) = U^\perp$, so there is a vector space isomorphism from $V/U^\perp$ onto U^*. To see that this is an FG-module isomorphism, it suffices to show that θ is an FG-module homomorphism, so we must prove that $(xg)\theta = (x\theta)g$ for all $x \in V$ and $g \in G$.

If $g \in G$, then $(xg)\theta = \lambda_{xg}$ and $(x\theta)g = \lambda_x g$, so we must show that $\lambda_{xg} = \lambda_x g$, or equivalently, that $\lambda_{xg}(u) = (\lambda_x g)(u)$ for all $u \in U$. We have

$$\lambda_{xg}(u) = \langle xg, u\rangle = \langle x, ug^{-1}\rangle = \lambda_x(ug^{-1}) = (\lambda_x g)(u),$$

as required. ∎

Next, we consider the homogeneous components of a semisimple symplectic module. In particular, we show that each homogeneous component is either nondegenerate or totally isotropic, and it is nondegenerate if and only if its type module is self-dual.

Lemma 10.23. *Let V be a semisimple symplectic FG-module, and let X be a homogeneous component of V. Then there exists a unique homogeneous component Y of V (possibly equal to X) that is not orthogonal to X.*

(a) *If $X = Y$, then X is nondegenerate.*

(b) *If $X \neq Y$, then X and Y are totally isotropic and $X + Y$ is nondegenerate.*

(c) *Y is the unique homogeneous component of V with type module dual to the type module of X.*

Proof. We show first that if A and B are simple submodules of V, where A and B are not dual to each other, then A and B are orthogonal. By Lemma 10.22, we know that $V/A^\perp \cong A^*$. Since A is simple, it follows by Lemma 10.14 that the dual module A^* is also simple, and thus $A^\perp$ is a maximal submodule of V. If B is not contained in $A^\perp$, therefore, then $A^\perp + B = V$, and since B is simple, we have $A^\perp \cap B = 0$. In this case, $B \cong V/A^\perp \cong A^*$, and thus B is dual to A, contrary to assumption. Then, $B \subseteq A^\perp$, and thus A and B are orthogonal, as claimed.

Now let S be the type of X, and suppose Z is a homogeneous component of V, with type T. Note that X is a sum of isomorphic copies of S and Z is a sum of isomorphic copies of T, so if S and T are not dual to each other, then by the result of the previous paragraph, the simple summands of X are orthogonal to the simple summands of Z, and thus X and Z are orthogonal. This shows that X is orthogonal to every homogeneous component of V with type not isomorphic to S^*. Since V is nondegenerate, X is not orthogonal to V, so there must exist at least one homogeneous component Y of V that is not orthogonal to X. Then Y is the component of type S^*, and in particular, Y is unique and (c) follows.

To prove (a), suppose that $X = Y$, and let R be the radical of X. Then R is orthogonal to X and also to all homogeneous components of V different from $Y = X$, and it follows that R is orthogonal to V. Since V is nondegenerate, we conclude that $R = 0$, and thus X is nondegenerate.

Now suppose that $X \neq Y$, so X is orthogonal to itself, or equivalently X is totally isotropic. Also, since X and Y are not orthogonal, we can interchange the roles of X and Y, and we deduce that Y is totally isotropic.

Now $X + Y$ is orthogonal to all homogeneous components of V different from X and Y. If R is the radical of $X + Y$, then R is orthogonal to X and Y, and also to all homogeneous components of V other than X and Y. Then R is orthogonal to V, so $R = 0$, and thus $X + Y$ is nondegenerate. This completes the proof of (b). ∎

By Lemma 10.21, we know that an anisotropic FG-module must be semisimple, and we begin now to investigate the question of when a semisimple symplectic FG-module can fail to be anisotropic. We begin with an easy observation.

Lemma 10.24. *Let V be a semisimple symplectic FG-module. Then V is anisotropic if and only if its homogeneous components are all anisotropic.*

Proof. It is immediate from the definition that every submodule of an anisotropic module is anisotropic, so it suffices to assume that every homogeneous component of V is anisotropic and to deduce that V is anisotropic. Otherwise, V has a simple totally isotropic submodule S, and since S is contained in one of the homogeneous components of V by Theorem 10.6, this is a contradiction. ∎

Because of Lemma 10.24, it suffices to study homogeneous anisotropic FG-modules, so we assume that V is a direct sum of some number n of isomorphic copies of a simple FG-module S. We shall see that if V is anisotropic and F is finite of odd order, then necessarily $n \leq 2$. Also, there are additional conditions that must be satisfied. We need a number of preliminary results.

Recall that an **anti-automorphism** of a ring E is a map $\tau : E \to E$ such that τ is an automorphism of the additive group of E, and in addition, $(\delta\varepsilon)^\tau = \varepsilon^\tau\delta^\tau$ for all $\delta, \varepsilon \in E$. Of course, if E is commutative, then an anti-automorphism of E is simply an automorphism.

Lemma 10.25. *Let S be a symplectic FG-module, and let E be the ring $\mathrm{End}_{FG}(S)$. Then there is a unique anti-automorphism τ of E such that $\langle x, y\delta\rangle = \langle x\delta^\tau, y\rangle$ for all $x, y \in S$ and all $\delta \in E$. Also, τ^2 is the identity map.*

Proof. Fix $\delta \in E$, and let $x \in S$. The map $y \mapsto \langle x, y\delta\rangle$ is a linear functional on S, and since S is symplectic, we know by Lemma 10.15(a) that every linear functional on S has the form $y \mapsto \langle v, y\rangle$ for some unique vector $v \in S$. Of course, v depends on x, and we let $\alpha : S \to S$ be the map defined by $x\alpha = v$. Then $\langle x, y\delta\rangle = \langle x\alpha, y\rangle$ for all $x, y \in S$, and it is easy to check that α is an F-linear operator on S.

We argue now that the map α lies in E. To see this, we must show that $(xg)\alpha = (x\alpha)g$ for all $x \in S$ and $g \in G$. We compute

$$\langle (xg)\alpha, y\rangle = \langle xg, y\delta\rangle = \langle x, y\delta g^{-1}\rangle = \langle x, yg^{-1}\delta\rangle = \langle x\alpha, yg^{-1}\rangle = \langle x\alpha g, y\rangle\,,$$

for all $x, y \in S$. Since y is arbitrary and the form is nondenerate, it follows for all $x \in S$ that $(xg)\alpha = (x\alpha)g$, as wanted, and thus $\alpha \in E$.

Now δ uniquely determines α, so we have a map $\tau : E \to E$, defined by $\delta^\tau = \alpha$. Then $\langle x, y\delta\rangle = \langle x\delta^\tau, y\rangle$ for $x, y \in S$, as required, and it is easy to check that τ is an additive homomorphism from E to itself.

Now for $x, y \in S$, we have

$$\langle x\delta^{\tau^2}, y\rangle = \langle x, y\delta^\tau\rangle = -\langle y\delta^\tau, x\rangle = -\langle y, x\delta\rangle = \langle x\delta, y\rangle ,$$

so reasoning as before, we conclude that $x\delta^{\tau^2} = x\delta$ for all $x \in S$. It follows that $\delta^{\tau^2} = \delta$, and thus τ^2 is the identity map on E. In particular, τ is both injective and surjective, and hence it defines an automorphism on the additive group of E.

Now let $\delta, \varepsilon \in E$. Then for all $x, y \in S$, we have

$$\langle x(\delta\varepsilon)^\tau, y\rangle = \langle x, y\delta\varepsilon\rangle = \langle x\varepsilon^\tau, y\delta\rangle = \langle x\varepsilon^\tau\delta^\tau, y\rangle ,$$

and it follows that $(\delta\varepsilon)^\tau = \varepsilon^\tau\delta^\tau$, and thus τ is an anti-automorphism of E, as required. ∎

We shall refer to the anti-automorphism τ of Lemma 10.25 as the **standard involution** on $E = \mathrm{End}_{FG}(S)$. (But note that we do not claim that τ has order 2, as the word "involution" suggests; τ can be the identity map.) Of course, by the defining property of the standard involution, we have $\langle x, y\delta\rangle = \langle x\delta^\tau, y\rangle$, for all $x, y \in S$ and $\delta \in E$, but note that since τ is its own inverse, it is also true that $\langle x\delta, y\rangle = \langle x, y\delta^\tau\rangle$ for all x, y and δ.

If S is a simple FG-module, where F is finite, then as we have seen, E is a finite division ring. In this case, E is a field, so the standard involution on E is a field automorphism.

Lemma 10.26. *Let F be a finite field, and let S and T be simple symplectic FG-modules. Write $E = \mathrm{End}_{FG}(S)$, and let τ be the standard involution on E. Finally, suppose that $\varphi : S \to T$ is an FG-isomorphism. Then there exists a unique nonzero element $\delta \in E$ such that $\langle x\varphi, y\varphi\rangle = \langle x\delta, y\rangle$, for all $x, y \in S$. Also,*

(a) *$\delta^\tau = \delta$;*

(b) *if $\mu \in E$, then for all $x, y \in S$, we have $\langle x\mu\varphi, y\mu\varphi\rangle = \langle x\delta\mu\mu^\tau, y\rangle$.*

Proof. For a fixed vector $x \in S$, the map $y \mapsto \langle x\varphi, y\varphi\rangle$ is a linear functional on S, and since the form on S is nondegenerate, it follows by Lemma 10.15(a) that there is a unique vector $v \in S$ such that $\langle v, y\rangle = \langle x\varphi, y\varphi\rangle$. Let $\delta : S \to S$ be the map defined by $x\delta = v$. Then

$$\langle x\varphi, y\varphi\rangle = \langle x\delta, y\rangle \qquad \text{for all } x, y \in S ,$$

and, since x determines v uniquely, δ is the unique map on S with this property. Also, δ is nonzero because φ maps onto T and the form on T is not identically zero. It is easy to check that δ is an F-linear operator on S.

We argue that in fact $\delta \in E$. To see this, we must show that $(xg)\delta = (x\delta)g$ for all $x \in S$ and $g \in G$. We have

$$\begin{aligned}\langle (xg)\delta, y\rangle = \langle xg\varphi, y\varphi\rangle = \langle x\varphi g, y\varphi\rangle = \langle x\varphi, y\varphi g^{-1}\rangle &= \langle x\varphi, yg^{-1}\varphi\rangle \\ &= \langle x\delta, yg^{-1}\rangle \\ &= \langle (x\delta)g, y\rangle\,,\end{aligned}$$

so $(xg)\delta = (x\delta)g$, and thus $\delta \in E$, as required.

Now

$$\langle x\delta^{\tau}, y\rangle = \langle x, y\delta\rangle = -\langle y\delta, x\rangle = -\langle y\varphi, x\varphi\rangle = \langle x\varphi, y\varphi\rangle = \langle x\delta, y\rangle$$

for all $x, y \in S$, and since S is nondegenerate, we have $\delta^{\tau} = \delta$, proving (a).

For (b), let $\mu \in E$. Then

$$\langle x\mu\varphi, y\mu\varphi\rangle = \langle x\mu\delta, y\mu\rangle = \langle x\mu\delta\mu^{\tau}, y\rangle$$

for $x, y \in S$. Since E is commutative because F is finite, μ and δ commute, and the proof is complete. ∎

Next, we need a fact from the theory of finite fields.

Lemma 10.27. *Let E be a finite field having an automorphism τ of order 2. Then the τ-fixed elements of E are exactly the elements of the form $\mu\mu^{\tau}$ for $\mu \in E$.*

Proof. Let K be the subfield consisting of the τ-fixed elements of E, and define the map $\theta : E \to E$ by $\theta(\mu) = \mu\mu^{\tau}$. Our goal, therefore, is to show that $\theta(E) = K$, or equivalently, that $\theta(E^{\times}) = K^{\times}$.

Since $o(\tau) = 2$, it follows by elementary Galois theory that E is a field extension of degree 2 over K. Writing $|K| = q$, we see that $|E| = q^2$, and thus the map $\mu \mapsto \mu^q$ is an automorphism of order 2 of E. Since $\mathrm{Aut}(E)$ is cyclic, this automorphism must be τ, and hence $\theta(\mu) = \mu\mu^{\tau} = \mu^{q+1}$.

Since $E^{\times}$ is cyclic, we can write $E^{\times} = \langle\alpha\rangle$, where the order $o(\alpha) = |E^{\times}| = q^2 - 1$. Clearly, θ defines a group homomorphism from $E^{\times}$ into itself, and thus $\theta(E^{\times}) = \langle\alpha^{q+1}\rangle$. Then

$$|\theta(E^{\times})| = o(\alpha^{q+1}) = (q^2 - 1)/(q + 1) = q - 1\,.$$

In fact, $\theta(E^{\times})$ is the unique subgroup of order $q - 1$ in $E^{\times}$ because $E^{\times}$ is cyclic. Since $K^{\times}$ is a subgroup of $E^{\times}$ and $|K^{\times}| = q - 1$, we have $\theta(E^{\times}) = K^{\times}$, as required. ∎

Lemma 10.28. *Let V be a symplectic FG-module, and suppose that $V = S \dotplus T$, where S and T are orthogonal simple submodules. Let $E = \mathrm{End}_{FG}(S)$,* *and let τ be the standard involution on E. Suppose $\varphi : S \to T$ is an FG-isomorphism, and assume that $\delta \in E$ has the property that*

$$\langle x\varphi, y\varphi\rangle = \langle x\delta, y\rangle$$

for all $x, y \in S$. Then V is anisotropic if and only if τ is trivial and δ is not the negative of a square in E.

In the situation of Lemma 10.28, observe that since by hypothesis the form on V is nondegenerate and S and T are orthogonal, we can conclude that each of S and T is nondegenerate. The standard involution on E is thus defined by Lemma 10.25, and the element $\delta \in E$ exists and is uniquely determined by Lemma 10.26. The statement of the lemma thus makes sense.

Proof of Lemma 10.28. Given $\mu \in E$, let

$$W_\mu = \{s + s\mu\varphi \mid s \in S\},$$

and observe that W_μ is a submodule of V, and $W_\mu \neq T$. Also, W_μ is isomorphic to S, and hence it is simple. Furthermore, since each element of V is uniquely a sum of the form $s + t$, where $s \in S$ and $t \in T$, it is easy to see that the submodules W_μ are distinct for distinct elements $\mu \in E$, and hence there are thus exactly q distinct submodules of the form W_μ, where $q = |E|$.

Lemma 10.10 tells us that the total number of simple submodules of V is $(q^2 - 1)/(q - 1) = q + 1$, and hence the submodules W_μ account for every simple submodule of V other than T. Since T is nondegenerate, we see that V fails to be anisotropic precisely when at least one of the submodules W_μ is totally isotropic. Our goal, therefore, is to show that this happens if and only if τ is nontrivial or $-\delta$ is a square in E.

Now W_μ is totally isotropic if and only if for all $x, y \in S$ we have

$$\begin{aligned} 0 = \langle x + x\mu\varphi, y + y\mu\varphi\rangle &= \langle x, y\rangle + \langle x\mu\varphi, y\mu\varphi\rangle \\ &= \langle x, y\rangle + \langle x\delta\mu\mu^\tau, y\rangle \\ &= \langle x(1 + \delta\mu\mu^\tau), y\rangle, \end{aligned}$$

where the second equality holds because S and T are orthogonal, and the third equality is a consequence of Lemma 10.26(b). We conclude that W_μ is totally isotropic if and only if

$$1 + \delta\mu\mu^\tau = 0.$$

We must prove, therefore, that there exists an element $\mu \in E$ satisfying this isotropy condition if and only if either τ is nontrivial or δ is the negative of a square in E.

If τ is nontrivial, there necessarily exists an element $\mu \in E$ satisfying the isotropy condition. This is because $\delta \neq 0$ and $-\delta^{-1}$ is τ-fixed by

Lemma 10.26, so by Lemma 10.27, there exists an element $\mu \in E$ such that $\mu\mu^\tau = -\delta^{-1}$, as required.

Now suppose that τ is trivial, so the isotropy condition becomes $\delta\mu^2 = -1$. This condition is satisfied for some element $\mu \in E$ precisely when δ is the negative of a square in E, and this completes the proof. ∎

We need one further elementary fact about finite fields.

Lemma 10.29. *Let ε and δ be nonzero elements of a finite field E. Then there exist μ and ν in E such that $1 + \delta\mu^2 + \varepsilon\nu^2 = 0$.*

Proof. If $|E|$ is even, then the multiplicative group $E^\times$ has odd order, and hence every element of $E^\times$ is a square. Since it is clear that μ and ν exist in this case, we can assume that $|E|$ is odd, and we write $q = |E|$.

Let $X = \{-\varepsilon\nu^2 \mid \nu \in E\}$. Each nonzero member of X is determined by two distinct nonzero elements ν and $-\nu$ of E, and it follows that X contains exactly $(q-1)/2$ nonzero members. Since also $0 \in X$, we have $|X| = (q-1)/2 + 1 = (q+1)/2$.

Now let $Y = \{1 + \delta\mu^2 \mid \mu \in E\}$. Observe that $1 \in Y$ and that each member of Y different from 1 is determined by two different nonzero elements of E. It follows that $|Y| = (q+1)/2$, and thus $|X| + |Y| > q = |E|$. We see, therefore, that X and Y must intersect nontrivially, and we conclude that there exist $\mu, \nu \in E$ such that $-\varepsilon\nu^2 = 1 + \delta\mu^2$, as required. ∎

Lemma 10.30. *Let V be a homogeneous anisotropic FG-module, where F is a finite field. Then either V is simple or else $V = S \dotplus T$, where S and T are orthogonal isomorphic simple FG-submodules.*

Proof. Since V is anisotropic, it follows by Lemma 10.21 that V is a direct sum of some number n of pairwise orthogonal simple FG-modules. Also, since V is homogeneous, all simple submodules of V are isomorphic, so it suffices to prove that $n \le 2$. Otherwise, V has a (necessarily anisotropic) submodule of the form $S \dotplus T \dotplus R$, where S, T, and R are isomorphic simple pairwise orthogonal submodules.

Fix isomorphisms $\varphi : S \to T$ and $\theta : S \to R$. Since V is anisotropic, each of the simple submodules S, T, and R is nondegenerate. We can thus apply Lemma 10.26 to deduce that there exist nonzero elements $\delta, \varepsilon \in E$ such that

$$\langle x\varphi, y\varphi\rangle = \langle x\delta, y\rangle \quad \text{and} \quad \langle x\theta, y\theta\rangle = \langle x\varepsilon, y\rangle$$

for all $x, y \in S$.

Let $E = \operatorname{End}_{FG}(S)$. Now $S + T$ is anisotropic, so Lemma 10.28 guarantees that the standard involution on E must be trivial, and thus by

Lemma 10.26(b), we have

$$\langle x\mu\varphi, y\mu\varphi\rangle = \langle x\delta\mu^2, y\rangle \quad \text{and} \quad \langle x\nu\theta, y\nu\theta\rangle = \langle x\varepsilon\nu^2, y\rangle$$

for all μ, ν in E.

By Lemma 10.29 we can choose μ and ν in E such that $1+\delta\mu^2+\varepsilon\nu^2 = 0$, and we let

$$W = \{s + s\mu\varphi + s\nu\theta \mid s \in S\}\,,$$

so W is a nonzero submodule of V. We will show that W is totally isotropic, and since V is anisotropic, this is a contradiction, and the proof will be complete.

Since S, T, and R are pairwise orthogonal, we have

$$\begin{aligned}\langle x + x\mu\varphi + x\nu\theta, y + y\mu\varphi + y\nu\theta\rangle &= \langle x, y\rangle + \langle x\mu\varphi, y\mu\varphi\rangle + \langle x\nu\theta, y\nu\theta\rangle \\ &= \langle x(1 + \delta\mu^2 + \varepsilon\nu^2), y\rangle \\ &= 0\,,\end{aligned}$$

so W is totally isotropic, as wanted. ∎

Theorem 10.31. *Let F be a finite field, and let V be a semisimple symplectic FG-module. Let $H \subseteq G$, and assume that the following conditions hold.*

(1) *Every simple FG-submodule of V is simple as an FH-module.*
(2) *For each simple FG-submodule T of V, the field extension $\mathrm{End}_{FG}(T) \subseteq \mathrm{End}_{FH}(T)$ has odd degree.*
(3) *The homogeneous FH-components of V are exactly the homogeneous FG-components of V.*

Then V is anisotropic as an FG-module if and only if it is anisotropic as an FH-module.

Note that since V is semisimple as an FG-module in Theorem 10.31, it is a sum of simple FG-submodules, and by (1), each of these summands is simple as an FH-module. Then V is semisimple as an FH-module, and thus it makes sense to refer to the homogeneous FH-components in (3).

Corollary 10.32. *Let F be a finite field of characteristic $p \neq 2$, and let V be a semisimple symplectic FG-module. Suppose that G/N is a p-group, and assume for all simple FN-submodules S of V and all elements $g \in G$ that S and Sg are FN-isomorphic. Then V is anisotropic as an FG-module if and only if V is anisotropic as an FN-module.*

Proof. It suffices to show that the three numbered hypotheses of Theorem 10.31 are satisfied if we take $H = N$, and we observe that we can apply Theorem 10.11 because the hypotheses of that result are being assumed. Condition (1) of Theorem 10.31 holds by Theorem 10.11(a), and (3) holds by Theorem 10.11(c). Also, by Theorem 10.11(d), the degree of the field extension $\mathrm{End}_{FG}(T) \subseteq \mathrm{End}_{FH}(T)$ is a power of p for each simple FG-submodule T of V, so (2) holds because we are assuming that $p \neq 2$. ■

Proof of Theorem 10.31. First, assume that V is anisotropic as an FH-module. Then V has no nonzero H-invariant totally isotropic subspace, and hence it certainly has no nonzero G-invariant totally isotropic subspace. Thus V is anisotropic as an FG-module, as required, so it suffices to prove the converse. We assume, therefore, that V is anisotropic as an FG-module, and we work to show that it also anisotropic as an FH-module.

By Lemma 10.24, each homogeneous FG-component of V is anisotropic, and since by (3), these components are the homogeneous FH-components of V, it suffices by Lemma 10.24 to show that each of these FH-components of V is anisotropic. It is no loss, therefore, to assume that V is homogeneous as an FG-module.

Since V is anisotropic as an FG-module, it follows by Lemma 10.30 that either V is simple as an FG-module, or else $V = S \dotplus T$, where S and T are orthogonal FG-isomorphic simple FG-submodules.

By (1), each simple FG-submodule of V is simple as an FH-module, so in the case where V is simple as an FG-module, it is also simple as an FH-module, and in that case, V is clearly anisotropic as an FH-module, as wanted. We can thus assume that $V = S \dotplus T$, where S and T are orthogonal, and they are simple and isomorphic both as FG-modules and as FH-modules.

Let $E = \mathrm{End}_{FH}(T)$ and $K = \mathrm{End}_{FG}(T)$, so $K \subseteq E$ is a field extension, and by (2), the degree of this extension is odd. Also, if τ is the standard involution on E, we see that the restriction of τ to K is the standard involution on K.

Since V is anisotropic as an FG-module, it follows by Lemma 10.28 that the standard involution on K is trivial, so if the standard involution on E is nontrivial, then the Galois group $\mathrm{Gal}(E/K)$ contains an element of order 2. This is not the case however, because the order of the Galois group is equal to the degree of the field extension, which is odd. We conclude that the standard involution on E is trivial.

Now let $\varphi : S \to T$ be an FG-isomorphism. By Lemma 10.26, there is a unique element $\delta \in K$ such that $\langle x\varphi, y\varphi\rangle = \langle x\delta, y\rangle$ for all $x, y \in S$. Also we note that δ is also the field element associated with φ when φ is viewed as an FH-isomorphism. Since V is anisotropic as an FG-module, it follows by Lemma 10.28 that $-\delta$ is not a square in K, and thus the polynomial $x^2 + \delta$

is irreducible over K. If this polynomial has some root ε in E, then the field extension $K \subseteq K[\varepsilon]$ has degree 2, and this is impossible since the degree of E over K is odd. It follows that δ is not the negative of a square in E, and we conclude by Lemma 10.28 that V is anisotropic as an FN-module. ∎

Lemma 10.33. *Let V be a symplectic FG-module, and let $N \lhd G$. Suppose that U is an FN-submodule of V, and let T be the stabilizer of U in the right-multiplication action of G on the set of FN-submodules of V.*

(a) *If V is anisotropic as an FG-module and U and Ug are orthogonal whenever $g \in G$ and $Ug \neq U$, then U is anisotropic as an FT-module.*

(b) *If V is hyperbolic as an FG-module and U is nondegenerate and is a sum of some of the homogeneous FN-components of V, then U is hyperbolic as an FT-module.*

Proof. For (a), we must prove that U is anisotropic, as an FT-module, so we consider a totally isotropic FT-submodule X of U, and we show that $X = 0$. Let $\mathcal{X}$ be the G-orbit of X, so $\sum \mathcal{X}$ is an FG-submodule of V. We will see that $\sum \mathcal{X}$ is totally isotropic, and since by assumption, V is anisotropic as an FG-module, it will that $\sum \mathcal{X} = 0$, and thus $X = 0$, as required.

To show that $\sum \mathcal{X}$ is totally isotropic, it suffices to prove that every two (not necessarily distinct) members of $\mathcal{X}$ are orthogonal. In fact, since $\mathcal{X}$ is a G-orbit and the form on V is G-invariant, it is enough to show that X is orthogonal to Xg for all $g \in G$. Now if $Ug \neq U$, then by assumption, U and Ug are orthogonal, and since $X \subseteq U$ and $Xg \subseteq Ug$, it follows that X and Xg are orthogonal, as wanted. We can assume, therefore, that $Ug = U$, so $g \in T$. Then $Xg = X$ because X is an FT-module, and since X is totally isotropic, it is orthogonal to $X = Xg$, as wanted.

For (b), we are assuming that U is a sum of some of the homogeneous FN-components of V, and thus we can write $V = U \dotplus Z$, where Z is the sum of the homogeneous FN-components of V that are not contained in U. We are assuming that U is nondegenerate, and we argue that Z is also nondegenerate. To see this, we will show that Z is orthogonal to U, and thus the radical R of Z is orthogonal to $U + Z = V$, and hence $R = 0$ because by hypothesis, V is nondegenerate.

To prove that Z is orthogonal to U, it suffices to show that if X is an arbitrary homogeneous FN-component of V contained in U, then X is orthogonal to Z. By Lemma 10.23, there exists exactly one homogeneous FN-component Y of V (possibly equal to X) such that X is not orthogonal to Y. If $Y \subseteq Z$, then X is orthogonal to U, and this is a contradiction since

U is nondegenerate. It follows that $Y \subseteq U$, and thus X is orthogonal to Z, as required, and we conclude that Z is nondegenerate.

By assumption, V is hyperbolic as an FG-module, so there exists a totally isotropic FG-submodule W with $\dim(W) = (1/2)\dim(V)$. By Corollary 10.7, we can write $W = (W \cap U) \dotplus (W \cap Z)$, and since U and Z are nondegenerate and $W \cap U$ and $W \cap Z$ are totally isotropic, Lemma 10.16 guarantees that

$$\dim(W \cap U) \leq (1/2)\dim(U) \quad \text{and} \quad \dim(W \cap Z) \leq (1/2)\dim(Z)\,.$$

Then

$$\begin{aligned}(1/2)\dim(V) = \dim(W) &= \dim(W \cap U) + \dim(W \cap Z)\\ &\leq (1/2)\dim(U) + (1/2)\dim(Z)\\ &= (1/2)\dim(U + Z)\\ &= (1/2)\dim(V)\,.\end{aligned}$$

Equality thus holds throughout, and we conclude that $\dim(W \cap U) = (1/2)\dim(U)$. Since $W \cap U$ is totally isotropic and T-invariant, it follows that U is hyperbolic as an FT-module, and this proves (b). ∎

We can now prove Dade's theorem.

Proof of Theorem 10.17. Recall that we are given a symplectic FG-module V of a p-solvable group G, where p is an odd prime and F is a finite field of characteristic p. We are also given a subgroup H of G having p-power index, and by assumption, V is hyperbolic as an FH-module. Our goal is to show that V is hyperbolic as an FG-module.

Assuming that the theorem is false, let G be a counterexample having minimum possible order, and among FG-modules that demonstrate that G is a counterexample, assume that $\dim(V)$ is as small as possible. Then V is anisotropic as an FG-module by Corollary 10.20, and hence V is semisimple by Lemma 10.21. Also, observe that we can assume that H is a maximal subgroup of G.

Let $N = \mathrm{core}_G(H)$. Then $N < G$, so we can choose a chief factor M/N of G, and we see that $M \not\subseteq H$. Since H is maximal, we have $MH = G$, and thus $|G : H| = |MH : H| = |M : M \cap H|$. By assumption, $|G : H|$ is a nontrivial power of p, and we conclude that p divides $|M : N|$. Then M/N is not a p'-group, and since G is p-solvable, it follows that M/N is an abelian p-group.

Next, we show that $\mathbf{C}_G(M/N) = M$. Write $C = \mathbf{C}_G(M/N)$, so $M \subseteq C$ because M/N is abelian, and since $MH = G$, we have $C = M(H \cap C)$. Now $C \triangleleft G$, so H normalizes $H \cap C$. Also, M normalizes $H \cap C$ because

$[H \cap C, M] \subseteq [C, M] \subseteq N \subseteq H \cap C$, and it follows that $H \cap C \triangleleft G$. Then $H \cap C \subseteq \operatorname{core}_G(H) = N \subseteq M$, and we have $C = M(H \cap C) = M$, as claimed.

Clifford's theorem (Theorem 10.8) guarantees that V is semisimple as an FN-module, and we let X be a homogeneous FN-component of V. By Lemma 10.23, there is a unique homogeneous FN-component Y of V such that X and Y are not orthogonal. If $X = Y$, we set $U = X$ and otherwise, we set $U = X + Y$. By Lemma 10.23, we see that in each case, U is nondegenerate and U is orthogonal to all homogeneous FN-components of V other than X and Y.

We saw in the discussion preceding Theorem 10.8 that G acts by right multiplication on the set of FN-submodules of V, and this action permutes the homogeneous FN-components of V. We argue that if $g \in G$ and $Ug \neq U$, then U and Ug are orthogonal. This is clear if $X = Y$ because in that case, $U = X$, so Ug is a homogeneous FN component of V different from X and Y, and thus U is orthogonal to Ug.

In the remaining case, $U = X + Y$, where X and Y are distinct homogeneous FN-components of V that are not orthogonal. Since $Ug = Xg + Yg$, and $Ug \neq U$, at least one of Xg or Yg is not contained in U. Now Xg and Yg are homogeneous FN-components of V, and we see that if neither Xg nor Yg is contained in U, then U is orthogonal to both Xg and Yg, and thus U and Ug are orthogonal, as wanted. We can thus assume that one of Xg or Yg is contained in U and that the other is not, and thus Xg and Yg are orthogonal. This is a contradiction since X and Y are not orthogonal, and this completes the proof that U and Ug are orthogonal if they are different.

Now let K be the stabilizer of U in G, so U is an FK-module. Since V is anisotropic as an FG-module, Lemma 10.33(a) guarantees that U is also anisotropic as an FK-module, and in particular, U is not hyperbolic as an FK-module.

Since $|G : H|$ is a power of p, it follows that H contains a full Hall p'-subgroup of G. Replacing H by a conjugate if necessary, we can assume that H contains a full Hall p'-subgroup of K, and thus $|K : H \cap K|$ is a power of p. Also, $H \cap K$ is the stabilizer of U in H, and since V is hyperbolic as an FH-module, Lemma 10.33(b) guarantees that U is hyperbolic as an $F(H \cap K)$-module. It follows that K is a counterexample to the theorem, so we conclude by the minimality of $|G|$ that $K = G$, and thus U is an FG-module. Also, by the minimality of $\dim(V)$, we have $U = V$, and thus V has at most two homogeneous FN-components.

In the case where $U = V$ has two distinct homogeneous FN-components X and Y, it follows by Lemma 10.23 that X and Y are totally isotropic, and since V is anisotropic as an FG-module, neither X nor Y can be G-invariant. Then X and Y have a common stabilizer R in G, and we have

$|G : R| = 2$. In the remaining case, where V has just one homogeneous FN-component, we set $R = G$. In both cases, we see that $N \subseteq R$.

Now R stabilizes each FN-homogeneous component of V, so if S is a simple FN submodule of V and $r \in R$, then Sr is a simple FN-submodule of the homogeneous FN-component that contains S, and thus S and Sr are FN-isomorphic. Since $|G : R| \leq 2$ and $|M : N|$ is a power of the odd prime p, it follows that $M \subseteq R$, and thus S and Sm are FN-isomorphic for all $m \in M$.

We can now apply Theorem 10.11 with M in place of G. By Theorem 10.11(a), every simple FM-submodule T of V is also simple as an FN-module, and thus there is a conjugation action of M on the field $\mathrm{End}_{FN}(T)$, and the fixed-point subfield of this action is $\mathrm{End}_{FM}(T)$.

Suppose that for every choice of a simple FM-submodule T of V we have $\mathrm{End}_{FM}(T) = \mathrm{End}_{FN}(T)$. Since V is hyperbolic as an FH-module, there is a totally isotropic FH-submodule W of V such that $\dim(W) = (1/2)\dim(V)$. Now W is N-invariant because $N \subseteq H$, and thus W is M-invariant by Theorem 10.11(e). Since $G = MH$ and W is both M-invariant and H-invariant, we see that W is an G-invariant, and thus V is hyperbolic as an FG-module. This is a contradiction, and we conclude that there exists a simple FM-submodule T of V, such that $\mathrm{End}_{FM}(T) < \mathrm{End}_{FN}(T)$.

It follows now that the conjugation action of M on the field $E = \mathrm{End}_{FN}(T)$ is nontrivial. Since $M \subseteq R$, Lemma 10.12 guarantees that the conjugation action of M can be extended to an action of R on E, and we let L be the kernel of this action of R. The conjugation action of N on E is trivial, so $N \subseteq L$, but since the action of M on E is nontrivial, we see that $M \not\subseteq L$.

Now R induces field automorphisms of E, and since the automorphism group of a finite field is cyclic, it follows that R/L is cyclic, and so $[M, R] \subseteq R' \subseteq L$. Then $N[M, R] \subseteq L$, and hence $M \not\subseteq N[M, R]$. Also $[M, R] \subseteq M$ since $M \lhd G$, and thus $N \subseteq N[M, R] < M$.

Recall that $|G : R| \leq 2$, so $R \lhd G$, and thus $N[M, R] \lhd G$. Since M/N is a chief factor of G, we deduce that $N[M, R] = N$, so $[M, R] \subseteq N$, and hence R centralizes M/N. We saw previously that $\mathbf{C}_G(M/N) = M$, so $R \subseteq M$, and thus $R = M$, and in particular $|G : M| \leq 2$.

Since V is hyperbolic as an FN-module, it is not anisotropic, and it follows by Corollary 10.32 with M in place of G that V is not anisotropic as an FM-module. We know, however, that V is anisotropic as an FG-module, and so we conclude that $M < G$, and thus $|G : M| = 2$.

We know now that V has exactly two homogeneous FN-components X and Y, and by Theorem 10.11(c), these are also the homogeneous FM-components of V. Also, $R = M$ is the stabilizer of X, and thus for every element $g \in G$ with $g \notin M$, we have $Y = Xg$. In particular, we can choose an element $h \in H$ such that $Y = Xh$.

Let T be a simple FM-submodule of X. Then Th is a simple FM-submodule of Y, and thus T and Th are not FM-isomorphic, and it follows by Lemma 10.13(a) that $T + Th$ is a simple FG-submodule of V, and we write $S = T + Th$. Now X is a sum of FM-isomorphic copies of T, and if J is any one of these, then by Lemma 10.13(c), we see that $J + Jh$ is FG-isomorphic to S. It follows that V is a sum of isomorphic copies of S, and thus V is homogeneous of type S as an FG-module. Also, T is simple as an FN-module by Theorem 10.11(a), and reasoning as before, we see that $T + Th = S$ is simple as an FH-module, and thus V is homogeneous of type S as an FH-module.

Now V is anisotropic as an FG-module, and we wish to apply Theorem 10.31 to deduce that V is also anisotropic as an FH-module. Since V is hyperbolic and not anisotropic as an FH-module, this contradiction will complete the proof of the theorem. It suffices, therefore, to check that the three conditions of Theorem 10.31 hold. For (1), let Z be an arbitrary simple FG-submodule of V. Then Z is isomorphic to S as FG-modules, and since S is simple as an FH-module, it follows that Z is also simple as an FH-module, as required. Condition (3) is trivially satisfied because V is homogeneous both as an FG-module and as an FH-module.

It remains to check that the degree of the field extension $\mathrm{End}_{FG}(S) \subseteq \mathrm{End}_{FH}(S)$ is odd. By Lemma 10.13(b), we have $|\mathrm{End}_{FG}(S)| = |\mathrm{End}_{FM}(T)|$ and $|\mathrm{End}_{FH}(S)| = |\mathrm{End}_{FN}(T)|$, so it suffices to observe that by Theorem 10.11(d), the degree of the field extension $\mathrm{End}_{FM}(T) \subseteq \mathrm{End}_{FN}(T)$ is a power of p, so it is odd because $p \neq 2$. ■

10D

In this section, we prove Theorem 10.1. The key step is an appeal to Dade's Theorem 10.17, and we show first how this result will be used.

Lemma 10.34. *Let $L = \mathbf{Z}(K)$, where L is cyclic and K/L is an elementary abelian p-group for some prime p. Then K/L can be viewed a symplectic F-vector space, where F is a field of order p. Under this identification, the F-subspaces of K/L are exactly the subgroups U/L of K/L, and an F-subspace U/L is totally isotropic if and only if U is abelian. Also, if H acts on K via automorphisms and H acts trivially on L, then K/L is a symplectic FH-module.*

Proof. By assumption, K/L is abelian, so for elements $x, y \in K$, the commutator $[x, y]$ lies in L. Since L is central in K, we see that $[x, y]$ depends only on the cosets of L that contain x and y, and thus we can define a function of two variables on $\overline{K} = K/L$ by setting $\langle \overline{x}, \overline{y} \rangle = [x, y]$. Again using the assumption that L is central in K, it is easy to check that this function is a homomorphism in each variable, and since K/L is an elementary abelian p-group, it follows that $\langle \overline{x}, \overline{y} \rangle$ has order dividing p for all $x, y \in K$.

We can certainly assume that K is nonabelian, so there exist elements $x, y \in K$ such that $[x, y]$ is nontrivial, and thus $\langle \overline{x}, \overline{y} \rangle$ has order p. Then $|L|$ is divisible by p, and since L is cyclic by hypothesis, there is a unique subgroup $Z \subseteq L$ of order p. We conclude that $\langle \overline{x}, \overline{y} \rangle$ lies in Z for all $x, y \in K$.

Identify Z with the additive group of a field F of order p, and view K/L as an F-vector space. From this perspective, $\langle \cdot, \cdot \rangle$ is an F-bilinear form on the space K/L. Also, since $[x, x] = 1$ for all $x \in K$, we have $\langle \overline{x}, \overline{x} \rangle = 0$, and hence the form $\langle \cdot, \cdot \rangle$ is alternating. This form is also nondegenerate because if $\langle \overline{x}, \overline{K} \rangle = 0$, then $[x, K] = 1$, so $x \in \mathbf{Z}(K) = L$, and thus $\overline{x} = 0$. The F-subspaces of K/L are the subgroups U/L of K/L, and U/L is totally isotropic precisely when $\langle \overline{u}, \overline{v} \rangle = 0$, or equivalently, $[u, v] = 1$ for all $u, v \in U$. In other words, U/L is totally isotropic if and only if U is abelian.

Now assume that H acts on K via automorphisms and that H centralizes L. Then H acts on K/L via F-linear transformations, and thus K/L is an FH-module. The form $\langle \cdot, \cdot \rangle$ on K/L is H-invariant since H centralizes Z, and thus K/L is a symplectic FH-module. ∎

In the situation of Lemma 10.34, a maximal abelian subgroup A of K must contain $L = \mathbf{Z}(K)$, and thus A/L is a maximal totally isotropic subspace of K/L. It follows by Lemma 10.16 that $\dim(A/L|) = (1/2)\dim(K/L)$, or equivalently, $|A/L|^2 = |K/L|$. Conversely, if A/L is a totally isotropic subspace of K/L with dimension $(1/2)\dim(K/L)$, then A/L is a maximal totally isotropic subspace, and thus A is a maximal abelian subgroup of K. In the case where H acts on K and centralizes L, it follows that K/L is a hyperbolic FH-module if and only if H normalizes some maximal abelian subgroup of K.

Corollary 10.35. *Let G be p-solvable with $p \neq 2$, and suppose that $K \lhd G$. Write $L = \mathbf{Z}(K)$, and assume that L is cyclic and central in G and that K/L is an elementary abelian p-group. Let $H \subseteq G$, where $|G : H|$ is a power of p, and assume that H normalizes a maximal abelian subgroup of K. Then some maximal abelian subgroup of K is normal in G.*

Proof. By Lemma 10.34, we can view K/L as a symplectic FH-module, where F has order p. By hypothesis, H normalizes a maximal abelian subgroup of K, and it follows that K/L is hyperbolic as an FH-module.

By Theorem 10.17, therefore, K/L is hyperbolic as an FG module, and thus some maximal abelian subgroup of K is normal in G. ■

Next, we recall some elementary group theory. For an arbitrary group G, the **lower central series**

$$G = G^1 \supseteq G^2 \supseteq G^3 \supseteq \cdots$$

of G is defined recursively by the formula $G^{r+1} = [G^r, G]$ for $r > 1$. We recall that G is nilpotent if and only if $G^n = 1$ for some integer n, and in that case, $G^r > G^{r+1}$ whenever $G^r > 1$. Also, a useful general fact is that $[G^r, G^s] \subseteq G^{r+s}$ for all positive integers r and s. (See Theorem 4.11 of [**FGT**].)

Lemma 10.36. *Let $L = \mathbf{Z}(K)$.*

(a) *If K/L is a p-group and every characteristic abelian subgroup of K is contained in L, then K/L is elementary abelian.*

(b) *If K/L is abelian and λ is a faithful linear character of L, then λ is fully ramified with respect to K/L.*

Proof. For (a), note that K is nilpotent. Let n be minimal such that $K^n = 1$, and suppose first that $n \geq 4$. Then $[K^{n-2}, K^{n-2}] \subseteq K^{2n-4}$, and since $2n - 4 \geq n$, we see that $K^{2n-4} = 1$. Then K^{n-2} is a characteristic abelian subgroup of K, so by hypothesis, $K^{n-2} \subseteq L = \mathbf{Z}(K)$. It follows that $K^{n-1} = 1$, and this is a contradiction. Then $n \leq 3$, so $[K', K] = [K^2, K] = K^3 = 1$, and thus $K' \subseteq \mathbf{Z}(K) = L$, so K/L is abelian.

To see that K/L is elementary, let E/L be the subgroup of K/L consisting of the elements with order dividing p. If $E < K$, we can choose an element $x \in K - E$ such that $x^p \in E$. Now let $y \in E$ be arbitrary, and observe that since commutators in K are central, we have $[x^p, y] = [x, y]^p = [x, y^p] = 1$, where the final equality holds because $y \in E$ and thus $y^p \in L = \mathbf{Z}(K)$. Then $x^p \in \mathbf{Z}(E)$, and since E is characteristic in K, it follows by hypothesis that $\mathbf{Z}(E) \subseteq L$, and thus $x^p \in L$, so $x \in E$. This is a contradiction, so $E = K$ and thus K/L is elementary, as required for (a).

For (b), let λ be a faithful linear character of L, and suppose that $\chi \in \mathrm{Irr}(K|\lambda)$. To show that λ is fully ramified with respect to K/L, it suffices to prove that χ vanishes on $K - L$. (See Lemma 7.2.) Let $x \in K$ with $x \notin L$. Then x is not central in K, so there exists an element $k \in K$ such that $x^k \neq x$, and thus since K/L is abelian, we can write $x^k = zx$ for some nonidentity element $z \in L$. It follows that $\chi(x) = \chi(x^k) = \chi(zx) = \lambda(z)\chi(x)$, where the final equality follows by considering a representation $\mathcal{X}$ affording χ and recalling that $\mathcal{X}(z)$ is the scalar matrix $\lambda(z)I$. Since λ is faithful and $z \neq 1$, we see that $\lambda(z) \neq 1$, and thus $\chi(x) = 0$, as required. ■

Lemma 10.37. *Suppose that a group M acts on an abelian group A, where $|M|$ and $|A|$ are relatively prime. Then*

(a) $\mathbf{C}_A(M) = 1$ *if and only if* $[A, M] = A$.

(b) *If* $[A, M] = A$ *and* $U \subseteq V \subseteq A$, *where* U *and* V *are* M*-invariant, then* $\mathbf{C}_{V/U}(M) = 1$.

Proof. We have $A = \mathbf{C}_A(M) \times [A, M]$ by Fitting's lemma (Theorem 4.34 of [**FGT**]), and (a) follows. Now assume that $[A, M] = A$ and that $U \subseteq V \subseteq A$, where U and V are A invariant. We have $\mathbf{C}_V(M) \subseteq \mathbf{C}_A(M) = 1$ by (a), and thus $[V, M] = V$ by (a) applied with V in place of A. It follows that $[(V/U), M] = V/U$, and another application of (a) yields $\mathbf{C}_{V/U}(M) = 1$. ■

Next, we present a bit of elementary character theory that will be used here and also in the proof of Theorem 10.2. Recall that if $\chi \in \mathrm{Irr}(G)$, then $\mathbf{Z}(\chi)$ is the largest subgroup Y of G such that χ_Y is a multiple of a linear character, and in particular, $\mathbf{Z}(G) \subseteq \mathbf{Z}(\chi)$.

Lemma 10.38. *Let $\alpha \in \mathrm{Irr}(H)$, where $H \subseteq G$, and suppose that $\alpha^G = \chi$, where χ is irreducible. Then $\mathbf{Z}(\chi) \subseteq H$, and thus $\mathbf{Z}(G) \subseteq H$.*

Proof. Let $Y = \mathbf{Z}(\chi)$, so χ_Y is a multiple of some linear character λ. Let $J = YH$ and $D = Y \cap H$, and note that α^J is irreducible and lies under χ. Then $(\alpha^J)_Y$ is a multiple of λ, and thus $(\alpha^J)_Y = \alpha^J(1)\lambda$. Then

$$|J : H|\alpha(1) = \alpha^J(1) = [(\alpha^J)_Y, \lambda] = [(\alpha_D)^Y, \lambda] = [\alpha_D, \lambda_D] \leq \alpha(1),$$

where the last inequality holds since λ_D is irreducible. Then $|J : H| = 1$, so $Y \subseteq J = H$. ■

The following theorem is a key component of the proof of Theorem 10.1.

Theorem 10.39. *Let $N \lhd G$, where G is p-solvable with $p \neq 2$, and let $L = \mathbf{Z}(N)$. Assume that every abelian normal subgroup of G that is contained in N is central in G, and suppose that there exists a monomial character $\chi \in \mathrm{Irr}(G)$, where $\chi(1)$ is a power of p and χ_N is faithful. Then N/L is an elementary abelian p-group, and if λ is a linear character of L lying under χ, then λ is invariant in G and λ is fully ramified with respect to N/L.*

Proof. If $L = N$, there is nothing to prove, so suppose $L < N$. Since L is an abelian normal subgroup of G contained in N, we have $L \subseteq \mathbf{Z}(G)$, and thus $L = N \cap \mathbf{Z}(G)$, so every abelian normal subgroup G that is contained in N is contained in L.

Now χ_L is a multiple of some linear character λ, so λ is invariant in G, as wanted. Also, λ is faithful because $\ker(\lambda) \subseteq \ker(\chi) \cap N = 1$.

Let $Y/L = \mathbf{O}_{p'}(N/L)$, and observe that since L is central in Y and Y/L is a p'-group, no irreducible character of Y can have degree divisible by p. Now $\chi(1)$ is a power of p, so all irreducible constituents of χ_Y have p-power degree, and it follows that these constituents are linear. Then $Y' \subseteq \ker(\chi) \cap N = 1$, and so Y is an abelian normal subgroup of G contained in N. Then $Y \subseteq L$, and thus $\mathbf{O}_{p'}(N/L)$ is trivial.

Now suppose that $K \triangleleft G$ and $L \subseteq K \subseteq N$, where K/L is a p-group. Every characteristic abelian subgroup of K is a normal abelian subgroup of G contained in N, so every such subgroup is contained in L, and in particular $L = \mathbf{Z}(K)$. It follows by Lemma 10.36(a) that K/L is an elementary abelian p-group, and thus Lemma 10.36(b) guarantees that λ is fully ramified with respect to K/L. If N/L is a p-group, we can take $K = N$, and there is nothing further to prove, so we can assume that N/L is not a p-group, and we work to obtain a contradiction.

Since N/L is not a p-group, there exists a subgroup $M \triangleleft G$ minimal such that $L \subseteq M \subseteq N$ and M/L is not a p-group. In particular, $M > L$, so we can choose a chief factor M/K of G with $L \subseteq K$. The minimality of M guarantees that K/L is a p-group, and thus M/K is not a p-group, and it follows that M/K is a p'-group because G is p-solvable. By the result of the previous paragraph, K/L is elementary abelian and λ is fully ramified with respect to K/L. Also, we note that $K > L$ because otherwise, M/L is is a nontrivial p'-group, and this is a contradiction since $\mathbf{O}_{p'}(N/L)$ is trivial.

We argue next that $[(K/L), M] = K/L$. To see this, write $J/L = [(K/L), M]$, and suppose that $J < K$. Since K/J is a central Sylow p-subgroup of M/J, we can write $M/J = (K/J) \times (U/J)$ for some subgroup U, where $M/U \cong K/J$, and thus M/U is a nontrivial p-group. Now $U \triangleleft G$ and $L \subseteq U < M$, so it follows by the minimality of M that U/L is a p-group. This is a contradiction because M/U is also a p-group but M/L is not a p-group. It follows that $J = K$, so $[(K/L), M] = K/L$, as claimed.

Since χ is monomial, there exists a subgroup $H \subseteq G$ and a linear character ν of H such that $\nu^G = \chi$. Then $L \subseteq \mathbf{Z}(G) \subseteq H$ by Lemma 10.38. Also, ν lies over λ because χ_L is a multiple of λ.

Now $|G : H| = \chi(1)$, and this is a power of p. Then KH/K has p-power index in G/K, and hence KH/K contains the normal p'-subgroup M/K. Then $K \subseteq M \subseteq KH$, and thus $M = K(M \cap H)$.

Let $D = K \cap H$, and note that $D \triangleleft H$. Also, $L \subseteq D \subseteq K$, so D/L is abelian. Write $\varphi = \nu_D$, so φ lies over λ and φ is linear and H-invariant. We have $D' \subseteq \ker(\varphi) \cap L = \ker(\lambda) = 1$, so D is abelian, and thus D is contained in the stabilizer E of φ in K. Also, since H stabilizes φ, it follows that H normalizes E.

We have

$$|\operatorname{Irr}(E|\varphi)| = |\operatorname{Irr}(K|\varphi)| \leq |\operatorname{Irr}(K|\lambda)| = 1\,,$$

where the first equality is a consequence of the Clifford correspondence, the inequality holds because φ lies over λ and the final equality is valid because λ is fully ramified with respect to K/L. Thus $|\operatorname{Irr}(E|\varphi)| = 1$, and since φ is invariant in E, it follows that φ is fully ramified with respect to E/D.

Let $R = EH$, and observe that H is a complement for E relative to D in R. We will apply Lemma 7.6 to conclude that every such relative complement is conjugate to H in R. To verify the hypotheses of Lemma 7.6, we must find a subgroup S with $E \subseteq S \lhd R$ and such that $|S : E|$ and $|E : D|$ are relatively prime and $\mathbf{C}_{E/D}(S) = 1$.

We check that the subgroup $S = M \cap R$ satisfies these conditions. Certainly $E \subseteq S \lhd R$. Also, since $M = K(M \cap H)$, we have $M = KS$, and thus

$$\mathbf{C}_{E/D}(S) = \mathbf{C}_{E/D}(M) = \mathbf{C}_{E/D}(M/K) = 1\,,$$

where the final equality holds by Lemma 10.37 because $[(K/L),(M/K)] = K/L$ and the orders of E/D and M/K are relatively prime. Finally, observe that $K \cap R = E$ because $K \cap H = D$ and H is a complement for E in R relative to D. It follows that $|S : E|$ divides $|M : K|$, and thus p does not divide $|S : E|$. Since $|E : D|$ is a power of p, we see that $|S : E|$ and $|E : D|$ are relatively prime, as required. Lemma 7.6 thus applies, so as claimed, every complement for E in R relative to D is conjugate to H in R.

We see now that we can apply Theorem 7.5 in the group R, with E/D in place of K/L and H in place of U. Now $\nu^G = \chi$, so ν^R is irreducible, and since ν lies over φ, it follows by Theorem 7.5(c) that $D = E$.

Recall now that D is abelian. In fact, D is a maximal abelian subgroup of K because any abelian subgroup of K containing D is contained in the stabilizer E of φ in K, and we know that $E = D$. Since D is H-invariant and H has p-power index in G, Corollary 10.35 guarantees that some maximal abelian subgroup A of K is normal in G. Since L contains every abelian normal subgroup of G that is contained in N, we have $A \subseteq L$, and thus $L = \mathbf{Z}(K)$ is a maximal abelian subgroup of K. Then $L = K$, and this is our final contradiction. ∎

Our next lemma handles the case of Theorem 10.1 where N is abelian.

Lemma 10.40. *Let $A \lhd G$, where A is abelian, and suppose that μ is a linear character of A that lies under some monomial character $\chi \in \operatorname{Irr}(G)$. Then the Clifford correspondent of χ with respect to μ is monomial.*

Proof. Since χ is monomial, we can write $\chi = \lambda^G$, where λ is a linear character of some subgroup $J \subseteq G$. Let $\gamma = \lambda^{AJ}$, and observe that $\gamma^G = \chi$, and thus γ is irreducible. The irreducible constituents of γ_A lie under χ, and hence they are conjugate in G to μ. There is no loss if we replace μ by a conjugate, so we can assume that μ lies under γ.

Let T be the stabilizer of μ in G, and write $S = T \cap AJ$. Then S is the stabilizer of μ in AJ, and we let $\beta \in \mathrm{Irr}(S|\mu)$ be the Clifford correspondent of γ, so $\beta^{AJ} = \gamma$ and $[\gamma_A, \mu] = [\beta_A, \mu]$.

Now $\beta_A = e\mu$ for some positive integer e, and thus,

$$e = [\beta_A, \mu] = [\gamma_A, \mu] = [(\lambda^{AJ})_A, \mu] = [(\lambda_{A\cap J})^A, \mu] = [\lambda_{A\cap J}, \mu_{A\cap J}] \leq 1\,,$$

where the final inequality holds because $\lambda_{N\cap J}$ and $\mu_{N\cap J}$ are irreducible since λ and μ are linear. It follows that $e = 1$, and thus $\beta_A = \mu$, and in particular, β is linear.

We have

$$(\beta^T)^G = \beta^G = (\beta^{AJ})^G = \gamma^G = \chi\,,$$

and thus β^T is irreducible. Also, β^T lies over μ and under χ, and it follows that β^T is the Clifford correspondent ψ of χ with respect to μ. Finally, since β is linear, we see that $\psi = \beta^T$ is monomial, as required. ■

Proof of Theorem 10.1. Recall that $N \triangleleft G$, where G is p-solvable and $p \neq 2$. We are assuming that $\theta \in \mathrm{Irr}(N)$ lies under a monomial character $\chi \in \mathrm{Irr}(G)$, where $\chi(1)$ is a power of p. Also, T is the stabilizer of θ in G, and $\eta \in \mathrm{Irr}(T|\theta)$ is the Clifford correspondent of χ. Our goal is to show that η is monomial, so we assume that this is false and that G is a counterexample having minimum possible order.

Since $G/(N \cap \ker(\chi))$ is also a counterexample to the theorem, the minimality of $|G|$ guarantees that $N \cap \ker(\chi) = 1$, and thus χ_N is faithful. Let $L = \mathbf{Z}(N)$, and let $\lambda \in \mathrm{Irr}(L)$ lie under χ. We wish to apply Theorem 10.39 to deduce that λ is invariant in G and that λ is fully ramified with respect to N/L, so it suffices to check that every abelian normal subgroup of G that is contained in N is central in G.

Given $A \triangleleft G$, where A is abelian and $A \subseteq N$, we argue that $A \subseteq \mathbf{Z}(G)$. Let $\mu \in \mathrm{Irr}(A)$ lie under θ. If μ is invariant in G, then since μ is linear and χ_A is a multiple of μ, we have $[A, G] \subseteq \ker(\mu) \cap N \subseteq \ker(\chi) \cap N = 1$, and thus A is central in G, as required. We can assume, therefore, that the stabilizer S of μ in G is a proper subgroup, and we work to obtain a contradiction.

Now η lies over θ, so η lies over μ, and we see that $T \cap S$ is the stabilizer of μ in T. Let $\gamma \in \mathrm{Irr}((T \cap S)|\mu)$ be the Clifford correspondent of η, so $\gamma^T = \eta$, and we write $\psi = \gamma^S$. We have

$$\psi^G = (\gamma^S)^G = \gamma^G = (\gamma^T)^G = \eta^G = \chi\,,$$

and hence ψ is irreducible.

Now ψ lies over γ, which lies over μ, and thus ψ lies over μ. Also, $\psi \in \mathrm{Irr}(S)$ and S is the stabilizer of μ in G. Since $\psi^G = \chi$, it follows that ψ is the Clifford correspondent of χ with respect to μ, and we conclude by Lemma 10.40 that ψ is monomial. Furthermore, $\psi(1)$ divides $\chi(1)$, and in particular, $\psi(1)$ is a power of p.

Recall now that $\gamma \in \mathrm{Irr}(T \cap S)$ and $\gamma^T = \eta \in \mathrm{Irr}(T)$. Since $T \cap S \subseteq \mathbf{N}(T \cap S) \subseteq T$, it follows that $\gamma^{N(T\cap S)}$ is irreducible character and lies under η. Also, η_N is a multiple of θ, so $\gamma^{N(T\cap S)}$ lies over θ. Writing $U = N \cap S$, we have $U = N \cap (T \cap S)$, and thus

$$0 < [(\gamma^{N(T\cap S)})_N, \theta] = [(\gamma_U)^N, \theta] = [\gamma_U, \theta_U],$$

and it follows that there exists a character $\xi \in \mathrm{Irr}(U)$ that lies under both γ and θ.

Since ψ lies over γ, we see that ψ lies over ξ. Also, ψ_A is a multiple of μ, and it follows that ξ lies over μ. Now $\xi \in \mathrm{Irr}(U)$ and U is the stabilizer of μ in N, and since ξ lies under $\theta \in \mathrm{Irr}(N)$, we see that ξ is the Clifford correspondent of θ with respect to μ.

Now $U \lhd S$ and $\xi \in \mathrm{Irr}(U)$, and we argue that $T \cap S$ is the stabilizer of ξ in S. First, observe that since ξ is the Clifford correspondent of θ with respect to μ, it is uniquely determined by θ and μ, and thus $T \cap S$ stabilizes ξ. Conversely, the stabilizer of ξ in S stabilizes $\xi^N = \theta$, and it follows that this stabilizer is contained in T. We conclude that $T \cap S$ is the full stabilizer of ξ in S, as claimed.

Since γ is an irreducible character of $T \cap S$ that lies under ψ and over ξ, we deduce that γ is the Clifford correspondent of ψ with respect to ξ.

Recall now that ψ is a monomial irreducible character of S having p-power degree. Also, $U \lhd S$ and ξ is an irreducible constituent of ψ_S having stabilizer $T \cap S$. Also, γ is the Clifford correspondent of ψ with respect to ξ, and since $S < G$, it follows by the minimality of $|G|$ that γ is monomial, and thus $\eta = \gamma^T$ is monomial, contrary to assumption, and this contradiction shows that $A \subseteq \mathbf{Z}(G)$, as wanted.

We conclude by Theorem 10.39 that λ is invariant in G and λ is fully ramified with respect to N/L. Now $\theta \in \mathrm{Irr}(N)$ lies over λ, so θ is the unique member of $\mathrm{Irr}(N|\lambda)$. Also, λ is invariant in G, and it follows that θ is invariant in G. Then $T = G$ and $\eta = \chi$, and this is a contradiction because χ is monomial and η is not. ∎

10E

We need a little more module theory for our proof of Theorem 10.2, which we present in this section.

Lemma 10.41. *Let $M \lhd G$, where $|G : M|$ is an odd prime number. Let V be a symplectic FG-module, where F is an arbitrary field, and suppose that U is an FM-submodule of V that contains no nonzero FG-submodule. Assume that V/U is simple as an FM-module, and suppose that V is semisimple and not homogeneous as an FM-module. Then V is anisotropic as an FM-module.*

Proof. Recall that G acts by right multiplication on the set $\mathcal{X}$ of FM-homogeneous components of V. (See the discussion immediately preceding Theorem 10.8.) The sum of the members of each G-orbit in this action is an FG-submodule, and since by hypothesis, U contains no such submodule, it follows that U cannot contain all of the members of any G-orbit on $\mathcal{X}$.

Now V is the direct sum of the members of $\mathcal{X}$, and by Corollary 10.7, we know that U is the direct sum of its intersections with the members of $\mathcal{X}$. It follows that V/U is isomorphic to the direct sum of the FM-modules $X/(X \cap U)$ for $X \in \mathcal{X}$. By hypothesis, however, V/U is a simple as an FM-module, and therefore there is a unique member T of $\mathcal{X}$ that is not contained in U. Since U does not contain all members of any full G-orbit on $\mathcal{X}$, we conclude that the G-orbit containing T is the unique G-orbit on $\mathcal{X}$, and hence G acts transitively on $\mathcal{X}$.

By assumption, V is not homogeneous as an FM-module, so $|\mathcal{X}| > 1$, and since the action of G is transitive, G stabilizes no member of $\mathcal{X}$. Since M stabilizes every member of $\mathcal{X}$ and M has prime index in G, we see that M is the full stabilizer in G of each member of $\mathcal{X}$, and in particular, $|\mathcal{X}| = |G : M|$.

We argue that $(U \cap T)g \subseteq U$ for all $g \in G$, where we recall that T is the unique member of $\mathcal{X}$ not contained in U. If $g \in M$, this is clear because U is an FM-submodule, and hence $(U \cap T)g \subseteq Ug = U$. Otherwise, g does not stabilize T, so Tg is a member of $\mathcal{X}$ different from T, and thus $(U \cap T)g \subseteq Tg \subseteq U$, as wanted. We see now that $\sum_g (U \cap T)g$ is an FG-submodule contained in U, and hence by hypothesis, it is the zero module, and thus $U \cap T = 0$. Also $U + T = V$ because $T \not\subseteq U$ and U is a maximal FM-submodule of V. Then $T \cong V/U$, and in particular, T is a simple FM-module. Since G acts transitively on $\mathcal{X}$, it follows that every member of $\mathcal{X}$ is simple as an FM-module.

Recall now that $\mathcal{X}$ is the set of FM-homogeneous components of V, so the members of $\mathcal{X}$ have pairwise nonisomorphic type modules. (Since each member $X \in \mathcal{X}$ is simple, X is its own type module, but this observation is not needed here.) By Lemma 10.23(c), there is associated with each member $X \in \mathcal{X}$ a member $Y \in \mathcal{X}$ (possibly equal to X) such that the types of X and Y are dual to each other, and thus X uniquely determines Y. There is thus a well-defined bijection $\sigma : \mathcal{X} \to \mathcal{X}$ such that the types of X and $(X)\sigma$ are duals, and we see that σ^2 is the identity map.

Now $|\mathcal{X}| = |G : M|$, which is odd. The involutory permutation σ, therefore, must have a fixed point, so some member $X \in \mathcal{X}$ has self-dual type. It follows by Lemma 10.23 that X is nondegenerate, and since G acts transitively on $\mathcal{X}$ and the bilinear form is G-invariant, every member of $\mathcal{X}$ is nondegenerate. By Lemma 10.23 once again, the members of $\mathcal{X}$ are pairwise orthogonal.

We must show that if $S \subseteq V$ is a totally isotropic FM-submodule, then $S = 0$. Corollary 10.7 guarantees that S is the sum of its intersections with the members of $\mathcal{X}$, and since $\mathcal{X}$ consists of simple FM-modules, we conclude that S is the sum of the members of the subset $\mathcal{Y} = \{X \in \mathcal{X} \mid X \cap S \neq 0\}$. Since S is totally isotropic, it is orthogonal to each member of Y. Since the members of $\mathcal{X}$ are pairwise orthogonal, S is also orthogonal to every member of $\mathcal{X}$ that is not in $\mathcal{Y}$, and it follows that S is orthogonal to $\sum \mathcal{X} = V$. Thus $S = 0$, as required. ∎

We need one more preliminary result.

Lemma 10.42. *Let $M \lhd G$, where G/M is a p-group for some prime p, and let $\chi \in \mathrm{Irr}(G)$ be primitive. Then χ_M is irreducible.*

Proof. There is nothing to prove if $M = G$, so we can assume that $M < G$, and we proceed by induction on $|G : M|$. Let X/M be a chief factor of G, and note that $|X/M| = p$ because G/M is a p-group. Write $\psi = \chi_X$, so ψ is irreducible by the inductive hypothesis. If θ is an irreducible constituent of χ_M, then θ is invariant in G because χ is primitive, and thus θ is the unique irreducible constituent of ψ_M. Since $|X : M|$ is prime, we can apply Lemma 1.8 to deduce that $\psi_M = \theta$. Then $\chi_M = \theta$, so χ_M is irreducible, as wanted. ∎

Proof of Theorem 10.2. We have $M \lhd G$, where G/M is a p-group, G is p-solvable and $p > 2$. Also $\chi \in \mathrm{Irr}(G)$, where χ is primitive and has p-power degree, and we want to show that χ_M is irreducible and primitive. We will prove this under the additional assumption that $|G : M| = p$. The full result then follows easily by repeated application of this special case.

Let $\psi = \chi_M$, so ψ is irreducible by Lemma 10.42. Our goal is to show that ψ is primitive, so working by contradiction, we assume that there exists a maximal subgroup H of M and a character $\alpha \in \mathrm{Irr}(H)$ such that $\alpha^M = \psi$.

Let $L = \mathrm{core}_G(H)$, so $L \subseteq H < M$. Since χ is primitive, χ_L is a multiple of some character $\lambda \in \mathrm{Irr}(L)$, and we will show that λ can be assumed to be linear and faithful.

The character triple (G, L, λ) is isomorphic to some triple (G^*, L^*, λ^*), where λ^* is linear and faithful, and thus $L^* \subseteq \mathbf{Z}(G^*)$. Let χ^* be the character in $\mathrm{Irr}(G^*|\lambda^*)$ that corresponds to $\chi \in \mathrm{Irr}(G|\lambda)$, and observe that $\chi^*(1) = \chi(1)/\lambda(1)$, so $\chi^*(1)$ is a power of p.

We claim that χ^* is primitive. Otherwise, $\chi^* = \tau^{G^*}$ for some irreducible character τ of a proper subgroup V of G^*. Now $L^* \subseteq \mathbf{Z}(G^*) \subseteq V$ by Lemma 10.38, and hence $V = U^*$ for some proper subgroup U of G with $L \subseteq U$. Also, τ lies over λ^* because $(\chi^*)_{L^*}$ is a multiple of λ^*. It follows that $\tau = \sigma^*$ for some character $\sigma \in \mathrm{Irr}(U|\lambda)$. Now

$$(\sigma^G)^* = (\sigma^*)^{G^*} = \tau^{G^*} = \chi^*,$$

and thus $\sigma^G = \chi$. This contradicts the primitivity of χ, and it follows that χ^* is primitive as claimed.

Recall that $\chi_M = \psi$, where ψ is irreducible and ψ is induced from a character α of H. Also, $L \subseteq H < M$ and $L = \mathrm{core}_G(H)$. All of this is reflected in G^*, so we can replace the triple (G, L, λ) by (G^*, L^*, λ^*), and thus we can assume that λ is linear and faithful and that L is central in G.

Suppose that $A \subseteq M$, where A is abelian and $A \triangleleft G$. Since χ is primitive and $\psi_A = \chi_A$, it follows that ψ_A is a multiple of some linear character, and thus $A \subseteq \mathbf{Z}(\psi)$. By Lemma 10.38, we have $A \subseteq H$, and thus $A \subseteq \mathrm{core}_G(H) = L$. It follows that L is maximal among abelian normal subgroups of G contained in M.

Now $L < M$, so we can choose a chief factor K/L of G with $K \subseteq M$, and thus $KH = M$ because $K \not\subseteq H$ and H is maximal in M. Now $|M : H|$ divides both $|K : L|$ and $\psi(1)$. Also, $|M : H| > 1$, and since $\psi(1) = \chi(1)$, which is a power of p, it follows that p divides $|K : L|$. Since G is p-solvable, the chief factor K/L is thus an elementary abelian p-group.

Let $C = \mathbf{C}_M(K/L)$, so $K \subseteq C \triangleleft M$ and C is normal in G. Write $D = C \cap H$, and note that $L \subseteq D \triangleleft H$. Now $L \subseteq \mathrm{core}_G(D)$, and in fact, equality holds here because $D \subseteq H$, and thus $\mathrm{core}_G(D) \subseteq \mathrm{core}_G(H) = L$. Also, since K/L is central in C/L, we see that K/L normalizes D/L and thus K normalizes D. Since H also normalizes D, we have $D \triangleleft KH = M$.

Now H is a complement for C relative to D in M, and since H is maximal in M, it follows that C/D is a chief factor of M. Also, $|C : D| = |M : H|$, which is a power of p, and thus C/D is an elementary abelian p-group.

It is easy to see that there is a unique normal subgroup E of C, minimal with the property that C/E is an elementary abelian p-group. Then E

is characteristic in C, and hence $E \triangleleft G$. Also, $E \subseteq D$ because C/D is an elementary abelian p-group, and thus $E \subseteq \mathrm{core}_G(D) = L$. It follows that C/L is an elementary abelian p-group. Also, $L \subseteq \mathbf{Z}(C)$ and $\mathbf{Z}(C)$ is an abelian normal subgroup of G contained in M, and it follows that $\mathbf{Z}(C) = L$.

Since C/L is abelian and λ is a faithful linear character of L, it follows by Lemma 10.36(b) that λ is fully ramified with respect to C/L. Let θ be the unique irreducible character of C that lies over λ, and observe that θ must lie under χ, and θ is invariant in G. Also, $|C : L| = \theta(1)^2$ because λ is linear.

Recall that C/D is an abelian chief factor of M and θ is invariant in M. By Corollary 7.4, there are just three possibilities: either θ_D is irreducible or θ is fully ramified with respect to C/D or θ is induced from D.

Suppose first that θ_D is irreducible, and write $\theta_D = \tau$. Then $\tau_L = \tau(1)\lambda$, and it follows by Lemma 7.2 that $\tau(1)^2 \leq |D : L| < |C : L| = \theta(1)^2$. This is a contradiction because $\tau(1) = \theta(1)$, and we conclude that θ_D is not irreducible.

Next, we show that θ is not fully ramified with respect to C/D. Recall that θ is induced from a character of H, and H is a complement for C relative to D in M. By Theorem 7.5, therefore, we will have a contradiction when we show that every complement for C relative to D in M is conjugate to H in M. This is obvious if $C = M$, so we suppose that $C < M$, and we observe that by Lemma 7.6, it suffices to find a subgroup B such that $C \subseteq B \triangleleft M$, where $\mathbf{C}_{C/D}(B)$ is trivial and $|B : C|$ and $|C : D|$ are relatively prime.

Let B/C be a chief factor of G, with $B \subseteq M$. Now $B > C = \mathbf{C}_M(K/L)$ so B does not centralize K/L. It follows that $\mathbf{C}_{K/L}(B)$ is trivial because $B \triangleleft G$ and K/L is a chief factor of G. Also, since C acts trivially on K/L, the action of B on K/L induces an action of B/C, and we see that B/C has no nontrivial fixed points on the p-group K/L. It follows that B/C is not a p-group, and thus it is a p'-group, as required. Finally, since $\mathbf{C}_{K/L}(B/C)$ is trivial, it follows by Lemma 10.37 that B/C has no nontrivial fixed points on $K/(K \cap D) \cong C/D$, and hence $\mathbf{C}_{C/D}(B)$ is trivial, as required.

We have now proved that θ does not restrict irreducibly to D and that θ is not fully ramified with respect to C/D, and we deduce that $\theta = \delta^C$ for some character δ of D.

By Lemma 10.34, we can view C/L as a symplectic FG-module, where F is the field of order p. Also, C/L is anisotropic as an FG-module because if A/L is a totally isotropic FG-submodule, then A is an abelian normal subgroup of G contained in C, and we have seen that implies that $A = L$.

It follows by Lemma 10.21 that C/L is semisimple as an FG-module, and thus by Theorem 10.8, it is also semisimple as an FM-module.

We argue next that not only is C/L anisotropic as an FG-module, but also, it is anisotropic as an FM-module. To see this, we propose to apply Lemma 10.41 with C/L as the module V, and with D/L in the role of the submodule U. To verify the hypotheses of Lemma 10.41, we must show that D/L contains no nonzero G-invariant subspace, or equivalently, that L is maximal among normal subgroups of G that are contained in D. This is true, however, because we have seen tha $\mathrm{core}_G(D) = L$. To apply Lemma 10.41, we must also show that the factor module $(C/L)/(D/L)$ is simple as an FM-module. This is clear, however, because C/D is a chief factor of M.

If C/L is not homogeneous as an FM-module, Lemma 10.41 guarantees that C/L is anisotropic as an FM-module, as wanted. If C/L is homogeneous as an FM-module, then all simple FM-submodules of C/L are isomorphic, and so we can apply Corollary 10.32. Since C/L is anisotropic as an FG-module, we deduce that it must also be anisotropic as an FM-module. In either case, therefore, we have shown that C/L is anisotropic as an FM-module, and thus C does not contain an abelian normal subgroup of M strictly larger than L.

We obtain our final contradiction by showing that $\mathbf{Z}(D) > L$. Recall that there exists a character $\delta \in \mathrm{Irr}(D)$ such that $\delta^C = \theta$, where θ is fully ramified with respect to C/L, and thus $\theta(1)^2 = |C : L|$. By Theorem 7.3, there exists a subgroup R with $L \subseteq R \subseteq D$, and such that the following two properties hold: δ is fully ramified with respect to D/R, and the unique character $\mu \in \mathrm{Irr}(R)$ that lies under δ restricts irreducibly to L.

Now μ is linear because μ_L is irreducible. Since μ is invariant in D, we have $[D, R] \subseteq \ker(\mu)$. Also, $[D, R] \subseteq L$ because D/L is abelian, and thus $[D, R] \subseteq L \cap \ker(\mu) = \ker(\lambda) = 1$, and we have $R \subseteq \mathbf{Z}(D)$.

Since δ is fully ramified with respect to D/R and the unique irreducible character of R lying under δ is linear, we have

$$\frac{|D|}{|R|} = \delta(1)^2 = \frac{\theta(1)^2}{|C : D|^2} = \frac{|C : L|}{|C : D|^2} = \frac{|D|^2}{|C||L|},$$

where the second equality holds because $\delta^C = \theta$, and so $\delta(1)|C : D| = \theta(1)$.

We deduce that $|R : L| = [C : D| > 1$, so $R > L$, and thus $\mathbf{Z}(D) > L$. This is a contradiction because $\mathbf{Z}(D)$ is an abelian normal subgroup of M that is contained in C. ∎

Notes and References

Theorem 10.1 is essentially due to Dade, and it appears in [**14**]. The main result of an earlier, preprint version of Dade's paper was that if χ is a monomial character of p-power degree of a p-solvable group G, where $p \neq 2$, then the irreducible constituents of χ_N are monomial for all normal subgroups N of G. When Isaacs saw the preprint of Dade's paper, he pointed out to Dade that the techniques of the paper could be used to prove Theorem 10.1 of this chapter, which is a somewhat stronger result. Dade incorporated Isaacs' remarks into the final version of his paper.

Isaacs' paper [**37**] contains what appears to be the first use of symplectic modules to study M-groups and related topics. In [**14**], Dade expanded on Isaacs' ideas to obtain much deeper results, including Theorem 10.17 of this chapter, on hyperbolic modules.

Theorem 10.2 appears as Theorem B in [**37**].

The idea underlying Lemma 10.10 and Theorem 10.11(a) was published in [**30**], where it appears in the proof of Theorem 2.2.3. The authors credit the argument to J. A. Green.

Bibliography

[AGC] I. Martin Isaacs, *Algebra: A graduate course*, Graduate Studies in Mathematics, 100, American Mathematical Society, Prov., RI, 2009.

[CTFG] I. Martin Isaacs, *Character theory of finite groups*, AMS Chelsea Publishing, Providence, RI. (Corrected reprint of the 1976 original, Academic Press, NY.)

[FGT] I. Martin Isaacs, *Finite group theory*, Graduate Studies in Mathematics, 92, AMS, Prov., RI, 2008.

[1] J. L. Alperin, The main problem of block theory, *Proc. of Conf. on Finite Groups* Academic Press, New York (1975) 341–356.

[2] J. L. Alperin, Weights for finite groups.*Proc. Sympos. Pure Math.* Amer. Math. Soc., Providence **47** (1987) 369–379.

[3] G. Amit and D. Chillag, On a question of Feit concerning character values of finite solvable groups, *Pacific J. Math.* **122** (1986) 257–261.

[4] T. R. Berger, Primitive solvable groups. *J. of Algebra* **33** (1975) 9–21.

[5] T. R. Berger, Characters and derived length in groups of odd order, *J. of Algebra* **39** (1976) 199–207.

[6] R. Brauer, Zur Darstellungstheorie der Gruppen endlicher Ordnung, *Math. Zeit.* **63** (1956), 406–444.

[7] P. Centella and G. Navarro, Correspondences between constituents of projective characters, *Archiv der Math.* **90** (2008) 289–294.

[8] I. A. Chubarov, Monomial characters and subgroups. *Uspehi Mat. Nauk* **33** (1978 191–192

[9] I. A. Chubarov, Normal subgroups of M-groups, *Trudy Sem. Petrovsk.* **4** (1978) 249–256.

[10] A. H. Clifford, Representations induced in an invariant subgroup, *Ann. of Math (2)* **38** (1937), no. 3, 533–550.

[11] J. P. Cossey, A construction of two distinct canonical sets of lifts of Brauer characters of a p-solvable group, *Arch. Math. (Basel)* **87** (2006) 385–389.

[12] E. C. Dade, Normal subgroups of M-groups need not be M-groups, *Math. Zeit.* **133** (1973) 313–317.

[13] E. C. Dade, A correspondence of characters, *Proc. Sympos. Pure Math.* **37** (1979) 401–404.

[14] E. C. Dade, Monomial characters and normal subgroups, *Math. Zeit.* **178** (1981) 401–420.

[15] F. R. DeMeyer and G. Janusz, Finite groups with an irreducible representation of large degree, *Math. Zeit.* **108** (1969) 145–153.

[16] L. Dornhoff, M-groups and 2-groups, *Math. Zeit.* **100** (1967) 226–256.

[17] W. Feit, Some consequences of the classification of finite simple groups, in: Finite Groups (Santa Cruz 1979), *Proc. Sympos. Pure Math.* **37** AMS, Providence (1980) 175–181.

[18] P. A. Ferguson and A. Turull, On a question of Feit. *Proc. Amer. Math. Soc.* **97** (1986) 21–22.

[19] P. Fong, On the characters of p-solvable groups. *Trans. Amer. Math. Soc.* **98** (1961) 263–284.

[20] P. Fong, Solvable groups and modular representation theory. *Trans. Amer. Math. Soc.* **103** (1962) 484–494.

[21] P. Fong, On decomposition numbers of J_1 and $R(q)$. *Symposia Mathematica, Vol. XIII* Academic Press, London, (1974) 415–422.

[22] H. Fukushima, Hall subgroups of M-groups need not be M-groups, *Proc. Amer. Math. Soc.* **133** (2005 671–675.

[23] D. S. Gajendragadkar, A characteristic class of characters of finite π-separable groups, *J. of Algebra* **59** (1979) 237–259.

[24] P. X. Gallagher, Group characters and normal Hall subgroups, *Nagoya Math. J.* **79** (1962) 223–230.

[25] P. X. Gallagher, The number of conjugacy classes inn a finite group, *Math. Zeit.* **118** (1970) 175–179.

[26] G. Glauberman, Fixed points in groups with operator groups, *Math. Zeit.* **84** (1964) 120–125.

[27] G. Glauberman, Correspondences of characters for relatively prime operator groups, *Canad. J. Math.* **20** (1968) 1465–1488.

[28] D. Gluck, Bounding the number of character degrees of a solvable group, *J. of London Math. Soc. (2)* **31** (1985) 457–462.

[29] P. G. Gres, Proof for p-solvable groups of the Alperin hypothesis on the number of characters of height 0, *Proc. 6th All-Union Symposium on Group Theory (Cherkassy,* (1978) 179–188.

[30] P. Hall and G. Higman, On the p-length of p-soluble groups and reduction theorems for Burnside's problem, *Proc. London Math Soc.* **6** (1956) 1–42.

[31] B. Huppert, Monomiale Darstellung endlicher Gruppen. *Nagoya Math. J.* **6** (1953) 93–94.

[32] B. Huppert, Lineare auflösbare Gruppen, *Math. Zeit.* **67** (1957) 479–518.

[33] B. Huppert, *Endliche Gruppen. I.* Springer-Verlag, Berlin-New York (1967).

[34] I. M. Isaacs, Characters of solvable and symplectic groups, *American J. of Math.* **95** (1973) 594–635.

[35] I. M. Isaacs, Character degrees and derived length of a solvable group, *Canad. J. of Math.* **27** 146–151.

[36] I. M. Isaacs, Lifting Brauer characters of p-solvable groups, II, *J. of Algebra* **51** (1978) 476–490.

[37] I. M. Isaacs, Primitive characters, normal subgroups, and M-groups, *Math. Zeit.* **177** (1981) 267–284.

[38] I. M. Isaacs, Character correspondences in solvable groups, *Advances in Math.* **43** (1982) 284–306.

[39] I. M. Isaacs, On the character theory of fully ramified sections, *Rocky Mtn. J. of Math.* **13** (1983) 689–698.

[40] I. M. Isaacs, Characters of π-separable groups, *J. of Algebra* **86** (1984) 98–128.

[41] I. M. Isaacs, Extensions of characters from Hall π-subgroups of π-separable groups, *Proc. of the Edinburgh Math. Soc.* **38** (1985) 313–317.

[42] I. M. Isaacs, Induction and restriction of π-special characters, *Canad. J. Math.* **38** (1986) 576–604.

[43] I. M. Isaacs, Fong characters in π-separable groups, *J of Algebra* **99** (1986) 89–107.

[44] I. M. Isaacs, Counting objects which behave like irreducible Brauer characters of finite groups, *J. of Algebra* **117** (1988) 419–433.

[45] I. M. Isaacs, Partial characters of π-separable groups, *Representation theory of finite groups and finite-dimensional algebras, Progr. Math.* **95** Birkhäuser, Basel (1991) 273–287.

[46] I. M. Isaacs, Characters and Hall subgroups of groups of odd order, *J. of Algebras* **157** (1993) 548–561.

[47] I. M. Isaacs, Characters and sets of primes for solvable groups. *Finite and locally finite groups*, NATO Adv. Sci. Inst. Ser. C **471** Kluwer Acad. Publ., Dordrecht, (1995) 347–376.

[48] I. M. Isaacs, Constituents of restricted and induced characters of odd order groups, *J. of Algebra* **178** (1995) 991–1001.

[49] I. M. Isaacs and G. Navarro, Weights and vertices for characters of π-separable groups, *J. of Algebra* **177** (1995) 339–366.

[50] I. M. Isaacs, G. Malle and G. Navarro, A reduction theorem for the McKay conjecture, *Invent. Math.* **170** (2007) 33–101.

[51] I. M. Isaacs and G. Navarro, New refinements of the McKay conjecture for arbitrary finite groups, *Annals of Math.* **156** (2002) 333–344.

[52] I. M. Isaacs and G. Navarro, Characters that agree on prime-power-order elements, *J. of Algebra* **263** (2003) 246–254.

[53] I. M. Isaacs and G. Navarro, A characteristic subgroup and kernels of Brauer characters, *Bull. Austral. Math. Soc.* **72** (2005) 381–384.

[54] I. M. Isaacs and T. Wilde, Primitive characters of maximal subgroups of solvable groups, *J. Algebra* **323** (2010 419–43.

[55] M. L. Lewis, Characters of maximal subgroups of M-groups. *J. of Algebra* **183** (1996) 864–897.

[56] G. Malle and B. Späth, Characters of odd degree. *Ann. of Math. (2)* **184** (2016) 869–908.

[57] J. McKay, Abstract 71T-A31, *Notices Amer. Math. Soc.* **18** (1971) 397.

[58] J. McKay, Irreducible representations of odd degree. *J. of Algebra* **20** (1972) 416–418.

[59] G. Navarro, Primitive characters of subgroups of M-groups, *Math. Zeit.* **218** (1995) 439–445.

[60] G. Navarro, Inducing characters and nilpotent subgroups, *Proc. Amer. Math. Soc.* **124** (1996) 3281–3284

[61] G. Navarro, New properties of the π-special characters, *J. of Algebra* **187** (1997) 203–213.

[62] G. Navarro, *Characters and blocks of finite groups*, London Mathematical Society Lecture Note Series, 250. Cambridge University Press, Cambridge, (1998).

[63] G. Navarro, Vertices for characters of p-solvable groups, *Trans. Amer. Math. Soc.* **354** (2002) 2759–2773.

[64] G. Navarro, Linear characters of Sylow subgroups, *J. of Algebra* **269** (2003) 589–598.

[65] T. Okuyama and M. Wajima, Irreducible characters of p-solvable groups, *Proc. Japan Acad. Ser. A* **55** (1979) 309–312.

[66] T. Okuyama and M. Wajima, Character correspondence and p-blocks of p-solvable groups. *Osaka J. Math.* **17** (1980) 801–806.

[67] A. E. Parks, Nilpotent by supersolvable M-groups, *Canad. J. of Math.* **37** (1985) 934–962.

[68] G. R. Robinson and R. Staszewski, On the representation theory of π-separable groups. *J. of Algebra* **119** (1988) 226–232.

[69] C.-H. Sah, Existence of normal complements and extension of characters in finite groups, *Illinois J. Math.* **6** (1962) 282–291.

[70] R. G. Swan, The Grothendieck ring of a finite group. *Topology* **2** (1963) 85–110.

[71] K. Taketa, Über die gruppen, deren Darstellungen sich sämtlich auf monomiale Gestalt transformieren lassen, *Proc. Acad. Tokyo* **6** (1930) 31–33.

[72] R. W. van der Waall, On the embedding of minimal non-M-groups, *Indag. Math.* **77** (1974) 157–167.

[73] W. Willems, On degrees of irreducible Brauer characters, *Trans. Amer. Math. Soc.* **357** (2004) 2379–2387.

[74] T. R. Wolf, Characters of P'-degree in solvable groups, *Pacific J. of Math.* **74** (1978) 267–271.

[75] T. R. Wolf, Character correspondences in solvable groups, *Illinois J. of Math.* **22** (1978) 327–340.

[76] T. R. Wolf, Variations on McKay's character degree conjecture, *J. of Algebra* **135** (1990) 123–138.

[77] H. Zassenhaus, *Lehrbuch der Gruppentheorie* Teubner, Leipzig and Berlin (1937).

Index

Selected Published Titles in This Series

189 **I. Martin Isaacs,** Characters of Solvable Groups, 2018
187 **John Douglas Moore,** Introduction to Global Analysis, 2017
186 **Bjorn Poonen,** Rational Points on Varieties, 2017
185 **Douglas J. LaFountain and William W. Menasco,** Braid Foliations in Low-Dimensional Topology, 2017
184 **Harm Derksen and Jerzy Weyman,** An Introduction to Quiver Representations, 2017
183 **Timothy J. Ford,** Separable Algebras, 2017
182 **Guido Schneider and Hannes Uecker,** Nonlinear PDEs, 2017
181 **Giovanni Leoni,** A First Course in Sobolev Spaces, Second Edition, 2017
180 **Joseph J. Rotman,** Advanced Modern Algebra: Third Edition, Part 2, 2017
179 **Henri Cohen and Fredrik Strömberg,** Modular Forms, 2017
178 **Jeanne N. Clelland,** From Frenet to Cartan: The Method of Moving Frames, 2017
177 **Jacques Sauloy,** Differential Galois Theory through Riemann-Hilbert Correspondence, 2016
176 **Adam Clay and Dale Rolfsen,** Ordered Groups and Topology, 2016
175 **Thomas A. Ivey and Joseph M. Landsberg,** Cartan for Beginners: Differential Geometry via Moving Frames and Exterior Differential Systems, Second Edition, 2016
174 **Alexander Kirillov Jr.,** Quiver Representations and Quiver Varieties, 2016
173 **Lan Wen,** Differentiable Dynamical Systems, 2016
172 **Jinho Baik, Percy Deift, and Toufic Suidan,** Combinatorics and Random Matrix Theory, 2016
171 **Qing Han,** Nonlinear Elliptic Equations of the Second Order, 2016
170 **Donald Yau,** Colored Operads, 2016
169 **András Vasy,** Partial Differential Equations, 2015
168 **Michael Aizenman and Simone Warzel,** Random Operators, 2015
167 **John C. Neu,** Singular Perturbation in the Physical Sciences, 2015
166 **Alberto Torchinsky,** Problems in Real and Functional Analysis, 2015
165 **Joseph J. Rotman,** Advanced Modern Algebra: Third Edition, Part 1, 2015
164 **Terence Tao,** Expansion in Finite Simple Groups of Lie Type, 2015
163 **Gérald Tenenbaum,** Introduction to Analytic and Probabilistic Number Theory, Third Edition, 2015
162 **Firas Rassoul-Agha and Timo Seppäläinen,** A Course on Large Deviations with an Introduction to Gibbs Measures, 2015
161 **Diane Maclagan and Bernd Sturmfels,** Introduction to Tropical Geometry, 2015
160 **Marius Overholt,** A Course in Analytic Number Theory, 2014
159 **John R. Faulkner,** The Role of Nonassociative Algebra in Projective Geometry, 2014
158 **Fritz Colonius and Wolfgang Kliemann,** Dynamical Systems and Linear Algebra, 2014
157 **Gerald Teschl,** Mathematical Methods in Quantum Mechanics: With Applications to Schrödinger Operators, Second Edition, 2014
156 **Markus Haase,** Functional Analysis, 2014
155 **Emmanuel Kowalski,** An Introduction to the Representation Theory of Groups, 2014
154 **Wilhelm Schlag,** A Course in Complex Analysis and Riemann Surfaces, 2014
153 **Terence Tao,** Hilbert's Fifth Problem and Related Topics, 2014
152 **Gábor Székelyhidi,** An Introduction to Extremal Kähler Metrics, 2014
151 **Jennifer Schultens,** Introduction to 3-Manifolds, 2014

For a complete list of titles in this series, visit the AMS Bookstore at **www.ams.org/bookstore/gsmseries/**.